ARSENAL SHIP OMNIBUS

ESSENTIAL PUBLIC DOCUMENTS ABOUT THE MOST INNOVATIVE NAVAL WEAPONS CONCEPT OF THE LATE 20TH CENTURY

EDITED BY W. FREDERICK ZIMMERMAN

NIMBLE BOOKS LLC

Nimble Books LLC
1521 Martha Avenue
Ann Arbor, MI, USA 48103
http://www.NimbleBooks.com
wfz@nimblebooks.com
+1.734-330-2593

Version 1.0; last saved 2011-02-26.

Printed in the United States of America

ISBN-13: 978-1-60888-117-8

The paper used in this publication meets the minimum requirements of the American National Standard for Information Sciences—Permanence of Paper for Printed Library Materials, ANSI Z39.48-1992. The paper is acid-free and lignin-free.

CONTENTS

NAVAL POSTGRADUATE SCHOOL
MONTEREY, CALIFORNIA

AN *ARSENAL SHIP* DESIGN

by

LT G. Baumann, USN
LCDR M. Chase, USN
LT B. Ellis, USCG
LT J. Gage, USCG
LT T. Heatter, USN
LT C. Mercer, USN
LT G. Null, USN
LT J. Sebastian, USCG
LT J. Brown, USCG
LT D. Florence, USN
LT M. France, USN
LT C. Holmes, USN
CAPT T. Langlois, USMC
LT A. Rowe, USN
LT J. Scrofani, USN

December 1996

Advisors: Assoc. Prof. C. N. Calvano
Assoc. Prof. R. C. Harney

NAVAL POSTGRADUATE SCHOOL

Report Documentation Page						Form Approved OMB No. 0704-0188
Public reporting burden for the collection of information is estimated to average 1 hour per response, including the time for reviewing instructions, searching existing data sources, gathering and maintaining the data needed, and completing and reviewing the collection of information. Send comments regarding this burden estimate or any other aspect of this collection of information, including suggestions for reducing this burden, to Washington Headquarters Services, Directorate for Information Operations and Reports, 1215 Jefferson Davis Highway, Suite 1204, Arlington VA 22202-4302. Respondents should be aware that notwithstanding any other provision of law, no person shall be subject to a penalty for failing to comply with a collection of information if it does not display a currently valid OMB control number.						
1. REPORT DATE **DEC 1996**		2. REPORT TYPE **N/A**				3. DATES COVERED **-**
4. TITLE AND SUBTITLE **An Arsenal Ship Design**						5a. CONTRACT NUMBER
						5b. GRANT NUMBER
						5c. PROGRAM ELEMENT NUMBER
6. AUTHOR(S)						5d. PROJECT NUMBER
						5e. TASK NUMBER
						5f. WORK UNIT NUMBER
7. PERFORMING ORGANIZATION NAME(S) AND ADDRESS(ES) **Naval Post Graduate School Monterey, CA 93943-5000**						8. PERFORMING ORGANIZATION REPORT NUMBER
9. SPONSORING/MONITORING AGENCY NAME(S) AND ADDRESS(ES)						10. SPONSOR/MONITOR'S ACRONYM(S)
						11. SPONSOR/MONITOR'S REPORT NUMBER(S)
12. DISTRIBUTION/AVAILABILITY STATEMENT **Approved for public release, distribution unlimited**						
13. SUPPLEMENTARY NOTES **The original document contains color images.**						
14. ABSTRACT						
15. SUBJECT TERMS						
16. SECURITY CLASSIFICATION OF:			17. LIMITATION OF ABSTRACT	18. NUMBER OF PAGES	19a. NAME OF RESPONSIBLE PERSON	
a. REPORT **unclassified**	b. ABSTRACT **unclassified**	c. THIS PAGE **unclassified**	**UU**	**263**		

Standard Form 298 (Rev. 8-98)
Prescribed by ANSI Std Z39-18

AN *ARSENAL SHIP* DESIGN

This report documents a systems engineering and design capstone project undertaken by students in the Total Ship Systems Engineering program at the Naval Postgraduate School. The project was performed under the direction of Prof. C. N. Calvano and Prof. R. Harney. The officer students who comprised the design team were: LT G. Baumann, USN; LT J. Brown, USCG; LCDR M. Chase, USN; LT B. Ellis, USCG; LT D. Florence, USN; LT M. France, USN; LT J. Gage, USCG; LT T. Heatter, USN; LT C. Holmes, USN; CAPT T. Langlois, USMC; LT C. Mercer, USN; LT G. Null, USN; LT A. Rowe, USN; LT J. Scrofani, USN; LT J. Sebastian, USCG.

ABSTRACT

The Navy's "*Arsenal Ship* Concept of Operations (CONOPS)" and "*Arsenal Ship* Capabilities Document (SCD)" address a need for the design of a large missile platform that can carry massive and precise firepower, accomplish long-range strike, and perform flexible targeting and multidimensional theater defense capabilities consistent with the policies of "Forward...From the Sea" and "Operational Maneuver from the Sea." The ship is designed to be a large missile magazine that receives its launch orders from remote air, land or sea forces. It is fully integrated into the joint command and control structure to assist current forces in the opening days of conflict. A major design goal of the *Arsenal Ship* is to limit the crew size to 50 personnel through the use of system automation, redundancy and equipment reliability, while imposing an additional constraint of limiting the sailaway price to 550 million dollars.

The Total Ship's Systems Engineering (TSSE) design team devised a design philosophy, functional analysis and flow diagrams, and a team specific *Arsenal Ship* CONOPS. These documents provided initial guidance in the selection of systems and procedures by identifying a prioritized list of design goals, and a detailed description of how the *Arsenal Ship* operates.

The preliminary design phase involved tradeoff studies to determine the optimal hull alternatives, combat system selection, and systems and procedures to reduce crew size. The *Arsenal Ship*'s hull is a modified repeat of the T-AO 201 class auxiliary ship with double hull. The selection of this hull is based partly on mass tonnage for survivability, carrying capacity for approximately 500 vertical launched missiles, and budget constraints that preclude a new hull design. The midship section of the ship is fitted with 64 8-cell Vertical Launch System (VLS) modules. The Cooperative Engagement Capability (CEC) was researched and implemented into the design to provide the controlling platform the ability to remotely fire the *Arsenal Ship*'s missiles. System automation, remote sensors and cameras are used in every situation to benefit the design for manning reduction.

INTRODUCTION

This is the final report for the Total Ship's Systems Engineering (TSSE) student design project of 1996. The report contains the compilation of work performed over a two quarter period from June through December of 1996. The various assignments and design products created have been integrated into this report to provide a detailed and comprehensive record of the work completed.

The Assistant Secretary of the Navy, (Research, Development, and Acquisition), Mr. John W. Douglass, approached the TSSE faculty with the task of using the TSSE program to provide a preliminary design for the *Arsenal Ship*. The *Arsenal Ship* acquisition program is unique in that civilian contractors team together and compete for the design, as well as the construction contract. The TSSE project is the only non-industry design being performed. Secretary Douglass and his staff will be given the TSSE report to be used as background information to assist them in understanding and assessing contractor submissions.

To provide a thorough investigation of the *Arsenal Ship* concept, the TSSE team was divided into three teams: Hull, Mechanical, and Electrical (HM&E), Operations, and Combat Systems. The project included the following major design phases:

(1) Design Philosophy
(2) Functional Analysis and Concept of Operations
(3) Hull, Mechanical, and Electrical Tradeoff Studies
(4) Combat System Concept of Design and Tradeoff Studies
(5) Preliminary Design, Crew Reduction and Cost Analyses
(6) Design Evaluation

The following report contains the TSSE design team's analysis and drawings of the *Arsenal Ship*'s preliminary design. We believe that the at-sea experience and thorough analysis from the officer's involved has resulted in a sensible design of a creative and highly realistic solution to the Navy's call for an *Arsenal Ship* of the 21st Century.

TABLE OF CONTENTS

LIST OF FIGURES

LIST OF TABLES

I. *ARSENAL SHIP* CONCEPT OF OPERATIONS

A. INTRODUCTION

1. Forward ...From the Sea

Projection of power from the sea to the land, sea control and maritime supremacy, strategic deterrence, strategic sealift, and forward naval presence: these describe the fundamental and enduring roles U. S. naval forces play in providing for our nation's security. In support of these fundamental roles, naval expeditionary forces are routinely forward-deployed, designed and trained with the objectives of preventing conflicts, controlling crises, and if called upon, fighting and winning wars. Forward-deployed naval expeditionary forces are essential elements of the fundamental roles of the U. S. Navy. These naval forces normally consist of aircraft carrier battle groups and/or amphibious readiness groups. Consistent with the Navy's strategic concept paper, *Forward ...From the Sea* [1], and the Marine Corps' concept of expeditionary warfare in *Operational Maneuver From the Sea* [2], it is envisioned that these forces will increasingly be called upon to play larger and larger roles in regional conflicts.

From peacetime presence to full joint and combined operations, the power-projection capabilities of forward naval forces must increase. As described in the Navy's paper on naval warfare in the 21st century, *2020 Vision* [3], the theater commander of the future will require massive and precise firepower, long-range strike, flexible targeting, and multi-dimensional theater defense capabilities that go beyond current levels. A concept which will provide these increased capabilities is the introduction of a new weapons platform into the existing expeditionary force structure which enables the concentration of massive firepower with netted targeting and weapons assignment. This new weapons platform, called an *Arsenal Ship*, will essentially be a massive, remote missile and fire support magazine, linked through Cooperative Engagement Capabilities (CEC) with off-board targeting and fire control platforms engaged in the battle space. Operating under the control and umbrella of Aegis surface combatants, the *Arsenal Ship* will provide the Joint Task Force (JTF) Commander with rapid response and firepower. This can be used to blunt the attack of regional aggressors and support the build-up and maneuver of coalition land-based air and ground forces through precision strike, naval surface fire support, and theater air and missile defense.

2. General Description of Mission and Threat

a. The Mission

The *Arsenal Ship* is a missile laden, forward-deployed, highly automated, optimally-manned ship that possesses a high degree of built-in protection and uses the most advanced communication networks available. In concert with traditional naval expeditionary forces, the *Arsenal Ship* provides an extremely potent forward presence in peacetime. It is used as a remote magazine, linked with the fire control and command platforms of the expeditionary forces. These *Arsenal Ship* augmented forces, operating under the concepts described in *Forward ...From the Sea* and *Operational Maneuver From the Sea*, are used by the JTF Commander for missions such as the following:

Halting Invasions. The *Arsenal Ship* provides massive quantities of advanced missiles, equipped with precision-guided munitions, used to stop attacking armored forces.

Long-Range Strike. The *Arsenal Ship* provides Tomahawk cruise missiles used to attack the enemy's center of gravity -- demolishing strategic targets, air defense sites and the enemy's military infrastructure.

Littoral Warfare. Using Standard Missile Lightweight Exoatmospheric Projectile (LEAP) or Theater Anti-Air Defense (THAAD) surface to air missiles, the *Arsenal Ship* and Aegis combatants provide tactical ballistic missile defense, defense against cruise missiles, and support to air operations. Using a naval version of the Army's Advanced Tactical Missile System (ATACMS) and an advanced naval gun system, the *Arsenal Ship* provides naval fire support to forces ashore, countering enemy artillery systems and suppressing second echelon forces and air defense sites.

Conventional Deterrence. The forward-deployed *Arsenal Ship* provides conventional deterrence against regional aggression in areas vital to U.S. national interests.

b. Threat

The *Arsenal Ship* would likely be employed in every major regional conflict and will be the primary means of delivering ordnance on target to slow and halt the advance of the enemy in all areas of the battlespace. Consequently, it will be considered an extremely high valued unit and most certainly a primary target of opposing forces. The threats posed by these opposing forces encompass all varieties of sea-, air-, and land-based weapon systems. The design and employment of the *Arsenal Ship* effectively counters the threats posed by these enemy forces.

The projected threat environment in which the mission is expected to be accomplished will range from natural environmental forces, such as heavy seas and storms, to operations in littoral environments congested with mines, small attack surface craft and coastal submarines, to environments contaminated by chemical, biological and radiological (CBR) weapons. However, the greatest challenge will be operation of the *Arsenal Ship* and its interface with the sensor and targeting platforms in an environment where the electromagnetic spectrum has been denied or degraded. The desired mission of the system, as designed, is fully realizable in such environments.

B. CRITICAL SYSTEM CHARACTERISTICS AND CONSTRAINTS

The Defense Advanced Research Project Agency (DARPA) provided two documents, which briefly describe the Navy's concept of operations (CONOPS) [4] and general ship capabilities (SCD) [5], that should be used as guides in developing detailed concepts of operations and designs for the *Arsenal Ship*. The goals established by these two documents require the *Arsenal Ship* design to be revolutionary in nature. The operational concept gleaned from a thorough review of the CONOPS and SCD challenges existing Navy culture and tradition. Developing a ship system that attains the goals of the CONOPS and SCD requires rethinking standard practices and beliefs, and implementing labor-saving technologies. The following are critical system characteristics and constraints which drive this revolutionary concept of operations:

- The *Arsenal Ship* has a crew numbering less than 50.

- The *Arsenal Ship* supplies massive firepower in the early phases of crisis response and control and provides naval surface fire support (NSFS) in direct tactical support of ground forces.
- The *Arsenal Ship* is not a targeting or fire control platform. It does not possess the ability to employ the offensive weapons it carries.
- The *Arsenal Ship* is capable of full-time communications with ships, aircraft, satellites and shore stations via responsive, reliable, clear and secure voice, tactical information distribution, and recorded communications. An over-the-horizon satellite link capability is provided.
- The *Arsenal Ship* possesses limited active self-defense capability. This is a function of simplicity, manning level and cost. Therefore, if unescorted, the *Arsenal Ship* would provide the enemy with a defenseless, high value target. Consequently, the *Arsenal Ship* will always be operated under the umbrella of escorts which can provide appropriate defenses.
- The *Arsenal Ship* is always available for rapid movement upon receipt of strategic warning, providing the JTF Commander flexibility in response to regional crises (total ship availability of 0.95).
- The *Arsenal Ship* is virtually unsinkable. It incorporates designs and systems which dramatically reduce susceptibility and vulnerability. These designs and systems are predominantly passive in nature and inherently make the ship difficult to detect, target and hit. In addition, if the ship is hit, it is designed to automatically, or inherently limit, and contain damage so that the ship can continue its mission.
- The *Arsenal Ship* can be fully integrated into the joint war fighting force structure, operating in both peacetime and war as an integral fleet unit within the chain of command under Joint Combatant Command (COCOM). Peacetime operational control is exercised by numbered fleet commanders. When operating under a joint task force, operational control will be exercised by the Joint Force Maritime Commander.
- A fleet of six *Arsenal Ship*s will be stationed continuously forward in the SW Asia/Persian Gulf, Western Pacific and Mediterranean theaters of operation. Like the Maritime Prepositioning Force (MPF) ships, they will remain on station in support of a Unified Combatant Commander for indefinite periods without

dependence on host nation support or permission. The maintenance, logistic and training concepts are consistent with the forward operating base (FOB) concept.

- The *Arsenal Ship* is designed with systems of high reliability and very low maintenance.
- The *Arsenal Ship* is able to refuel underway via connected replenishment (CONREP) and take on stores via vertical replenishment (VERTREP).
- The *Arsenal Ship* does not possess the capability to rearm vertical launch systems underway.
- The *Arsenal Ship* is capable of transiting the Panama and Suez canals.
- The *Arsenal Ship* has a minimum sustained speed of 22 knots.

C. *ARSENAL SHIP* CONCEPT OF OPERATIONS

1. Key Employment Elements

The discussions of mission, threats and critical ship characteristics and constraints provide the foundation for the development of a concept of operations. This concept of operations addresses the various key elements necessary to operate and maintain three squadrons of two ships, forward-deployed, with near-constant availability, capable of performing the stated missions within the constraints placed on the ship design. The requirement to remain forward-deployed for indefinite periods of time with near-constant availability and radically reduced manning demands revolutionary concepts and innovative approaches to many ship functions and operations. The CONOPS addresses the following key employment elements:

- Basing and ship movement.
- Integrated logistics support.
- Manpower, personnel and training.
- Command and Control.
- Security.

- Interoperability and integration with carrier battle groups, amphibious
- Readiness groups, and joint task forces.

2. Basing and Ship Movement

In order for the *Arsenal Ship* to perform its mission effectively, it must be forward deployed and continuously available for the majority of its operational life. By necessity, forward operating bases within the required geographic regions must be established. They must be capable of providing the required support services to achieve the availability goals and response time, and also provide the strategic reach. The *Arsenal Ship* will utilize the forward operating bases already established for the three MPF squadrons: Diego Garcia, Guam and Ascension Island. Figure (I-1) illustrates the locations of these bases and the strategic reach and response times associated with the sustained speed requirement of 22 knots. The circles represent closure times of 7 and 14 days.

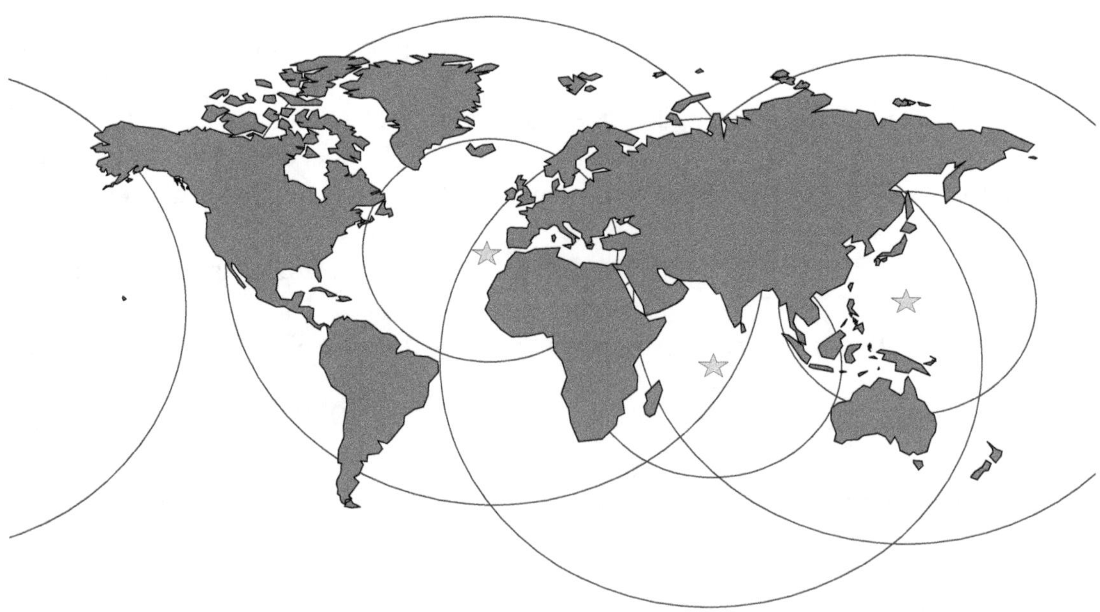

Figure I-1. *Arsenal Ship* Forward Operating Bases and Strategic Reach

The forward operating bases will have facilities for the maintenance and support of hundreds of missiles, to include lift, rearming and storage capabilities. All required inport logistics will be provided at these bases. All three bases have airlift capabilities that will facilitate the procurement, distribution, maintenance and replacement of materiel and personnel while the ships are inport. All organizational and intermediate level maintenance will be supported from these bases.

The *Arsenal Ship* will never be underway without appropriate escorts. Due to its limited self-defense capabilities and the high military value of its payload, the unescorted *Arsenal Ship* would present an irresistible target to potential foes. Consequently, the *Arsenal Ship* must be escorted by other fleet assets when it is moved into, out of, or within its area of operation.

3. Integrated Logistics Support

a. Maintenance Concept

Maintenance requirements for the entire ship system are kept to a minimum. The requirements for a near constant availability and dramatically reduced ships force demand use of radically different systems and procedures. Organizational repair capability is minimal. Shore and afloat intermediate maintenance activities (IMA) and detachments will be used to the maximum extent possible. Ship systems are redundant, and are standardized and modular in order to capitalize on the ease and speed of "unit replacement" vice "repair-in-place" maintenance concepts. The *Arsenal Ship* is designed and will be operated such that regularly scheduled depot level maintenance is infrequent.

The *Arsenal Ship* will return from forward operations every five years for dry docking and depot level maintenance.

The upkeep of the weapons payload is minimal. Missile certifications will be performed during the regularly scheduled depot level maintenance availabilities. Routine maintenance and operational checks are eliminated and/or automated providing the readiness goals inherent in the mission concept.

b. Logistics Considerations

The provisioning strategy inport is much the same as for the Maritime Prepositioning Ships. The forward operating bases support airlift operations so that stores can be flown in and transported to the ship. Underway, the *Arsenal Ship* will utilize standard fleet logistic requests for provisioning via VERTREP and for underway refueling.

Storage and handling facilities for massive amounts of missiles will be required at each forward operating base. Due to the massive amount of missiles stored on both ships and the storage/handling facilities required ashore, security and safety issues are of concern.

4. Manpower, Personnel and Training

The normal crew size for the *Arsenal Ship* is less than 50 personnel. To achieve this, the *Arsenal Ship* employs a manning concept that parallels a typical Merchant Marine tanker, augmented with the personnel required for the military operations of the vessel. The ship makes maximum use of automation for monitoring and control of all ship functions. Damage control techniques are radically different, requiring minimal personnel. The *Arsenal Ship* relies on passive measures incorporated into the design to inherently reduce susceptibility and vulnerability. The *Arsenal Ship*, by design, is virtually unsinkable. Fire fighting, flooding and stability control are largely automated and centrally monitored. This allows the *Arsenal Ship* to operate in combat situations with dramatically reduced numbers of personnel.

The driving factor in crew size is the manpower surge required in certain types of operations. Although it would not be unusual for a merchant vessel the size of the *Arsenal Ship* to be operated with a crew of about a dozen people, the *Arsenal Ship* is a warship and must be able to operate as such. Operations for extended periods at Condition 1 and Condition 3, refueling underway, precision anchoring, small boat operations and helicopter flight operations have all traditionally been manpower intensive. The *Arsenal Ship* is designed with innovative systems and will be operated with innovative procedures which will allow the ship to perform these functions safely and reliably with very few personnel.

The human element has not been discounted in the employment of the *Arsenal Ship*. A well trained, motivated, professional crew is the cornerstone of any weapons system. The general manning concept is closely tied to ship system design and to the maintenance concept. The ship is manned primarily by "operators." Since most of the maintenance functions are shore-based, a minimum number of maintenance-specific ratings are part of the crew. The crew focuses on training to operate and fight the ship. Constant readiness, intensive mission-specific training, and the constant routine of ship's business will be demanding with such a small crew. Personnel selected for duty in the *Arsenal Ship* program will require thorough screening, intense training and qualification, and a tireless work ethic.

The ships remain forward-deployed, with crew rotations normally taking place at the forward operating bases. In order to maintain a high level of readiness, and to respect the spirit of the Navy's personnel management goals, the entire crew rotates as a unit. This allows for the crew to train as a unit in the continental United States (CONUS) and then deploy as a unit. The crew cycle for the *Arsenal Ship* is approximately 18 months long. Approximately six months is spent assembling and training the crew. This detailing and training phase is followed by a twelve month deployment to one of the forward-deployed ships. Due to the intense duty, as well as the forward- deployed nature of the *Arsenal Ship*, assignment to the *Arsenal Ship* program is considered a hardship assignment. Personnel are assigned to the program for approximately 18 months. This would allow for a single complete tour on an *Arsenal Ship*. A crew member will either voluntarily extend or be reassigned at the end of their tour in the *Arsenal Ship* program.

In order to achieve maximum readiness with a crew of less than 50, several conditions must be met. First, when personnel are assigned to the *Arsenal Ship* program, they must complete the cycle. Second, it is paramount that the crew members arrive at the *Arsenal Ship* proficient in their rating and qualifications. With a small crew, the individual crew members must arrive at the ship ready to do their job with a minimum amount of on-ship training. These needs are incorporated into a CONUS based "*Arsenal Ship* School." Ideally, this school would not be a new training pipeline, but a crew assembly and integration headquarters. Personnel arrive at the "school," are assigned to a crew, and are sent to existing service schools for rating specific training. *Arsenal Ship*-specific training is done at the "school" in order to qualify officers and enlisted personnel for duty on the *Arsenal Ship*.

As stated before, the crew is primarily composed of "operators." At the "*Arsenal Ship* School," the training is simulator and mockup intensive. This philosophy is carried over to forward ports where additional simulation-based training is integrated into the ship's routine to augment the infrequent live fire exercises with other fleet assets.

The ship remains forward-deployed. Therefore, it is not available to participate in group work-ups in preparation for deployment. However, the crew, or at least a portion of the crew, is available. Through remote connectivity and existing fleet simulators, the crew links and trains with other fleet units, as if the Arsenal Ship is actually with the battle group. This type of training is necessary because the *Arsenal Ship* must operate as an integral part of a battle group in order to perform her mission.

5. Command and Control

The *Arsenal Ship*'s ability to interface with command and control systems, world wide, is the center of gravity for this platform. The most advanced communication systems are employed to create a network of sensor systems that link fire control information, in real time, back to the *Arsenal Ship*. When called upon by the targeting platforms, the *Arsenal Ship* will deliver overwhelming amounts of ordnance to a multitude of targets. Electromagnetic compatibility and frequency spectrum assignment issues are vital to the ship's ability to perform her mission. In addition, the communication and information security systems are impenetrable.

The level of connectivity required to enable the *Arsenal Ship* concept is revolutionary. This connectivity revolution is enabled through advances in Command and Control technologies married with intensive training and exercises with naval expeditionary forces. The Manpower, Personnel and Training section of the CONOPS addresses how the forward-deployed *Arsenal Ship* maintains its proficiency at interfacing with the Command and Control architecture of the expeditionary forces which routinely work-up and deploy as integrated groups.

6. Security

Due to the nature of the *Arsenal Ship*'s mission and its cargo, maintaining physical security is a paramount concern. At sea, the ship must be protected from all threats that can endanger the ship and its payload. Escort ships, coupled with a sound shipboard security plan provide the measures necessary to safeguard the *Arsenal Ship* and its missiles.

The forward operating bases which support and maintain the *Arsenal Ship*s require innovative security measures to protect the vital capability of these platforms. Opposing forces and terrorists on missions to damage U. S. warfighting capabilities and erode the resolve of forward-deployed forces will view these ships and the bases which support them as prime targets. The port facilities developed to service these missile laden ships must facilitate the enhanced levels of physical security required for such high valued assets.

7. Interoperability and Integration with Carrier Battle Groups and Amphibious Readiness Groups

The *Arsenal Ship*'s interoperability and integration with the regularly deploying naval expeditionary forces is of concern due to its forward-deployed nature of employment. The following passages and figure (I-2) describe the basic operational cycle of two *Arsenal Ship*s, stationed at a designated forward operating base, in six month blocks:

- Block 1 (months 1-6):

  - *Arsenal Ship* "A," with Crew A1, is underway in her area of operation with Group A1 (the group Crew A1 trained with prior to flying out and relieving the previous crew of ship "A").

  - *Arsenal Ship* "B," with Crew B1, is inport at the forward operating base conducting routine maintenance on redundant ship systems and preparing for crew turnover. The ship remains available to the Unified Combatant Commander for most of her time inport.

  - Back in CONUS, Crew B2 is assembled and is training with Group B2. Prior to the end of Block 1, Crew B2 will fly to the forward operating base and relieve Crew B1.

  - Detailers are writing orders and assembling members for Crew A2.

- Block 2 (months 7-12):

  - *Arsenal Ship* "B," with Crew B2, is underway in her area of operation with Group B2.

  - *Arsenal Ship* "A," with Crew A1, is inport at the forward operating base conducting routine maintenance and preparing for crew turnover.

  - Back in CONUS, Crew A2 is assembled and is training with Group A2. Prior to the end of Block 2, Crew A2 will fly to the forward operating base and relieve Crew A1.

  - Detailers are writing orders and assembling members for Crew B3.

- Block 3 (months 12-18):

  - *Arsenal Ship* "A," with Crew A2, is underway in her area of operation with Group A2.

  - *Arsenal Ship* "B," with crew B2, is inport at the forward operating base conducting routine maintenance and preparing for crew turnover.

  - Back in CONUS, Crew B3 is assembled and is training with Group B3. Prior to the end of Block 3, Crew B3 will fly to the forward operating base and relieve Crew B2.

  - Detailers are writing orders and assembling members for Crew A3.

This operating cycle supports the operational concepts outlined above, providing two trained *Arsenal Ships* continuously available to integrate with regularly deploying naval expeditionary forces in each of the three operating areas.

	Block 1 (months 1-6)		Block 2 (months 7-12)		Block 3 (months 12-18)
***Arsenal Ship* A**	Underway with Crew A1 and Group A1.		Inport with Crew A1 at FOB conducting routine maintenance and preparing for crew turnover.	Crew Turn-over	Underway with Crew A2 and Group A2.
***Arsenal Ship* B**	Inport with Crew B1 at FOB conducting routine maintenance and preparing for crew turnover.	Crew Turn-over	Underway with Crew B2 and Group B2.		Inport with Crew B2 at FOB conducting routine maintenance and preparing for crew turnover.
***Arsenal Ship* "School" (in CONUS)**	Crew B2 is assembled and training with Group B2.	Crew B2 Flies to FOB/Group B2 Deploys	Crew A2 is assembled and training with Group A2.	Crew A2 Flies to FOB/Group A2 Deploys	Crew B3 is assembled and training with Group B3.
Detailing	Detailers are writing orders and assembling Crew A2		Detailers are writing orders and assembling Crew B3.		Detailers are writing orders and assembling Crew A3

Figure I-2: Basic Operational Cycle for an *Arsenal Ship* Squadron

D. SUMMARY

In an era that now finds the majority of U. S. military might based solely in the United States, vice overseas in regions of vital national interest, the forward-deployed naval expeditionary forces will certainly play a much larger role in the early stages of future engagements. These expeditionary forces must be shaped to meet the demands of future warfare. The forward-deployed forces of today must be augmented with larger numbers of precision-guided weapons and more robust command and control systems if they are to succeed in their roles as outlined in the Navy's vision of future naval warfare. The *Arsenal Ship*, and the revolutionary warfare concept it enables, brings such capabilities to the fleet.

The CONOPS described above envisions a fleet of six *Arsenal Ship*s capable and ready to integrate with regularly deploying expeditionary forces. The concept is fully consistent with the Navy and Marine Corps' strategic concepts of expeditionary warfare as described in the papers *Forward ...From the Sea* and *Operational Maneuver From the Sea.* The *Arsenal Ship*s, together with traditional naval expeditionary forces, will provide the rapid, robust, combat-ready response critical to thwarting the advance of future regional aggressors and to shaping the battlefield so that follow-on forces can quickly and decisively win wars.

II. DESIGN PHILOSOPHY

A. OVERVIEW

"A top level statement of guidance, for the design team, to

assist in carrying out design tradeoffs in a consistent manner"

Professor Charles N. Calvano, CAPT, USN (Ret)

This design philosophy provides a prioritized list of factors to be used by the design team throughout the duration of the project. Specific issues and systems in the *Arsenal Ship* design are considered and trade-offs made. In order to ensure this is a logical and consistent process, the following prioritized list of the major design issues is developed. This section outlines and provides justification of the issues considered important enough to be incorporated in the design philosophy and thereby incorporated into the design.

B. PRIORITY OF DESIGN CONSIDERATIONS

1. Acquisition Cost and Life Cycle Cost
2. Mission Effectiveness
3. Survivability and Self Defense
4. Reduction in Manning
5. Reliability, Maintainability and Availability
6. Commonality: Other Platforms, Commercial off the Shelf (COTS), and Exploiting DoD Investments
7. Upgradeability and Modularity
8. Minimized Maintenance
9. Environmental Impact
10. Habitability

1. Acquisition Cost and Life Cycle Cost

The acquisition cost is the only concrete requirement placed on the design team. The acquisition cost cannot exceed 550 million dollars. All other requirements are secondary and tradeable.

Cost is given the highest priority because of the "cost cap" of the program. Some may argue that placing cost ahead of military effectiveness is unimaginable, but it is the driving factor in future procurement and getting the most effective platform for the least cost.

Life cycle cost is viewed as important as acquisition cost. The life cycle cost, if not considered in early stages of design, could make operation of the vessel uneconomical in the future. For example: The number of crew is a fixed operating cost for the life of the vessel. Also, fuel economy is designed into the hull and power plant and once finalized is only a function of how many miles the ship is to be driven. Decision trade-offs are to be made considering life cycle cost in mind, to provide a less costly ship today and in the future.

2. Mission Effectiveness

The primary role of the *Arsenal Ship* is to deliver a large number of missiles for other platforms to use. The priority of mission effectiveness is considered secondary only to cost. It is a focus of the design team to provide the most mission "bang" for the 550

3. Survivability and Self Defense

Survivability is defined as the capability of a ship to avoid or withstand a man-made hostile environment [6]. The survivability capabilities of the *Arsenal Ship* require critical design attention. Design considerations that increase survivability should be incorporated.

The ship design should exploit the use of passive self-defense measures by reducing infrared (IR) and electromagnetic (EM) signatures and employing electronic warfare measures and decoys. Additional active weapon systems may need to be incorporated to provide additional protection. Extensive hull compartmentation, hardening and automated systems were included to help in controlling damage, keeping the ship afloat, and providing graceful degradation of systems.

These features provide a very survivable platform. However the *Arsenal Ship*, also depends on the ability of the escort ships to provide a protective umbrella and early warning of threats coming within the area of operations.

4. Reduction in Manning

The design goal of the TSSE team is to limit personnel manning to 50. Manning reduction is primarily achieved through function automation in all aspects of ship operations, including ship control, engineering plant operations, damage control and warfighting operations. Good ship designs take advantage of technological innovations which replace or assist a crew member in making decisions, thereby reducing inconsistencies and human errors.

Watchstation manning levels are greatly reduced by relying on the automated systems to carryout routine, manpower intensive, duties. Manning levels for watchstation training are also reduced by automated systems, since fewer personnel are required. A reduction in training effort is a hidden benefit of automation. An adequately trained person should be able to operate the ship effectively and efficiently with the assistance of automated services.

Many skills cannot be replaced or done in a cost effective manner. The major effort will be to balance the two and create the best solution. Manning assignments will be addressed on a person by person basis. Every crew member will have to be justified and balanced against the cost of automation. Reductions in manning also provide hidden and intangible savings in the areas of reduced casualties.

5. Reliability, Maintainability and Availability

The requirement to operate forward-deployed, with minimal manning, for extended periods of time, make reliability, maintainability, and availability paramount in meeting the mission. Use of onboard equipment monitoring systems to provide Condition Based Maintenance (CBM) information predicts maintenance based upon need. Reliability Based Maintenance (RBM) techniques incorporated with typical Preventative Maintenance (PMS) practices cut maintenance time, reduces the number of spare parts, improves reliability, improves system performance and reduces manning. The use of easily interchangeable spares and redundant systems also help provide additional reliability and availability of the *Arsenal Ship*.

6. Commonality: Other Platforms, Commercial off the Shelf (COTS) and Exploiting other DoD investments

Systems that are already in use in the Navy or other services were selected for use on the *Arsenal Ship*. The supporting infrastructure is already in place and can be tapped without additional cost. COTS items are to be chosen where DoD systems do not exist. This provides fast procurement, of spares and replacements, easy upgrading and interoperability. Systems and equipment from Aegis platforms are to be selected, wherever possible, because of the ease of obtaining spares from these escort ships.

Other DoD programs, such as "Smart Ship" are to be relied upon for information about new systems and programs to be incorporated into the *Arsenal Ship*, to reduce the need to invest in research and development funding.

7. Upgradeability and Modularity

Future growth and entire system upgrades are to be designed into the vessel. The ship is designed to be upgraded quickly, in order to significantly cut the cost and minimize the time and effort to upgrade. Modular designs also allow quick repairs by allowing change out vice repair in place. Locations for allowing removal of large equipment were included to limit down time for maintenance, ensuring the ability to meet the design goal of 95% availability.

A design goal of the ship is to provide space for a future gun system in the near future. The ship is designed to allow for this installation and already provides the needed supporting systems to speed the installation and prevent loss of availability.

8. Minimize Maintenance

The duel challenge of increased system sophistication and decreased manning in the *Arsenal Ship* is to be accomplished by using new maintenance technology. These systems can incorporate monitoring and analysis and replace the current program of preventative and corrective maintenance. The systems identify degraded or degrading performance. Unattended embedded sensors continuously monitor the condition of all equipment. Monitoring stations record and analyze the information. Appropriate warnings and equipment shutdowns are controlled by the monitoring station. Critical maintenance items are attended to by the limited personnel onboard. Other items may be automatically scheduled for an upcoming port call and shore side technicians attend to the maintenance needs. Wherever possible all major maintenance is designed to be conducted "dockside" in short mini-availabilities, thereby lengthening the time between required drydockings to 5 + years. The reduction of longer maintenance time (drydockings) by performing needed maintenance dockside will allow the availability of the *Arsenal Ship* to remain above 95%.

It should be noted the design allows for the maintenance work to be completed by either Navy personnel or by a commercial contractor. Systems and required maintenance will be selected on their easy of use and ease of repair.

9. Environmental Impact

Environmental regulations presently in force, today and those expected to be in place in the near future, are to be adhered to in the design. Great effort should go into balancing the cost and need of creating an environmentally friendly warship as reasonably and cost effectively as possible. Systems are considered by their use of environmentally friendly materials and processes, prevention of typical waste disposal problems (trash dumping), and prevention of negative image related incidents (oil spills, smoke generation, etc.).

10. Habitability

Sufficient habitability standards are considered as to provide adequate living space for all crew members without compromising morale. The need for higher habitability standards over conventional naval ships arises due to the smaller, mixed gender, and isolated crew of this minimally manned ship. The attention given to this area can be seen in large living quarters and recreational areas.

III. DESIGN FOR REDUCED MANNING

A. INTRODUCTION

The design of the *Arsenal Ship* must incorporate innovative measures to reduce manning. The requirements are specified as follows:

- "To meet mission goals at affordable cost, ship design will be based on commercial practices and rely extensively on automation in engineering, damage [control], ship [control] and weapon systems to achieve a crew size of no more than 50." [4]
- "Life Cycle Considerations. The ships are to be manned, if at all, by a Navy crew to be as small as practicable, but in any event not to exceed 50 people." [5]

The rationale for this concept is primarily based on cost. An analysis of operating and support (O&S) costs for a typical destroyer [7] reveals that personnel costs have the most significant impact on ship affordability. This is depicted in figure (III-1), where personnel costs are shown to represent 65% of O&S expenditures. Although this percentage may not be accurate for the *Arsenal Ship*, its relative proportion is considered a reasonable approximation. Therefore, to design an affordable *Arsenal Ship*, major emphasis must be place on manning reduction.

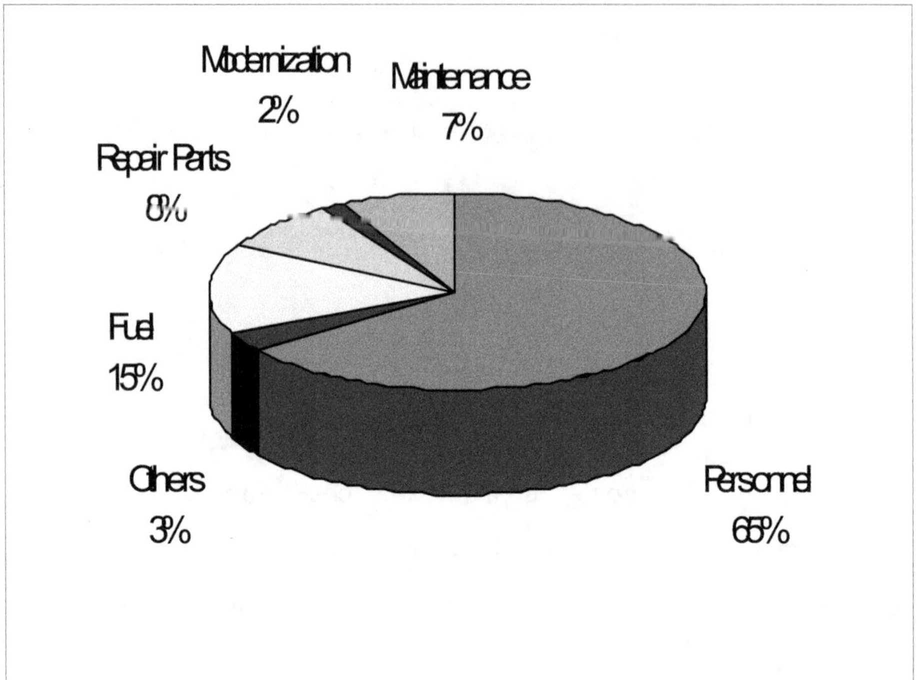

Figure III-1. O&S Costs for a Typical Destroyer

An additional benefit of reduced manning is the reduction of personnel placed in harm's way. By minimally manning the *Arsenal Ship*, personnel density is reduced and therefore, personnel loss due to casualty is lower.

This manning reduction is accomplished by extensive use of automation, electronics and computer technologies, as well as enacting concepts that challenge current Navy customs and traditions. These concepts or core themes are summarized as follows:

- Automation, Electronics and Computer Technologies
- Cross Functionality of Personnel
- Experience-- Professional Maturity
- Top Performers
- Selective Screening

In the design for reduced manning, extensive implementation of automation, electronics and computer technologies is required. The availability and affordability of advanced shipboard sensors and actuators, coupled with increased computer processor speeds, provide the means to effectively and safely reduce shipboard manning.

Secondly, personnel aboard the *Arsenal Ship* are required to be versatile. Cross-functionality is a crucial element to this concept of reduced manning. Crew members must not only demonstrate expertise in their area of specialty, but will be called upon to perform additional tasks, that may require additional training prior to deployment.

Additionally, a minimally manned ship requires a crew that is highly experienced. Personnel must report aboard fully trained, qualified and ready to perform at full capacity. Only self-motivated professionals are suitable for assignment to this ship.

Finally, discriminating medical, dental and psychological evaluations are required. In a manner similar to selection for submarine duty, *Arsenal Ship* personnel must be "hand-picked" for assignment. With a small, highly trained crew, personnel losses must be kept to a minimum. Selective screening is required to ensure that only "top-flight" sailors are assigned to the *Arsenal Ship*.

B. TRAINING CONCEPT

1. Overview

The *Arsenal Ship* reduced manning concept demands the ship be staffed with a fully mission capable crew requiring little or no onboard training. To satisfy this goal, crew members must arrive to the ship fully trained and qualified, immediately capable of performing the ship's mission. To do this, an *Arsenal Ship* training command must be established. This does not require a new training pipeline, but is a supplement to existing service schools with *Arsenal Ship*-specific simulator-based training.

2. The *Arsenal Ship* Training Command (ASTC)

At ASTC, crews are formed and trained, and then sent as a unit to the forward-deployed ship. This command serves as administrative headquarters for the *Arsenal Ship* program, as well as the homeport for the crews. The ASTC should be located in the vicinity of existing fleet infrastructure (i.e. Norfolk, VA). This reduces the requirement for construction of administrative and support facilities.

3. Crew Training Cycle

The *Arsenal Ship* training cycle is designed to build and deploy the crew as a cohesive unit. The *Arsenal Ship* cycle is approximately eighteen months in length. The first six months of the cycle are spent training the crew both as individuals and as a unit. During the next 12 months of the cycle, the crew is assigned to one of the forward-deployed *Arsenal Ship*s.

4. Individual Training Concept

a. Rating/Designator Training

It is paramount to the reduced manning concept that each crew member on the ship is fully qualified according to billet. To accomplish this, existing service schools are used to provide initial and refresher training and qualification. This may require creation of additional *Arsenal Ship*-specific courses suited to the special needs of the program.

Additionally, existing schools may need to provide instructional and facility support to the program.

b. Cross-training

Crew cross-training is another cornerstone of the reduced manning concept. Crew members will perform numerous interdisciplinary tasks. Cross-training is performed along departmental lines. For example, a crew member reporting to the engineering department must be familiar with the operation and maintenance of every aspect of the engineering plant, not just his rating area. Service schools may need to develop courses to support cross-training personnel in out of rating area specialties.

5. Crew Training Concept

a. Special Detail Training

Like any other ship, special evolutions, such as underway refueling, boat operations, and flight quarters, are manning and training intensive. By locating the ASTC near an existing fleet infrastructure, land-based trainers and mockups can be used to perform team training for these special evolutions.

b. Fleet Integration

The *Arsenal Ship* is envisioned as a supporting platform requiring dedicated fire control support from fleet assets. Because the *Arsenal Ship* is forward-deployed and unavailable to participate in traditional battle group workups (i.e. FLEETEX, BGE), its crew must be trained in a manner similar to inport team training presently conducted at Fleet Combat Training Centers. Through remote connectivity and existing fleet simulators, the crew is able to link and train with other fleet units, as if the *Arsenal Ship* was actually steaming with the battle group. This method of fleet integration enables the *Arsenal Ship* to remain forward-deployed while its relief crew is fully trained and integrated into the deploying battle group.

6. Effect on Manning

a. *Arsenal Ship* Manning

The concept of deploying trained people to forward-deployed equipment is not new, and is routinely performed in the U.S. Marine Corps. The *Arsenal Ship* adopts this concept in order to meet the requirements of steaming a minimally manned crew and an availability of 95%. Presently, crew members on naval warships are sent to schools when certain training requirements onboard are delinquent. The remaining crew members perform not only their own assignments, but also help take up the slack of their missing shipmate. The *Arsenal Ship* is already minimally manned and crew members will not be able to leave the ship for extra training once forward-deployed.

C. MAINTENANCE CONCEPT

The *Arsenal Ship* is designed to operate forward-deployed for extended periods of time with minimal maintenance requirements. New methods of maintenance must be developed for this to be performed by a reduced crew.

1. Depot-Level Maintenance

The *Arsenal Ship* is required to be available 95% of the time [4]. This requires no more than 18 days of unavailability for combat operations per year, or 90 days every five years. To accomplish this, most intermediate- and depot-level repairs are performed at the forward operating bases (FOB). Access soft patches in the aft superstructure allow equipment to be easily removed and replaced requiring little more than crane service. For example, the normal lengthy overhaul of a diesel engine is reduced to a few days by removal and reinstallation of a new diesel engine, vice repair of the engine in place.

Drydocking is conducted in 90 day availabilities every five years. The driving factor for this five-year periodicity is the missile recertification requirements of the ship's payload. The *Arsenal Ship* returns to CONUS for missile offload at a Naval Weapons Station (i.e., Seal Beach, CA or Yorktown, VA) prior to entering drydock.. Upon completion of drydocking, the ship is outfitted with a fully-certified missile payload.

2. Intermediate-Level Maintenance

Each FOB must have an intermediate-level maintenance activity (IMA) capable of supporting the forward-deployed *Arsenal Ship*s. These IMAs are crucial in maintaining 95% availability because the ship is far removed from depot-level maintenance activities. Additionally, IMA personnel are needed to augment ship's force in performing periodic and corrective maintenance during inport periods.

3. Organizational-Level Maintenance

Three different, but linked, maintenance systems are implemented in *Arsenal Ship*: Periodic Maintenance System (PMS), Reliability-Based Maintenance System (RBMS) and Condition-Based Maintenance System (CBMS) . These systems reduce unneeded maintenance. The PMS and RBMS determine maintenance requirements based upon known characteristics of installed equipment. The CBMS is a computer-based monitoring system integral to the Engineering Control and Monitoring System (ECMS).

Sensors are attached to major pieces of equipment and transmit status to the monitoring system. Baseline information recorded at equipment installation is used to determine required maintenance intervals. Significant time-intensive PMS items are replaced by the use of the CBMS. Maintenance is driven by need, vice a preset time interval. When used effectively, coordinated use off these three maintenance systems maximize availability of the ship and decrease maintenance man-hours and costs.

4. Redundant systems

A major consideration in the design of the *Arsenal Ship* is the use of redundant systems. These systems allow a shift to offline equipment in the event of failure vice immediate repair. The failed equipment can remain idle, pending repair. The equipment can then be repaired by ship's force, a technical assistance team, or by the IMA upon returning to port. Sufficient redundant systems are installed to prevent loss or degradation of any primary mission area.

Redundant systems also add to the overall survivability of the vessel. Multiple redundant systems give the ability to circumvent battle damage and restore lost functions quickly, greatly increasing survivability and ship availability.

5. Modularity

Modularity encompasses several concepts. This design incorporates two of them. The first is the use of commonality, interchangeable components in various systems. By pre-selecting common components (i.e., pumps and motors), the ability to use a single spare for multiple systems reduces required logistics infrastructure and parts storage. An added advantage is that you can take a component from a non-vital system to replace a damaged vital system component in an emergency.

The second is the use of total modular systems. For example, each auxiliary system can be mounted on an individual skid. This concept supports the 90 day depot-level maintenance period by allowing entire systems to be swapped out, vice repaired in place. This creates a rotatable pool of well-maintained equipment to help maintain the required 95% availability of the *Arsenal Ship* fleet. Another benefit is the ability to perform configuration changes easily and quickly, keeping *Arsenal Ship* systems on the cutting edge of technology.

6. Industrial Coatings

Commercially available coatings are used to the fullest extent possible to prolong the time between topside paintings. These coatings include those used on industrial machinery and submarines. This is coupled with new methods of application that make underwater coating possible, thereby extending the time between drydockings for hull inspections.

D. SPECIAL EVOLUTIONS

1. Overview

Special evolutions are manpower intensive aboard naval ships. The purpose of this section is to describe how the *Arsenal Ship* is designed to reduce manning for these evolutions when compared to current practices. *Arsenal Ship* special evolutions are listed below.[1]

- Underway Replenishment (UNREP)
- Helicopter Operations
- Boat Operations
- Sea and Anchor Detail

2. Underway Replenishment

The Navy's current methods of underway replenishment are far too manpower intensive for the *Arsenal Ship* [8]. Table III-1 shows the total number of personnel required for UNREP on a Arleigh Burke destroyer is approximately 80, depending on the number of stations involved and whether the evolution involves refueling at sea (RAS) and/or connected replenishment (CONREP) [9]. The large manpower requirement is primarily due to the need for linehandlers to pull receiving lines aboard.

1. Gunnersmate							
2. Fuel Sampler							
3-8. Phone and Distance Line							
9-22. Linehandlers for Refueling							
23-36. Linehandlers for Highline							
37-52. Rig Team							
52-62. Ready Lifeboat							

Table III-1. Typical Naval Combatant Manning for Underway Replenishment [8]

[1] NOTE: The ideas presented in this section are based on the at-sea experience of the design team, and are developed in response to the *Arsenal Ship*'s reduced manning concept. These ideas require exceptions to [8], [10] and [11].

The *Arsenal Ship* carries enough fuel for a 90 day mission such that there is no need for RAS. If RAS becomes necessary, the *Arsenal Ship* has capstans located at each refueling station in order to reduce the number of line handlers to four (Table III-2). The primary method of stores transfer to the *Arsenal Ship* is vertical replenishment (VERTREP). Wireless communications and laser rangefinders replace the phone and distance line and station-to-station phonetalkers to further reduce the manning requirement for this evolution.

Phone and Distance Line							
1. Phone Talker							
2. Linehandler							
3. Linehandler							
4. Linehandler							
5. Linehandler							
6. Gunnersmate/Signals							
7. Fuel Sampler							
8. Winch Operator							
9. Safety Observer for Winch							
10. Linehandler							
11. Linehandler							

Table III-2. Projected *Arsenal Ship* Manning for Refueling at Sea Evolutions.

3. Helicopter Operations

Helicopter landing capabilities are critical. The helicopter is the primary means to transfer personnel (i.e., technical support personnel, medical evacuation). Since the helicopter lands on the *Arsenal Ship*, the ship has landing deck lights, tie-downs, safety nets, and a glideslope indicator.

In order to minimize manning and maximize on-station time, the *Arsenal Ship* relies on VERTREP, vice CONREP, for rapid stores transfer. The *Arsenal Ship* has a strikedown elevator in the vicinity of the flight deck to strike below stores. The longitudinal passageways that link the fore and aft sections of the ship provide a path between the strikedown elevator and the freezer, chill box, dry provisions storeroom and ready storage spaces located forward. To move the stores from the flight deck to these storage locations, a conveyance system is installed along these longitudinal passageways.

The *Arsenal Ship* has a class-3 flight deck certification for flight deck operations [10]. Class-3 certification means the ship only provides a landing platform with no services. The decision not to carry JP-5 is based on the personnel required to maintain an additional fuel oil service and transfer system and process the fuel. Should the need arise for helicopter refueling, the escort ship or VERTREP delivery ship is responsible for providing the necessary services.

To further reduce manning during this evolution, a remotely-operated, integrated firefighting system is installed. This system includes sensors, video monitors and AFFF flight deck sprinklers. The helicopter control officer, with full visibility of the flight deck, controls activation of this system from the helicopter control station. Since this system is used to provide the initial response to a flight deck fire, the number of hoseteams is reduced to one (Table III-4).

Typical Naval Combatant
1-2. HIFR/Hotpump/Grounders (static discharge)
3. Landing Signalman
4. Helo Control Officer/comms/AFFF sprinkler operator
5-6 Chocks/Chainmen
7. Fire Party Scene Leader
8. Sound Powered Phone Talker
9-11. Plugmen/Reel Tenders
12-17. Hosemen
18-20. Team Leaders/Nozzlemen
21-22. Hotsuitmen
23-32. Ready Lifeboat Crew
Arsenal Ship
1. Landing Signalman
2-3. Chocks/Chainmen
4. Fire Party Scene Leader/AFFF sprinkler activator
5. Sound Powered Phone Talker
6. Plugman/Reel Tender
7-8. Hosemen
9. Team Leader/Nozzleman
10-11. Hotsuitmen

Table III-3. Typical Navy Combatant vs. Projected *Arsenal Ship* Manning for Helicopter Operations [10].

4. Boat Operations

The rigid-hull inflatable boat (RHIB) is essential for the *Arsenal Ship*. It provides additional flexibility to an already minimally manned ship. The RHIB not only serves to transport additional personnel to the ship for corrective maintenance, but also provides a ready lifeboat during certain evolutions. Requirements for the *Arsenal Ship*'s RHIB are as follows:

- The *Arsenal Ship* carries the smallest RHIB in the Navy's inventory.
- The RHIB burns F-76 diesel fuel, marine (DFM). The *Arsenal Ship* does not carry aviation fuel (i.e., JP-5), onboard due to the additional storage and maintenance requirements, and does not carry gasoline due to the intolerably low flash point that presents a fire hazard to the ship.
- The RHIB is stowed on the main deck internal to the ship, to prevent an increase in radar cross-section (RCS).
- The *Arsenal Ship* reduces the manning for lowering and operating the RHIB as shown in Table III-4.

Navy Combatant (lower small boat)	*Arsenal Ship* (lower small boat)
1. Safety Observer	1. Safety Observer
2. Davit Captain	2. Davit Operator
3. Davit Operator	3. Forward Linesman
4. Forward Linesman	4. Aft Linesman
5. Aft Linesman	
Navy Combatant (operate boat)	***Arsenal Ship* (operate boat)**
1. Coxswain	1. Coxswain/Bowhook
2. Engineman	2. Engineman/Swimmer
3. Bowhook	
4. Swimmer	
5. Boat Officer	
TOTAL MANNING: 10	**TOTAL MANNING: 6**

Table III-4. Typical Navy Combatant vs. Projected *Arsenal Ship* Manning for Small Boat Operations [9].

During helicopter operations and underway replenishment special evolutions, the *Arsenal Ship* will not normally man a ready lifeboat. The standard procedure for these evolutions is to use a battle group asset as the primary man overboard recovery unit. This requirement stems from the fact that the *Arsenal Ship* is minimally manned, lacking extra personnel to man special evolution watch stations and a ready lifeboat simultaneously.

5. Sea and Anchor Detail

The sea and anchor detail evolution is also a manning intensive operation due to the significant number of linehandlers required at each station. The *Arsenal Ship* reduces this number by positioning retractable capstans in close proximity to each station. This decreases the number of personnel required from ten per station to merely four. The four consist of a safety observer, capstan operator, and two linehandlers. The total number of personnel involved in this evolution is eight; two four-man teams, one fore and one aft. These teams, working in concert, make up two lines at a time until the ship is tied up.

A keel anchor is installed for several reasons. By removing topside equipment, the radar cross section, maintenance and preservation requirements are reduced. This reduction in maintenance and preservation requirements significantly reduces man-hours The evolution watch is reduced from six personnel to one [9].

6. Maintenance

Table III-5 contains the special evolutions PMS requirements performed at a periodicity not greater than quarterly by Deck Department personnel [11].

Boat Operations	7 PMS checks
UNREP	12 PMS checks
Helo Ops	14 PMS checks
Anchoring	2 PMS checks
Other (lifeboats, lifevests etc.)	4 PMS checks

Table III-5. PMS requirements for Deck personnel on topside equipment.

The majority of these PMS checks are inspections and, by shielding the equipment from the weather, the volume of corrective maintenance based on these checks is drastically reduced. The recommended deck force personnel to accomplish topside PMS and corrective maintenance is three [11].

Figure III-2 contrasts the projected *Arsenal Ship* manning requirements for special evolutions against the current Navy combatants.

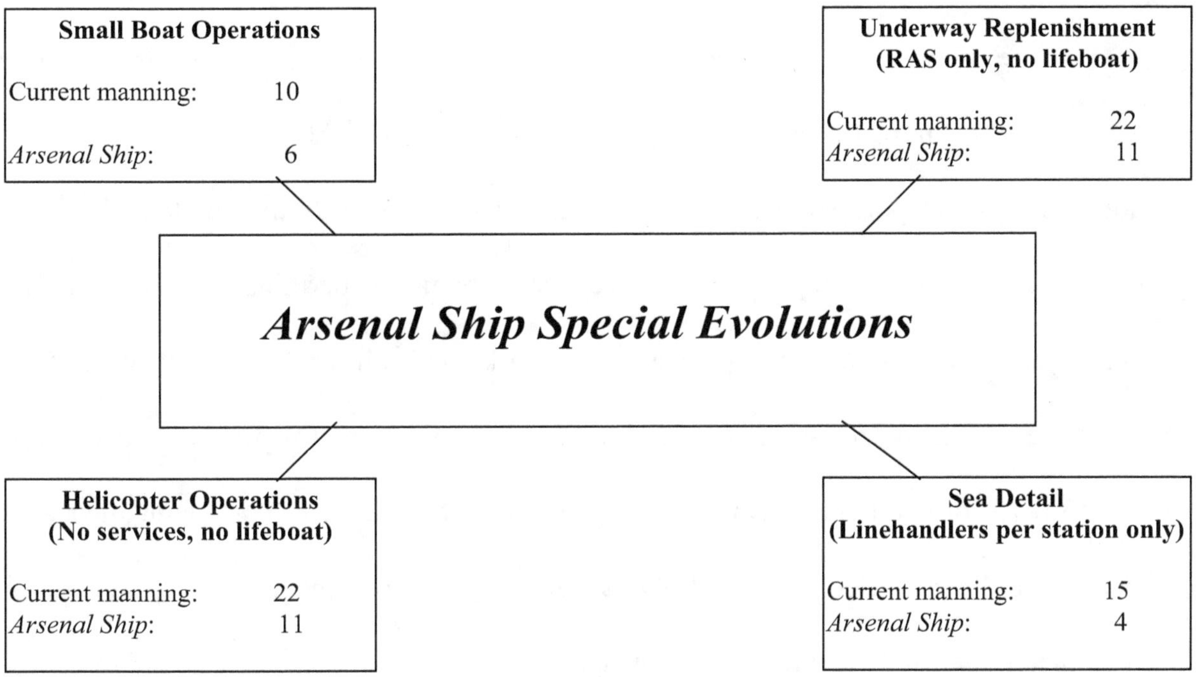

Figure III-2. Typical Navy Combatant vs. Projected *Arsenal Ship* Manning for Special Evolutions.

E. ADMINISTRATIVE CONCEPT

1. Overview

In the past, shipboard administrative functions (i.e., service records, disbursing) have had a significant impact on manning requirements. This situation has been further exacerbated each time a new program is developed. The purpose of this section is to describe how the *Arsenal Ship* is designed to reduce manning for administrative duties onboard.

2. Administrative Concept

The *Arsenal Ship* relies on computer networks to handle all administrative duties. The goal is to make the *Arsenal Ship* a "paperless" ship. The ship does not have any administrative support personnel onboard. All personnel records are maintained using the COMPASS computer program [12]. Every stateroom and berthing compartment has a computer for the ship's crew to access the network. This allows easy exchange of fitness reports, evaluations, and training records; ordering of parts; and work order generation to all be performed on the network. A hard copy of each record is maintained at the Personnel Support Detachment (PSD) at the respective home port. All records are periodically updated via wireless transmission.

3. Postal Services

A ship's post office and postal clerk are not assigned to the *Arsenal Ship*. This is possible because electronic mail can replace paper letters. Furthermore, each crew member has access to a personal computer located in their living quarters. Additionally, all personnel receive indoctrination training on methods of electronic financial management during their six-month pre-deployment period at the ASTC. A crew member is assigned the collateral duty of mail handling.

4. Medical and Dental Services

The ship has an Independent Duty Corpsman (IDC) assigned and sickbay facilities onboard. The IDC administers emergency care to the crew. All routine medical and dental examinations and procedures are performed inport at appropriate clinics. The

Arsenal Ship screening process discussed previously minimizes the need for medical and dental services underway.

5. Disbursing Office

In addition to being paperless, the *Arsenal Ship* is also a cashless ship. The ship does not have vending machines or a traditional ship's store. All snacks and necessary laundry items are provided for each crew member. Since the *Arsenal Ship* is not expected to visit any foreign ports due to physical security requirements, there is no need for a Disbursing Officer.

F. HABITABILITY

1. Overview

The purpose of this section is to describe how the *Arsenal Ship* is designed to maximize the size and comfort of living arrangements for a minimal crew. This section also describes the messing concept and miscellaneous crew services.

2. Messing

Two Mess Specialists (MS) are assigned to the *Arsenal Ship* to prepare and cook the meals for the entire crew. A mess attendant is assigned from the junior enlisted personnel onboard to assist the MS in preparation and cleanup of the mess and galley. The MS can use the computer in his berthing compartment to prepare the menu and automatically track the inventory of all necessary items. The entire crew pass through the buffet-style line, but dine in their own messing areas. These dining spaces include a general mess, chief petty officer' (CPO) mess, and officer's wardroom. The location of the dining facilities and food storerooms is forward in close proximity to the primary habitability area.

3. Berthing

The berthing spaces onboard the *Arsenal Ship* are designed such that all officers have individual staterooms. CPOs are assigned to two-person quarters, and enlisted personnel (E-6 and below) are assigned to six person berthing spaces. All rooms include complete head facilities. An additional benefit of this layout is that female sailors are easily accommodated. All rooms also include a television, VCR and carpeting. This "home away from home" setup is created to provide sailors a relaxing environment for their time outside normal working hours. Mess attendants are not assigned for the officer or CPO quarters due to reduced manning. All personnel are responsible for cleanliness of their living quarters.

The forward berthing spaces are designated the primary habitability area. This area includes sufficient berthing to accommodate 12 officers, 12 CPOs, and 24 other enlisted personnel. Additional berthing accommodations are located aft. This area includes sufficient berthing for five officers, four CPOs, and 24 enlisted personnel. The additional berthing not only provides for overflow, but also provides the ability to distribute their crew during wartime operations. Distributing the crew increases their survivability mitigating the effects of a single missile strike in a berthing area.

4. Crew Services

a. Laundry

The ship is designed to have a separate laundry space that includes two washing machines and two dryers. Ship's personnel are responsible for doing their own laundry. In order to reduce the need for pressing services, the standard uniform underway is fire retardant coveralls [13]. Laundry detergent and dryer sheets are provided to the crew as needed.

b. Ship's Store

The *Arsenal Ship* does not have a ship's store onboard. The crew may purchase the standard ship's ball caps and collared shirts at their home port. Crew members are required to bring any necessary items that they may need underway. This concept is the standard mode of operation for the submarine community. The *Arsenal Ship* uses this mode operation due to the reduced manning concept.

c. Barber Services

The *Arsenal Ship* does not have a barber shop onboard. Underway, personnel are not required to keep their grooming standards within regulations until the week before the ship returns to port. A crew member is then assigned as the ship's barber and cuts all

d. Crew Recreation/Physical Fitness

The ship is designed with a crew recreation room. This room is available during working hours for meetings and classroom training, and after working hours for socializing. The ship also has a physical fitness space onboard. The space includes all of the standard state of the art equipment (i.e., weights, bicycle and stair climbers), and any desired Nautilus equipment.

e. Divine Services

All divine services are conducted by ship's crew members assigned as lay leaders.

G. AUTOMATION

1. Overview

The *Arsenal Ship* takes advantage of the latest technology to perform routine watchstanding procedures. The installation of a central, automated system, similar to the one used in the Smart Ship Program, is the baseline for the *Arsenal Ship* [14]. Several additional subsystems are added to improve overall system performance.

2. Voyage Management System

The Sperry Marine Vision 2100 Voyage Management System (VMS), with an integrated bridge, is used for navigation [15]. It is a commercially available, Windows NT based, computer navigational system, commonly referred to as the Integrated Bridge System (IBS). This system allows only two watchstanders to safely navigate the ship. The operator has full ship control, or he can allow the system to automatically control the ship by inputting selected waypoints at the keyboard. The ADG3000 autopilot subsystem keeps the ship on track by controlling the ship's course and speed. The MK 37 Gyro, SRD-421 Doppler Radar, Global Positioning System (GPS), and depth and weather sensors provide the VMS with necessary information to automatically correct the ship's track. In autopilot, the bridge operators are only needed to monitor alarms, control casualties, establish communications, and serve as lookouts.

The VMS uses electronic charts that can be created in the chartroom using a chart digitizer, or purchased prior to deployment. Additionally, a RASCAR VT radar system provides the autopilot with a collision avoidance feature and displays radar over-layed images on the IBS console .

Additional modules provide voyage recording (i.e., blackbox recorder), docking displays, precision anchoring, man overboard monitoring, and engineering and damage control status. These features give the two watchstanders adequate information to properly handle abnormal situations without the requirement for additional personnel on the bridge.

3. Two Wire Automatic Remote Sensing Evaluation System

The Two Wire Automatic Remote Sensing Evaluation System (TWARSES), coupled with remote television cameras, provide the bridge, engineering, and combat system watchstanders with complete all around visibility and interior space monitoring [16].

4. Standard Monitoring Control System

The Standard Monitoring Control System (SMCS) is an integrated control and monitoring system for all shipboard machinery [17,18]. Watchstanders in the engineering control station (CCS), combat information center (CIC) and the pilothouse have displays depicting the engineering plant status. The enginerooms are unmanned. The engineering watchstanders carry a personal information pad (PIP). The engineering watchstanders are free to roam throughout the ship until the PIP beeps, notifying the watchstanders of an engineering casualty.

The SMCS provides graphical interface, as well as a hardcopy printout. This allows the engineering watchstander to make quick decisions affecting the status of the engineering plant. The SMCS, in conjunction with the ICAS (Integrated Condition Assessment System) and Damage Control System (DCS), is capable of complete control of the engineering plant including casualty control procedures [17-21]. The SMCS can shutdown and startup systems. It can also cross-connect systems and prevent loss of any vital equipment. The SMCS uses an open-architecture design that allows for future equipment upgrades and changes.

5. Damage Control System

The DCS is a computer system that monitors and initiates controlling actions in the event of a casualty. The DCS provides real-time graphical information (i.e., stability, electrical isolation) to the control stations allowing rapid decision making. In the fully-automated mode the information is passed to the SMCS. This allows the SMCS to automatically isolate the compartment and activate installed damage control systems. The DCS does not require a dedicated watchstander. The engineering watchstander are responsible for monitoring the system's displays. The bridge and CIC watchstander can also monitor the damage control status of the ship locally, via the DCS monitor on their respective consoles.

6. Integrated Condition Assessment System

The Integrated Condition Assessment System (ICAS) monitors, tracks, and provides a complete machinery-condition assessment. It is a diagnostic tool for maintenance management. By monitoring the equipment, a maintenance requirement estimate can be predicted. This reduces unneeded time-based maintenance. No watchstander are required for this system. Pending problems are reported to the engineering watchstander for maintenance scheduling.

7. Local Area Networks

Zonal Local Area Networks (LANs) are interconnected with the ship's administrative LAN and provide engineering and ship control status to the consoles in CIC, CCS and the pilothouse. Each of these multi-purpose control centers are capable of duplicating the entire functions of the other two stations (i.e., CIC can monitor engineering plant status), thus increasing the survivability of the overall system. The control stations are also capable of displaying multiple system information on several displays

The automated systems described above provide for ship control and engineering monitoring with only three watchstanders.

H. DAMAGE CONTROL

The *Arsenal Ship* uses a combination of automated systems to drastically reduce damage control party manning. The Damage Control System (DCS), in conjunction with the Standard Machinery Control System (SMCS), control the installed engineering and damage control systems automatically. The Two Wire Automatic Remote Sensing

Evaluation System (TWARSES) is the remote-sensing system that monitors all shipboard compartments.

The DCS continually monitors the TWARSES for problems. TWARSES sensors are located in every compartment, tank and void. The DCS can be monitored from CIC, CCS, and the bridge. This removes the requirements for a dedicated damage control watchstander.

The SMCS and DCS are capable of enacting a predetermined sequence of events in response to damage. These systems isolate the damaged compartment by closing hydraulic watertight doors surrounding the damage. Electrical and electronic equipment and ventilation to the affected space are secured in accordance with preprogrammed doctrine. Depending on the damage detected, the SMCS activates installed damage control systems. The compartment is isolated within seconds of damage detection, with no crew member action.

A highly capable, six-man damage control party arrives on scene to take additional action, if required. The damage control party is comprised of personnel from the non-watchstanding maintenance force.

I. COMBAT SYSTEM MANNING REQUIREMENTS

1. Overview

The *Arsenal Ship* concept strives for functionality and simplicity with minimum required manning. Integrated sensors, computers and electronics have allowed the automation of functions that previously required human monitoring and response. At the same time, solid state and digital electronics have become more reliable, longer-lasting and require less maintenance. The assumption, based on this trend, is that electronics will only continue to improve and can be relied on for monitoring and response functions and require little intervention. The largest mechanical component of the *Arsenal Ship*'s combat suite is the Vertical Launching System (VLS). A preliminary study has been completed to determine the minimum number of personnel required to perform preventative maintenance on the MK 41 VLS [20].

The combat system watchstanding concept is two-fold, with combat system watchstanders (CSWS) and combat system maintenance personnel (CSMP). Watchstations consist of the Combat System Officer of the Watch (CSOW), Communications (COMMS), and Ship Defense (SD) (Table III-6). CSMP consist of the combined rates of Electronics Technicians (ET) and Gunner's Mate Missile (GMM). The ET rate brings electronics expertise and the GMM rate the missile, launcher and small arms expertise.

2. Command

The Commanding Officer (CO) is ultimately responsible for the safety of the ship. He also ensures that the *Arsenal Ship* can carry out its combat mission. The CO can personally exercise his missile launch authority from the Combat Information Center (CIC). The CO is the only officer who can take the ship from the "weapons-tight" to "weapons-free" condition. Once the ship is placed in the "weapons-free" condition, the CSOW can be delegated the authority to launch all defensive weapons.

COMMAND					
		CO			
		XO			
CSOW		COMMS		SD	
O4	CSO	LDO	RM	E6	EW
O3	WO	E6	RM	E5	EW
COMBAT SYSTEM MAINTENANCE					
E6	ET	E5	GMM		
E6	ET	E4	GMM		
E5	ET	E4	GMM		
E5	ET	E4	GMM		
E6	GMM	E4	GMM	TOTAL: 16	

Table III-6. Combat Systems Manning Requirements

3. Combat Systems Officer of the Watch (CSOW)

CIC and the Communications Center are fully manned with a CSOW, COMMS and SD when the *Arsenal Ship* is in a combat area. The CSOW operates from the Advanced Tactical Weapons Control System Console (ATWCS), monitoring system and weapon statuses, selection requests, and post-fire assessment. Combat system maintenance and damage control parameters are monitored in CIC on the combat system control console. This console is identical to the SMCS in CCS and the IBS on the bridge. The CSOW may be thought of as a Local Area Network (LAN) manager, monitoring the entire combat system with the capability to access it to respond to individual casualties. The CSOW uses the system and determines the required level of casualty control. He can task maintenance personnel remotely via the personal information pad (PIP). The CSOWs stand a six-hour watch during combat operations. During peacetime operations, the CSOWs conduct system diagnostics and review prioritized maintenance work lists, and are not required to remain in CIC.

4. Communications (COMMS)

The COMMS watchstander is responsible for ship connectivity. The COMMS watchstander ensures the proper communications plan is established, setting up and monitoring long-haul, tactical and link communications through the SACCS (Ships Automated Communications Control System). He conducts message delivery and communications system diagnostics. The COMMS watchstander stands a six-hour watch for one week, rotating with CSMP ETs to maintain a proficient operator/maintainer force. Since the COMMS watch is automated, the watchstander mainly monitors the system, such that a typical six-hour watch is not physically or mentally demanding.

5. Self Defense (SD)

The SD operator controls the ship's self-defense system. SD watchstanders stand a six-hour watch as required for ship self-defense. When manning of the self-defense system is not required, the SD watchstanders augment the CSMP. Similar to the COMMS watchbill, two CSMP ETs are also trained to stand the SD watch. The self-defense system includes the Advanced Integrated Electronic Warfare System (AIEWS), Decoy Launcher Control and the Rolling Airframe Missile (RAM) Weapon Control Panel. The SD operator receives engagement orders from the CSOW.

6. Combat System Maintenance Personnel (CSMP)

CSMP have day work hours and prioritized maintenance schedules, repairing out of commision equipment first and then completing preventative maintenance. CSMP are on call 24 hours for emergent work or emergency response to casualties. CSMP are contacted by the PIP. The PIP notifies the CSMP of the system casualty and location or the need to contact with Officer of the Deck (OOD) or CSOW. CSMP receive damage control and security force training as a ready response force for ship damage and protection.

J. ENGINEERING MANNING REQUIREMENTS

1. Overview

In support of reducing manning, the engineering spaces are unmanned during normal steaming. The *Arsenal Ship* has an Engineering Officer of the Watch (EOOW) and an Engineering Operator (EO) in a four-section watch rotation (Table III-7). These watchstanders are not required to remain in the engineering spaces. The watchstanders carry a personal information pad (PIP). The watchstander is free to roam throughout the ship until the PIP beeps, thus notifying the watchstander of an engineering casualty. Four additional personnel are available for troubleshooting and repair of engineering casualties.

Maintenance Officer			
EOOW		EO	
E7	EN	E6	EN
E7	EN	E6	EN
E7	EM	E5	EN
E7	EM	E5	EM
ENGINEERING MAINTENANCE			
E6	DC	E5	EM
E6	EN	E5	DC
		TOTAL: 13	

Table III-7. Engineering Manning Requirements

2. Engineering Watchstanders

The EOOW operates from the Standard Monitoring Control System (SMCS) console in the engineering control station (CCS) during engineering casualties. The EOOW uses the system to determine the required level of casualty control. He can task the EO and maintenance personnel remotely via the PIP. The EOOW may be thought of as a Local Area Network (LAN) manager, monitoring the entire engineering system with the capability to access it to respond to individual casualties. The SMCS allows the entire engineering plant to be monitored and controlled from any of the consoles in CIC, CCS, the pilothouse; all Local Operating Panels (LOP); and all Control and Collection Units (CCU). The digital monitor and control capability of this system is the primary justification for leaving all engineering spaces unmanned. The EO is primarily an assistant to the EOOW, and is available to provide local casualty control on engineering equipment in case the SMCS fails to take the proper controlling action.

3. Engineering Maintenance Personnel (EMP)

EMP have day work hours and prioritized maintenance schedules, fixing broken equipment first and then completing preventative maintenance. EMP are on call 24 hours for emergent work or emergency response to casualties. EMP are contacted by the PIP. The PIP notifies the EMP of the system casualty and location or the need to contact the EOOW. EMP comprise the damage control party, and are involved in all special evolution details.

K. PHYSICAL SECURITY

1. Overview

The *Arsenal Ship* is a truly high value asset, thereby making physical security a significant concern with a reduced crew size. The *Arsenal Ship* relies mainly on passive design characteristics, augmenting traditional security practices, in the protection of the ship and its payload.[2]

2. Threat Assessment

- At Sea Physical Security Threat
  - Hostile boarding and small boat attacks from terrorist organizations or special operations forces.

- Inport Physical Security Threat
  - Sabotage and deliberate attack from terrorist organizations or special operations forces.

[2] Note: A comprehensive analysis of these threats is beyond the scope of this report.

3. Passive Design Measures

The central feature in the design for passive security is limited and controlled access. This is accomplished by minimizing the number of access points between the weatherdecks and the interior of the ship. The fore and aft superstructures each have two accesses The doors can be remotely closed and locked from the bridge. The internal watertight doors are also remotely actuated and lockable, thus providing enhanced security and compartmentalization. The fore and aft armories are located in close proximity to the starboard accesses.

4. At Sea Security Concept and Tactics

The *Arsenal Ship* relies on tactical evasion, high freeboard, and the formation of the security alert team (SAT) to combat at sea security threats. The SAT is comprised of the non-watchstanding maintenance personnel. They immediately man the armories during security alert, and arm themselves with small arms and stinger missiles. The escort ship's five inch and 50 caliber guns provide the first layer of defense against potential small boat and helicopter attacks. The *Arsenal Ship*'s rolling airframe missiles (RAM) are the second layer of attack against helicopter attacks. The final layer of defense against incoming boat and helicopter attacks is the SAT.

5. Inport Security Concept and Tactics

Inport, the *Arsenal Ship* is protected by existing shore-based security forces. TWARSES assists the security forces by monitoring interior spaces using infrared (IR) sensors. There are existing security forces in place to provide pierside and port security at the three forward operating bases. It is beyond the scope of this report to estimate the effect of basing *Arsenal Ships* at these locations.

6. Active Design Explorations

The design team explored additional active security measures. They are used to give the SAT sufficient time to obtain weapons from the armories, and provide additional protection to support the reduced manning concept. The systems **were not** included in this overall design because they are drastically different from current security alert procedures and are untested.

a. Weapons Positions

Armored gun tubs are placed on the weatherdecks surrounding the fore and aft superstructures. The SAT mans the tubs with small arms, 50-caliber machine guns, and stinger missiles. These tubs are positioned to place grazing fire over the weatherdeck of the ship and to provide close-in fields of fire around the superstructures.

b. Border Suppression System (BSS)

A systems of command-detonated anti-personnel mines (i.e., naval variants of the M18A1 and M14 anti-personnel land mines) are placed flush with the weatherdecks (M14) and recessed into the superstructure (M18A1). These mines are positioned to cover the weatherdecks with overlapping casualty producing zones. The mines are controlled from the bridge and provide immediate response to a hostile boarding, giving the SAT time to react to the threat.

c. Intruder Detection System (IDS)

A key feature to the physical security system inside the ship is an IDS [22]. The IDS is integrated with TWARSES. An intruder is sensed by the TWARSES IR sensors. A signal is sent to the IDS. The IDS signals the damage control system (DCS) to flood the selected space with a damage control agent (i.e., CO_2).

L. WATCHSTANDING CONCEPT

The *Arsenal Ship* operates with a reduced manning concept. The crew is divided into operational watchstanders and a maintenance force. The maintenance personnel augment the normal watchstanders for all special evolution stations, and serve on damage control parties and security alert teams. Tables III-8 and III-9 show that 44 personnel are needed to man the *Arsenal Ship*. This manning reduction is accomplished by extensive use of automation, electronics and computer technologies, as well as enacting concepts that challenge current Navy customs and traditions. The crew must be highly experienced and versatile. Selected screening is required to ensure that only "top-flight" sailors are assigned to the *Arsenal Ship* program.

	Crew	Billet Number	Title	Rank	Quals	Normal Steaming	Combat Steaming	Bridge Evolutions	UNREP	Helo Operations	Boat Operations	Damage Control Physical Security
Command	1	Ops - 100	Captain	O-5/6	SWO							
	2	Eng - 100	Maintenance Officer	O-4	LDO							
Watch	3	Ops - 201	OOD	O-3	SWO	OOD #1	OOD #1	OOD #1	OOD #1	OOD #1	OOD #1	OOD #1
	4	Ops - 202	OOD	O-3	SWO	OOD #2	OOD #2	OOD #2	OOD #2	OOD #2	OOD #2	OOD #2
	5	Ops - 203	OOD	O-3	SWO	OOD #3	OOD #3	OOD #3	OOD #3	OOD #3	OOD #3	OOD #3
	6	Ops - 204	OOD	O-3	SWO	OOD #4	OOD #4	OOD #4	OOD #4	OOD #4	OOD #4	OOD #4
	7	Ops - 301	JOOD/lookout	E-7/8	QM	JOOD #1	JOOD #1	JOOD #1	JOOD #1	JOOD #1	JOOD #1	JOOD #1
	8	Ops - 302	JOOD/lookout	E-7/8	QM	JOOD #2	JOOD #2	JOOD #2	JOOD #2	JOOD #2	JOOD #2	JOOD #2
	9	Ops - 303	JOOD/lookout	E-7/8	QM	JOOD #3	JOOD #3	JOOD #3	JOOD #3	JOOD #3	JOOD #3	JOOD #3
	10	Ops - 304	JOOD/lookout	E-7/8	QM	JOOD #4	JOOD #4	JOOD #4	JOOD #4	JOOD #4	JOOD #4	JOOD #4
	11	Com - 201	CSO	O-4	SWO	CSO #1	CSO #1	CSO #1	CSO #1	CSO #1	CSO #1	CSO #1
	12	Com - 202	CSO	O-3	SWO	CSO #2	CSO #2	CSO #2	CSO #2	CSO #2	CSO #2	CSO #2
	13	Com - 301	COMMS	O-3	RM	COMMS #1	COMMS #1	COMMS #1	COMMS #1	COMMS #1	COMMS #1	COMMS #1
	14	Com - 302	COMMS	E-6	RM	COMMS #2	COMMS #2	COMMS #2	COMMS #2	COMMS #2	COMMS #2	COMMS #2
	15	Com - 401	SD	E-6	EW	SD #1	SD #1	SD #1	SD #1	SD #1	SD #1	SD #1
	16	Com - 402	SD	E-5	EW	SD #2	SD #2	SD #2	SD #2	SD #2	SD #2	SD #2
	17	Eng - 201	EOOW	E-7	EN	EOOW #1	EOOW #1	EOOW #1	EOOW #1	EOOW #1	EOOW #1	EOOW #1
	18	Eng - 202	EOOW	E-7	EN	EOOW #2	EOOW #2	EOOW #2	EOOW #2	EOOW #2	EOOW #2	EOOW #2
	19	Eng - 203	EOOW	E-7	EM	EOOW #3	EOOW #3	EOOW #3	EOOW #3	EOOW #3	EOOW #3	EOOW #3
	20	Eng - 204	EOOW	E-7	EM	EOOW #4	EOOW #4	EOOW #4	EOOW #4	EOOW #4	EOOW #4	EOOW #4
	21	Eng - 205	EO	E-6	EN	EO #1	EO #1	EO #1	EO #1	EO #1	EO #1	EO #1
	22	Eng - 206	EO	E-6	EN	EO #2	EO #2	EO #2	EO #2	EO #2	EO #2	EO #2
	23	Eng - 207	EO	E-5	EN	EO #3	EO #3	EO #3	EO #3	EO #3	EO #3	EO #3
	24	Eng - 208	EO	E-5	EM	EO #4	EO #4	EO #4	EO #4	EO #4	EO #4	EO #4

Table III-8. Watch, Quarter and Station Bill for Watchstanders

		Billet				Normal	Combat	Bridge		Helo	Boat	Damage
	Crew	Number	Title	Rank	Quals	Steaming	Steaming	Evolutions	UNREP	Operations	Operations	Control
												Physical
												Security
Admin	25	Adm - 501	Corpman	E-6	HM							
	26	Adm -502	Cook	E-6	MS							
	27	Adm -502	Cook	E-4	MS							
Labor	28	LF - 501	Deck Force	E-6	BM	Normal Work	Normal Work	Normal Work	Safety	LSE	Safety	SAT
Force	29	LF - 502	Deck Force	E-5	BM	Normal Work	Normal Work	Normal Work	Winch	Tiedown	Coxswain	SAT
	30	LF - 503	Deck Force	E-5	BM	Normal Work	Normal Work	Normal Work	Linehandler	Tiedown	Boat	SAT
	31	LF - 504	HME Force	E-6	DC	Normal Work	Normal Work	Normal Work	Fuel Sampler	OSL/AFFF		OSL/AFFF
	32	LF - 505	HME Force	E-6	EN	Normal Work	Normal Work	Normal Work	Linehandler	Team Leader	Engineman	Hoseman
	33	LF - 505	HME Force	E-5	EM	Normal Work	Normal Work	Normal Work	Linehandler	Hoseman	Linehandler	Hoseman
	34	LF - 506	HME Force	E-5	DC	Normal Work	Normal Work	Normal Work	Linehandler	Hoseman	Linehandler	Team Leader
	35	LF - 507	Combat Force	E-6	ET	Normal Work	Normal Work	Normal Work	Signals	Hotsuitman		Hoseman
	36	LF - 508	Combat Force	E-6	ET	Normal Work	Normal Work	Normal Work	Phonetalker	Hotsuitman		Plugman
	37	LF - 509	Combat Force	E-5	ET	Normal Work	Normal Work	Normal Work		Plugman		SAT
	38	LF - 510	Combat Force	E-5	ET	Normal Work	Normal Work	Normal Work				SAT
	39	LF - 510	Combat Force	E-6	GMM	Normal Work	Normal Work	Normal Work				SAT
	40	LF - 510	Combat Force	E-5	GMM	Normal Work	Normal Work	Normal Work				SAT
	41	LF - 510	Combat Force	E-4	GMM	Normal Work	Normal Work	Normal Work	Linehandler	Phonetalker		SAT
	42	LF - 510	Combat Force	E-4	GMM	Normal Work	Normal Work	Normal Work	Linehandler			SAT
	43	LF - 510	Combat Force	E-4	GMM	Normal Work	Normal Work	Normal Work				SAT
	44	LF - 510	Combat Force	E-4	GMM	Normal Work	Normal Work	Normal Work				SAT

Table III-9. Watch, Quarter and Station Bill for Special Evolutions

IV. COMBAT SYSTEMS

This chapter provides an detailed description of the *Arsenal Ship* combat system. First, a brief overview of the entire combat system is provided followed by sections covering the launcher system, Anti-Air Warfare, Strike Warfare, Theater Ballistic Missile Defense, ship self defense, Naval Surface Fire Support, Command and Control, Communications, topside design, and underway missile replenishment.

A. COMBAT SYSTEMS DESIGN OVERVIEW

The selection of combat systems for the *Arsenal Ship* is driven by the single requirement that it must complete the assigned missions within the constraints of minimum acquisition and life cycle cost [4,5]. The *Arsenal Ship*'s combat system is one of low maintenance and high reliability. The total ship inherent availability goal is 0.95 and the operational availability is 0.85. The weapon control subsystem reliability goal is 0.99. Therefore, combat system material selection, equipment, arrangement, built-in-test equipment, redundancy, and equipment reliability is heavily dependent on currently available and proven technologies.

The *Arsenal Ship* deploys under the escort of an Aegis warship equipped with the Cooperative Engagement Capability (CEC). The Aegis escort ship provides Command and Control over the weapons onboard the *Arsenal Ship.* The *Arsenal Ship* contains minimum equipment required to support remote launching of its weapons. The capability exists for control to be extended to platforms other than the Aegis escort, such as an Airborne Warning and Control System (AWACS) platform or even a CONUS station through satellite communication (SATCOM) links, but the concept of operations requires an Aegis *Control Ship* for protection as well as connectivity. The *Arsenal Ship* is effectively a remote magazine for the Aegis escort and does not have an autonomous engagement capability.

The *Arsenal Ship* has connectivity with Army, Air Force, and Marine Corps personnel, for launch of strike weapons and Naval Surface Fire Support (NSFS), through the *Control Ship*. The combat systems onboard the *Arsenal Ship* are kept to a minimum to reduce cost, manning, maintenance, training, and logistics. Tactical Aircraft Control and Navigation (TACAN) and search radars are not included in order to reduce the required avionics support personnel, combat information architecture, and manning. The *Control Ship* has responsibility for planning and reporting any weapon firing...All post launch missile communications are performed by the *Control Ship.*

CEC provides the primary link between the Aegis escort ship and the *Arsenal Ship* for digital targeting information and weapon data. UHF Link 16 is the secondary link and HF link 22 (improved link 11) is the tertiary backup for providing digital data. This is shown in Figure IV-1. SATCOM connectivity includes the future Global Broadcast System (GBS). This permits CONUS based planning for weapon systems. Support equipment is networked through a dual ring LAN that provides data from CEC/Link-16/link-11 to the Advanced Tactical Weapon Control System commonly referred to as big “T” ATWCS (A”T”WCS). A”T”WCS is currently being designed to communicate with SM-2, TLAM, and ATACMS.

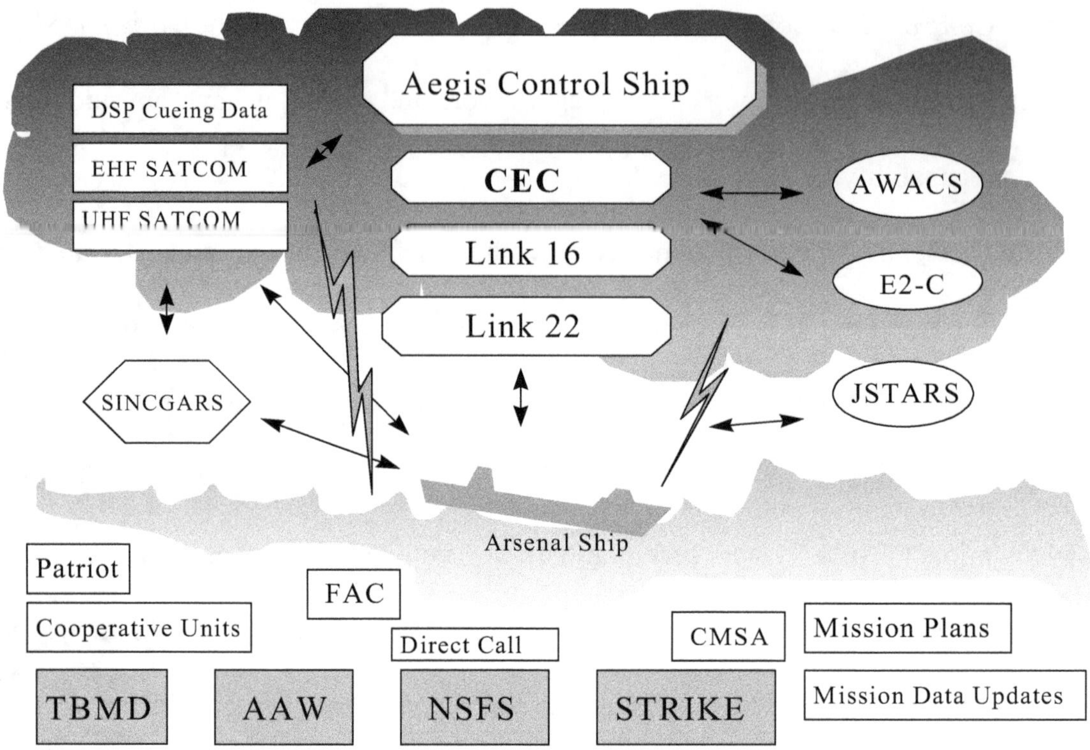

Figure IV-1: Communications For The *Arsenal Ship*

Required missions for the *Arsenal Ship* are Anti-Air Warfare (AAW), Strike Warfare, Theater Ballistic Missile Defense (TBMD), ship self defense, and NSFS. The *Arsenal Ship* employs Standard Missile (SM-2), Tomahawk Land Attack Missile (TLAM), Army Advanced Tactical Missile (naval version ATACMS), and SM-2 Lightweight Exoatmospheric Projectile (LEAP), all launched from the Mk 41 Vertical Launcher System (VLS). The Rolling Airframe Missile (RAM), Advanced Integrated Electronic Warfare System (AIEWS) and various decoys provide ship self defense. The future Vertical Gun for Advanced Ships (VGAS) is a preplanned improvement to the *Arsenal Ship* for increased NSFS capability. This is conceptualized in Figure IV-2.

The Arsenal Ship Combat System

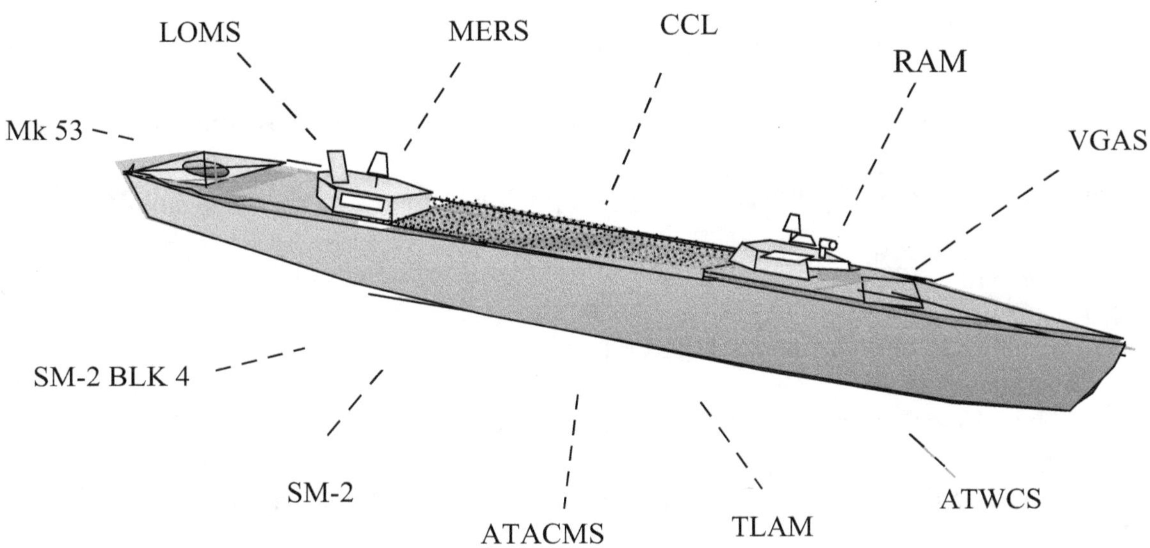

Figure IV-2: ***Arsenal Ship*** **Combat System Summary**

B. LAUNCHER SYSTEMS

The MK-41 VLS and Concentric Canister (CCL) launcher systems were considered for the *Arsenal Ship* design. A description of the operation, advantages and disadvantages of each system is provided along with a justification for final launcher selection.[23,24]

1. MK-41 Vertical Launch System

The MK-41 has tremendous reliability and support and is capable of launching all the desired weapons. The MK-41 is presently carried on 22 ships of the Ticonderoga Class (CG 52), 24 ships of the Spruance Class (DD 963), and 24 ships of the Arleigh

Burke Class (DDG 51). It is also in use by our allies on the Tribal Class, Brandenberg Class, Kongo Class, and Mursame Class. The MK-41 program has delivered over 900 Modules over the past 12 years with almost 7000 cells presently in use.

The MK-41 VLS is fully supported by the existing military and industrial infrastructure from acquisition, installation, in-service engineering, and complete life-cycle support. The use of the MK-41 VLS for the *Arsenal Ship* has minimal impact on existing shore support facilities, training support, and shore manning requirements. It meets the requirements for prevention of Hazards of Electromagnetic Radiation to Personnel (HERP) and Ordnance (HERO).

The MK-41 is capable of launching the Tomahawk Anti-ship and Land Attack Missile, Standard Missile-2, Standard Missile-2 Block 4, Seasparrow and Evolved Seasparrow Missile (ESSM), and the Vertical Launch Anti-submarine Rocket (VLA). It has the flexibility to be adapted for ATACMS as a NSFS launch platform. It integrates existing technology and is available for a FY00 installation.

2. Concentric Canister Launcher

The Concentric Canister Launcher (CCL) is an alternative to the MK-41 VLS. CCL has several distinct advantages over the MK-41 VLS. There are significant considerations for ship survivability, hull mechanical and electrical (HM&E) to weapons systems integration, safety, producibility, affordability, and combat system integration. The CCL is depicted in Figure IV-3.

"Concentric Canisterized" Launcher

- •Concentric Tubes Launcher
- •Anti-fragmentation Shields
- •Concentric Plenum
- •COTS/Open Electronics
- •Under development

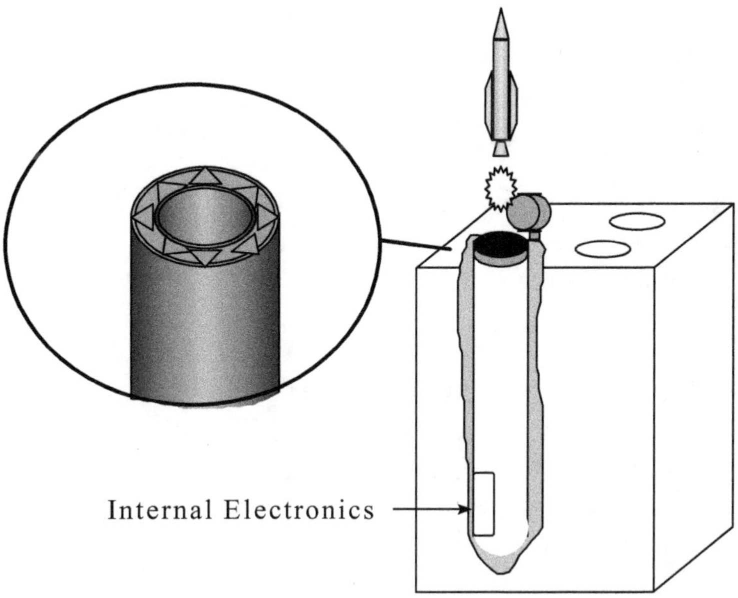

Figure IV-3: CCL Summarized

The MK-41 vertical launcher uses a shared-exhaust gas management system, encapsulated missiles, and shares a magazine. The MK-41 is a low risk, but expensive option for the *Arsenal Ship* launcher. The CCL is a second option for the *Arsenal Ship*. The CCL concept employs a concentric launcher that can be used for diverse munitions types and sizes. It is an All-Up-Round (AUR) that is stowed ready to fire and provides passive armor and a self contained Gas Management system.

The CCL is a self-contained launcher mechanically and electrically. The electronic assembly can be packaged within a protected annular space. It can include anti-fragmentation shields between concentric tubes to provide increased protection between launchers than is available in the MK 41.

Shock effects cause an impulsive motion of the ship up and away from the charge. Accelerations typically measured in a ship are much greater below the surface for an underwater blast than are experienced topside. CCL incorporates a shock collar that mitigates the larger G- forces experienced on lower decks. Care needs to be taken during ship design to ensure sufficient GMT due to the pendulum type effects of mounting weight at a high attach point.

3. Launcher Selection

The CCL open architecture provides the framework to cost-effectively capture existing and future technology. The costs of the CCL are projected to be less than the MK 41 with the increased benefits of simplicity, increased survivability, and an open-electronic architecture. Should the CCL evolve as anticipated by NSWC, and prove reliable, it would make a better choice for the launcher of the *Arsenal Ship* than the MK 41. But the CCL concept is far from existing technology and we have chosen to design the ship around the MK 41, due to the practical nature of the Total Ship System Engineering approach.

C. AAW, STRIKE WARFARE AND TMBD

The *Arsenal Ship* acts as a force multiplier for the AAW, Strike Warfare and TMBD mission by increasing the number of weapons available in theater for these warfare areas. The ability of the *Arsenal Ship*'s missiles to be remotely launched by the *Control Ship* is the cornerstone of this design. The specific functional breakdown used to obtain this capability is government sensitive information. Therefore, the details of the interrelationships between the *Arsenal Ship* and the *Control Ship* for SM-2 and TLAM engagements are given in Appendix A, which is a government only distribution document. For completeness, it should be noted, that the remote launch capability is achievable with the designed combat system.

D. SHIP SELF DEFENSE

Determination of the required *Arsenal Ship* self defense capability begins with an investigation of the possible threats, which the ship will encounter. System selection is based on expected threats and operating conditions, while maintaining a balance between active and passive defense capabilities.

1. Threat Analysis

The *Arsenal Ship* by the nature of its concept of operations will be presented with a variety of threat postures worldwide. While the nature of threat remains multi-dimensional, the specific threats will vary over time and location. Our design incorporates a qualitative look at the threat, rather than a quantitative look, as a guideline for a survivable platform. This assessment is summarized in Figure IV.4.

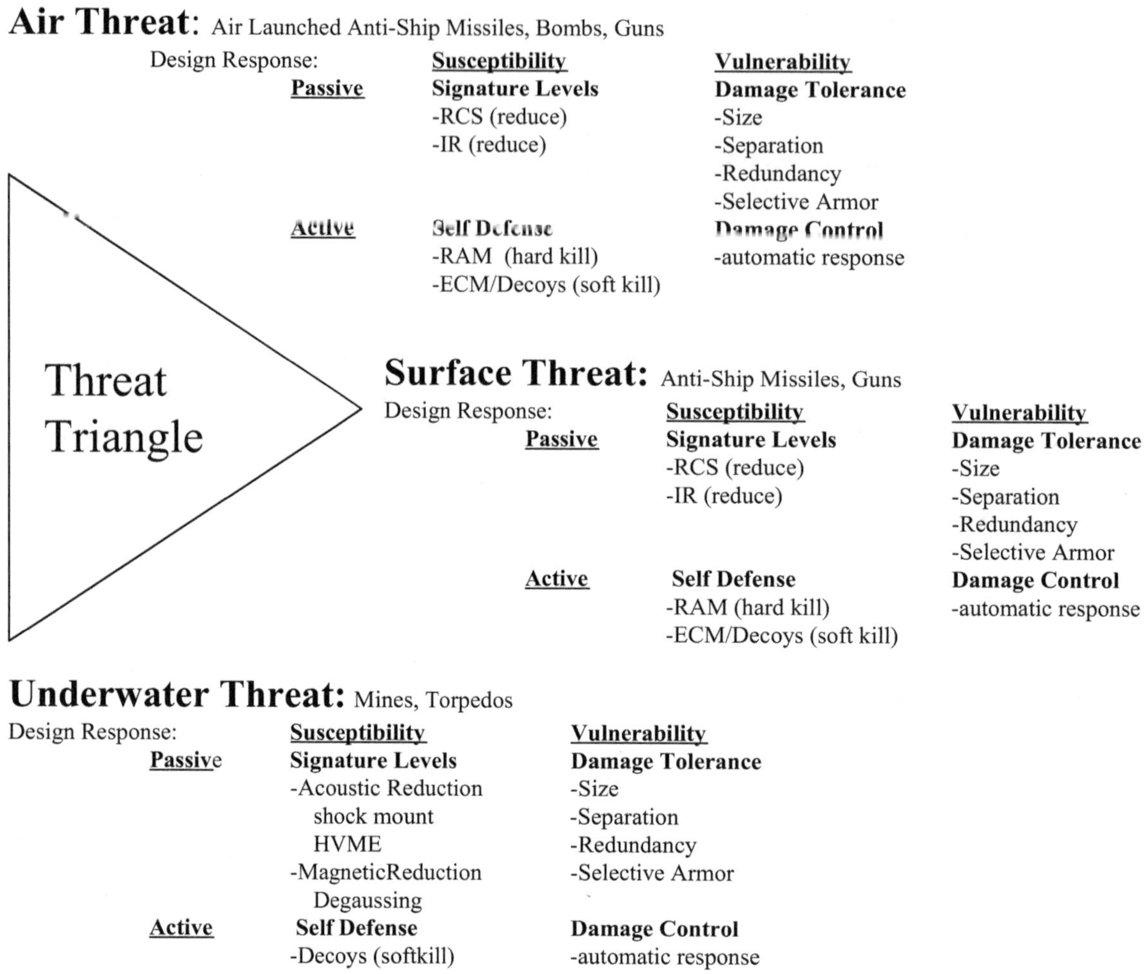

Figure IV-4. ***Arsenal Ship*** **Threat Triangle**

In our assessment of the threat to the *Arsenal Ship*, we determined that a majority of the defense for the vessel will be offboard. The ship itself has little active defense and relies on passive capability. However, it must be kept in mind how the ship works as part of a larger system when considering threat and defense. The *Arsenal Ship* does not operate alone. As a remote magazine with an Aegis *Control Ship*, it has the protection of the *Control Ship* systems. As a battlegroup member, the *Arsenal Ship* has a further umbrella of protection. Thinking of the *Arsenal Ship* as a subsystem of a larger system, it is not defenseless.

2. Detection Elements

a. AAW/ASCM Threat

Except for a short-range navigation radar, the *Arsenal Ship* possesses no active surface- or air-search radar. Therefore, the primary detection of AAW/ASCM threats rests upon surface escorts.

b. Electronic Warfare

The *Arsenal Ship* uses the new Advanced Integrated Electronic Warfare System (AIEWS) Phase II as its Electronic Support Measures (ESM) detection system. The initiation of Phase I of this system into the fleet is scheduled for FY00, with Phase II to follow soon thereafter. This system possesses passive radio frequency (RF) detection and identification capabilities, and active jamming capabilities. Since this program is still in the design stages, the existing hardware of SLQ-32 (V)3 is used as a model for design purposes. AIEWS is made up of the following elements:

- Port and Starboard Antenna/Receiver Units
- Signal Processing Units
- Display and Control Console

c. ASW Threat

Based on the manning and fiscal constraints, the *Arsenal Ship* depends on surface and submarine escorts to provide underwater threat detection and protection.

3. Control

a. Ship Self-Defense System (SSDS)

The SSDS is an integration and control system for the *Arsenal Ship*'s self-defense suite. It provides a quick reaction combat capability to the *Arsenal Ship*. The SSDS receives target-data inputs from CEC, and ESM information from AIEWS. The capabilities of the SSDS are summarized as follows:

- Correlates CEC tracks and ESM detections for tracks which meet the proper criteria, passes necessary missile-initiation information to RAM MK 31 Block 0/1, and receives and displays decoy recommendations provided by AIEWS.
- Integrates the use of hardkill and softkill defenses.
- Designates whether systems operate in an automatic or semi-automatic mode.
- Automatic mode allows for the defense systems to be fired with no operator intervention.
- Semi-automatic mode recommends the system to be used for the engagement and a time to fire, but the operator is required to physically fire the weapon or decoy [25].

b. AIEWS

The AIEWS supports the various types of softkill decoys launched from the *Arsenal Ship*. It gives decoy selection, tube selection, re-seed time calculations, and reload recommendations [26].

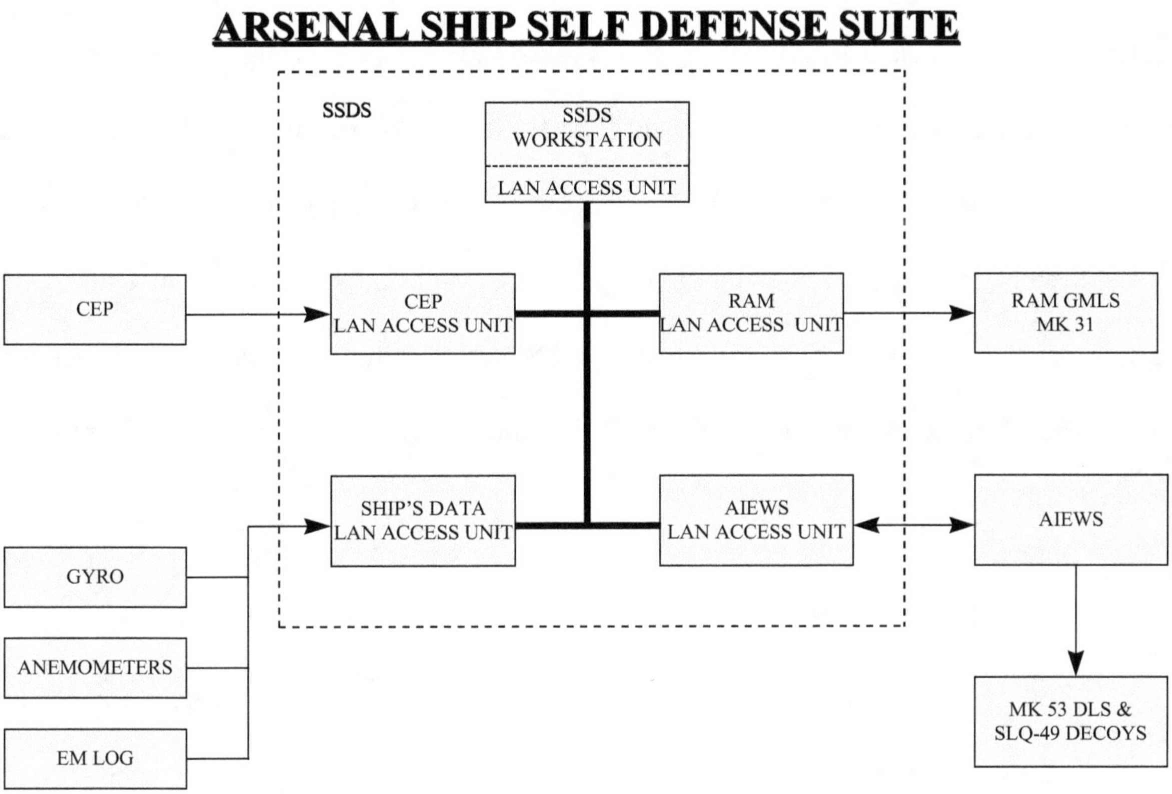

Figure IV-5: *Arsenal Ship* Self-Defense Suite

4. Engagement

a. Missiles

The *Arsenal Ship* uses the RAM MK 31 Block 0/1 Guided Missile Weapon System (GMWS) for its hardkill self defense. The RAM MK 31 GMWS consists of the following elements:

- Weapon Control Panel (WCP)
- Launcher Control Interface Unit (LICU)
- Launcher Servo Control Unit (LSCU)
- (1) MK 144 Launcher (missile capacity 21)

The block 0 RAM system uses RF midcourse and infrared (IR) terminal phase guidance. The block 0 RAM system receives 2-D range and bearing information from CEC, and correlates bearing and RF Band (I or J) information from AIEWS. The block 1 RAM system uses IR for both its midcourse and terminal phase guidance. The Block 1 RAM system must receive 3-D inputs of range, bearing, and elevation from CEC. All inputs are passed to the RAM system via SSDS for missile initiation. A distinct advantage of the RAM missile is that it's a fire-and-forget, lock-on after launch, missile [27].

b. Decoys

The *Arsenal Ship* possesses an assortment of softkill decoys to defeat a possible variety of threats. It employs chaff, IR, NULKA and SLQ-49 (rubber duck) decoys. The chaff decoys are launched into the air, and explode to create a chaff cloud of aluminum strips. The IR decoys are launched and float on the water surface producing a large thermal signature. The NULKA decoy is launched into the air, and follows a controlled flight path. While in flight, the NULKA decoy receives, amplifies, and re-radiates signals as specified by the EW operator. The chaff, IR, and NULKA decoys are supported by the MK 53 Decoy Launching System (DLS). The MK 53 DLS consists of the following elements:

MK 24 Decoy Launch Processor (DLP)

MK 174 Processor Power Supply (PPS)

(2) MK 137 Mod 7 Launchers

(2) Ready Service Lockers (RSL)

(2) MK 164 Mod 2 Bridge Launcher Control (BLC)

(1) Combat Launcher Control (CLC) [28]

The Rubber Duck decoy consists of two tethered un-inflated radar retro-reflectors which are stored in a lifeboat-like canister, ready to deploy over-the-side, launcher. Once launched and inflated, the pair of retro-reflectors produce a large ship-like radar return. This type of return is broadband in nature, and does not limit the recognition of narrow-band RF energy used by the incoming weapon [29]. Therefore, this type of decoy is useful in combating RF weapons operating in a frequency band outside the active ESM frequency band, and outside NULKA's capabilities. Rubber Ducks are remotely launched from the Combat Information Center (CIC), or manually at the launcher.

E. NAVAL SURFACE FIRE SUPPORT

As the Navy's threat focus shifted from blue water to the littorals, the Naval Service has shifted its operational focus as highlighted in *"From the Sea..."* and *"Forward, From the Sea..."*. *Operational Maneuver from the Sea* has become the predominant concept for the projection of naval power ashore, of which Naval Surface Fire Support is a major component.

The *Arsenal Ship* Concept of Operations (CONOPS) establishes the requirement for the *Arsenal Ship* to provide NSFS to Joint Ground Forces. CONOPS describes representative Surface Fire Support target sets to be countered by *Arsenal Ship* and recommends the Surface Fire Support weapon capabilities desired. This is summarized in Tables IV-1 and IV-2. [4].

	Halt Invasion	Long Range Strike	Battlespace Dominance	Surface Fire Support
Complex Adaptive Armed Forces	Air Land Maneuver Battle Groups (e.g., OMGs)	National / Regional C4I Space Control	Manned A/C TBMs, UAVs Cruise Missiles. SAM/AAA	Long-Range Artillery TBMs Logistics Assets
Armored Mech Armed Forces	Armor-Heavy Comb. Arms Formations Divisions/BDEs	National and Regional C4I	Manned A/C TBMs SAM/AAA	Long-Range Artillery
Infantry Based Armed Forces	Armor/Mech "Pure" units (BDEs/BNs)	Military Region District C4I	Manned A/C SAM/AAA	Medium-Range Artillery Logistics Assets
Internal Security Light Force	Transportation Railroads Trucking, Light Vehicles	National CMD Authority Military Concentrations	OP Bases Light A/C Coastal Patrol Craft	Logistics Assets Economic Asset Local Forces

Table IV-1. *Arsenal Ship* Representative NSFS Targets.

	Halt Invasion	Long Range Strike	Battlespace Dominance	Surface Fire Support
Complex Adaptive Armed Forces	ATACMS-BAT SLAM TLAM-TSTAR TLAM-C	TLAM	ATACMS TLAM-C/D SM-2 Blk III A/B and Blk IVA	ATACMS, SLAM, STRIKE-SM TLAM-C/D NAVAL GUNFIRE (VGAS/SCRAM)
Armored Mechanized Armed Forces	ATACMS-BAT TLAM-TSTAR SLAM STRIKE-SM	ATACMS-BAT TLAM-TSTAR SLAM STRIKE-SM	ATACMS TLAM-C/D SM-2 Blk III A/B and Blk IVA	ATACMS, SLAM, STRIKE-SM TLAM-C/D NAVAL GUNFIRE (VGAS/SCRAM)
Infantry Based Armed Forces	ATACMS SLAM STRIKE-SM	TLAM-D ATACMS-ER	ATACMS	ATACMS, SLAM, STRIKE-SM TLAM-C/D NAVAL GUNFIRE (VGAS/SCRAM)
Internal Security Light Force	NAVAL GUNFIRE (VGAS/SCRAM)	TLAM-C	ATACMS NAVAL GUNFIRE (VGAS/SCRAM)	ATACMS NAVAL GUNFIRE (VGAS/SCRAM)

Table IV-2. Suggested *Arsenal Ship* NSFS Weapon Capabilities.

CONOPS addresses NSFS by requiring the *Arsenal Ship* to carry 500 missiles and have space, weight and support system capacity for future installation of an extended range gun system. The Navy currently has a number of programs, projects and Advanced Technology Demonstrators (ATD) that provide solutions to the NSFS problem. The

NSFS systems is broadly separated into Guns/Projectiles and Missiles. Connectivity for targeting and command and control is similar across the spectrum of NSFS.

1. Guns/Projectiles

The Navy is currently pursuing a 5-inch gun modification and the VGAS, along with numerous projectile options for NSFS. VGAS. VGAS is the future gun system for the *Arsenal Ship*.

As a current Advanced Technology Demonstrator within the NSFS Program, VGAS embodies the Navy's plan to use a 155-mm gun for fire support and deep strike applications. This answers the NSFS Cost and Operational Effectiveness Analysis (COEA) in which the need for larger payloads and extended range is identified. [30] The larger gun has a longer range than the 5-inch, covering missions beyond 150 nautical miles. This range is only reachable by ATACMS, within the NSFS 10 minute response limit.

VGAS employs a completely automated gun system packaged in a Ship System Engineering Standard (SSES) B-module which is set flush with the ships' deck [31] (Figure IV-5).

Two 155mm guns per module use an automated loader with multi-ram modular advanced projectiles. An automated system of propelling charge buildup accommodates a wide range of projectile weights. The program piggy-backs off of Navy/Army/Marine Corps technological base to provide a high payoff . The payoff in technology is electro-thermal ignition of high energy propelling charges, high packing density of 700 rounds

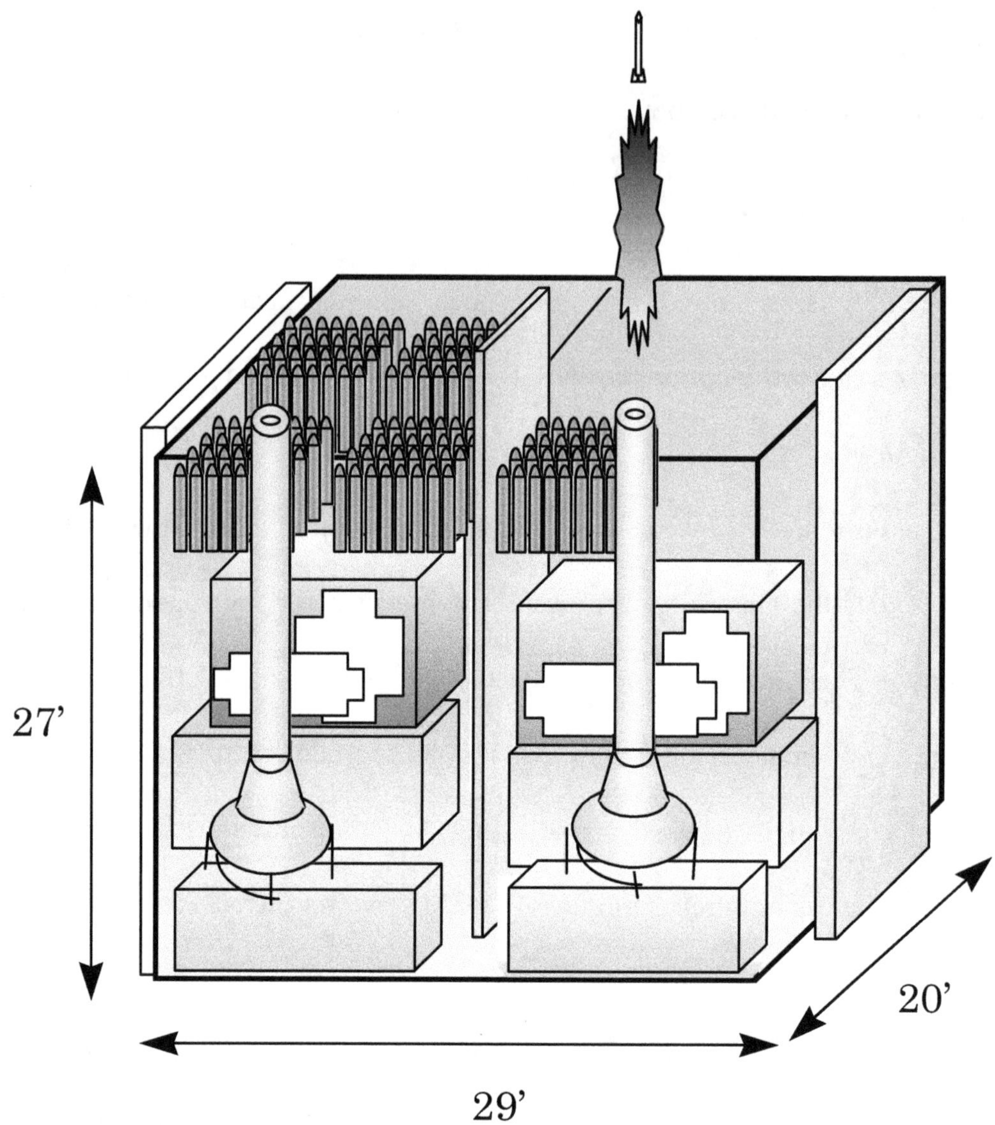

Figure IV-6. The Vertical Gun for Advanced Ships (VGAS)

per gun, 15 rounds per minute per gun, and a gun which can support not only NSFS, but also deep strike applications. The advantages and disadvantages of the 155-mm VGAS are briefly summarized in Table IV-3. [32] The physics of guns and projectiles is well understood and the risk defined within the initial development of a new system is minimal.

Vertical Gun for Advanced Ships	
Advantages	**Disadvantages**
•SSES-B Module Design	•Development Risk
•Reduced Manning (0 to operate, 1-3 to maintain)	•Advanced metallurgy for key gun components
•Reduced Topside RCS (Low, Protected Location)	•High rate-of-fire with the rotating chamber
•Extended Performance (300 lbm warhead to 275 NMI)	•Thermal management
•Commonality of Munitions with Army/Marine Corps	•Multiple ram operations for guided munitions
•No Train or Elevation Mechanisms	•Overpressure requires elevated firing

Table IV-3. Vertical Gun for Advanced Ships Advantages and Disadvantages

The near term solution (IOC 2001) is 5-inch MK 45 gun mount modifications to allow the firing of the Extended Range Guided Munitions (ERGM). This is a major rework of the current gun system requiring a new 60-70 cal barrel, breech and breech operating mechanism, empty case tray, and modified slide, gun barrel housing, recoil, counter-recoil system, data communication interface, stand, carriage and upper hydraulics. These changes increase the gun energy from 10 Megajoules to 18 Megajoules, and nearly doubles the ballistic range. This also allows for a sustained rate of fire of the ERGM (double ram) of 10 rounds per minute and maintains an AAW/ASUW and Shore Fire Ballistic Round Capability of 20 rounds per minute [33].

To achieve the required NSFS ranges, the gun fires the Extended Range Guided Munitions (ERGM-EX 171). The ERGM is a rocket-assisted projectile equipped with a submunition warhead and Global Positioning System (GPS) and Inertial Navigation

System (INS) coupled- guidance. A pre-planned product improvement of the ERGM calls for a terminal homing-seeker and unitary payload to extend coverage to point targets [34]. The advantages and disadvantages of the 5-inch gun using the ERGM are listed in Table IV-4.

5-inch MK 45 Gun Mount Modifications with ERGM	
Advantages	**Disadvantages**
•Current Program Status	•Near Term Fix-Just Meets NSFS Requirements
•IOC 2001	•No Reduction in Manning to Operate (Requires 10+)
•Range of 63 NMI with ERGM	•Inherent Above Deck RCS
	•Requires Train and Elevation (Gun Fire Control System)
	•Requires Major Rework of Gun System

Table IV-4. Advantages and Disadvantages of the 5-inch Gun modifications with ERGM

The 5-inch gun with ERGM make it a viable candidate for the future gun system. The disadvantages, however, are significant and the VGAS program is concurrently researched.

2. NSFS Missiles

The Tomahawk missile is used for non-time critical NSFS. Despite high per-missile replacement costs, the TLAM has already covered development costs and continue to decrease in per-unit cost. The ATACMS) is under investigation as a joint candidate for a fire support missile. The viability of ATACMS rests on industry funded MK 41 VLS integration and test firing in December 1996. The challenge of ATACMS is to fit a 24-inch diameter missile, with folded fins and modified-umbilical connections, in a 25-inch VLS cell.

A modified-Harpoon variant named the Sea SLAM, and a Standard Missile Strike variant have been investigated as fire support missile options [35]. These missiles are not selected for the *Arsenal Ship* due to lack of current and projected program funding [36].

3. NSFS Weapon Control

NSFS weapon control on the *Arsenal Ship* is performed by the Advanced Tactical Weapon Control System (A"T"WCS). This system provides an open architecture, flexible-control interface to NSFS precision munitions. It is currently configured for TLAM and is being configured through software modules for VGAS, ERGM and ATACMS. Additional modules are added for follow-on NSFS munitions (i.e., Strike-SM, SLAM, etc.) [37].

4. Discussion

NSFS Cost and Operational Effectiveness Analysis (COEA) completed in December 1994, addresses the need for naval surface forces to provide credible NSFS for expeditionary forces [30]. This support is required at a minimum 43 nautical miles, with a goal of 63 nautical miles, within 10 minutes. This range is based on the firing platform stationed 25 nautical miles offshore. Ranges in excess of 63 nautical miles provide additional flexibility and extended lethality. The NSFS COEA is a complete assessment which considers 30 guns, 100 projectiles and 7 missiles. The Center for Navy Analysis

(CNA) along with weapon system development program offices at NAVSEA and warfare centers at Dahlgren, Louisville, and China Lake have estimated the cost for candidate weapon systems. The lethality estimated for each candidate weapon system, is based on number of rounds required to strike each target as a function of range, target location error and desired effect. Headquarters Marine Corps (HQMC) and the Marine Corps Combat Development Command (MCCDC) selected three different representative scenarios, developed operational concepts and produced time-phased target lists for each scenario consistent with the Defense Planning Guide to establish baseline effectiveness comparisons. Operational effectiveness is established as the cost to achieve a given effectiveness for each candidate weapon system and combination of systems. System integration considerations determine the feasibility of integrating the most promising system(s) onto a ship. The COEA also addresses follow-on efforts, highlighting longer range possibilities and NSFS as a complement to TACAIR.

The COEA team identified the most cost-effective solution as a combination of 155-mm guns and Tomahawk missiles. The major driving factors for NSFS weapon options were the requirement that the system concept be technically feasible, achieve IOC within 10 years (of 1994) and be considered effective.

The combination of guns and missiles provides a cost-effective balance of systems for NSFS. In terms of cost-effectiveness not one of the missile concepts considered could compete with guns as a low-cost delivery platform of munitions. The one gun concept considered could fire a sufficiently large unitary charge far enough inland, with sufficient anti-jam capability, to take care of certain elements of a target set. However, a combination of guns and missiles, with guns taking out the majority of the required targets, is modeled to defeat scenario target-sets at minimum total life-cycle cost. Guns and missiles are usually complementary in range, warhead size and cost. Guns usually have a shorter range, smaller warhead and smaller replacement cost. Advanced gun concepts with increased ranges allow projectiles to engage targets previously reachable only with a missile option.

The NSFS COEA options are listed in Table IV-5 in order of increasing cost.

NSFS Cost Effective Options
(1) New 155mm/60 cal gun with advanced propellant and a family of rocket assisted precision guided munitions AND TLAM
(2) The above gun with TLAM plus ATACMS (deck-launched derivative)
(3) The above gun with TLAM plus ballistic missile sized to fit short VLS cell (i.e. SMASHR, Standard Missile Autonomous Strike Homing Round)
(4) The above gun with TLAM plus boost/glide missile sized to fit short VLS cell (i.e. JSOW, Joint Stand Off Weapon derivative)

Table IV-5. NSFS Cost Effective Options.

The Operational Analysis (OA) criteria examines which rounds can attack each given target for the smallest replacement cost. Various warhead selections are required for both gun and missiles to defeat a target set. This is summarized in Table IV-6.

•Unitary HEU (conventional high explosive unguided without terminal seeker)
•Unitary HEG (conventional high explosive guided with or without terminal seeker)
TARGET SETS: radar complex, vans, troops in open, truck convoy, buildings
•DPICM (Dual Purpose Improved Conventional Munitions i.e. submunitions)
TARGET SETS: distributed targets (dispersed troops/foxholes)
•Penetrating/Shaped Charge
TARGET SETS: command bunkers, pill boxes, hardened sites, tanks, armored vehicles
•ATGSM (Autonomous Terminally Guided Submunitions)
•BAT (Brilliant Anti-Tank)
•SADARM (Sense And Destroy Armor)
TARGET SETS: same type targets, tanks and armored vehicles

Table IV-6. Gun and Missile Warhead Selections.

These warheads provide extended range as RAP (Rocket Assisted Projectiles) and provide the necessary precision through GPS and INS, with or without terminal seekers. GPS and INS are adequate for most HE unitary and DPICM rounds. Penetrating and shaped-charge rounds usually require better accuracy than is achievable with GPS alone.

F. COMMAND AND CONTROL SYSTEMS

1. Command and Control Goals

The *Arsenal Ship* must have complete connectivity with the *Control Ship* and other Commanders who may need to use the weapons within its magazine. It must have interoperability with Joint and Service C4 systems. It has commonality with existing systems that do not require further development costs that the *Arsenal Ship* would have to be responsible for. Non-developmental items and existing commercial off the shelf (COTS) technologies keep costs down and bring the most advanced electronics equipment in the shortest time to the *Arsenal Ship* Program.

The C4 Architecture onboard the *Arsenal Ship* is redundant and provides reliable, back-up paths for information to flow despite battle damage or malfunctioning equipment. It is adapted from existing equipment and technologies and further automated with Smart Ship efforts to reduce manning and maintenance. A fiber optic network and distributed/redundant equipment provides rapid information distribution subsystems for the "Combat Commander."

The communications equipment onboard the *Arsenal Ship* takes advantage of a range of frequency regions in order to guarantee data transmission under various meteorological, oceanographic, and operational conditions. It is imperative that the *Arsenal Ship* have survivable, protected, and sustainable command, control, communications, computer, and Intelligence systems.

2. C^4 Employment

Different conflict levels will impose different, and perhaps contentious, requirements on the C4 systems that support the *Arsenal Ship*. Peacetime C4 systems support daily operations, attack warning, and transition to warfare. During a crisis, the *Arsenal Ship* will provide an opportunity for conventional deterrence. Critical C2 connectivity is needed between the military units and their combatant commander. In a conventional war, the combatant commander may take control of C4 forces that are not organic to tactical forces. System control provides network status and supports reconfiguration and reconstitution. As a crisis expands, additional nodes may be brought on line.

The Joint Maritime Communications Strategy (JMCS) is the communications architecture the Navy will use to implement *Copernicus...Forward*, the C4I vision for the future. It is technical as well as a strategy using nondevelopmental item communication systems. JMCS consists of three technical elements the Arsenal Ship will employ: the Automated Digital Network System (ADNS), the Digital Network System (DMR System), and the Integrated Terminal Program (ITP). ADS will increase the efficiency of information transfer by pooling communication resources. DMR uses modular radio components for HF, VHF, and UHF communications and covers the spectrum of tactical comm. It will field the integrated, multiband antenna known as the Multifunction Electromagnetic Radiating System which will reduce topside space and weight and reduce the Arsenal Ship cross section. ITP will provide SHF, EHF, and commercial SATCOM band terminals. It will leverage commercial terminal systems and services such as Direct Broadcast Satellite Service and Global Broadcast Service. ITP will also develop and implement multifunction antennas such as the Low Observable Multifunction Stack, that integrates SATCOM antennas into a single lightweight structure. This will significantly reduce the Arsenal Ship radar cross section and IR signature. (ref. E1, p 18-19)

Command can take three very different forms in peacetime operations: Combatant Command (COCOM), Operational Control (OPCON), or Tactical Control. Combatant Command means owning the forces. The commander has the full range of authority and responsibility inherent in the concept of military command. Operational Control allows for maximum control without full command or the burden of support. [38]

Presidential Decision Directive 25 notes that "Operational control is a subset of command. It is given for a specific time frame or mission and includes the authority to assign tasks to US forces already deployed by the President and assign tasks to US units led by US officers." Tactical control is defined by JCS Pub 1 as "the detailed and, usually local direction and control of movements or maneuvers necessary to assign missions and assigned tasks." [39]]

Major US military operations are going to be joint for a long time! They will include forces from more than one Service and will participate under a single commander. The National Command Authority (NCA) consists of the President and his advisors provides the overall guidance. One CINC, usually a theater commander such as CINCPAC or CINCCENT, provides the operational guidance in close cooperation with the commander of the Combined Joint Task Force (CJTF). The CJTF is usually made of up force components that are functionally differentiated. [40]

The *Arsenal Ship* will function within the existing command and control infrastructure. It will not be staffed such that it can assume Composite Warfare Commander responsibilities, but will be an integral ship whose weapons are under the control of the Composite Warfare Commanders who will report to the CJTF. Existing C4 equipment and procedures will be adapted to include the *Arsenal Ship*.

3. Naval Telecommunications Automation Programs

Major initiatives have been undertaken since the 1960s to automate shipboard communications. The Naval Tactical Data System (NTDS), Naval Modular Automated Communication System (NAVMACS), Naval Telecommunications System (NTS) Architecture, Integrated Communications System/Shipboard Communications Area Net (ICS/SCAN), Survivable Adaptable Fiber Embedded Network (SAFENET), Unified Networking Technology (UNT), Communication Support System (CCS) and Copernicus Architecture have all been developed as major steps in the modernization of shipboard suites. [41]

The *Arsenal Ship* takes advances in technology to provide a survivable, integrated, redundant, effective, and automated system. The overall goals of reduced manning and cost through application of existing technologies still apply!

G. COMMUNICATIONS

This section explores the exterior and interior communication requirements for the *Arsenal Ship*. The shipboard systems required are identified and the interrelationships between the systems is explained.

1. 1. External Communications Equipment

The *Arsenal Ship* Concept of Operations requires the ship to be capable of full-time communications with ships, aircraft, satellites and shore stations via responsive, reliable, clear and secure voice, tactical information distribution, and recorded communications. A survey of current primary communications equipment was conducted to set a baseline for the *Arsenal Ship*. Current Advanced Technology Demonstrators (ATDs) were reviewed to see what concepts will be available close to, or soon after the *Arsenal Ship* is built to ensure infrastructure is in place to facilitate integration.

ATDs and concepts with applicability to the *Arsenal Ship* include the Low Observable Multi-Function Stack, Multi-Function Electromagnetic Radiation System (MERS), Advanced Enclosed Mast Sensor System (AEM/S), and the Integrated HF Broadband Antenna, Figure IV-7, [42]. The *Arsenal Ship* superstructure was designed not only with reduced radar cross-section in mind using a ten degree slope, but also with the intention of integrating these advanced concepts as they become available. Our design incorporates current available equipment as the baseline communications suite with the capability and intent to upgrade.

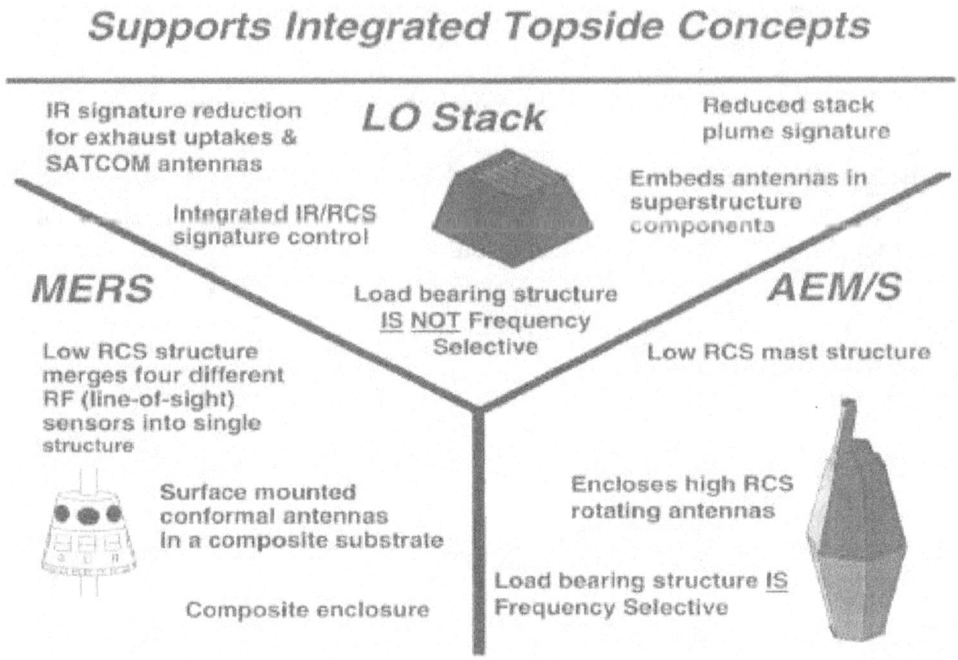

Figure IV-7. Communication Related Advanced Technology Demonstrations

The Low Observable Multi-Function Stack will provide exhaust suppression and imbedded satellite antennas as described in Figures IV-8 and IV-9, [42].

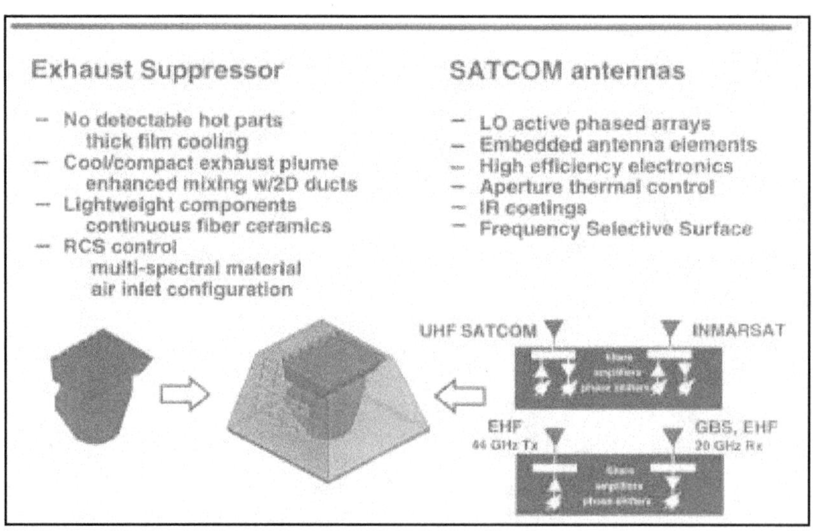

Figure IV-8. Low Observable Multi-Function Stack Exhaust Suppresser and SATCOM Antennas

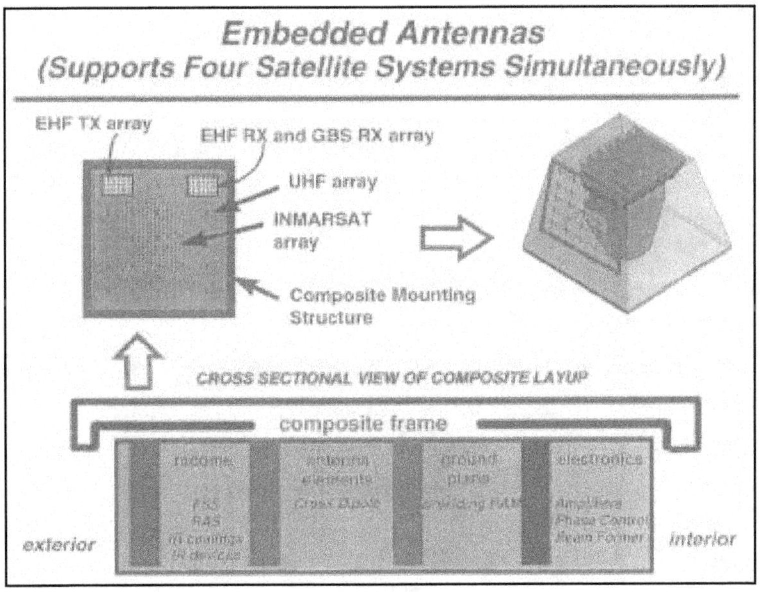

Figure IV-9. Expanded View of Low Observable Multi-Function Stack Embedded Antennas

The Multi-Function Electromagnetic Radiation System (MERS), will provide UHF and Link 16 Line-of-Sight (LOS) communication antennas as well as Identification Friend or Foe (IFF) and Combat Direction Finding Equipment, Figure IV-10 [42].

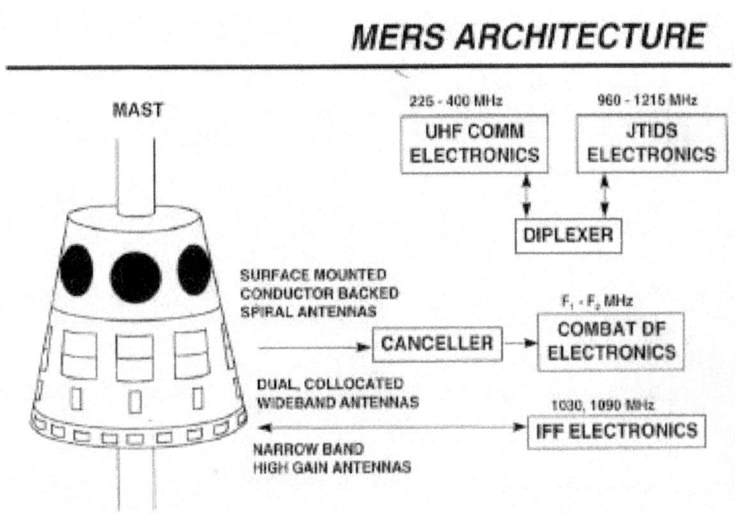

Figure IV-10. Multi-Function Electromagnetic Radiation System (MERS) Architecture.

Our interest in the Advanced Enclosed Mast Sensor System (AEM/S), which was designed to cover existing mast mounted equipment on present surface ships, is the integrated VHF/UHF/IFF communications antenna and integrated HF Broadband Antenna concept. The top of the mast holds an extended integrated antenna for VHF, UHF and IFF communications.. This mast is near the top to optimize line-of-sight distance, Figure IV-11, [43]. The Upper half of the enclosed mast is hexagonal with a ten degree angle. Investigations are currently ongoing at NRaD, San Diego to see if an integrated HF broadband antenna, 2-30 MHz can be imbedded, possibly in the corner structures, Figure IV-12, [43].

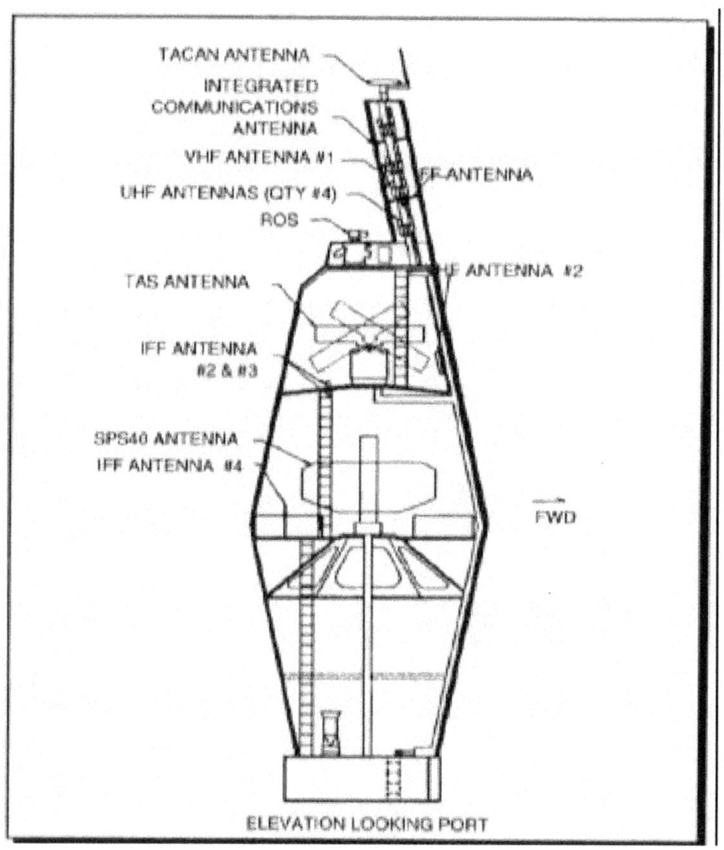

Figure IV-11. Advanced Enclosed Mast

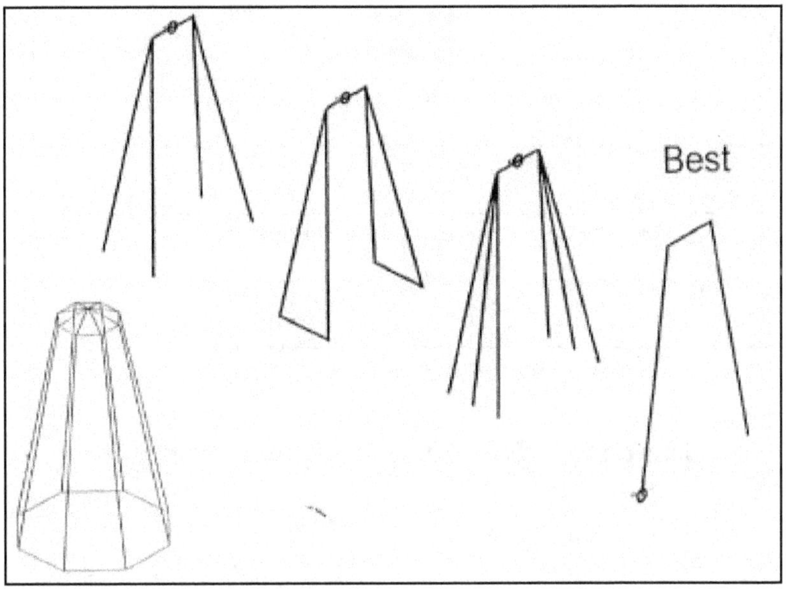

Figure IV-12. Example of Integrated HF Broadband Antenna, Octagonal Orientation.

a. Long Haul Communications [44]

Satellite communications have become the primary global means of communication with HF communications as a back-up. Satellites have inherent disadvantages including electronic vulnerabilities, high costs, coverage limitations, and bandwidth/load limitations. No single long-range communications medium can meet all shipboard communications requirements and these limitations have spurred improved HF communications programs, Table IV-7, [42].

• HF Improvement program (HFIP)
• Link 11 Improvement Program (LEI - Link 22)
• HF Automatic Link Establishment System
• Communications Support System (CSS)
• HF multimedia communications systems

Table IV-7. HF Communications Programs

HF communications operate in the 2-30 MHz frequency range primarily for ship-to-ship and ship-to-shore connectivity. Typical bandwidths are 3 kHz for voice, 75bps for teletype (TTY) and 2400 bps for data. HF propagation is subject to ionospheric fluctuations varying with the time of day, season, sunspots, multipaths and fading, but is relatively free from atmospheric noise. Optimum HF performance has been attained through improved antenna performance, use of single-side-band (SSB) modulation to preserves bandwidth, efficient use of transmitter power, and propagation prediction methods. HF improvements are striving for higher data rates, improved point-to-point link establishment, adaptive frequency selection, better shipboard radio resource applications, Anti-Jam (AJ) capabilities, and Low Probability of Intercept (LPI), [42]. Table IV-8 is a survey of required HF communication equipment for the *Arsenal Ship* and the associated antenna systems.

ARSENAL SHIP HF EQUIPMENT
• AN/URC-131A(V) High Frequency Radio Group (HFRG)
• AN/TRQ-42 Tactical Frequency Management System (TFMS)
• AN/URT-23(E) HF Transmitters
• R-2368B/URR HF Receivers
• Broadband Three-Wire, Twin-Fan, Light Weight Communications Antenna (LWCA)
• AS-2537A/SR HF Narrowband Monopole Antennas
• HFRG Receive Antennas
• HF Transmitter Multicouplers
• HF Receive Multicouplers
• AN/SSQ-33 Ships Automated Control System (SACCS)

Table IV-8. *Arsenal Ship* HF Communications Equipment

b. HF Communication Component Descriptions:

(1) AN/URC-131A(V) High Frequency Radio Group (HFRG), [45]

The AN/URC-131A(V) High Frequency Radio Group (HFRG) replaces existing shipboard manually tuned/controllable and rigid HF communications systems with a modern, agile communications system that is remote controllable/tunable, data capable and easier to operate and maintain. The HFRG forms part of the *Arsenal Ship* Exterior Communications System (ECS), and when interfaced with other equipment supports interrupted continuous wave (ICW), voice, teletype and digital data communications services. The HFRG supports tactical and Long-Haul (LH), full duplex, half duplex, and simplex, ship-to-ship, ship-to-submarine, ship-to-aircraft and ship-to-shore communications. The HFRG is capable of receiving signals in the frequency range 14 kHz to 30 MHz and transmitting signals in the frequency range 2-30 MHz. The HFRG is controlled remotely by the Ship Automated Communications Control System (SACCS) over a MIL-STD-1553B digital time division command/response multiplex data bus, and RS-422 buses, or locally from the equipment front control panels.

The HFRG consists of three subsystems: The transmit subsystem, the receive subsystem and the control/monitor subsystem. These subsystems replace the individual narrowband strings of receive and transmit equipment. The HFRG supports various configurations and quantities of transmit/receive circuits. The HFRG can support various transmit powers (4,8,12 kW), antenna numbers, different peak envelope power (PEP) rated Broadband Power Amplifiers (BBPA), and 1kW narrowband transmit circuits.

Modes of Operation: The modes of operation for the transmission and reception of user information are Lower Side Band (LSB), Intermediate Side Band (ISB), Upper Side Band (USB), Frequency Shift Keying (FSK), Continuous Wave (CW), Amplitude Modulation (AM), LINK-11-ISB and LINK-11-USB. The transmission of FSK, CW,AM, and LINK-11-USB are in the USB mode. Full carrier is used in the AM mode and the carrier is fully suppressed when transmitting in any other mode of operation. Reduced carrier, when required, is selectable. The reception of FSK, CW, AM, and LINK-11-USB are in the USB mode. The reception of AM is band-limited by the USB filter. The appropriate receive subsystem automatic gain control (AGC) is used for each mode of operation.

Transmit Subsystem: The *Arsenal Ship* (information source) generates a baseband signal which is routed by the ships transmitter switchboards to the HFRG transmit subsystem, where the signal is modulated, in the USB, LSB or ISB mode, by an exciter to provide an output in the 2-30 MHz frequency range. Each transmit circuit has its own exciter. The output signal of each exciter is amplified by the Broad Band Power Amplifier (BBPA) and routed to the appropriate broadband transmit antenna. The baseband signals are also routed to the narrowband antenna tuner-coupler which matches the output of the amplifier to the 35 foot whip antenna. The transmit subsystem operational configuration is established in response to commands from the SACCS, and the transmit subsystems reports when the system is configured and also reports detected failures. The *Arsenal Ship* is configured with 4kW allocated transmit power for 4 broadband transmitters and 2 1kW narrowband transmitters. (Comparison: YORKTOWN is configured with 4kW transmit power for 7 broadband transmitters and 2 1 kW narrowband transmitters.)

Receive Subsystem: Radio Frequency (RF) signals between 14 kHz and 30 MHz are received by small antennas provided with the system. These antennas are physically smaller than the transmit antennas and electrically small at the lower frequencies. The RF signals from the antenna are distributed to the appropriate receiver, and the signal is demodulated to provide a baseband output which is routed by the surface ship receiver switchboards to the appropriate user (information sink) within the ECS. The receive subsystem is established in response to commands from the SACCS and the receive subsystem reports when the system is configured and also reports detected failures.

(Comparison: YORKTOWN is equipped with 3 racks of 6 receivers (18 channels total). Two receivers cover 14 kHz-30MHz and 16 receivers cover 2-30MHz.)

Control/Monitor Subsystem: Remote control, performance monitoring, test, operator interface and display of the operational characteristics of the HFRG are the functions of the CMS. The transmit and receive subsystems are capable of configuring and reporting operational characteristics in response to commands from the CMS over dual-redundant RS-422 multi-drop buses. The remote control, operator interface, and display functions are performed by the CMS. With SACCS installed, it performs the remote control, operator interface and display functions. The CMS automatically assumes those functions if the SACCS fails.

The HFRG provides a major improvement in time required for circuit activation/deactivation and significantly reduces man-hours required for system maintenance and repair. The HFRG significantly reduces the workload of communication and maintenance personnel by providing remote-controlled, automatically-tuned circuit configuration and activation, extensive self-test capabilities and diagnostic programs. Using previous manually tuned HF systems, each circuit required approximately 5-7 minutes to tune. The HFRG requires 30 seconds for one person to reconfigure one or multiple circuits using the CMS/SACCS.

(2) AN/TRQ-42 Tactical Frequency Management System (TFMS), [42]

The Tactical Frequency Management System (TFMS) uses a Chirpsounder which conducts HF sounding of the ionosphere to detect the best HF communications frequencies, Figure IV-13, [42]. The TFMS improves the quality and reliability of HF circuits by providing continuous measurement and display of the best HF communications frequencies as the ionosphere changes.

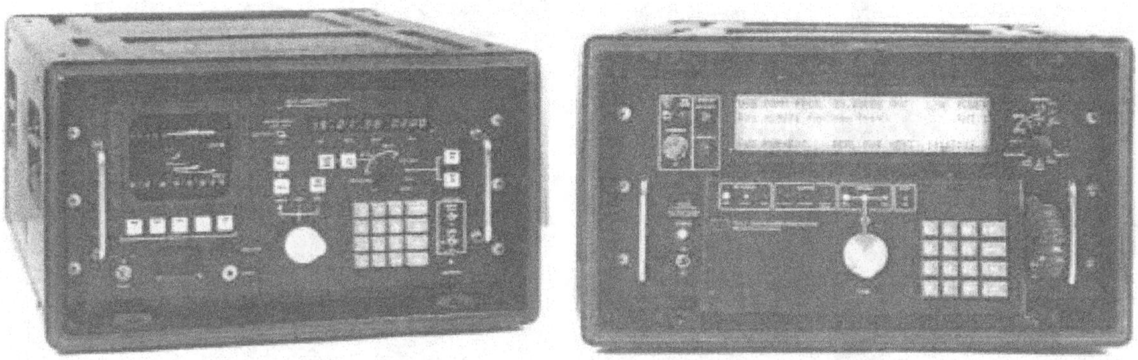

Figure IV-13. AN/TRQ-42 Transmitter and Receiver

(3) AN/URT-23(E) HF Transmitters, [42]

The AN/URT-23E transmitter is an ISB HF radio set designed for reduced and suppressed carrier applications in the 2-30MHz frequency range. It operates in AM voice, CW, RATT (radio teletype) and ALE (Automatic Link Establishment) modes with the automatic antenna coupler group AN/URA-38. The AN/URT-23(E) has improved coding and waveforms and a MIL-STD-1553B bus. ALE is a method of establishing HF point-to-point communications using selective calling, hand-shaking, scanning, sounding, polling, and coordination among HF networks conforming to MIL-STD-188-141A for HF interoperability and ALE. Figure IV-14 shows the physically similar AN/URT-23(D) transmitter, [44].

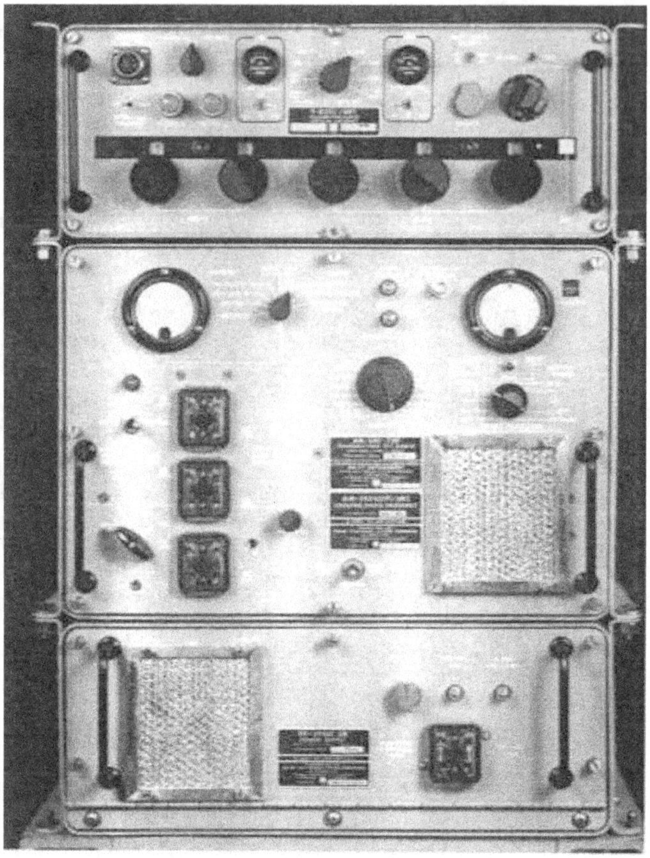

Figure IV-14. AN/URT-23(D) Transmitter

(4) R-2368B/URR HF Receivers, [42]

The R-2368B is the standard LF/MF/HF receiver operating in the frequency range 10 kHz-30 MHz, tunable in 1 Hz increments, Figure IV-15, [44]. Its modes of operation include USB/LSB/ 2-ISB, CW, AM and FM with fault isolation to a replaceable module/board. The receiver includes a serial two-wire databus remote control with MIL-STD188C, RS-232C or RS-422 formats. LINK 11 operability and internal FSK demodulation with selectable shift and center frequency are standard. The receiver size is 13.3 cm x 48.3 cm x 19.5 cm, weighing 18 kg, with a power requirement of 115 VAC +-20%, 47-400 Hz.

Figure IV-15. R-2368B/URR HF Receiver

(5) Broadband Three-Wire, Twin-Fan, Light Weight Communications Antenna (LWCA) [42]

Fan-type antennas are designed to be nearly omnidirectional operating over a 4-to-1 frequency range with power loss of only 20% due to mismatch. Efficiency is provided by multicouplers used to combine signals from transmitters and receivers for one or more antennas. The LWCA is the antenna for the HFRG transmit subsystem. The LWCA uses a thermoplastic polymer instead of ceramic insulators to improve shock resistance and the overall weight is reduced by using 4.8mm diameter phosphor-bronze wire instead of 7.9mm diameter wire. The *Arsenal Ship* will use the LWCA in a twin fan setup similar to Figure IV-16, [44].

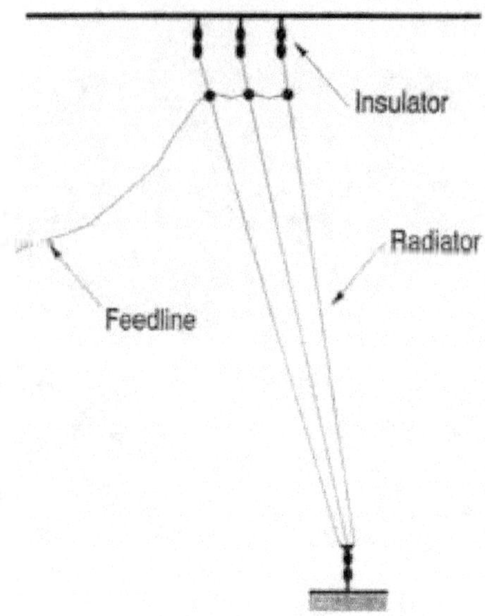

Figure IV-16. Broadband Three-Wire Light Weight Communications Antenna

(6) AS-2537A/SR HF Monopole Whip Antennas, [42]

The *Arsenal Ship* has two 35 foot monopole antennas located forward of the aft stack separated by 35 feet to prevent excessive mutual coupling interference between adjacent base tuners if both are transmitting. The normal setup is for one HF monopole to work with the HFRG transmitting the narrowband baseband signal and the other to operate in a receive mode for additional HF applications. The AS-2537A can receive from 14kHz-32MHz and transmit from 2-32 MHz depending on the antenna coupler used with the antenna, [46].

(7) HFRG Receive Antennas, [45]

The HFRG receive antennas are described as physically smaller than transmit antennas, however, no picture descriptions were obtained. Based on the size of the vessel, it was concluded that the incorporation of this receive group would not present any difficulties.

(8) AN/URA-38 Antenna Coupler Group, [44]

The AN/URA-38 Antenna Coupler Group provides the narrowband tuning for the AS-2437A/SR monopole antenna when using the AN/.URT-23(E) HF transmitter.

(9) CU-1772A/SRA-56, CU-1774/SRA-57 and CU-1776A/SRA-58 Shipboard HF Transmitter Multicouplers [44]

HF transmitter multicouplers provide the capability of simultaneously connecting up to four 1kW transmitters into a single broadband antenna, (e.g. LWCA), thereby reducing the number of HF antennas required for effective communication. The CU-1772A/SRA-56 can connect up to four broadband transmitters simultaneously to a single antennas in the 2-6 MHz band, the CU-1774/SRA-57 in the 4-12 MHz band and the CU-1776A/SRA-58 in the 10-30 MHz band.

(10) CU-1789/SRA-49B Shipboard HF Receiver Multicoupler [44]

The CU-1789/SRA-49B HF Receiver Multicoupler provides the capability of simultaneously connecting HF receivers such as the R-2368 into a single broadband antenna such as the HF monopole antenna AS-2537A/SR or LWCA.

(11) AN/SSQ-33 Ships Automated Control System (SACCS), [47]

SACCS provides an advanced ship radio communications control system for the *Arsenal Ship* that features centralized operator control with automated monitoring and reconfiguration. As the local radio network manager installed on the *Arsenal Ship*, SACCS interfaces with the USN multi-ship radio network manager, the Automated Integrated Communications System (AICS). The AICS supports a global secure-level, relational database-serviced computer and communications network manager for the Navy's ship and shore communications control system.

The AN/SSQ-33 uses TAC-3 open architecture, Oracle and UNIX technologies to provide centralized operator control over ship radio communications. SACCS provides automated, reliable, and robust circuit connectivity, HF, UHF, and VHF radio network monitoring and reconfiguration.

c. VHF/UHF Line-of-Sight Communications [44]

VHF/UHF communications support ship-to-ship, ship-to-shore, ship-to-air, line-of-sight (LOS) connectivity. The VHF band is between 30 MHz - 300 MHz and the UHF frequency spectrum is 300 MHz - 3 GHz. Navy shipboard communications in the VHF/UHF bands are attractive because smaller antennas can be used than for HF, LOS propagation results in low detectability, the wider bandwidth allows spread spectrum applications, and very low transmission power can be used for limited transmitter range.

Shipboard tactical VHF in the 30 MHz - 88MHz frequency range uses the Single Channel Ground and Airborne Radio System (SINCGARS), while the 108 MHz - 156 MHz spectrum supports ship-shore communications in amphibious operations and land-mobile shore communications. Principal tactical ship-ship communications and ship-aircraft ECCM circuits operate in the overlapping VHF/UHF 225 MHz - 400 MHz band, while LINK 16, HAVEQUICK II, and some satellite relays occupy 950 MHz - 1150 MHz , Figure IV-17, [44].

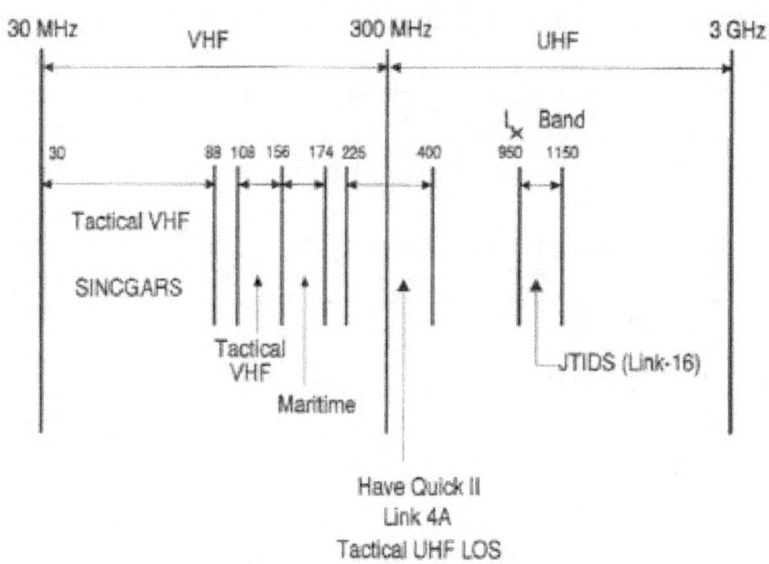

Figure IV-17. VHF/UHF Spectrum

In the VHF and UHF bands, it is necessary to pay considerable attention to noise reduction within the equipment itself, since for frequencies above 100 MHz, the noise generated by the receiver front-end amplifier becomes dominant. Receivers are designed so as to perform at as high a S/N ratio as possible. VHF/UHF propagation is not affected by ionospheric reflection and travels in straight lines. Small wavelengths (below 1 meter for frequencies above 300 MHz) may reflect from a ship's structure. Reflected signals interfere with direct LOS signals and cause multipath effects that may weaken the signal. In low altitude applications, reflected multipath signals on the water vary receiver signal strength. Above 400 MHz, transmission resembles transmission of light with shadowing by obstacles, reflection from terrain features, structures and vehicles with sufficient reflectivity. Wider signal bandwidth than HF allows spread-spectrum techniques and higher data transmission rates (UHF is often used for digital data transmission). Shipboard VHF and UHF receivers are designed for high sensitivity, for a very high range of signal strengths, and for combating interfering signals by utilizing spread spectrum technology. LOS ranges are increased by elevating shipboard VHF/UHF radio antennas. To extend the LOS range, airborne relays are used such as aircraft or UAVs. Table IV-9 is a comparison of LOS ranges and elevation.

Antenna Height	Range
Ship (30m) to Ship (30m)	25-50km
Ship (30m) to Helicopter (300m)	60-90km
Ship(30m) to Aircraft (3000m)	200-240km

Table IV-9. LOS VHF/UHF Range versus Antenna Height

(1) AN/VRC-90 SINCGARS - Single Channel Ground and Airborne Radio System [42]

SINCGARS is a joint tactical VHF radio which is a frequency hopping (FH), frequency modulated (FM), spread spectrum system covering the 30 MHz-88MHz frequency band in 25 kHz wide discrete bands. SINCGARS is designed to provide secure voice and data communications in jamming environments. The shipboard VHF SINCGARS is a modified version of the ground-based SINCGARS. The shipboard VHF radio is integrated into the ships exterior communications system, available antenna system and the Single Audio System (SAS), which integrates secure and nonsecure voice interior communications, Figure IV-18, [44].

SINCGARS radios consist of transmitter-receiver equipment, antenna/antenna coupler equipment, Communications Security (COMSEC) equipment, Transmission Security (TRANSEC) equipment, and baseband equipment. The *Arsenal Ship* contains the Type II SINCGARS VHF radio configuration. Type II SINCGARS consists of four circuits with multiple antennas for the heavy traffic anticipated for NSFS. When operated in the single-channel (SC) mode, the four antennas are connected to a Multicoupler (TD-1289). When operated in frequency hopping (FH) mode, the four antennas are connected to four separate power sensors in order to support the FH spread spectrum operation.

The shipboard SINCGARS includes four RT-1523 receiver-transmitters as part of the AN/URC-90 SINCGARS radio system, Figure XX. The COMSEC/TRANSEC equipment includes a communications encryption device for user traffic and a transmission encryption device for frequency hopping ECCM. The baseband equipment includes the AN/PSC-2 Digital Communication Terminal (DCT), a portable communications message processor for Marine Corp gunfire support, the H-250 handset for voice communications and the Single Audio System. Baseband equipment is connected to the RT-1523 via a 1553B data bus or an RS-232C interface.

The SINCGARS System Improvement Program (SIP) of 1996 provides Time Division Multiplexing (TDM), voice and data networking, automatic net radio interfacing, packet radio, forward error correction, reduced synch times and an interface with the Global Positioning System (GPS). The technical specifications of the SINCGARS radio are shown in Table 4, [47].

SINCGAR RADIO Technical Specifications
• Modes: (single channel, frequency hopping, cypher text, plain text, digital data, analogue data, FSIC, retransmission)
• Frequency Range: 30-88 MHz
• Number of Channels: 2,320
• Data Capability: 600 bps - 16 kbps digital plus FSK
• Dimensions: 51x119x269mm, weight 2kg
• Power Output: 500 mW, 160 mW, 4.5 W, 50 W (with power amplifier)

Table IV-10. SINCGARS Technical Specifications.

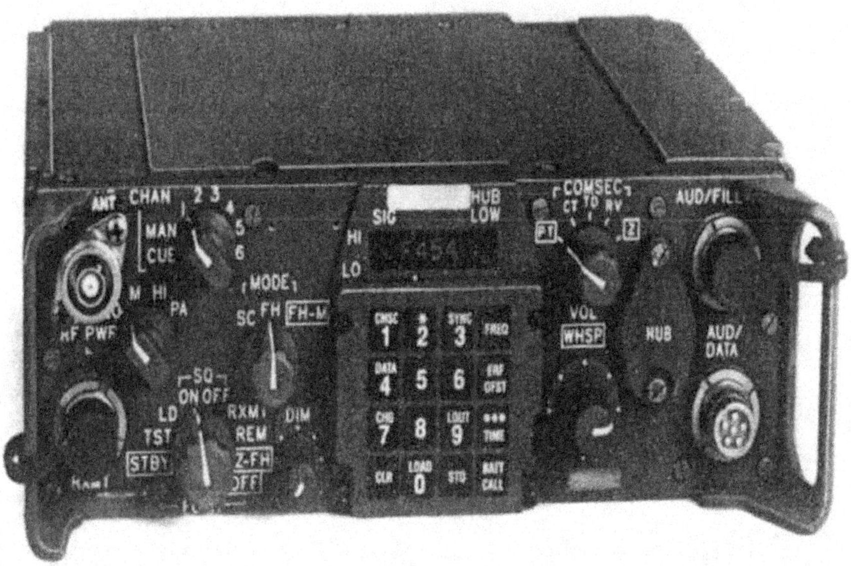

Figure IV-18. SINCGARS RT-1523 Transceiver

(2) AN/ARC-182 VHF/UHF Radio 30Mhz - 400 MHz [42]

The AN/ARC-182 VHF/UHF multiband/mulitmode tactical radio for close air support, air traffic control, maritime radio-telephone, military and NATO use. It has 25 kHz spacing in VHF/UHF modes supporting 11,960 channels and can be interfaced via a 1553B data bus or RS-232 serial data channel, Figure IV-19, [44].

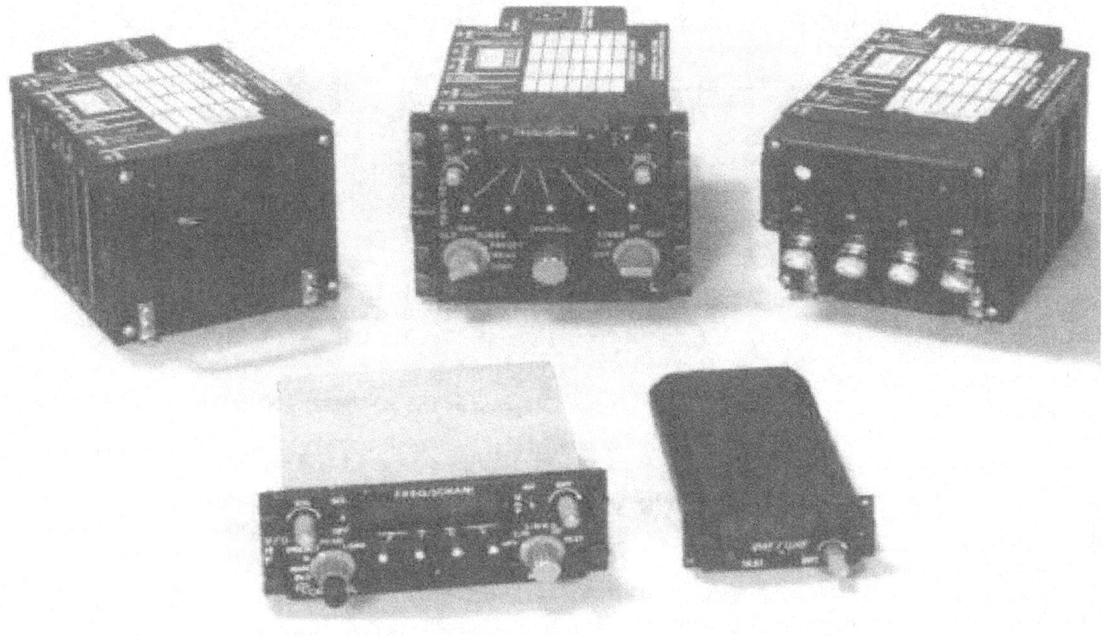

Figure IV-19. AN/ARC-182 VHF/UHF Radio

(3) AN/WSC-3(V)6 UHF LOS Radio [42]

The AN/WSC-3 is a shipboard UHF LOS radio in the 225 Mhz-400 MHz range. The WSC-3 radio group consists of an antenna coupler, an RT-1217/WSC-3 transceiver and a Quadratic Phase Shift Keying (QPSK) modem which can handle voice, data, TTY LOS and UHF SATCOM communications, Figure IV-20, [44].

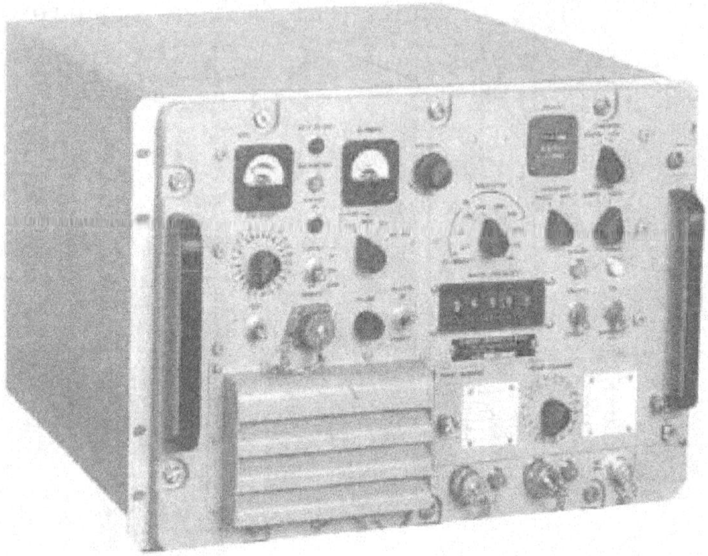

Figure IV-20. AN/WSC-3(V)6 UHF Transceiver

(4) AN/URC-93 VHF/UHF LOS Radio [42]

The AN/URC-93 is a Navy shipboard VHF/UHF radio operating in the 225-400 MHz frequency band and consists of an antenna coupler, OR-176(V) transceiver, and a remote controller. The radio group can handle local or remote voice, TTY, data and wideband communications in ECCM and Low Probability of Intercept (LPI) modes, Figure IV-21, [47].

Figure IV-21. AN/URC-93 VHF/UHF Transceiver

(5) LINK-16 (JTIDS - Joint Tactical Information Distribution System) [42]

Link 16 is a high capacity Time Division Multiple Access (TDMA) system providing integrated communications, navigation, and identification of friend or foe (IFF). The Link 16 waveform consists of a series of 6.4 us pulses in the 960 MHz - 1215 MHz band with anti-jamming capability using frequency hopping techniques. Link 16 provides ECCM capability for airborne and surface ship platforms, extended range of communications and OTH communications between surface ships using an airborne relay platform. The *Arsenal Ship* contains the Class I terminal for surface ships and uses LINK 16 as a backup to CEC for data communication with the Control Ship.

(6) Global Position System (GPS)

The *Arsenal Ship* uses GPS for precision navigation. The *Arsenal Ship* will have a GPS flush mounted antenna at each corner of the forward upper deck superstructure, tilted at 10 degrees from vertical and electronically switched for the maximum receive signal, Figure IV-22, [48]

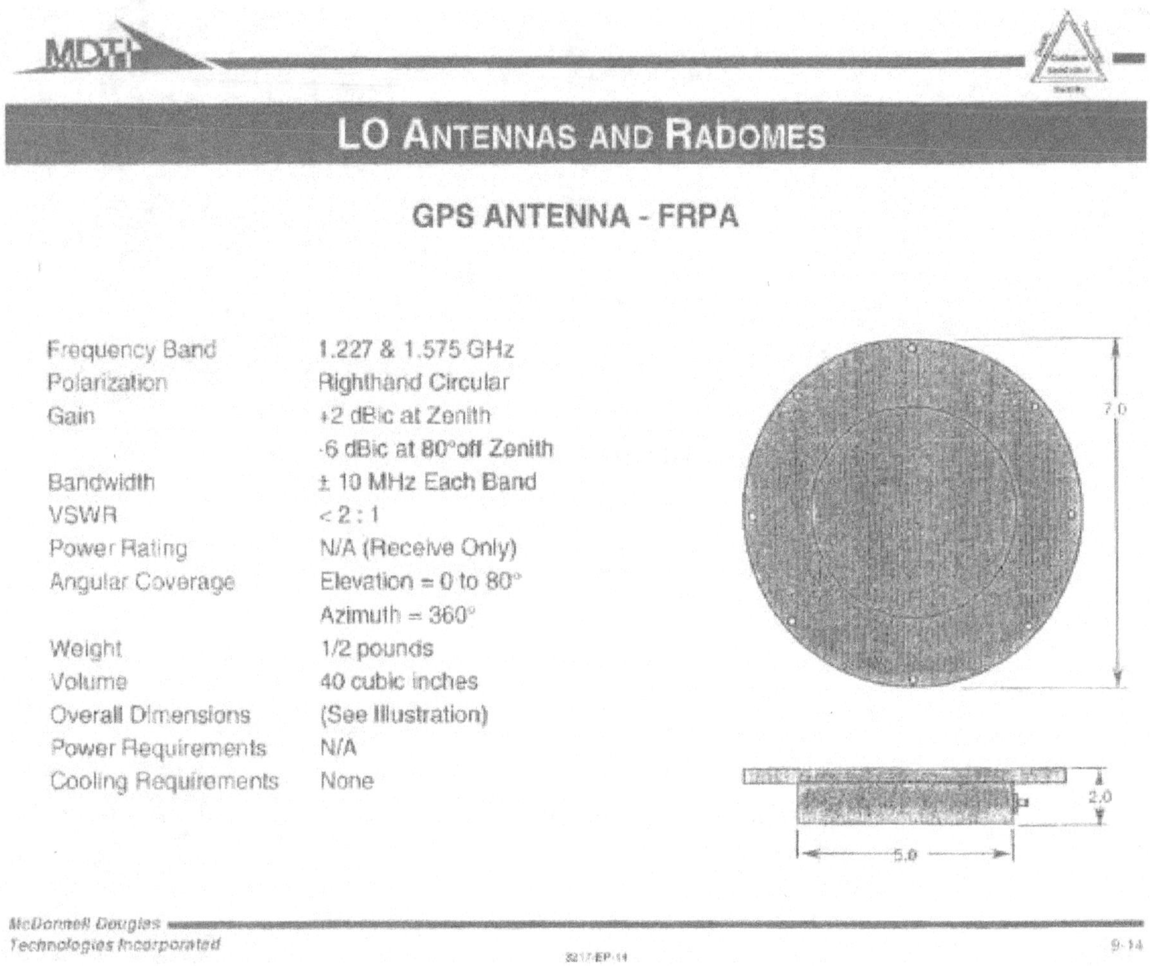

Figure IV-22. GPS antenna

(7) AN/USQ-123 Common High Bandwidth Data Link (CHBDL) [44]

The Common High Bandwidth Data Link is a microwave LOS link supporting imagery data communications from airborne platforms such as a reconnaissance aircraft to ships. The CHBDL provides full duplex microwave digital data links between shipboard and airborne terminals in the X and Ku band frequencies (9.7 GHz - 10.5 GHz and 14.4 GHz - 15.5 GHz). Currently used by aircraft carriers for acquisition of sensor signals form airborne reconnaissance aircraft and other data link-equipped aircraft, the system has a downlink capability of 10.71 Mbps, 137 Mbps, and 274 Mbps and an uplink rate of 200 kbps. This system is included as a consideration for a possible high data rate link capability between the control ship and the *Arsenal Ship*, perhaps as a backup for CEC. The higher data rate from the control ship to the *Arsenal Ship* could provide the weapon initialization data and updates while the lower data rate could provide the *Arsenal Ship* status messages. This system was not placed on our design because of the backfit cost required to fit all ships that might act as Control Ships with CHBDL. However, with the *Arsenal Ship* as part of a battlegroup with the capability to use aircraft as the downlink media, this system may bear consideration. The technical characteristics of the CHBDL AN/USQ-123(V) Transceiver are shown in Table IV-11.

• Frequency: X-band, Ku-band
• Data Rate: Uplink - (200 kbps) Downlink - (10.71 Mbps, 137 Mbps, and 274 Mbps)
• Modulation: Uplink - (BPSK-binary phase-shift-keying) Downlink - (Offset QPSK-quadrature phase-shift-keying)
• Antenna: Two 1-meter dish antennas

Table IV-11. Technical Characteristics of CHBDL AN/USQ-123(V) Transceiver

d. VHF/UHF Antennas

VHF/UHF LOS antennas are small, with VHF antennas ranging from 1.5-3 meters and UHF antennas sized from 60 cm - 1.5 m. These VHF antennas are placed topside, mounted on the *Arsenal Ship*'s two masts supporting the CEC antennas. They are internal, enclosed within a frequency selective structure, mounted around the load bearing support. The UHF LOS communication antennas are representative of low observable UHF antennas available from McDonnell Douglas Technologies Incorporated. Eight 32 inch diameter, flush mounted antennas symmetrically located around the deck house will provide UHF LOS connectivity, Figure IV-23, [48].

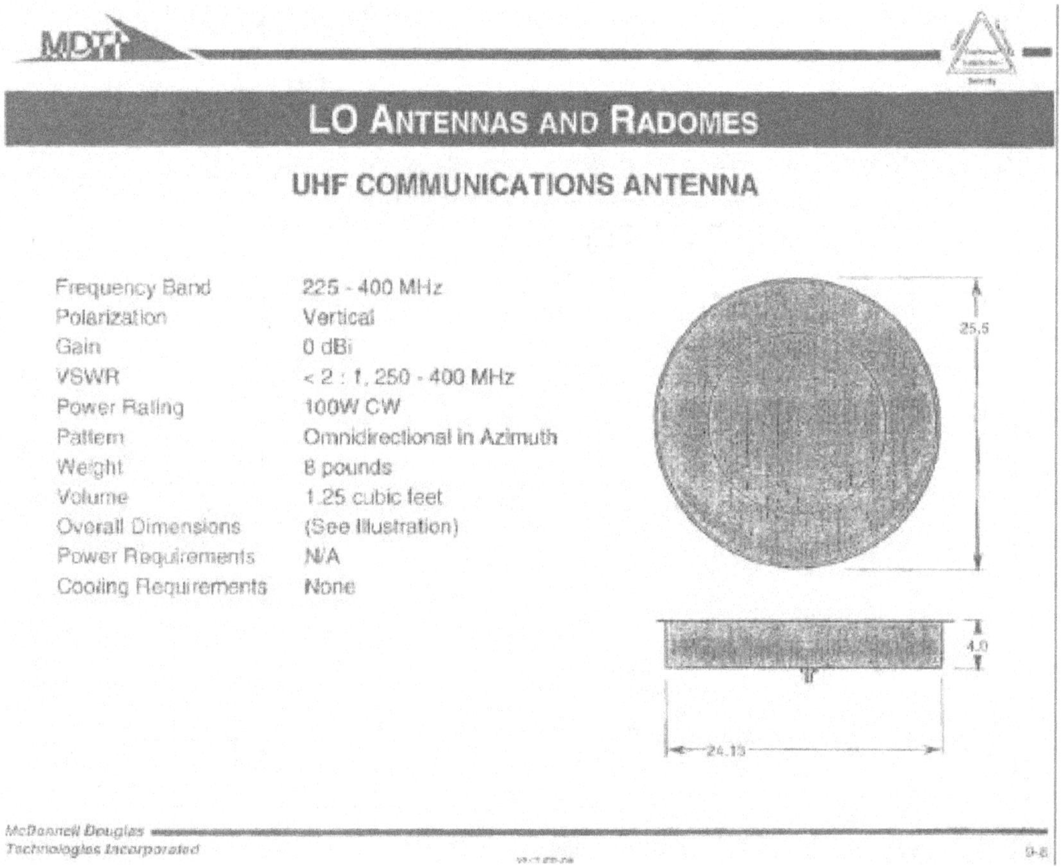

Figure IV-23. Low Observable UHF Communications Antenna

The LINK 16 and IFF antennas are flush mounted and located adjacent to the eight UHF antennas. The eight antennas are 12 inches in diameter and time share between Link 16 and the 1030 MHz, 1090 MHz IFF frequency, Figure IV-24, [48].

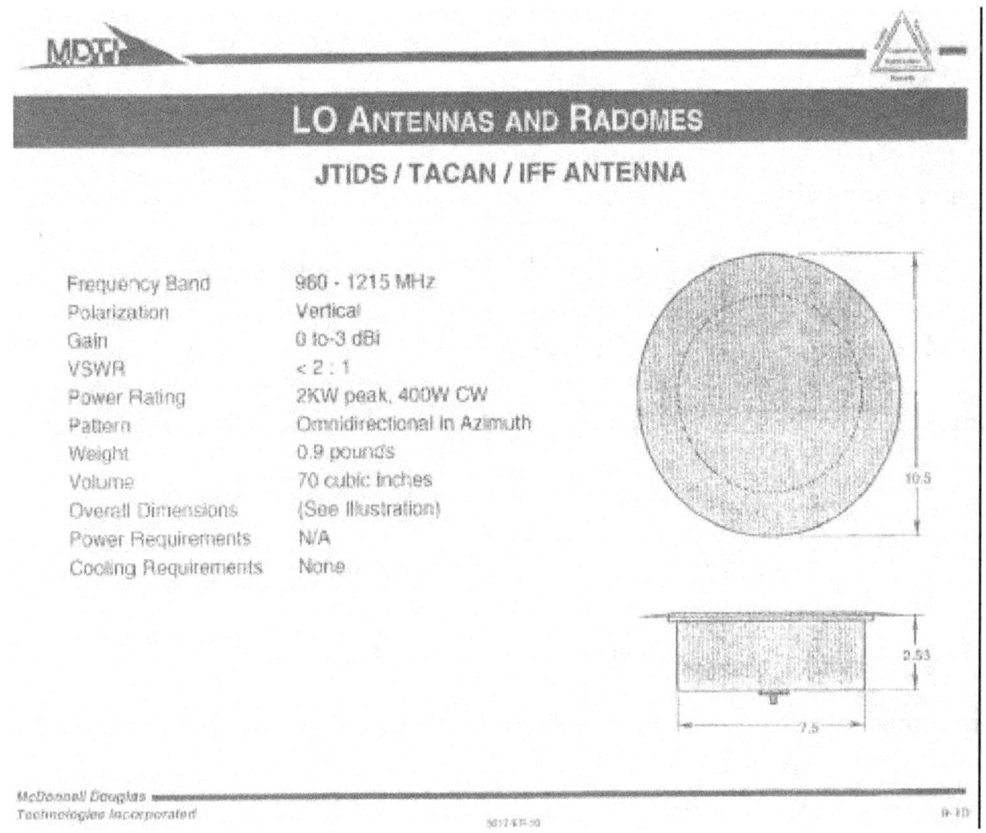

Figure IV-24. Low Observable LINK 16/IFF Communications Antenna

(1) VHF/UHF Antenna Coupler CU-1559 [44]

The CU-1559 antenna coupler is a two port impedance matching device and isolation network that allows duplex operation of two VHF/UHF transceivers. The CU-1559 may be used to couple two UHF transmitters to one antenna and two VHF/UHF receivers to a second antenna.

e. UHF Fleet Satellite Communications (FLTSATCOM) [44]

UHF satellite communications (SATCOM) are provided by the combination of FLTSAT (Gapfiller Satellites), and LEASAT (Leased Satellites), to provide global FLTSATCOM coverage between 76° N and 76° S. Four FLTSAT equatorial satellites provide coverage with ten 25 kHz channels dedicated for Navy use. Each 25 kHz channel has a separate transmitter with channel one carrying Fleet Broadcast transmissions. Four LEASAT satellites, in service since the 1980s, each provide one 500kHz, seven 25kHz, and five 5kHz repeater channels. The seven 25 kHz channels are dedicated to Navy use with channel one providing Fleet Broadcast and the other six providing UHF SATCOM with or without demand assigned multiple access (DAMA) equipment.

The UHF follow-on (UFO) Satellite Communications System is the replacement system for the FLTSATCOM network. The FLTSATCOM satellite network was originally intended to have a useful service life of five years, but has more than doubled its intended design life with the first launched in 1978. The Navy requires UHF satellite circuits to support the Navy's everyday administrative and tactical communications. The UHF follow-on program is designed to provide reliable and redundant tactical satellite communications well into the future. Eight satellites with one orbital spare will provide near-global coverage between 71° N and 71° S and are completely compatible with the existing FLTSATCOM satellites. The last three UFO satellites will include the Navy's FLTSAT EHF Package (FEP) which will add survivable and protected communications in the next higher band of the spectrum, extremely high frequency, (EHF) [49].

UHF FLTSATCOM subsystems will provide a significant part of the communications for the *Arsenal Ship* C^3I (Command, Control, Communications and Intelligence), Table IV-12.

FLTSATCOM Subsystems
• Fleet Satellite Broadcast (FLTBROADCAST)
• Common User Digital Information Exchange Subsystem / Naval Modular Automated Communications Subsystem (CUDIXS/NAVMACS)
• Officer-in-Tactical Command Information Exchange Subsystem (OTCIXS)
• Tactical Intelligence Subsystem (TACINTEL)
• Tactical Data Information Exchange Subsystem (TADIXS)
• Secure Voice Subsystem (SECVOX)
• Teletypewriter Subsystem (TTY)
• Submarine Satellite Information Exchange Subsystem (SSIXS)

Table IV-12. Fleet Satellite Communications Subsystems

Fleet Broadcast supports the transmission of broadcast one-way message traffic normally operating on channel one. The channel has 15 subchannels each operating at 75 bps with subchannels integrated into a time-division multiplexed (TDM) data stream of 1200 bps. Primary transmission from the shore station to the satellite is a direct-sequence spread-spectrum SHF uplink to provide jamming protection with a UHF downlink. A second channel operating on UHF for both the uplink and downlink provides a backup capability. This backup channel is normally assigned to secure voice transmission. Shipboard subscribers receive the UHF broadcast signal that is then demodulated and demultiplexed. The individual data streams are sent to the Naval Automated

Communications System (NAVMACS) and the Tactical Intelligence (TACINTEL) processors for screening prior to teletypewriters. Weather data and General Service (GENSER) data are sent directly to teletypewriter equipment. Messages in the FLTBROADCAST use formats that adhere to the Joint Interoperability of Tactical Command and Control System (JINTACCS) standards, the U.S. Message Text Format (USMTF), the Joint Army, Navy, Air Force Publications (JANAP) and the Allied Communications Policies (ACP) 121 and 127.

Common User Digital Exchange Subsystem (CUDIXS) is a shore-based processor suite that provides message traffic processing and transmission control. The Naval Modular Automated Communications Subsystem is the ship-based subscriber terminal and processing system that transmits, receives and processes messages. NAVMACS processes the CUDIXS messages and operates interactively with the ship message processing and distribution system (MPDS). CUDIXS/NAVMACS is a functional replacement for ORESTES TTY network. It provides improved operational message throughput rates, increased traffic volume and better link reliability. Each subscriber to a CUDIXS net is assigned an identification number (ID) that is recognized by the CUDIXS and NAVMACS processors. The NAVMACS processor forwards only screened messages addressed to the ID subscriber. The processor automatically sends the messages to the proper end equipment, either a teletypewriter, printer, or recorder. The NAVMACS processor is shown in Figure IV-25, [44].

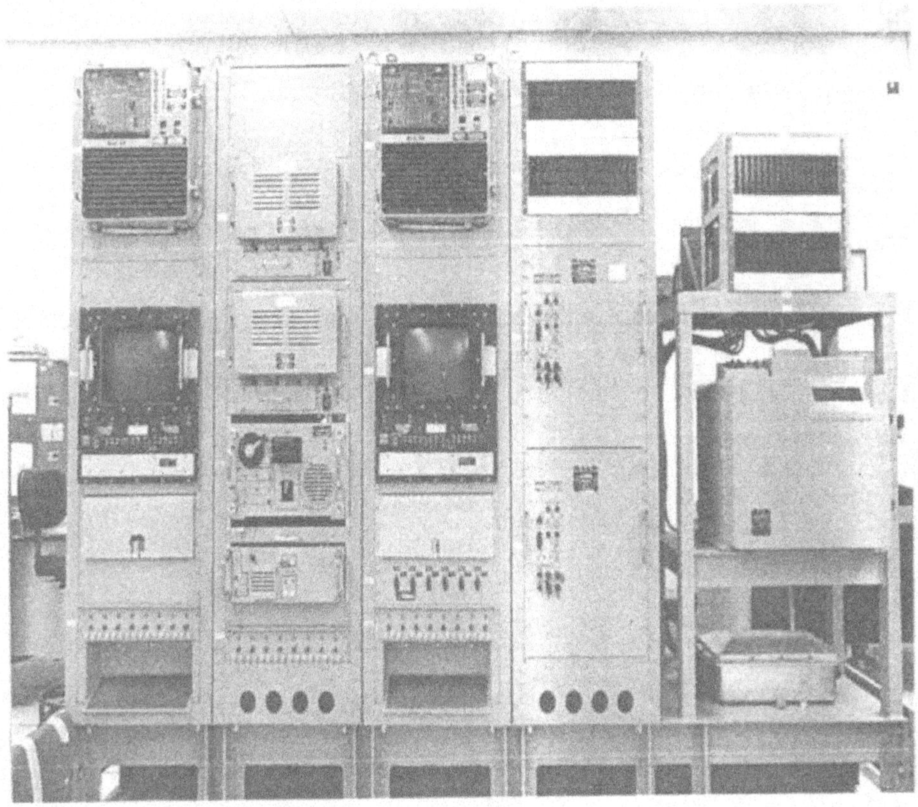

Figure IV-25. NAVMACS Processor

The Officer-in-Tactical Command information Exchange Subsystem (OTCIXS) supports full duplex communications to the Battle Group (BG) for command and control (C^2) and over-the-horizon targeting (OTH-T). The OTCIXS circuit carries TTY message traffic and formatted messages for tactical data processor (TDP) operation. The TDP capability supports the exchange of surveillance formatted track data and targeting data among BG platforms.

The Tactical Intelligence Subsystem (TACINTEL) supports the transmission of special intelligence (SI) information. This subsystem has an automated message processing system and is used for the transfer of information via satellite in a controlled environment. A 25 kHz channel is allotted to TACINTEL which is structured similar to OTCIXS with up to 23 subscribers in each satellite footprint.

The Tactical Data Information Exchange Subsystem (TADIXS) supports broadcast transmission of TDP data from shore sites to the fleet. TADIXS B is raw data that uses the Tactical Receive Equipment (TRE). The TRE is a computerized message processing system that demodulates, decodes, decrypts, and processes the TADIXS-B transmissions. The information is data or messages in the Over-the-Horizon Rainform Gold format. The TRE screens duplicate messages and filters them according to user-specified parameters

The Secure Voice Subsystem (SEVOX) supports shore-to-ship, ship-to-ship, and ship-to-shore voice transmission. The narrowband voice (3 kHz channel) is digitized using a vocoder, producing a 2400 bps data stream that is encrypted. The SECVOX subsystem uses push-to-talk half-duplex transmission over a 2400 bps FLTSATCOM channel on each of the four satellites. SECVOX channels on the satellites are dedicated and when DAMA is used, a dedicated time slot is available.

The Teletypewriter Subsystem (TTY) ORETES uses the FLTSATCOM to provide a reliable beyond-line-of-sight (BLOS) TTY capability via satellite. The ORESTES subsystem provides a full period and on-call tactical and report-back TY circuits that serve as backup for the other networks (e.g. CUDIXS/NAVMACS and TACINTEL).

f. UHF FLTSATCOM Equipment [44]

There are currently two shipboard UHF FLTSATCOM receivers in use. One is the stand-alone AN/SRR-1A and the other is the receiving section of the AN/WSC-3 UHF transceiver. The AN/SSR-1A receiver is installed aboard most surface ships and receives the FLTBROADCAST message traffic.

The AN/SSR-1 receiver drives TTY equipment and consists of antennas, amplifier-converters, the MD-900 combiner-demodulator, and the MD-1063 demultiplexer.

(MD-900 Demodulator, TD-1063 Demultiplexer, SRR-1 antenna and receiver equipment)

The AN/WSC-3 is used for receiving UHF LOS communications and SATCOM. The AN/WSC-3 transceiver has two built-in demodulators for DPSK data rates from 75-9600 bps, FSK at 75 bps and AM or FM voice modulation. The transmitter segment generates 30 W RF output for AM and 100W for FM, FSK, DPSK. It can support AM narrowband and wideband as well as FM narrowband operations, AM voice and SECVOX, PSK data at 75, 300, 1200, 2400, 4800 and 9600 bps, and FSK at 75 bps. The AN/WSC-3 is tunable to 7000 channels at a 25 kHz spacing.

(1) UHF FLTSATCOM Antennas

The AS-2815/SSR-1 antenna is used for receiving the Fleet Broadcast and is combined with the AN/SRR-1 receiver. A set of four antennas is used to provide full hemispherical coverage and space diversity reception, Figure IV-26, [44]. The *Arsenal Ship* uses a replacement for the SSR-1 antenna, the AS-3439/G low gain antenna with complete hemispherical coverage. The AS-3439/G is compatible with low data rate and low radar cross section installations, Figure IV-27, [48].

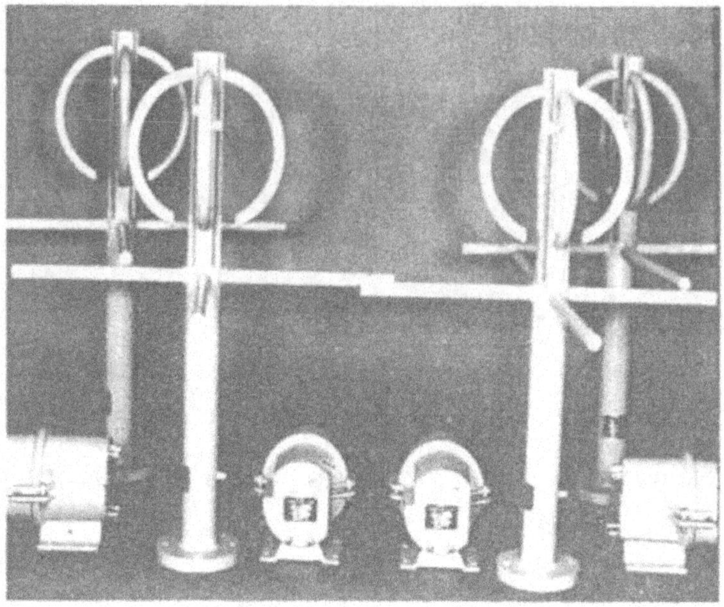

Figure IV-26. SRR-1 Antenna System

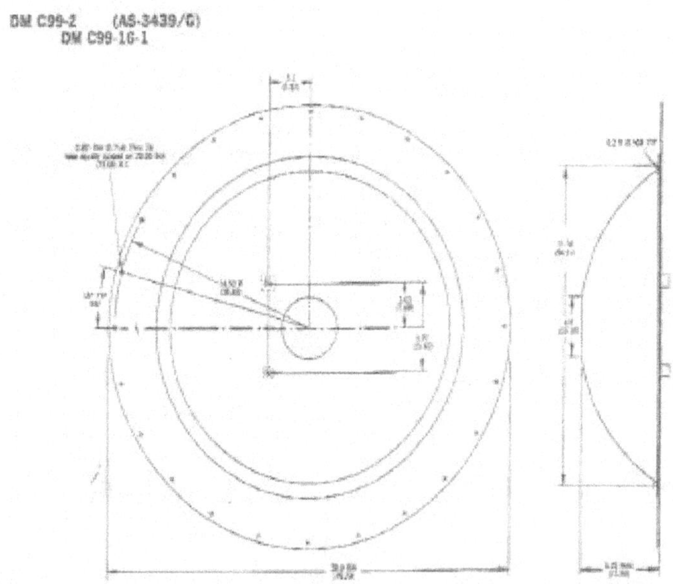

Figure IV-27. AS-3429/G

The AS-3018/WSC-1(V) antenna is part of the antenna group OE 82C/WSC-1(V), interfacing with the AN/WSC-3 equipment. The antenna group consists of the antenna and additional equipment for tracking , control, sensing, switching, and preamplification. The AS/3018A/WSC-1(V) antenna consists of a single crossed dipole element recessed within a barrel-shaped drum structure. The antenna size is 1.37 x 0.85m (54 x 33 in) (diameter x depth) and weighs 133 kg (295 lbs). The antenna mount permits 360° azimuthal and 0°-90° elevational orientation with the antenna tracking the satellite automatically. Two AS-3018A/WSC-1(V) antennas are used onboard the *Arsenal Ship* to eliminate superstructure blocking and provide redundancy, Figure IV-28, [44].

Figure IV-28. AS-3018/WSC-1(V)

g. INMARSAT

The commercial International Maritime Satellite (INMARSAT) constellation provides point-to-point phone calls ship-ship or ship-shore, on demand, for voice or data connectivity up to 19.2 kbps. Current service is expensive at more than $5.00 per minute.

Future digital enhancements (INMARSAT B) are projected to improve the quality of service and to lower the cost of a phone call to nearly a dollar a minute by enabling the sharing of a 64 kpbs access among several callers. INMARSAT equipment is installed aboard the *Arsenal Ship*.

h. SHF Satellite Communications

SHF Satellite Communications are provided by the Defense Communications Satellite System (DSCS) consisting of eight geostationary satellites. The SHF X-band frequency region offers certain benefits for SATCOM operations as outlined in Table IV-13, [44].

SHF SATCOM Advantages
- Substantial bandwidths are available for allocation, which enables the operation of several 50-, 60- and 85 MHz repeaters on the DSCS satellite.
- High data rates can be supported.
- Propagation is very stable and propagation-related outages (e.g. fading or scintillation) are minimal.
- High-gain antennas with narrow beamwidth can be used, which reduces the probability of interception and provides advantages against jamming.
- Band spreading can be used to provide jamming-resistant communications.
- Interoperability among the services is possible.

Table IV-13. SHF SATCOM Advantages

DSCS terminals are capable of 2.0 Mbps throughput and typically get allocated 64 kbps-1.0 Mbps depending on the C4I requirements and the priority of the assigned mission. DSCS provides video-teleconferencing, telemedicine, and data lines for Joint

Deployable Intelligence Support System (JDISS), Joint Maritime Command Information Strategy (JMCIS) subcomponent of the Global Command and Control System (GCCS) and imagery transmission needs. The last three UHF-Follow-On (UFO) satellites will carry a Ku-band SHF high-data-rate broadcast service providing a 23 Mbps service in three spot beams called the Global Broadcast Service (GBS).

SHF shipboard equipment includes the AN/WSC-6 SATCOM terminal, the OE-279/WSC-6 antenna group and the AS-3399/WSC antenna. The WSC-6 antenna is a 1.2m (4-ft) reflector antenna with a Cassegrainian feed. The antenna (including the radome) has a diameter of 1.83 m (72 in) and is 2.20 m (87 in) high weighing 263 kg (580 lbs.), Figure IV-29, [47]. Commercial wideband SATCOM capabilities have been demonstrated by the Navy as Challenge Athena (An NPS Thesis) providing primary imagery, multiple phone lines, teleconferencing, telemedicine, and multiple data lines for C4I systems. Commercial wideband terminals have been programmed for all command ships (LCCs/AGFs, CV/Ns, LHA/Ds, LPDs), and would benefit the *Arsenal Ship*.

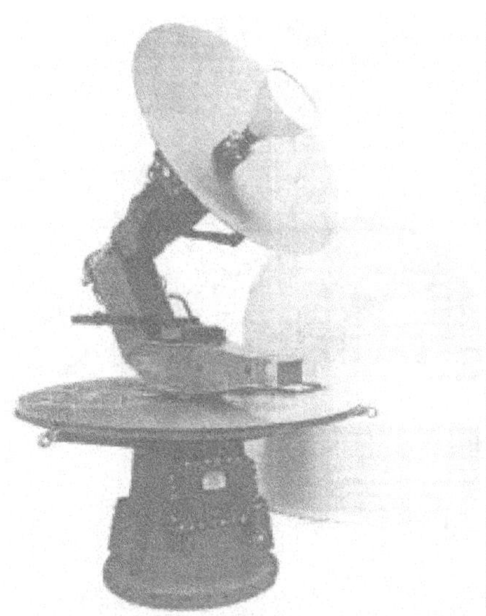

Figure IV-29. AN/WSC-6 SHF SATCOM Antenna

i. Cooperative Engagement Capability (CEC)

The Cooperative Engagement Capability is the *Arsenal Ship*s primary method for transferring data for remote firing. The antenna that accomplishes the data distribution is a phased array antenna system that provides omnidirectional coverage. It operates in the high S-band region and consists of a 1000 element antenna providing elevation and azimuth coverage, Figure IV-30, [43]. The *Arsenal Ship* will have two mast mounted antennas for redundancy, one forward and one aft. The antennas are 44 inches in diameter and 14 inches high, Figure IV-31, [50].

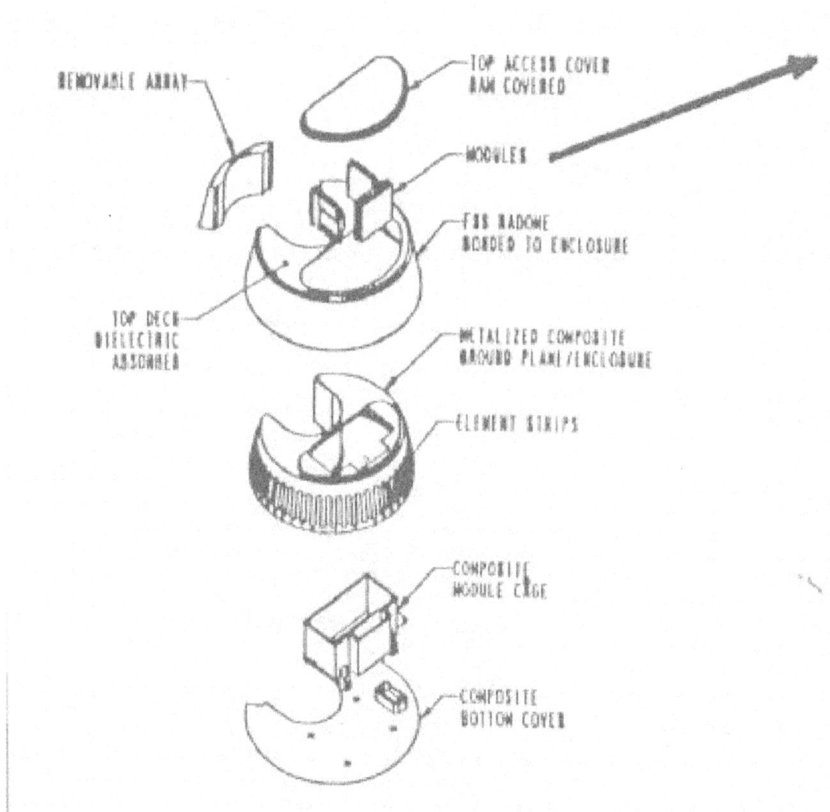

Figure IV-30. CEC Antenna Schematic

Figure IV-31. CEC Antenna Location

j. EHF Satellite Communications [44]

The extremely high frequency band (EHF) provides large bandwidths and allows narrow spot beams that are utilized for substantial anti-jamming protection for military communications. In addition, man-made electromagnetic disturbances (e.g. scintillation's resulting from high altitude nuclear bursts) are not as harmful as at lower frequencies and can be counteracted more easily. The Air Force managed Military Strategic Satellite Relay (MILSTAR) along with the Navy FLTSAT EHF Package (FEP) and UHF Follow-On Satellites-EHF (UFO-E) is part of the modernization of US C^3 (Command, Control and Communications) and will provide worldwide coverage by a constellation of geostationary and inclined orbit satellites with satellite crosslinking. EHF brings the advantage of protected services. The current implementation of EHF service provides low-data rate channels (2.4 kbps) whose link cannot be interrupted by jamming or nuclear effects. The high frequency provides extremely narrow signal beamwidths enabling users to communicate with a very low probability of intercept or detection (LPI/LPD). The same attributes that make the signal very robust also make the signal very clean, translating into fewer (if any) errors for each transmission. This clean signal is critical for passing TLAM mission data updates (MDUs) or system software updates.

The shipboard EHF satellite terminal is the AN/USC-38 Terminal Equipment which includes a high power amplifier (HPA) and communications equipment group (CEG). The HPA uses 2.6 kW three phase power, 60 Hz, 440 V AC with dimensions 54 in. high, 19 in. wide and 24 in. deep weighing 615 lbs. The CEG cabinet is 72 in. high, 24 in. wide and 30 in. deep weighing 1050 lbs requiring 2.3 kW, three phase, 60 Hz, 440 V, AC power. The ship MILSTAR antenna is a 35 in. diameter reflector with a dual band feed and a rotatable subreflector to provide a conical scan for downlink frequencies, Figure XX. The antenna is enclosed in radome, 56 in. diameter by 60 in. high. The waveguide run from the HPA to the antenna should not exceed 75 feet. One antenna mounting is sufficient for full hemispherical coverage with no superstructure blockage, Figure IV-32, [44].

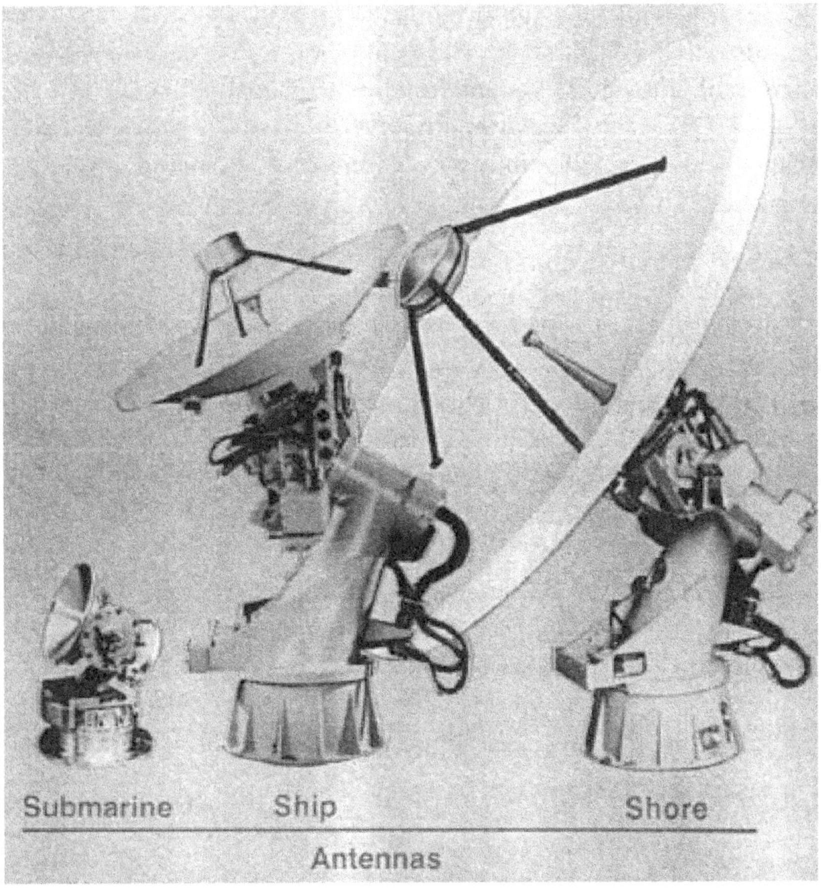

Figure IV-32. EHF MILSTAR Antennas

2. Naval Telecommunications Automation Programs

Major initiatives have been undertaken since the 1960s to automate shipboard communications. The Naval Tactical Data System (NTDS), Naval Modular Automated Communication System (NAVMACS), Naval Telecommunications System (NTS) Architecture, Integrated Communications System/Shipboard Communications Area Net (ICS/SCAN), Survivable Adaptable Fiber Embedded Network (SAFENET), Unified Networking Technology (UNT), Communication Support System (CCS) and Copernicus Architecture have all been developed as major steps in the modernization of shipboard suites [43].

The *Arsenal Ship* takes advances in technology to provide a survivable, integrated, redundant, effective, and automated system. The overall goals of reduced manning and cost through application of existing technologies still apply!

3. Shipboard Technical Control Systems

The typical Shipboard Technical Control Facility provides automated technical coordination, testing, activation and deactivation, and report generation. It contains COMSEC equipment, RF and base-band equipment, and a Technical Control room. A typical Shipboard Interior Communications System consists of a Radio Room, Communications Center, and a User Area [43].

The *Arsenal Ship* should take advantage of an automated communications control and switching system for controlling and monitoring the operation of the EXCOMM system and routing communications and weapon control signals. The AN/SQQ-33 SSECMCS and OJ-631 Technical Control Console provides the capabilities necessary to provide full supervisory control over the exterior communications system by two operators. The first performs normal technical control functions, and the second performs back-up functions. The SSECMCS is controlled by the control console and the data terminal set [44].

The AN/SSQ-88 QMCS provides the equipment necessary for a radio operator to monitor the quality of communications circuits. It is a system for on-line testing and is comprised of 11 test devices housed in a single CY-4516 electronic cabinet. The AN/URQ-23 is commonly used as shipboard frequency standard. Tactical Frequency Management Systems such as the AN/TRQ-35 AND AN/TRQ-42 TFMS consist of a "Chirp-sounder" transmitter, receiver, and a spectrum monitor [44].

Antenna multi-couplers can facilitate the use of a single antenna as a signal source for several receivers. It is an accepted practice to install multi-couplers in a cascaded "normal-through" arrangement. The CU-1772A/SRA-56, CU-1774/SRA-57, and CU-1776A/SRA-58 shipboard HF transmitter multi-couplers can simultaneously connect up to four 1kw transmitters into a single broadband antenna [44].

RF patch panels, such as the SB-4249 distribute antenna outputs, multi-coupler inputs and outputs, and receiver RF inputs.

The Black Communications Switch SA-2112A (V)6/STQ provides non-secure automated analog radio-telephone switching system and the Red Communications Switch SA-2112(V)8/STQ provides digital control and switching of plain and cipher audio and data terminals and external radio circuits. The MCS-2000 Interior Communications System provides control and switching of the interior audio and data terminals and the exterior communications system similar to the above mentioned switches [44].

Data switching systems are important to the weapon system in the *Arsenal Ship*. There are several shipboard systems available today (i.e., the SA-4176 Data transfer Switch, the NAVMACS-V AN/SYQ system, and the ON-143 Interconnecting Group). These functions for tactical data links and exterior communications radios, automated message processing and distribution [44].

The *Arsenal Ship* does not have watches dedicated to providing 1MC service. Personnel can carry personal information pads (PIP) that can be remotely-beeped throughout the ship.

4. Interior Communications

The mind boggling variety of C^4 equipment and procedures found on major Combatants and CJTF staffs requires numerous personnel with advanced training to operate and maintain. The *Arsenal Ship* does not have the luxury of collections of complex systems, but needs smart initiatives to automate the links, circuits, processors, and interfaces required to support the various combat weapon systems.

"Communications automation devices can transfer information and process communications to support information transfer. The shipboard resources available for implementing communications services are radios, modems, encryption equipment, and a Communications Support Service (CSS) controller. The CSS architecture is designed to implement services and resources by using a building block approach and interoperable interfacing component hardware and software systems. CSS can be implemented using current Navy shipboard systems in transferring information among several platforms." The users are then connected using a LAN [44] (Figure IV-33).

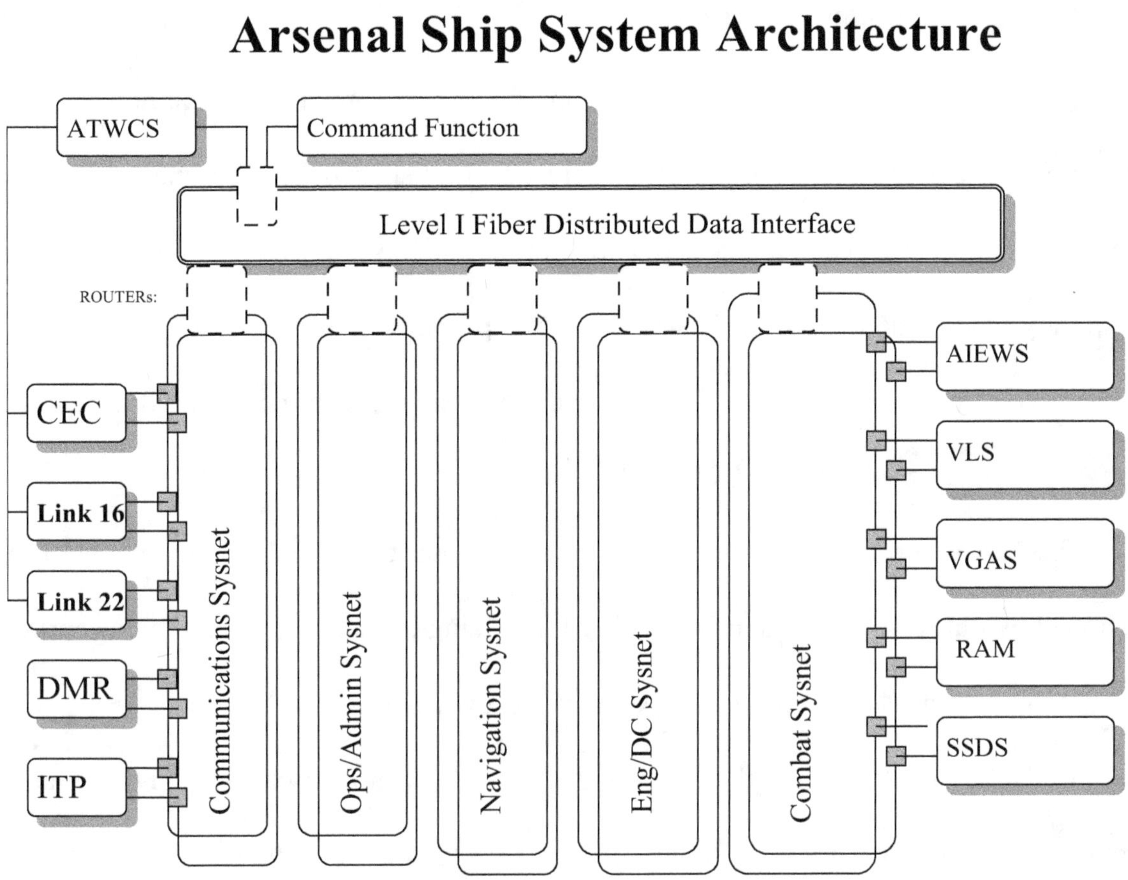

Figure IV-33. CCS Architecture

5. Copernicus Architecture

Integral to the development of the *Arsenal Ship* command and control architecture is the Navy initiative called the Copernicus Architecture. The U.S. Navy has developed the Copernicus Architecture for communications systems ashore and afloat. It consists of four entities: Global Information Exchange Systems (GLOBIX), CINC Command Complex (CCC), Tactical Data Information Exchange System (TADIX), and Tactical Command Center (TCC) [44].

6. Communication Links

A "Link" is a complete facility and system encompassing equipment and procedures through which communications are accomplished on a permanent basis as opposed to a circuit, which is a communications path established for message transfer on a temporary path. The *Arsenal Ship* supports three permanent links (i.e., CEC, Link-16, and Improved Link-11 (Link-22)). Communications is still maintained through a variety of systems connecting the *Arsenal Ship* to the Information Infrastructure.

CEC is the primary link for sharing composite targeting quality track data. It is envisioned by our team to be the primary link for passing digital information between the Control Ship and the *Arsenal Ship* for weapons employment.

Link 22 is the name of the improved Link 11. It is also called TADIL A and operates are 2.25 Kbps for maintaining the NTDS picture. It is an HF frequency operating between 2-30 MHz. HF has anti-jam capabilities and a low probability of intercept and can incorporate automatic Link establishment techniques. Link 11 operates in a line-of-sight (LOS) mode using UHF. Typical equipment includes HF radios such as the AN/SNC23 or AN/URC-109 Integrated Communication System, UHF radios like the AN/URC-83 Tactical Data Link System, Data Terminal Set such as the MX-512P, and external antenna (i.e., the HF monopole whip antenna AS-2537) [46].

Link 16, previously called Joint Tactical Information Distribution System (JTIDS) and known as TADIL J, operates at 28.8, 57.6, and 115.2 Kbps on UHF between 95 and 115 MHz. Link 16 is a high capacity Time Division Multiple Access (TDMA) system providing integrated communications, navigation, and Identification of Friend or Foe (IFF) capabilities. The waveform consists of a series of 6.4 us pulses in the 969 MHz to

1215 MHz band with anti-jamming capability using frequency hopping techniques The link provides electronic protection capability, extended rang of communications, and over-the-horizon communications between surface ships with an airborne relay platform [46].

7. Beyond-line-of-sight Communications

Beyond-line-of-sight communications (BLOS) can be provided to the *Arsenal Ship* using Wideband SHF SATCOM such as the jam-resistant secure communications (JRSC), FLTSATCOM Broadcast, UHF FLTSATCOM (OTCIX, CUDIXS, and TADIX), and long-haul communications equipment (Commander's Ship-to-shore, HICOM, and Fleet primary). Examples of airborne and shipboard VHF/UHF radios include the 30 watt AN/ARC-159A ship-to-aircraft and the 100 watt AN/URC-93 ship-to-ship radio [46].

8. SINGARS

The U.S. Army, Air Force, Navy, and Marine Corps utilize a tactical VHF radio called the Single Channel Ground and Airborne Radio System (SINGARS), which is a frequency-hopping (FH), frequency-modulated (FM) spread-spectrum system covering the 30 MHz to 88 MHz frequency band in 25 kHz wide discrete channels. The SINGARS is designed to provide secure voice and data communications in jamming environments and is a candidate for the *Arsenal Ship* C^4 for Naval Surface Fire Support missions [46].

The shipboard SINGARS is a modified version of the ground-based SINGARS. In the ECCM Mode, the AN/PRC-119 (man-pack), the AN/VRC-88 (vehicular long-range dismountable), the AN/VRC-90 (vehicular long-range), and the AN/VRC-92 (vehicular dual long-range for retransmit) radios can be used ashore. In the airborne relay mode, the family of radios that are compatible with the airborne relay mode SINGARS radios are the AN/ARC-201 and the AN/ARC-210 transceiver. Shipboard SINGARS (type II) radios have up to four antennas (AS/3226A) that can support four channels, three for transmitting and one for receiving. When frequency hopping using spread-spectrum operation, the antennas are connected to four separate RF power sensors [46].

9. *Arsenal Ship* Combat Systems Communications Protocol

The *Arsenal Ship* combat system is a complex system consisting of data-processing and communications subsystems, weapon systems, and ESM subsystems. These systems are interconnected connected through a local area network (LAN). Processing systems use standardized data transfer methods such as the Navy Digital System Interconnect Standard (MIL-STD-1397A), which defines the interface characteristics (i.e., physical, functional, and electrical) and the standard interfaces between digital equipment (i.e., computer-to-peripheral, computer-to-computer, and peripheral-to-peripheral). There are three types of NTDS formats to accommodate different data transfer rates (i.e., NTDS fast, slow, and serial) [46].

The *Arsenal Ship* LAN is structured for use by more than one protocol. This balances the conflicting needs of interoperability and performance. EXCOMM systems such as Link-16 and Link 22 use standard message formats for the exchange of digital information.

Modern warships interconnect subsystems with two types of LANs--high capacity token rings and Ethernet (IEEE standard 802.3) which uses a random access method, and high capacity token rings. The military versions are the Survivable Adaptable Fiber-Optic Embedded Network (SAFENET) and the Fiber Distributed Data Interface (FDDI), a high-speed fiber-optic cable token ring using IEEE compatible protocol. Currently, shipboard LANs predominantly utilize rings wired by coaxial or fiber-optic cables. The standard governing the ring LAN is the Token Ring protocol. Fiber-optic cables can accommodate data rates of 100 Mbps, supports up to 240 taps, and have unrelayed transmission ranges of 2 km [46].

"A token ring connects elements of a network in a ring. A token (i.e., data frame with a special bit pattern) moves along the ring in a round robin fashion. When the token arrives at a node, the node can seize the token and send data which passes until the intended receiver intercepts the token, but this method is not as efficient as Early Token Release, where the token is released after transmitting. This would allow the nodes on the LAN to send data and access the ring simultaneously. The Fiber-Distributed Data Interface is a high speed token ring that uses a complex capacity allocation method to accommodate both long data streams and data bursts using synchronous and asynchronous modes [46]."

The communications protocol standards are contained in the International Standard Organization's (ISO) Open Systems Interconnection (OSI) reference model. The government's standard for Open System Interconnection is the Government Open System Interconnection Profile (GOSIP), which allows commercial-off-the-shelf (COTS) products to be used in shipboard communications design [46].

H. TOPSIDE DESIGN

The discussion of the topside design begins with an overview of the factors which control the topside equipment placement, including electromagnetic compatibility, antenna integration, and electromagnetic interference control. The details of the topside design are presented taking into account the controlling factors

1. Electromagnetic Compatibility

a. Topside

The *Arsenal Ship* cannot afford to be a perfectly stealthy platform. It radiates intermittently through data and voice circuits on various antennas and continually pass information onto the links that it is a part of. It has HF, UHF/VHF, Satcom, and other communications antennas that are transmitters and receivers of RF energy. These antennas interact with each other and with the topside superstructure of the ship at particular wavelengths and signal strengths. Careful planning is required to maintain acceptable levels of electromagnetic interference between antennas onboard the *Arsenal Ship*. Physical placement and proximity between antennas is very important when designing the topside area. Design changes and corrections can be costly in both the financial and temporal domains [46].

Advanced Mast and integrated and embedded antenna concepts greatly reduce EMI problems for the *Arsenal Ship*. But the systems may not be developed sufficiently for use on the Demonstrator. The *Arsenal Ship* uses existing antenna and component systems in order to develop a design with a viable "fall back" option.

"Electromagnetic interference (EMI) is an electromagnetic disturbance that degrades the performance of electronic and electrical equipment due to atmospheric conditions, enemy electronic warfare, or unintended intermodulation effects. Electromagnetic compatibility (EC) refers to the ability of electronic and electrical systems to operate under their intended circumstances without undue degradation due to unwanted EMI. Electromagnetic radiation hazards exist to personnel, fuels, and ordnance (i.e., HERP, HERF, and HERO) [46]."

The main interference sources to the HF band comes from man-made noise followed by atmospheric and cosmic conditions, while the major contributors to VHF, UHF, SHF, and EFE band noise comes from galactic, solar, and atmospheric sources. Man-made noise is a mixture of impulse noise from machinery-- motors and generators, engines, and impulse noise from switch arcing, circuit breakers, relays, and florescent lighting. Machinery generates noise from arcing between the brush contacts and the slip rings. Gasoline engines that drive generators, pumps, and other equipment produces ignition noise in the HF and VHF range. Florescent lights generate noise similar to a discharge tube, which couples into and is radiated from the power distribution system onboard the ship. Electrical components, microwave ovens, data processing equipment, and appliances onboard the ship contribute to the EMI noise environment [46].

Transmitters onboard the *Arsenal Ship* contribute to topside EMI for communication equipment throughout the RF spectrum. Transmitters should be designed so that they radiate in as narrow a frequency band as possible. Interference occur due to harmonic frequencies, sideband splatter, intermodulation between signals, cross-modulation between transmitters, parasitic oscillations from antennas, leakage at waveguide joints, and hull-generated intermodulation due to ground loops. Corrosion produces non-linear effects at grounding points for electronic equipment. Finally, one of the largest contributors to topside EMI is reflected energy which creates multipath effects, constructive and destructive interference, and standing wave patterns [46].

The *Arsenal Ship* topside is arranged such that the weapon system elements and superstructures do not contribute greatly to reflections. Preferring an integrated advanced antenna system to the traditional stacked yardarm, emphasis on geometric angles for radar signature reduction (stealthy) and a lack of yardarms, booms, davits and masts assist in minimizing EMI problems.

The Navy initiated a Shipboard Electromagnetic Compatibility Improvement Program (SEMCIP) to help design ships with electronic systems that are electromagnetically compatible, identify existing fleet EMI problems, and train personnel involved in ship design how to understand EMC problems. Every EMI problem has a source, a coupling, and a victim. Front-door EMI results when the victim receives interference directly from antenna signal. Back-door EMI results when the victim gets it interference internally, such as a power supply or feed line. Intersystem coupling can occur between two different systems and intersystem interference can occur internally within a single system [46].

The Navy convenes an EMC Advisory Board (EMCAB) composed of EMI and EMC experts from Naval Laboratories, universities, and industry to support the *Arsenal Ship* project manager in system design analysis, EMC predictions, test plans, system installation and construction planning. The EMCAB generates an EMI control plan for routing of cables, wavequides, shielding, grounding, bonding, quality control inspections, and testing to drive EMI out of the *Arsenal Ship* [46]!

b. Electromagnetic Compatibility below Topside

The communications systems inside the skin of the *Arsenal Ship* include transmitters, receivers, digital processing equipment, RF distribution networks, and encryption equipment. RF signals are either generated below deck for transmission, sent below deck following reception, or else they represent data used in system operation. The power levels of signal sent out are several magnitudes larger than signals received, which must compete with noise from cosmic sources and man-made machinery. This lopsided balance requires shielding and filtering [40].

Shielding to reduce the electric field strength helps isolate signals and protects victims from unwanted EMI. Shields have to be properly terminated and grounded to prevent leakage. The most effective shield is a complete metal enclosure (i.e., Faraday cage). These need to be open to ventilation for cooling, usually with perforations that are effectively frequency dependent wavequides. Honeycomb shielding is very effective at high frequency and reasonably effective at low frequency [46].

Grounding presents a unique challenge for the combat system and communications engineers. Galvanic corrosions leads to nonlinear interference at grounding bolts and wires (which also need to be shielded). The *Arsenal Ship* hull is the ground reference for all electronic equipment. All grounds should be welded to avoid corrosion and particular attention should be paid to topside connections [46].

Filters protect equipment by limiting the frequency band of signals reaching protected equipment. A typical application of low-pass filters is the use of the control of harmonic radiation at cable connectors. Both 400 Hz and 60 Hz ship power supply systems can receive interference from one piece of equipment and carry it to another. Filtering at the source reduces EMI [44].

TEMPEST methods isolate encrypted signals from plain ones. Cabling between red and black (i.e., plain and cipher) zones must be entirely separate and equipment racks are different. There is significant efforts to shield and isolate the network that ties the communications systems and combat system nodes onboard the *Arsenal Ship*. EMC engineering is going to be a series of compromises between cost and convenience, performance and practicality. Reasonable levels of EMI can be found and accepted during the eventual construction phase [44].

2. Antenna Integration

Specific problems that the *Arsenal Ship* encounters are blockage (i.e., shadowing) of electromagnetic waves by the superstructure, coupling (i.e., resonant effects) between antenna and superstructure, unwanted RF emission, and hazards of high level radiation. The *Arsenal Ship* has omni-directional receiver antennas such as the HF whip, and directional transmitter and eceiver antennas for receiving directional EM waves from satellites. The HF whip antenna length is of the same order of the superstructure length. It is a broadband antenna with couplers for several frequency regions to reduce the number of antennas needed. This lessens the EMI problem. Shipboard receiving HF antennas are usually designed for low efficiency so that the atmospheric noise and the receiver noise generally match [46].

The Navy maintains a test range at the Naval Command and Control and Ocean System Center (NCCOSC) RTD & E Division where measurements are made on 1/48-scale brass model to determine optimal HF antenna placement. The best location for the

VHF/UHF antennas is high on the yardarms for maximum line-of-sight range covering 360 degrees around the *Arsenal Ship*. The trade-off is between antennas and yardarm size (i.e., yardarms increase signature). SATCOM antennas should have hemispherical-coverage. UHF SATCOM transmitter antenna are big, heavy, and often mounted in multiple orders of decreasing coverage. The SHF SATCOM terminal (AN/WSC-6) can be placed high as a single antenna or low as a dual set. Trade-off studies ensure optimal placement [46].

3. EMI Control in HF and VHF Systems

The six categories of EMI sources for HF communications are internally-generated distortions between transmitters, externally-generated nonlinear products, intermodulation due to non-linearities in the receiver, sideband noise generated by transmitters that fall within the receiver's processing band, reciprocal mixing of the receiver local oscillator noise caused by an interfering signal in the receiver bandpass, and front-end gain compression caused by the passband of the receiver [46].

Modern EMI control techniques include the use of nonmetallic topside fixtures whenever possible, antenna separation, narrowband receiver pre-selector filters, self-steering filters, and time division multiplexed transmission [40].

The SINCGARS VHF radio may experience antenna co-site interference due to simultaneous operation of up to four antennas. Most interference sources for VHF radios are the same ones as for HF radios. Some EMI control features include using frequency hopping filters and linear power amplifiers [46]."

The Table IV-34 is an example of anticipated EMI between sources (left) and victims (right) [46]:

	IFF	AIR RAD	SUR RAD	UHF SAT	UHF LOS	HF	VHF LOS	BRG	SAT NAV	OM	SHP AIR
IFF	⊕			⊕							
AIR RAD	⊕		⊕	⊕							
SUR RAD		⊕									
UHF SAT				⊕	⊕				⊕		
HF						⊕					
VHF							⊕				
BRG											⊕
SHP AIR							⊕				

Figure IV-34. EMI Between Sources and Victims

4. Topside Design Method

Once the *Arsenal Ship* TAO modified repeat hull was selected, topside design began. The initial step was to remove all the existing topside structures down to the main deck. This removal included the huge aft deckhouse superstructure and all masts, booms and rigging. What remained was a large smooth main deck with a slightly elevated forward section. The priority for topside design was simplicity and function. The goal was to maintain a low radar cross section (RCS) for topside superstructure, limit ingress/egress points, and reduce topside equipment while providing sufficient superstructure for ship sensors and integration of future sensors discussed in the communications section. The *Arsenal Ship* topside is divided into five areas; the forecastle gun deck, forward superstructure, missile deck, aft superstructure, and helicopter deck.

With the realization that the selected hull design had an inherently large RCS, the topside design was selected to minimize additional RCS. The removal of dihedral scattering surfaces, two adjacent flat surfaces which meet at nearly perpendicular angles, would eliminate multi-bounce energy reflection paths and reduce large radar cross section (RCS) spikes on the main deck. Shaping of the topside superstructure was used to direct reflected energy to a non-threatening area in space by tilting large surfaces ten degrees. The two superstructures were primarily selected to provide elevated line-of-sight (LOS) and mounting surfaces for ship sensors. The hexagon shape of the superstructures was selected for the addition of a future integrated HF antenna utilizing the six intersection points. Ship ingress and egress points were kept to a minimum for ship protection and monitoring. Located forward and aft on the superstructures, each ship access passes through video/sound monitored double door decontamination points which provide a means of determining identification and automatic containment if desired. To reduce exposure to the elements and to maintain the reduced RCS goal, all non-essential topside equipment was removed or moved inside the ship. The topside of the *Arsenal Ship* is "clean". With little topside equipment, maintenance associated with equipment exposed to the harsh sea environment is reduced. All essential topside equipment is flush mounted with appropriate covers or is accessible from inside.

The forecastle gun deck is simplified by the exchange of a bow mounted anchor for a keel anchor. The forward windlass was moved below deck for line handling through removable hull panels. The gun mount space is provided for an advanced 155 mm flush mounted vertical gun. The forward superstructure provides LOS elevation for the bridge, EW equipment mount, flush mounted communications antennas and CEC antenna. Until an integrated hull mounted CEC, the use of a low RCS mast provides the CEC antenna mount. With ship self defense a concern, the top of the forward superstructure was selected as the location for the ship's RAM launcher. This location gave the best area coverage for a bow-on threat sector. With such a large beam RCS, the intent for ship placement in a high threat environment is bow-on. This posture is hypothesized to provide the *Arsenal Ship's* lowest RCS toward the threat and allow the ship self defense system the optimum coverage for response. The missile deck is flush mounted with eight cell armored Mk 41 VLS launchers. The aft superstructure is similar in design to forward. It maintains space available for flush mounted advanced communications antennas and contains the main engine IR exhaust suppresser, satellite antennas, mast mounted CEC antenna. Port and starboard underway replenishment sites are located behind sealed retractable enclosures. The aft helicopter deck provides a platform for aircraft landing only.

I. UNDERWAY MISSILE REPLENISHMENT

Underway replenishment (UNREP) of the *Arsenal Ship*'s missile payload is not a requirement in the Ship Capabilities Document [5]. Due to this, the *Arsenal Ship* does not have this capability installed. However, the need to replenish a ship's payload during combat operations is a necessity.

Underway replenishment via connected replenishment (CONREP) is typically an extremely manpower intensive evolution. The manning level required under current doctrine is unacceptable. The primary challenge remaining is how to perform the CONREP of missiles within the framework of the *Arsenal Ship* reduced manning concept.

A method has been developed to not only reduce the number of personnel required for general CONREP, but has also been developed to reduce the number of personnel required to handle missile payloads during UNREP [51]. This design incorporates a rail and trolley system to handle the missiles and a retractable kingpost and storage for the trolley.

The manning requirement for safely conducting UNREP of missiles at sea onboard the *Arsenal Ship* is four: safety observer, trolley driver and two linetenders. This manning level is compatible with the reduced manning concept developed for the *Arsenal Ship*. The other piece of equipment that enables this manning reduction is the winch. The winch is developed to assist the *Arsenal Ship* crew in bringing the UNREP rig over from the replenishment ship, and it reduces the number of linehandlers required for CONREP to three.

The uses for this UNREP system onboard the *Arsenal Ship* are numerous. The most significant of them are the ability to replenish expended ordnance during combat operations, tailor the missile loadout in theater and conduct consol operations between relieving *Arsenal Ship*s to maintain valuable missile assets in theater.

With only 512 missiles installed in the *Arsenal Ship*, a limited number of missile mixes may be developed for a specific mission profile. Without UNREP capability for ammunition, once one type of missile is completely expended, the *Arsenal Ship* is required to leave the area of combat operations to return for missile reload at the forward operating base (FOB). This situation causes the *Arsenal Ship* to be out of theater for a period of up to four weeks plus the time required to perform the missile reload. Tailoring the missile loadout in theater gives the Battle Group (BG) Commander the flexibility to ensure the *Arsenal Ship* has exactly the payload required for its mission.

The *Arsenal Ship* can also be used as a source of replenishment missiles for Aegis cruisers and destroyers. With an *Arsenal Ship* on one side of the replenishment ship and an Aegis on the other, the missiles may be transferred ensuring that the combatant has the missiles required for its mission.

Conducting a consol between relieving *Arsenal Ship*s enables the next *Arsenal Ship* the proper loadout required by the BG Commander. This allows the ships to share valuable weapons in our limited missile arsenal. By sharing missiles between two *Arsenal Ship*s, the Navy can limit the number of missiles stored at the FOB.

The missile UNREP system is the appropriate equipment to effect missile reload at sea with the *Arsenal Ship*. The estimated cost of installing this system as currently designed is $6 million. This includes four UNREP stations, two port and two starboard; eight trolleys, one for each 64-cell vertical launching system (VLS) launcher; and the associated storage required for the trolley and other UNREP gear. This is an insignificant cost compared to the potential benefits of having the capability of conducting missile reload and transfer at sea.

V. HULL, MECHANICAL, AND ELECTRICAL

A. HULL ALTERNATIVES

The design philosophy and Concept of Operations (CONOPS) are the driving factors for hull selection. The Arsenal Ship's hull selection is based on overall cost, survivability, manning and the warfighter's needs. This report considers the following hull alternatives:

- Warship
- Small Waterplane Twin Hull (SWATH)
- Catamaran, Trimaran, and Slice
- Specialty Hulls
    a. Hydrofoils
    b. Surface Effect Vessels
    c. Planing Hulls
    d. Semi-submersible
    e. Tanker/Cargo Carrier
        1. Single Hull
        2. Double Hull

1. Warship

The warship hull-form has an advantage over other hull alternatives in speed and appearance (i.e., it looks like a combatant). It also has a psychological advantage that the other hull-forms do not. A fast hull, however, has only a marginal benefit, because the CONOPS requires a minimum sustained speed of only 22 knots.

This hull-form has many disadvantages that outweigh its benefits. The disadvantages are cost, inefficient volume for a deep 500 vertically launched system

(VLS) load, and poor seakeeping. The warship hull-form was not selected for the *Arsenal Ship* because of these disadvantages.

2. Swath, Catamaran, Trimaran, and Slice

These hull-forms have excellent seakeeping in high sea states. The hull-forms, however, are much more expensive than other alternatives. Additionally, these platforms have never been tested for this type of application. Hence, the overall acquisition costs are also significant. The twin-hulled vessels need a significant amount of freeboard to keep the centrally-located VLS cells out of the water. This high freeboard creates a large and undesirable radar cross-section (RCS). The *Arsenal Ship* design does not use these hull-forms because of these disadvantages.

3. Specialty Hulls

The planing, hydrofoil, and surface effect ship (SES) hulls are considered specialty-hulls. Each hull-form offers unique benefits. These alternatives are expensive and not practical for the *Arsenal Ship* design.

4. Semi-Submersible

The semi-submersible hull offers reduced RCS and excellent seakeeping in the semi-submersed state. The concept behind the semi-submersible is that the platform can submerse itself during combat operations and make itself nearly invisible to surface-search radars. The submersed state also adds excellent seakeeping for missile launch in high states.

A significant disadvantage for the semi-submersible hull-form in this design is the large design and engineering costs. These costs result from the amount of design analysis and testing that must be performed for this nontraditional application. The *Arsenal Ship* design does not use this hull-form because of this disadvantage.

5. Tanker/Cargo Carrier

The tanker/cargo carrier is relatively inexpensive and has a larger volume for the 500 VLS missiles in comparison to the other hull-form alternatives. An additional advantage is that the tanker/cargo carrier hull-form can easily be designed with a double-hull. Furthermore, the hull-form has proven reliability, and most shipyards have extensive experience in its construction. The disadvantages of the hull-form are large RCS and slow speed through the water. Despite these disadvantages, the large-bodied tanker hull-form is selected as the best alternative for the *Arsenal Ship*. The decision is based on the CONOPS requirements for a hull-form with large volume, and relatively slow speed.

The CONOPS requires the *Arsenal Ship* to be virtually unsinkable. Special attention must therefore be paid to minimizing the progressive damage and flooding, maximizing reserve buoyancy, and optimizing compartmentation. These factors can be achieved with a double-hull design. Double-hull design offers a "sacrificial outside layer" that absorbs the energy of a hostile weapon or other platform better than a single-hull design of the same material. The *Arsenal Ship* uses a double-hull design for these reasons.

Cost minimization is important in the *Arsenal Ship* design. A new ship design has significant engineering costs, especially in the case of small production. A cheaper alternative is to use a modified-repeat design. This principle entails selecting a previously designed and built hull and modifying the internal arrangements to satisfy the current requirements. Savings can be further achieved if the alterations to the hull are minimal and the same propulsion plant is used. The *Arsenal Ship* uses a modified-repeat design.

The T-AO 201 class auxiliary ship is a double-hulled ship that lends itself well to this modified-repeat concept for the *Arsenal Ship*. This design can be modified to accommodate 500 VLS cells and allow room for internal attributes that make the ship virtually unsinkable. In addition, the ship-class is primarily built to commercial standards and the Navy already owns the design rights. These two factors provide a considerable cost savings. The delivery cost of the T-AO 201 is $166 million (1995) [52]. A modified-repeat design, with installed VLS cells and CEC capability, can definitely be built within the $550 million sailaway cost [5]. For these reasons, the *Arsenal Ship* uses modified-repeat design of the T-AO 201 class auxiliary ship with double-hull.

B. PROPULSION PLANT SELECTION

The principle hull characteristics of the T-AO 201 class auxiliary ship, with a modified- repeat design, remain mostly unchanged. The propulsion system, however, is redesigned to satisfy CONOPS requirements for the *Arsenal Ship*.

The *Arsenal Ship* propulsion system is comprised of two Colt Pielstick 4.2V18 diesel engines. The primary factors in the selection of the diesel engines are available engineroom volume, reduced manning requirements, reliability, costs, and infrared (IR) signature. Diesel engines can be operated and maintained by Enginemen (EN), who are also responsible for the auxiliary systems. Gas turbine engines are not selected for the *Arsenal Ship* because they require additional qualified gas turbine specialists (i.e., GSE, GSM), and considerable engineroom volume for intakes and exhaust trunks, and support systems and electronics. The nuclear propulsion plant is not selected for the *Arsenal Ship* because it also requires additional qualified nuclear power personnel, considerable engineroom volume, and is extremely expensive. Finally, the fuel cell is not selected for the *Arsenal Ship* primarily because it is still in the developmental stages of design and the costs of engineering testing is expensive.

The *Arsenal Ship*'s Colt Pielstick diesel engines are larger and more powerful than the existing T-AO diesel engines. These larger engines are necessary to provide mechanical propulsion to the ship, as well as, electrical power to the 500 vertically launched system (VLS) cells. This electrical power is taken from the 2500 kW power take-off (PTO) generators that are located aft of the reduction gears. The increase in horsepower is needed to meet the sustained speed requirement of 22 knots. Figure V-1 shows a horsepower versus speed comparison between the *Arsenal Ship* and the T-AO. The *Arsenal Ship* clearly requires a larger amount of horsepower to obtain the higher sustained speed requirement. The *Arsenal Ship*'s diesel engines provide a combinV-ed 58,000 shaft horsepower (SHP). The engines provide 46,400 SHP at the sustained speed requirement which exceeds the previous installed T-AO propulsion plant's horsepower of 46,203 SHP at 22 knots.

The *Arsenal Ship* has a mechanical transmission, controllable reversible pitch (CRP) propeller and two shafts. The CRP system is installed to eliminate the need for a reversing gear or reversible diesels in the drive train. The mechanical transmission is installed because it provides a cost savings over electric drive for the modified-repeat design. The *Arsenal Ship* has twin screws as an inherent survivability feature.

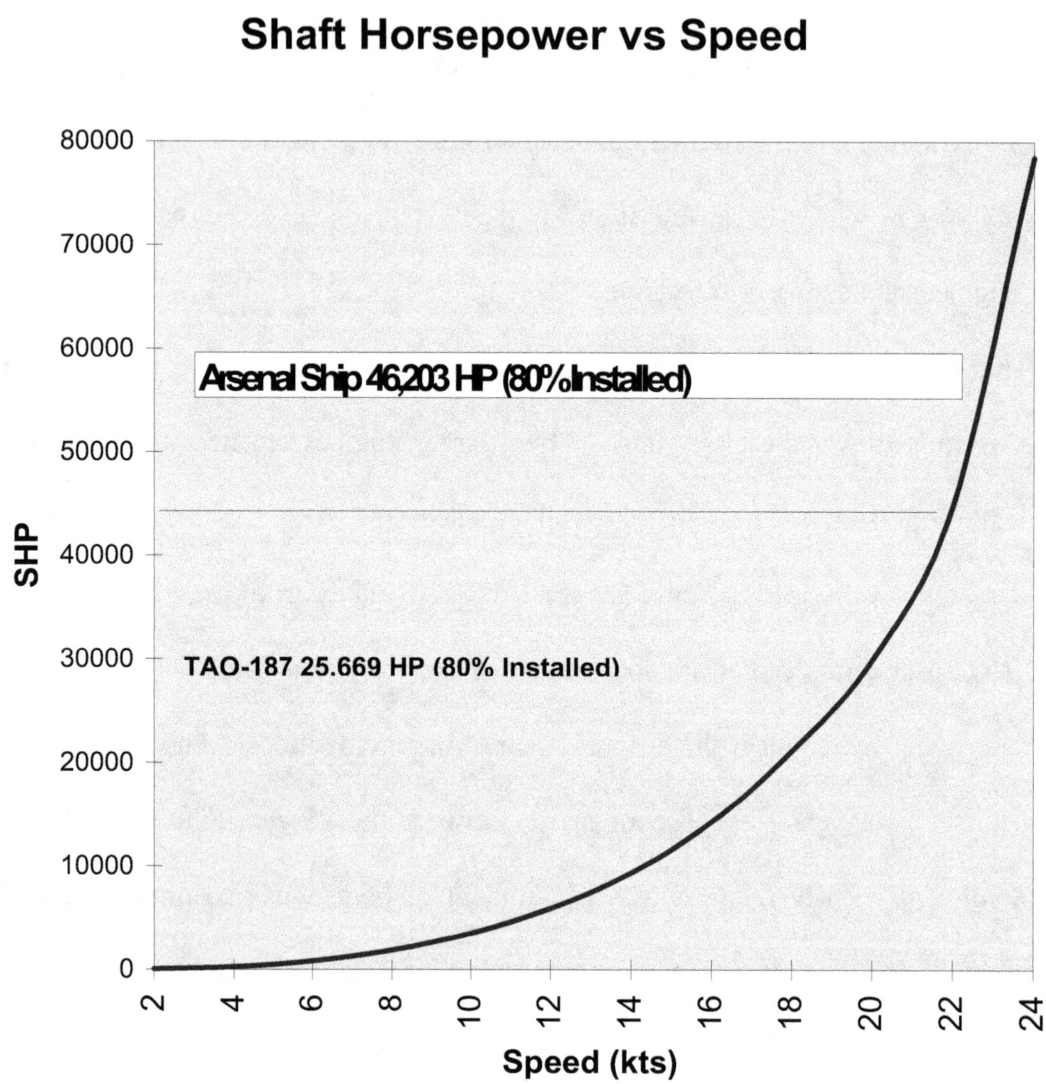

Figure V-1: Shaft Horsepower vs. Speed

C. PROPULSION PLANT LAYOUT

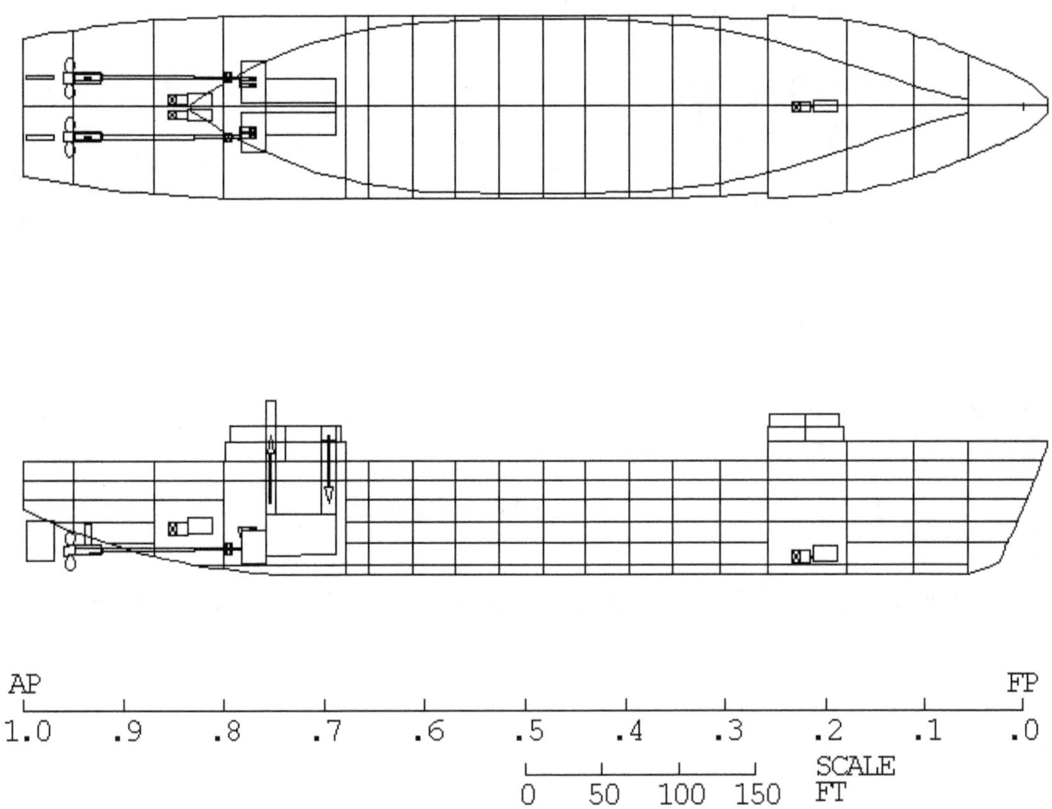

Figure V-2: ***Arsenal Ship*** **Machiner y Layout**

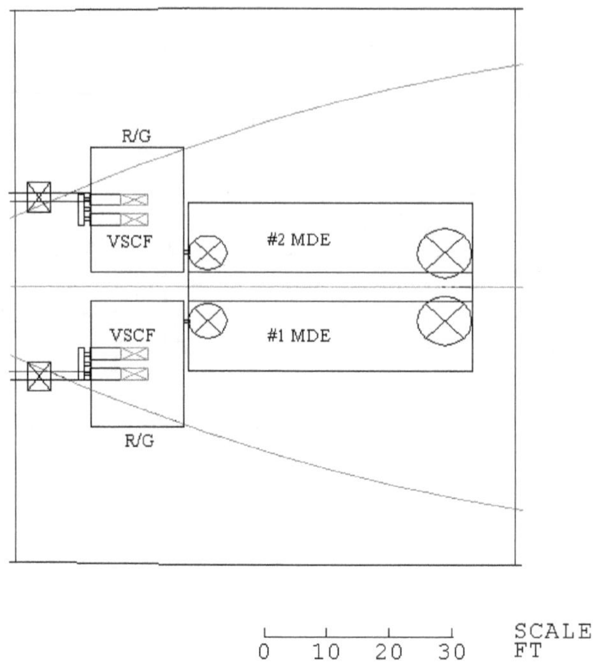

Figure V-3: *Arsenal Ship* Main Engineroom

MDE: Main Diesel Engines R/G - Reduction Gears VSCF - Variable Speed, Constant Frequency Generator

D. ELECTRICAL PLANT SELECTION

The original electrical plant configuration of the T-AO 201 class auxiliary ship is two take-off generators rated at 2500 kW each, two ship's service diesel generators (SSDG) rated at 2500 kW each and an emergency diesel generator (EDG) rated at 600 kW. In exploring alternative configurations for the electrical plant, several options were considered, including steam turbine generators, fuel cells and gas turbine generators.

Steam turbine generators require either a nuclear power plant or steam boilers to provide the prime mover. Neither of these options was seriously considered; nuclear power plants have a significant upfront cost that is beyond the budget of this design and steam boilers are no longer being designed into ships for fuel economy reasons.

Fuel cells take standard DFM fuel and reformulate it to produce hydrogen. It reacts with oxygen in the air to produce electricity directly in the fuel cells. Each cell produces 1 volt and a certain amperage per square foot. The cells are stacked to produce higher voltages. The fuel consumption of the plant is very low, but at the price of very expensive equipment that weighs more than a conventional SSGTG system. This system has possibilities for future implementation onboard ship, but is not developed enough to be ready for installation in the *Arsenal Ship*.

Gas turbine generators offer distinct advantages over diesel generators. These include a much higher power-to-weight ratio, a much higher power density and shorter time from start-up to full power. In designing the modified-repeat of the T-AO 201, these issues were of lesser importance. The mission profile of the *Arsenal Ship* does not indicate a need for rapid deployment from port and there is little need for the advantage of greater power density as the sustained speed requirement is only 22 knots. The two most significant reasons for not selecting gas turbine generators are the large intake and uptake volumes required and the high exhaust temperature, significantly increasing the *Arsenal Ship*'s IR signature.

For the reasons outlined above, the electrical plant configuration of the *Arsenal Ship* is of a similar design to the original T-AO 201. The modifications that were made to the plant are to upgrade the EDG to a full 2500 kW SSDG and to design in a zonal electrical distribution system with three buses to provide sufficient redundancy to provide ship's power for the *Arsenal Ship*'s vital payload. The additional 2500 kW SSDG is installed below the forward living spaces to provide increased survivability of the electrical plant. The four other generators, two take-off and two SSDGs, are installed aft in the machinery spaces.

The *Arsenal Ship*'s electrical distribution system is an AC zonal distribution (Figure V-4). There are 15 electrical zones, eight of which are in the VLS section of the ship with 64 VLS cells each. Zonal electrical distribution provides complete interconnectivity of the generators, switchboards and load equipment. The immediate advantage in construction is the weight reduction of the cabling since significantly less cabling is required to implement three buses than to implement the spider web of electrical cabling that is prevalent on today's Naval ships. Three buses provide easier isolation of spaces and rerouting of power in the event of damage. Each electrical zone incorporates three load centers, providing primary, secondary and tertiary sources of power for all vital loads within the zone. Non-vital loads will be supplied from the nearest load center, reducing the cable length required.

Figure V-4: Electrical Distribution System: 3 Primary Busses, 15 Zones

E. EXTERNAL SHIP VIEWS

Figure V-5: Plan View

Figure V-6: Forward Port Bow View

Figure V-7: Forward Stbd Bow View

Figure V-8: Aft Port Qtr View

Figure V-9: Aft Starboard Qtr View

Figure V-10: Bow View

Figure V-11: Stern View

VI. ARRANGEMENTS

A. INTERNAL ARRANGEMENTS

1. Deck Level Plans

The internal arrangement of spaces for the *Arsenal Ship* is based on the following list of priorities:

- Survivability
- Reduction in radar cross section by hiding topside equipment inside the hull
- Physical Security
- Habitability

a. Forward Internal Arrangements

(1) Superstructure

The goals for the design of the forward superstructure are to:

- Provide the Officer of the Deck (OOD) all of the necessary elements to totally command the ship from the bridge
- Optimize the field of vision for the OOD
- Provide spaces for forward combat system equipment and maintenance
- Minimize radar cross section

The superstructure is designed to reduce the amount of topside equipment by eliminating the construction of bridge wings. Windows completely surround the forward superstructure except in the extreme port and starboard sides that contain the SLQ-32s. A passageway is built into the forward superstructure to give bridge watchstanders a 360 degree view of the horizon, while still keeping them within the skin of the ship.

Additional spaces have been provided in the superstructure for SLQ-32 and AEWIS equipment and maintenance. Space is also provided for a Captain's at-sea cabin and a chartroom.

(2) Habitability

The goal for the design of the habitability spaces is to locate all of the crew berthing and messing facilities in the forward section of the ship. This minimizes the need for the crew to transit the entire length of the ship during peacetime operations. NOTE: Additional berthing for officers, chief petty officers (CPOs), and enlisted personnel is provided in the aft section of the ship for combat operations. This enhances survivability by separating personnel living quarters. In addition, this provides additional overflow berths for temporary duty personnel.

Officer's country is located on the main deck. Twelve staterooms are located on this deck to accommodate the estimated ship's complement of officers. Space is also provided on this deck for the Captain's inport cabin and a Flag cabin.

The CPOs' quarters are located on the second deck. Six berthing spaces are located on this deck to accommodate the estimated ship's complement of CPOs.

The enlisted berthing spaces are located on the fourth deck. Four berthing spaces are located on this deck to accommodate the estimated ship's complement of enlisted personnel. The ship's laundry is also located on this deck.

The galley, messdecks, chief's mess, and wardroom are located on the second deck. This deck is also the damage control deck. These spaces are all located within close proximity of one another to facilitate the use of only one galley area for all food preparation and serving. The location of these facilities on the second deck provides the crew a direct access to the aft section of the ship along the port and starboard longitudinal passageways. This is especially important for the onload of food and offload of trash from these spaces. The food storerooms and reefers are also located on this deck to minimize the distance between the galley and these frequently used spaces.

(3) Gun

Space has been provided in the forward section of the ship to accommodate a future gun system. This space is three decks deep. Due to safety considerations, the spaces directly below the gun have been designated as voids, which can also accommodate a substantial structural reinforcement.

(4) Anchor

The *Arsenal Ship* has a keel anchor to reduce topside equipment and the number of personnel needed for anchoring evolutions. The anchor windlass is located on the main deck, directly above the chain locker. By keeping this equipment within the skin of the ship, the radar cross section is once again reduced from that of a standard naval combatant.

(5) Combat Center

The combat, communications, and computer centers are located in an armored enclave on the third deck. This armored enclave technology is similar to that of the Arleigh Burke Class destroyer. The location was chosen to provide survivability from topside missile damage, while keeping it above the waterline to limit damage due to underwater threats and flooding. A remote Standard Monitoring Control Station is located in this well protected space.

(6) Combat Support Spaces

Space is provided in the forward section of the ship for electronic support spaces. Currently these spaces include: Forward IC/Gyro Room; Electronics Repair Eight; Electronics Cooling Equipment Room; and a Data Processing Systems room. The 400 Hz electrical distribution spaces are located on the main deck starboard side and the third deck port side. The separation of these spaces is to enhance their effectiveness should one become damaged in a casualty.

(7) Engineering Spaces

The fifth deck of the forward section contains auxiliary engineering spaces. These spaces include: CHT room; Reverse Osmosis Room; Auxiliary Generator Room; Forward A/C room; and two Forward Pumprooms. The pumprooms contain space for firepumps and miscellaneous equipment. These spaces are designed to provide primary engineering services to the habitability area. In addition, the separation of these spaces from their aft counterparts provides sufficient distance for maximizing survivability in the event of damage.

(8) Physical Security

The *Arsenal Ship* is designed with limited access to the internal spaces. The forward section of the ship has two accesses located transversely and aft of the main deck. The starboard access enters into the forward decontamination station. The port access is located in close vicinity to the forward armory. This provides the security force a primary exit to combat any hostile boardings. Both accesses also provide for an overpressurization zone.

(9) Damage Control

The forward repair locker and AFFF station are located on the starboard side of the second deck (damage control deck). This location is chosen due to its close location to the port and starboard longitudinal passageways. Should these lockers be needed for damage control in the after engineering spaces, the crew, who are assumed to be concentrated in the forward section of the ship, can rapidly don equipment and then transverse down these passageways to the after section of the ship. In addition, the repair locker is placed adjacent to the messdecks to provide a staging area for personnel and equipment.

(10) Small Boat

The rigid inflatable boat (RHIB) is stowed in a shielded space on the port side of the main deck. The boat is deployed and recovered by a hydraulic extendable arm, 360 degree azimuthing crane. Both the crane and the RHIB are hidden within the skin of the ship for purposes of reducing radar cross section. Access to this space is gained by opening a watertight enclosure mechanism.

Figure VI-1: Forward Superstructure

Figure VI-2: Main Deck Forward

Figure VI-3: Second Deck Forward

Figure VI-4: Third Deck Forward

Figure VI-5: Fourth Deck Forward

Figure VI-6: Fifth Deck Forward

Figure VI-7: Sixth Deck Forward

b. Aft Internal Arrangements

(1) Physical Security

The after section of the ship also has two accesses to internal spaces of the ship. The port side access enters the aft decontamination station. The starboard access enters the quarterdeck area and is in close proximity to the after armory. Both accesses provide for an overpressurization zone.

(2) Refueling Stations

The *Arsenal Ship*'s refueling stations are located in the after superstructure. The stations are shielded by two weathertight doors that separate at the port and starboard corners. A winch and padeye are located at each station to provide for refueling at sea capability. The stations are located within weathertight enclosures to minimize radar cross section and system degradation due to the environment (ie. Less maintenance) and for safety purposes.

(3) VERTREP Stations

The after superstructure also provides a staging area for VERTREP operations. The firefighting team are protected inside a firefighting staging area located next to the port access. Additionally located next to this access is the entrance to the VERTREP pallet staging area. This space provides a large area for the rapid and efficient clearance of the flight deck during VERTREP operations. A strikedown elevator is located in this space, and it connects the staging area to the supply storeroom, engineering machinery spaces, and the two passageways leading to the forward section of the ship. The elevator is designed to enable the minimally manned crew a system that they can use to strikedown/remove small equipment or stores directly from the replenishment area. A VERTREP control station is located on the second level of the aft superstructure to enable the VERTREP/Helicopter Control Officer a safe lookout location to supervise the helicopter evolutions.

(4) Battle Berthing Spaces

Additional berthing spaces are located in the aft section of the ship to segregate the crew during combat operations. These spaces accommodate 24 enlisted personnel, 4 CPOs, and 5 officers.

(5) Damage Control

The aft repair locker and AFFF station are located on the damage control deck. Also on this deck is the sickbay facilities. Sickbay is located centerline to provide a stable ride for medical emergency procedures.

(6) Engineering Spaces

The main machinery spaces, after steering, and aft A/C machinery spaces are all located in the aft section of the ship. The engineering control station is located on the damage control deck. The aft section also contains a large void directly above the engineering spaces to accommodate a soft patch for engine removal.

(7) Combat Support Spaces

Space has been saved in the after superstructure for future electronic rooms that may serve additional communications systems and/or a phased array radar system.

Additionally, the aft IC/Gyro room is located on the third deck, which provides separation and redundancy for increased survivability.

(8) Crew Support Spaces

A large recreation room and physical fitness room is located in the aft section of the ship. The need for these spaces forward was not considered a priority. Sufficient space is available in the after section of the ship to accommodate these facilities.

Figure VI-8: Main Deck Aft

Figure VI-9: Second Deck Aft

Figure VI-10: Third Deck Aft

Figure VI-11: Fourth Deck Aft

B. DETAILED ARRANGEMENTS

1. Bridge Layout Drawings

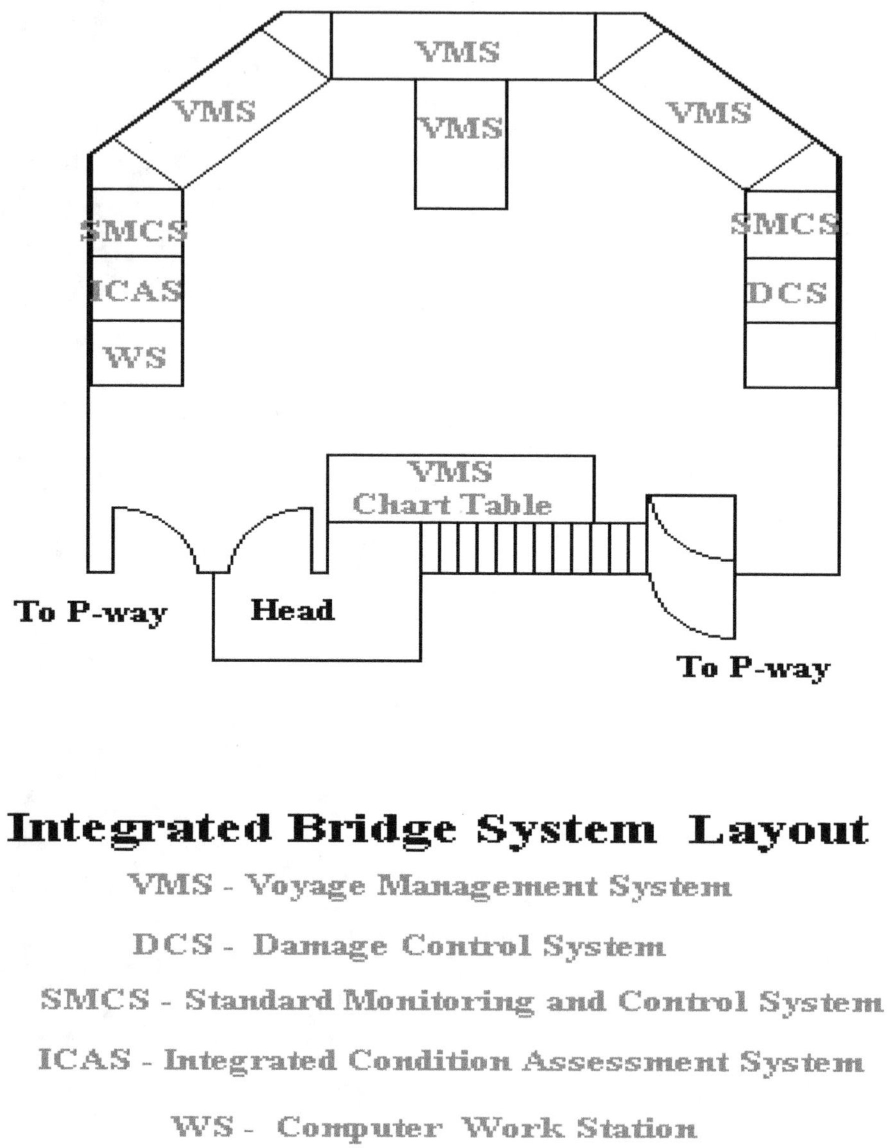

Figure VI--12: Integrated Bridge System Layout

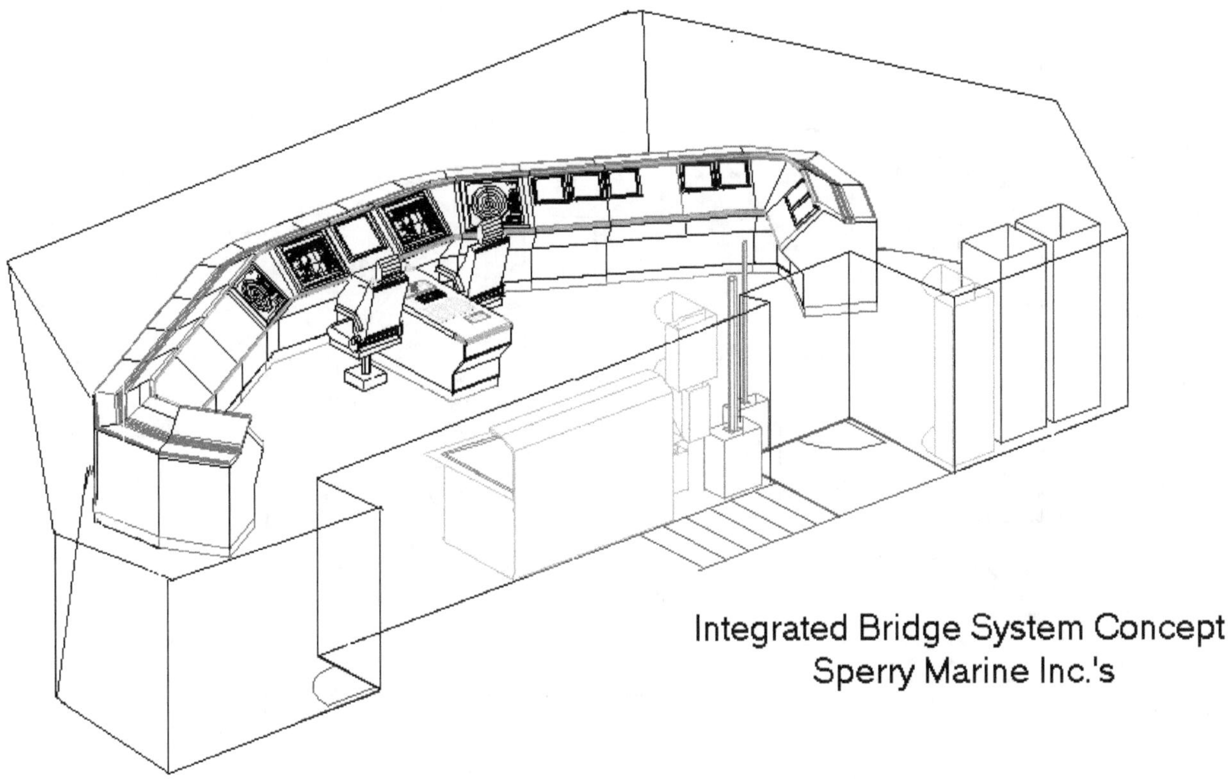

Figure VI-13: Integrated Bridge System

2. Engineering Central Control Station Drawings

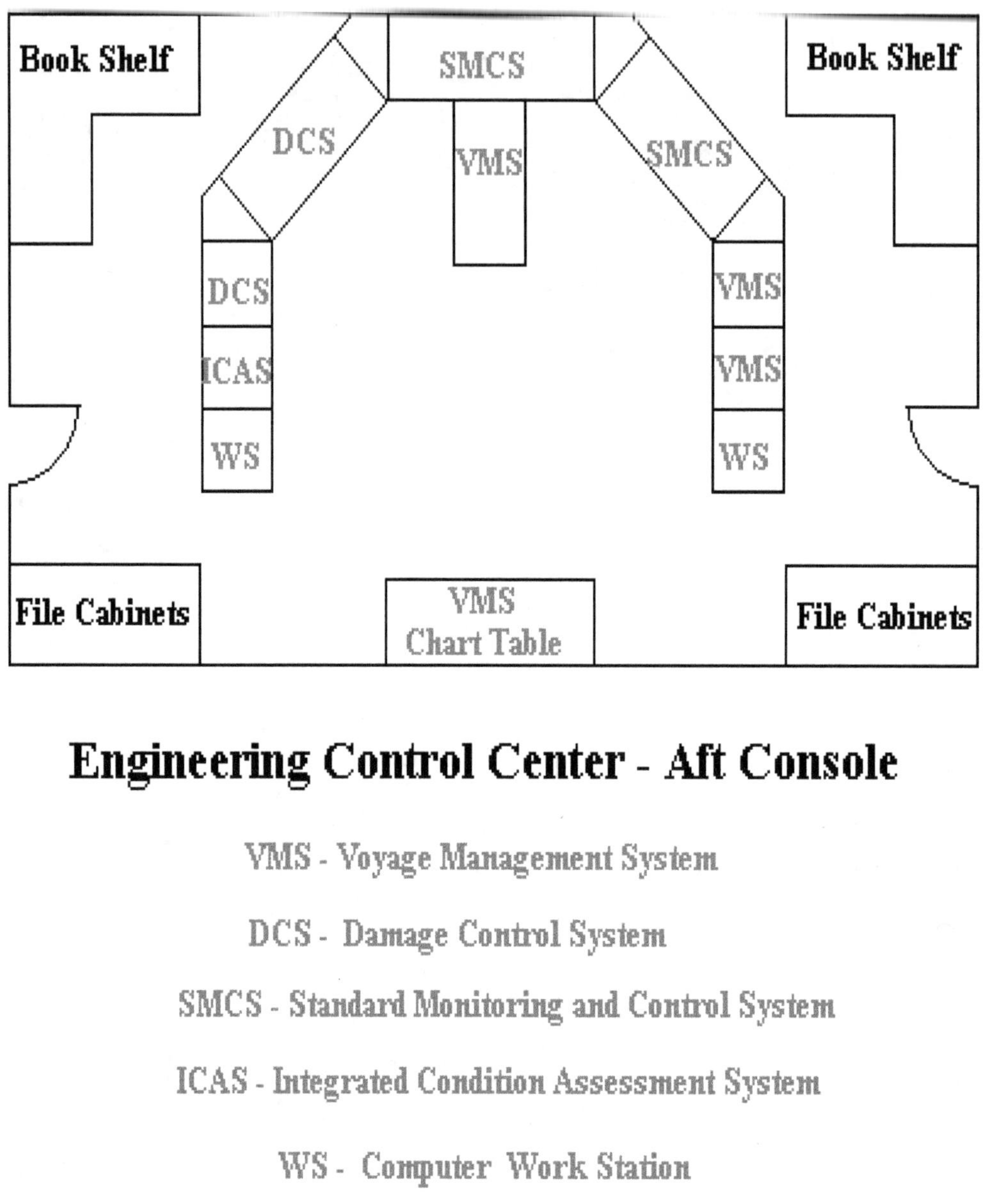

Figure VI-14: Engineering Central Control Station Drawings

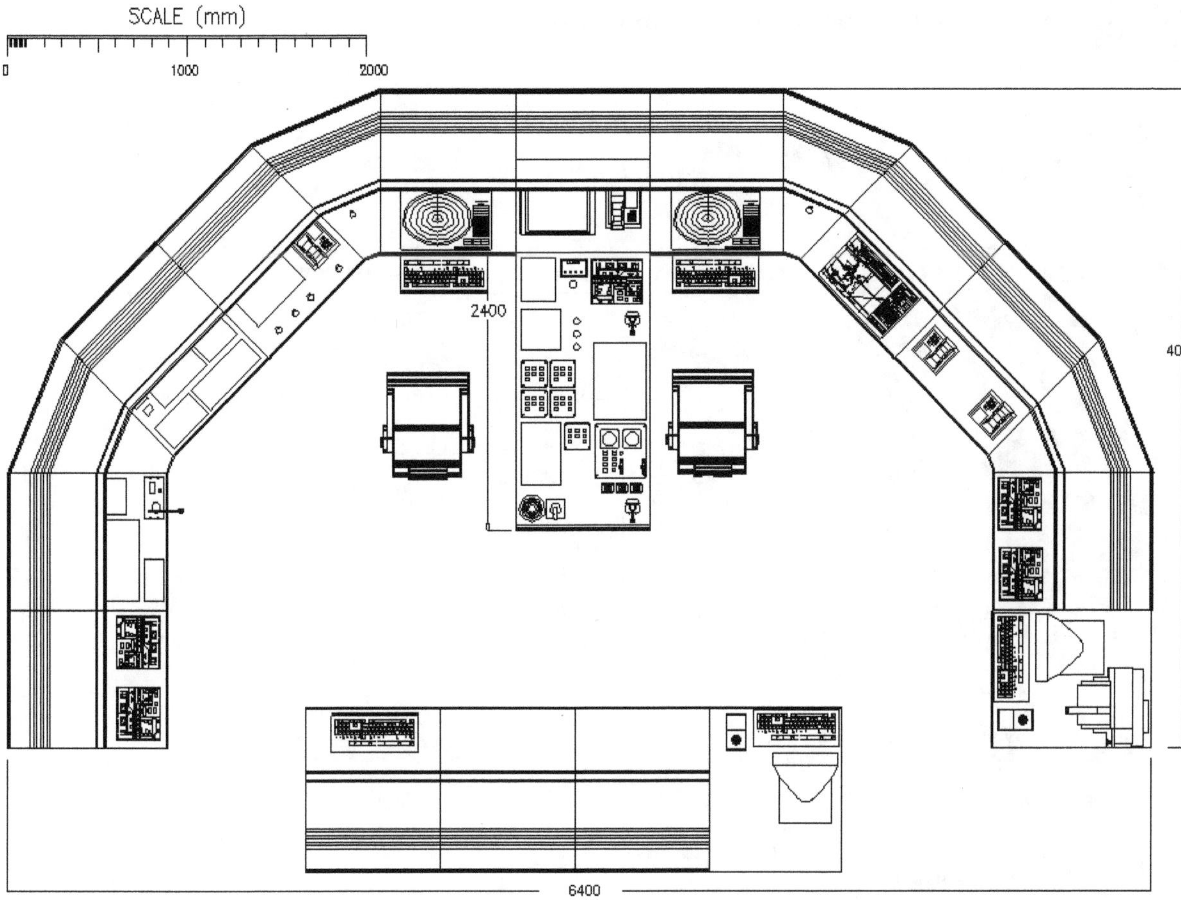

Figure VI-15: Engineering Central Control Station

3. Combat Information Center Drawings

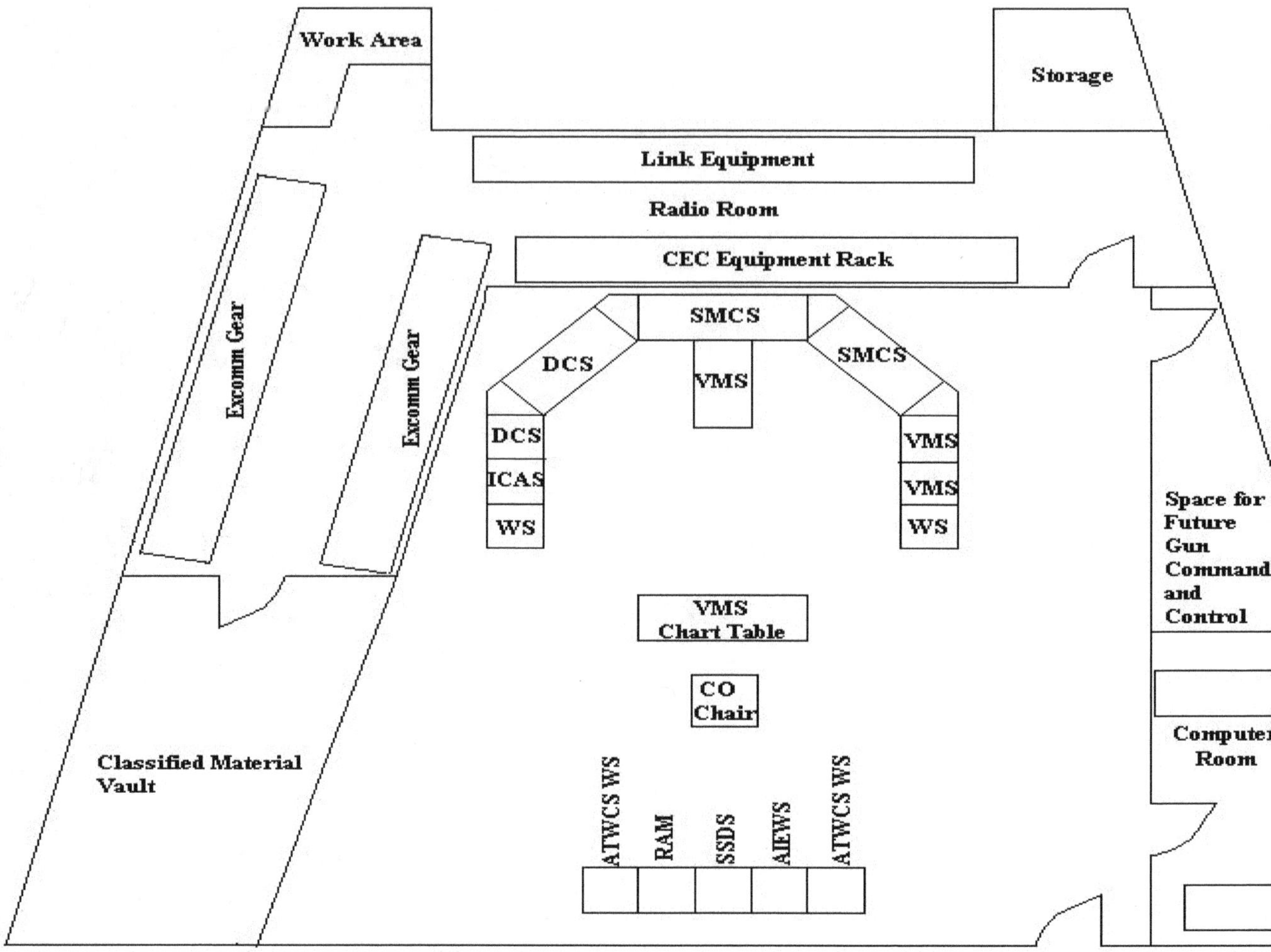

Figure VI-16: Combat Information Center

Combat System Control Consoles:

ATWCS WS-Advanced Tactical Weapons Control System Work Station
RAM-Rolling Airframe Missile Console
SSDS-Ship's Self Defense System Console
AIEWS-Advanced Integrated Electronic Warfare System Console

Integrated Bridge Control Consoles:

VMS-Voyage Management System
DCS-Damage Control System
SMCS-Standard Monitoring and Control System
ICAS-Integrated Condition Assessment System
WS- Computer Work Station

VII. SURVIVABILITY

A. INTRODUCTION

The *Arsenal Ship* is an extremely high value unit and undoubtedly faces the most powerful and sophisticated anti-surface weaponry our enemies have to offer. The sinking or incapacitation of an *Arsenal Ship* carrying hundreds of missiles, not only serves as a terrible tactical blow to U.S. forces, but also fuels our enemies with hope. It is therefore paramount that this ship is designed with a high level of survivability.

Survivability is defined as the "capability to avoid and/or withstand a man-made hostile environment. [6]" Survivability is divided into two categories:

- Susceptibility - the inability to avoid being hit
- Vulnerability - the inability to withstand a hit

The *Arsenal Ship* is designed to sufficiently decrease susceptibility and vulnerability, thereby increasing the overall survivability of the ship.

B. APPROACH

Because it is a high value asset, the *Arsenal Ship* is designed with Level III protection [53]. Level III protection means that the *Arsenal Ship* is placed in the same protection category as aircraft carriers and battle force surface combatants, operating in the most severe warfare environments.

A series of survivability comparisons between the T-AO and the *Arsenal Ship* were conducted using a trade-off analysis program [54]. Attributes such as percent of radar cross-section (RCS) reduction, number of escorts, relative amount of topside hardening and self defense features (i.e., RAM, CIWS, SM-2, etc.) were varied for each run. The algorithm was based on all of the past warship casualties that resulted from missile hits (i.e., Falkland War, Arab-Israeli War). The program confirmed the concept that a balanced approach to protecting the ship was the most survivable alternative.

The *Arsenal Ship* incorporates a reasonable amount of signature reduction, hardening, redundancy, separation, active damage control and self defense systems. Exhaustive efforts in stealth technology, and hardening from missile and torpedo damage, are not in the *Arsenal Ship* design due to excessive costs. The *Arsenal Ship* survivability features are designed to reduce the risk of damage caused by air, surface, and sub-surface forces. Automation is added into the survivability features to increase response time and decrease manning levels.

C. SUSCEPTIBILITY

The initial step in thinking about survivability is to prevent being hit. Therefore, it is believed that sufficient effort should be placed in providing *Arsenal Ship* with designs and systems which reduce susceptibility.

The *Arsenal Ship* is designed with an increase in passive self-defense systems and minimal active self-defense systems[5]. Since the *Arsenal Ship* has no active self-defense capability, it always operates under the protective umbrella of an AEGIS escort ship.

Due to the possible saturation of the escort's anti-air active self-defense umbrella, the possibility of an incoming anti-ship missile leaking past these defenses is high. Applying sufficient stealth technology to prevent targeting from an anti-ship missile is impractical due to the *Arsenal Ship's* large size. Consequently, the *Arsenal Ship* is designed with sufficient mass tonnage to absorb anti-ship missile explosions. A Rolling Airframe Missile (RAM) launcher is designed to provide the *Arsenal Ship* a limited active self-defense capability.

D. VULNERABILITY

A key design feature for the *Arsenal Ship* is that one shot does not keep the ship from performing its mission. It is designed for rapid damage localization and minimization. The minimization and localization of damage is enhanced by improvements in redundancy, separation of common systems, hull design, damage control, and hardening.

1. Redundancy and Separation

The most significant implementation of redundancy and separation on *Arsenal Ship* is the use of sixty-four individually armored 8-cell vertical launching system (VLS) launchers [55]. This design feature reduces the effects of a single mass detonation of a large 64-cell VLS launcher, by replacing it with smaller 8-cell launchers with armored separation. Each of the eight-cell launchers has an independent power source and is protected by a reinforced bulkhead on all four sides. Furthermore, these launchers are transversely arranged in rows of four with the two outboard, most vulnerable, cells separated from the inboard two by 11.5 feet. This 11.5 feet separation includes five inches of steel provided by the two interior bulkheads.

In addition to the separation of the VLS launchers, *Arsenal Ship* has several other redundant and/or separated systems. The remote-launch capability employs a Cooperative Engagement Capabilityt (CEC) antenna on the forward mast as the primary system, and

Link-16 and Link-22 antennas on the aft mast as the secondary and tertiary systems. The electrical system has five equally-rated ship service generators located in three separate compartments. Each generator can power the entire ship via any one of the three separate power distribution busses. Each bus feeds the ship's 15 electrical zones, which are completely isolated from one another. This architecture allows *Arsenal Ship* to withstand damage in one zone without degrading the power to the others. The firemain system is also designed using zonal architecture. Eight fire pumps feed into a firemain loop which provides firefighting to eight firemain zones.

The *Arsenal Ship* is designed with a single engineroom due to cost constraints and reduced manning levels. The impact of this on the *Arsenal Ship*'s mission is minimal since the ship is less dependent on propulsion than on electrical power for VLS and CEC. The ship is designed with twin screws that provides additional survivability to the ship's ability to maneuver.

2. Passive Damage Suppression

The *Arsenal Ship* uses increased structural strength, armor, watertight compartmentation and reserve buoyancy to limit damage to manageable levels for the automated systems and shipboard personnel.

The *Arsenal Ship* has sufficient reserve buoyancy to take on large amounts of water and remain afloat. It is highly compartmentalized with seventeen transverse watertight bulkheads and three longitudinal watertight bulkheads. The three compartment floodable length standard is a conservative estimate due to its double-hull design, added longitudinal bulkheads and permeability reduction features. One such feature is the foam-in salvage system[62]. This system "foams" a flooding compartment, thus decreasing permeability and restoring reserve buoyancy. The foam is used only in extreme emergency situations, because the solidified foam renders the space untenable.

The *Arsenal Ship* does not have any access penetrations through watertight bulkheads, below the damage control deck. Watertight closures are hydraulically-activated. The hydraulically-activated doors and hatches are remotely-triggered, and have a "fail to close" default. This system ensures complete material conditions are set.

In addition to maintaining the watertight integrity of the ship, damage due to missile attacks is reduced by side-impact structures. These structures, located between the inner- and outer-hulls, extend from the main deck to five feet below the waterline, and run the entire length of the VLS section of the ship. They are constructed of four layers of six-inch thick concrete and six-inch thick honeycomb polymer material. This system is designed to distribute the impact of the missile over the effected panel and dissipate the kinetic energy of the missile by redirecting the energy in directions away from the launchers.

3. Component Location and Shielding

The missiles are protected on all sides by 2.5-inch thick armor-plating. The armor provides additional resistance to missile impact and contains the exploding ordnance within the launcher. This, coupled with separation between launchers, aids in preventing mass detonation of all missiles onboard. The 2.5-inch steel thickness is limited by the maximum thickness that can currently be welded with a single-pass electroslag weld.

The command and control spaces are located deep within the ship. Crew habitability spaces are dispersed fore and aft. This dispersion avoids the potential of losing a majority of the crew to a single missile hit. Half of the VLS cells are located near the centerline, providing them with added protection. Additionally, the electrical distribution busses are arranged to provide maximum survivability. The port and starboard busses are located high near the main deck, while the centerline bus is located near the keel. All three busses are shielded by a steel enclosure to reduce the likelihood of damage.

E. DAMAGE CONTROL

The *Arsenal Ship's* Damage Control System (DCS), Standard Machinery Control System (SMCS) and Integrated Condition Assessment System (ICAS) responds automatically to any damage control situation. The ship uses remote-sensors to detect fire, flooding, access to spaces, toxic gasses, and monitor equipment. This information is sent to the DCS for evaluation. SMCS monitors and controls all machinery systems. ICAS monitors the major machinery for maintenance problems that effects plant operation, and interacts with SMCS for control procedures on machinery casualties.

DCS require no full-time manning. It has two modes of operation. The first is the fully-automated mode, and the primary mode used. The second is the prompting-mode. In this mode, the system suggests the proper response and awaits for directions from the console operator. These automated systems control propulsion, electrical distribution, ventilation, drainage, firemain and sprinklers, CO_2 flooding, AFFF, material condition, cooling water, fuel and oil systems.

Hydraulically-activated watertight closures are automatically programmed to close in the event of damage. The closure signal is sent to the doors from DCS, upon detection of a casualty in the effected zone. Ventilation fittings are isolated with electrically-operated valves. SMCS reroutes vital services around the damaged area, and places additional equipment online, as needed. The electrical distribution panels in the zones are secured as required. Two separate distribution panels, 110 volts and 440 volts, are used to allow lighting to be left energized, while electrical power is secured in the effected area.

All spaces are fitted with an overhead sprinkler system. The sprinkler system is activated when a fire is sensed in the space. High velocity fog, carbon dioxide (CO_2), or aqueous film-forming foam (AFFF) are selected based upon equipment in the space and degree of protection required. The installed drainage system is then activated by the sewage vacuum system or bilge drainage system, to remove accumulated firefighting water.

Flooding situations are handled along the same lines as fire casualties. The additional foam-in-salvage system is used in extreme casualties, since the space is left untenable after activation. The foam is sprayed in through an installed sprinkler system to restore lost buoyancy.

Three variations are used to control space ventilation. First, positive ventilation is set in adjacent spaces near the damaged area. Second, negative ventilation is set in the damaged compartment. Finally, ventilation fittings are secured in the space.

F. CHEMICAL, BIOLOGICAL AND RADIOLOGICAL (CBR)

The entire ship is protected by a Collective Protection System (CPS). All ventilation intakes are routed through filter rooms to remove contaminates before they can enter the ship. The ventilation systems pressurize the interior of the ship to two inches of water, to provide a positive seal. The four weatherdeck accesses have hydraulically-actuated watertight doors, that are remotely or automatically opened or closed. A decontamination station is located on the starboard side of the fore and aft weatherdeck access.

The external skin of the ship is protected by a water washdown system. This system prevents airborne contaminates from adhering to the ship. The *Arsenal Ship* is designed to reduce the accumulation of topside contaminates by minimizing the amount of topside equipment. Calcium hypochloride can be injected directly into the water washdown system. This cleaning solution is sprayed onto the deck in the event of a biological or chemical threat, thus eliminating the need for personnel exposure to the agents.

VIII. NAVAL ARCHITECTURE

The *Arsenal Ship* naval architecture design is tailored to an existing computer model of the T-AO 201 class auxiliary ship. Advanced Surface Ship Evaluation Tool's (ASSET) MONOLA is used to model monohull landing and amphibious ships. Although *Arsenal Ship* is neither of these, the basic hull form is more similar to them than it is to the traditional surface combatant. Using the T-AO as baseline, the *Arsenal Ship* model does not include the weights and volumes associated with cargo (fuel, water, stores, etc.), cargo pumps and underway replenishment equipment. The weights, volumes and center of gravity locations are added to the model for equipment and systems peculiar to *Arsenal Ship*, (i.e., VLS launchers, concrete side impact system and the future gun system). A complete list of these additions is located in the Payload and Adjustments report of Appendix C. Most of the other weight groups, identified by three-digit Ship Worklist Breakdown Structure (SWBS) numbers, also require adjustments from the original model. Some of these adjustments are scaled from the T-AO using built-in parametric equations. Others variables are calculated by ASSET, or require user input. ASSET performs detailed analyses (i.e., floodable length, resistance, power requirements, endurance, longitudinal strength, etc.), after the weight and space allocations are entered

A. HULL GEOMETRY

Since *Arsenal Ship* is a modified-repeat design, some of the traditional naval architecture are fixed. The hull geometry for the *Arsenal Ship*, specifically hull offsets, enclosed volume and appendages, are identical to the T-AO. Below is a summary of the important hull characteristics of the *Arsenal Ship*.

Principal Characteristics

• LBP (ft)	650.0
• LOA (ft)	666.3
• Beam at DWL (ft)	97.7
• Beam at Weather Deck (ft)	97.7
• Depth at Station 10 (ft)	60.8
• Draft to Keel from DWL (ft)	34.5
• Draft to Keel from LWL (ft)	33.3
• Freeboard at Station 3 (ft)	37.4
• Transverse Metacentric Height, GMT (ft)	11.3
• Prismatic Coefficient, Cp	0.653
• Midship Section Coefficient, Cx	0.980
• Full Load Displacement (LT)	38706.0

The body plan of the *Arsenal Ship*, shown in Figure VIII-1, gives an excellent perspective of the high (0.98) midship section coefficient.

S

Figure VIII-1: *Arsenal Ship* Body Plan

B. HYDROSTATIC CURVES

The *Arsenal Ship*'s hydrostatic curves, or curves of form, are identical to the T-AO. These curves, shown in Figure VIII-2, show how the *Arsenal Ship* responds to various static conditions, dependent on mean draft. The individual curves are::

- A. Displacement (5000 LT/division)
- B. Moment to Trim 1-inch, MT1" (500 ft-LT/inch per division)
- C. Tons per Inch Immersion (TPI) (20 LT/inch per division)
- D. Transverse Metacentric Radius, BM_T (5 ft per division)
- E. Longitudinal Metacentric Radius, BM_L (100 ft per division)
- F. Vertical Center of Buoyancy, KB (2 ft per division)
- G. Change in Displacement per unit trim by stern (10 LT/ft per division)
- H. Wetted Surface Area (10000 sq ft per division)
- I. Longitudinal Center of Buoyancy (5 ft per division)
- J. Longitudinal Center of Flotation (5 ft per division)

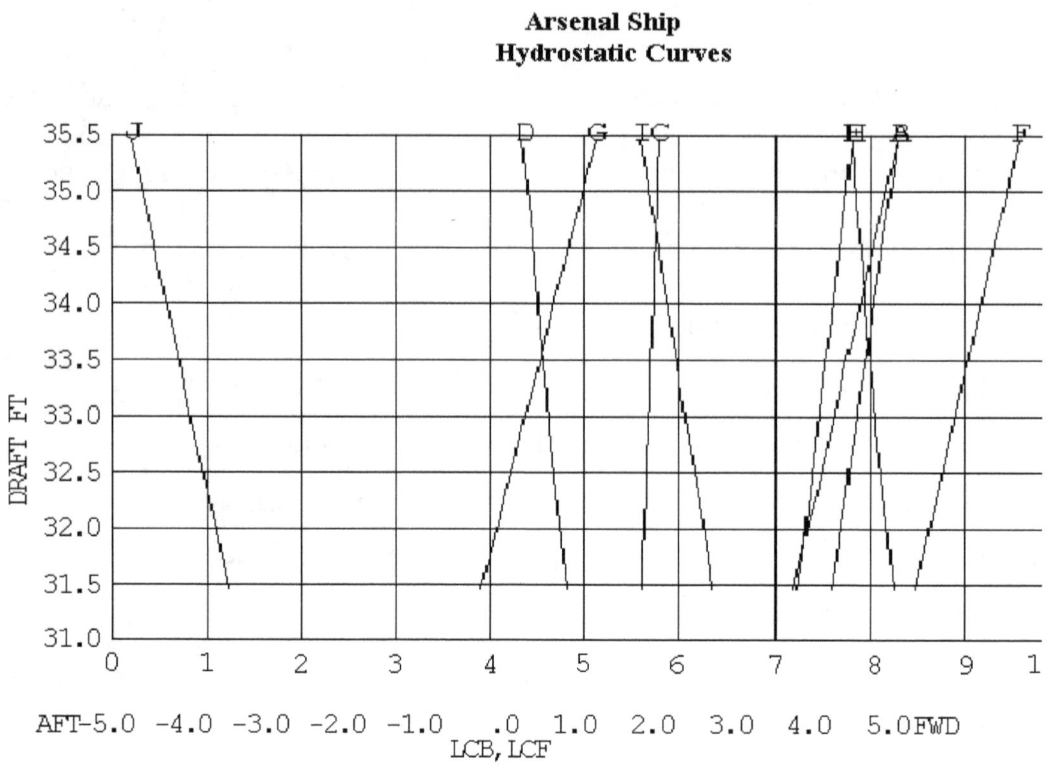

Figure VIII-2: Hydrostatic Curves

C. SECTIONAL AREA CURVE

The sectional area curve shown in Figure VIII-3 illustrates the extremely large sectional areas in the midsection of the ship. The flattened portion at the top of the curve is of particular importance. This large, nearly parallel, midbody allows maximum flexibility in the VLS arrangement, and thus optimizes survivability.

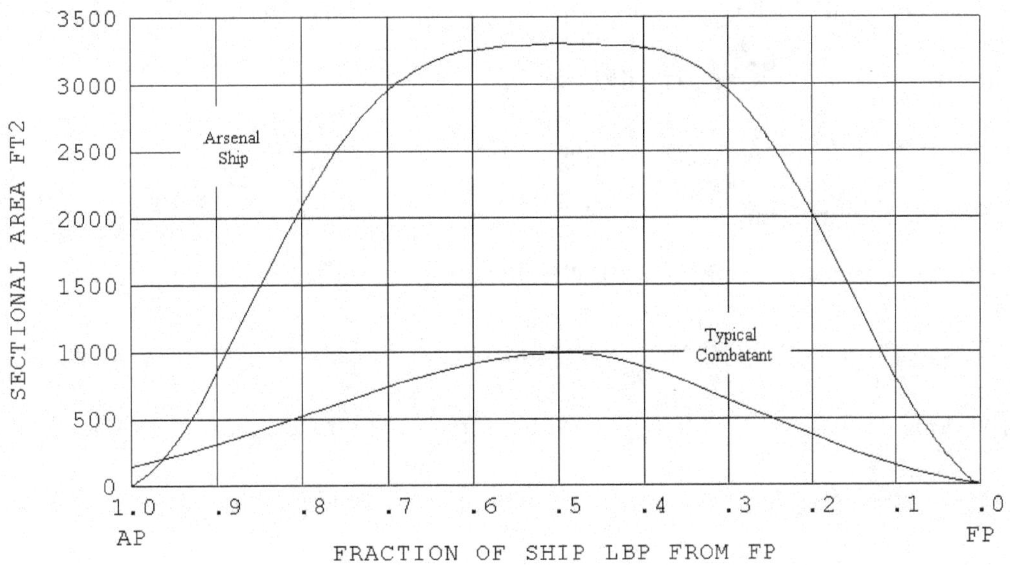

Figure VIII-3: Sectional Area Curve

D. HULL SUBDIVISIONS AND FLOODABLE LENGTH

Arsenal Ship has seventeen transverse watertight bulkheads to maintain reserve buoyancy in the event of flooding. Bulkhead locations are first determined by using the existing engineroom and auxiliary machinery bulkheads. Next, transverse bulkheads are placed between every two rows of VLS cells. And finally, the remaining bulkheads are placed such that the three compartment floodable length standard is achieved. This is a conservative estimate of *Arsenal Ship*'s ability to withstand flooding however, because it does not include several important factors. First, *Arsenal Ship* is designed with a double-hull with six feet of spacing between the two hulls. The entire midbody of the ship has longitudinal watertight bulkheads to prevent the spread of damage transversely. These two features reduce the likelihood of flooding the entire volume between transverse bulkheads to the waterline. Second, the three compartment floodable length is based on using the standard permeability factors recommended in Principles of Naval Architecture [56]. These permeability factors are as follows:

- Habitability Spaces: 0.95
- Machinery Spaces: 0.85
- Cargo/Storage Spaces: 0.60

These permeability factors are applied to the entire section of the ship, as is typically done. Features, such as the side impact system contained in the double-hull, raises the *Arsenal Ship*'s permeability and improves floodable length. Further iterations can implement a closer, yet more complex, approximation of these permeability factors. This can account for such spaces as the forward auxiliary machinery room (μ=0.85) that is actually located in a subdivision that is otherwise habitable (μ=0.95) on the T-AO. Finally, the existence of a foam-in-salvage system has the potential to drastically reduce the permeability of a compartment and thus restore reserve buoyancy. For these reasons, it is reasonable to predict that *Arsenal Ship*'s floodable length is greater than three (Figure VIII-4).

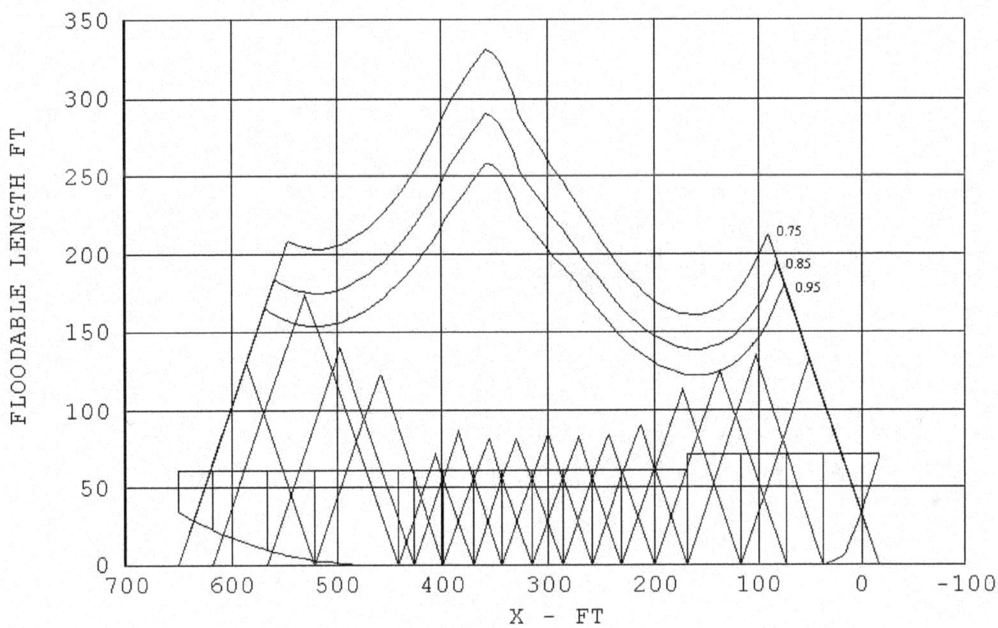

Figure VIII-4: ***Arsenal Ship*'s Floodable Length Curves**

E. STATIC STABILITY

Static stability is the measure of whether a ship possesses sufficient righting energy to withstand various types of upsetting and healing moments, such as beam winds (especially when combined with rolling), lifting of heavy weights over the side, high-speed turning and topside icing [57]. A comparison is made between the ship's corrected righting arm curve versus various heeling arm curves to determine *Arsenal Ship*'s stability. Figure VIII-5 is the static stability curves for a 100-knot beam wind, showing the righting arm and heeling arm. Stability is considered satisfactory if:

- The heeling arm at the intersection of the two curves is less than 60-percent of the maximum righting arm; and
- The area between the curves to the right of the intersection is at least 1.4 times the area between the curves to the left of the intersection.

It is apparent from this diagram that *Arsenal Ship* is capable of withstanding such winds.

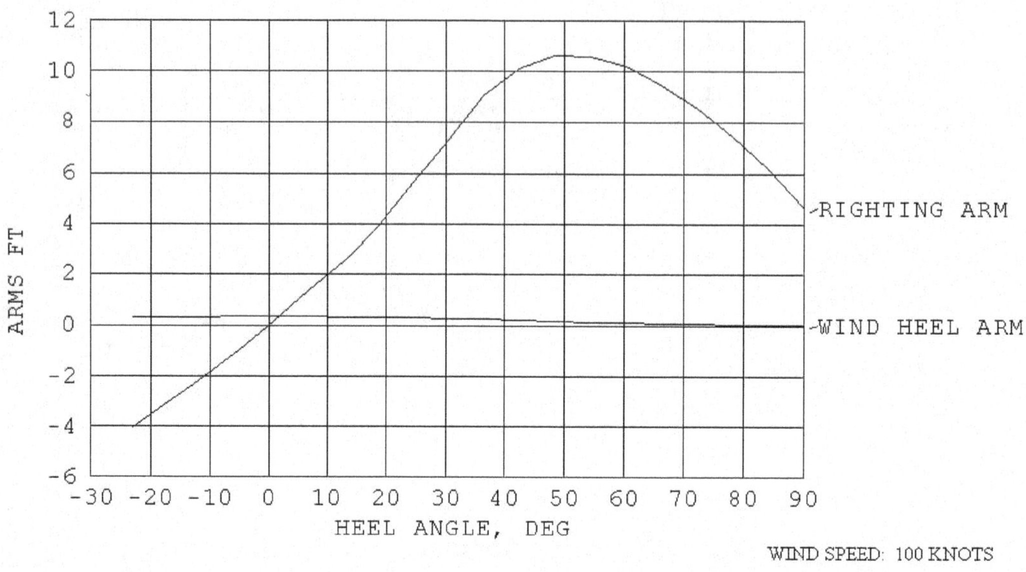

Figure VIII-5: Static Stability Curve for 100 kt Beam Winds

Figure V III-6 shows the same comparison for a high speed turn heeling effect. In this case, stability is considered satisfactory if:

- The angle of steady heel does not exceed 15 degrees;
- The heeling arm at the intersection of the two curves is less than 60-percent of the maximum righting arm; and
- The reserve dynamic stability (the area between the curves to the right of the intersection) is at least 40-percent of the total area under the righting arm curve.

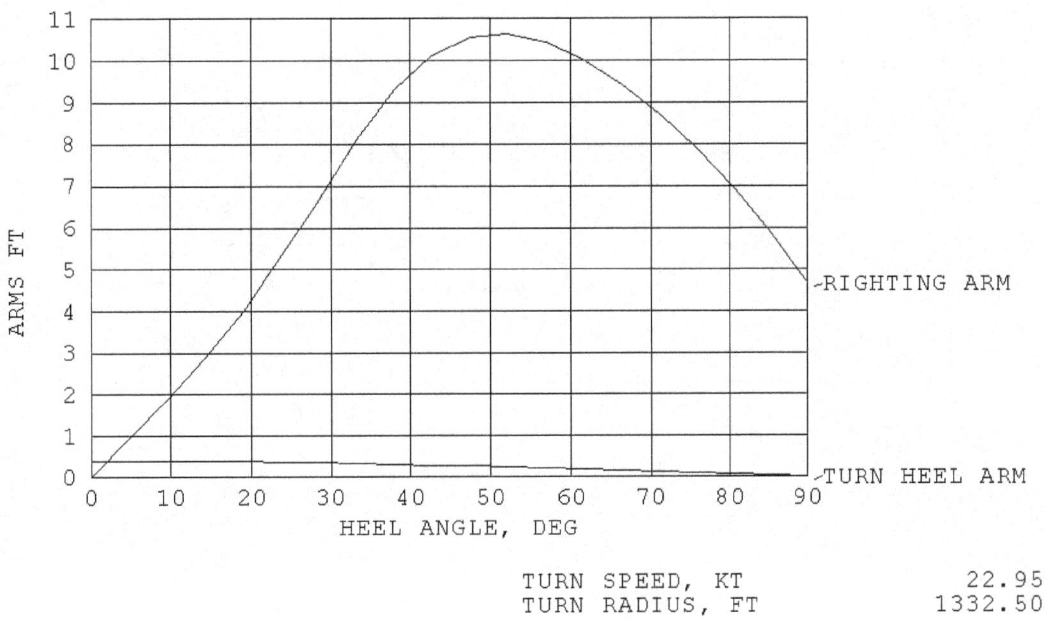

Figure VIII-6: Static Stability Curve for High Speed Turn

Once again, the diagram shows that these criteria are easily met.

Finally, the damaged static stability curve is shown in Figure VIII-7. This shows the effect of beam winds after damage is suffered. For this analysis, stability is considered satisfactory if:

- The static trimmed-heeled waterline after damage does not submerge the margin line;
- The static heel angle without wind effects does not exceed 15 degrees;
- Adequate dynamic stability (the area between the curves to the right of the intersection) exists to absorb the energy imparted on the ship by moderately rough seas in combination with beam winds.

This diagram is evidence that *Arsenal Ship* is highly stable, even after damage.

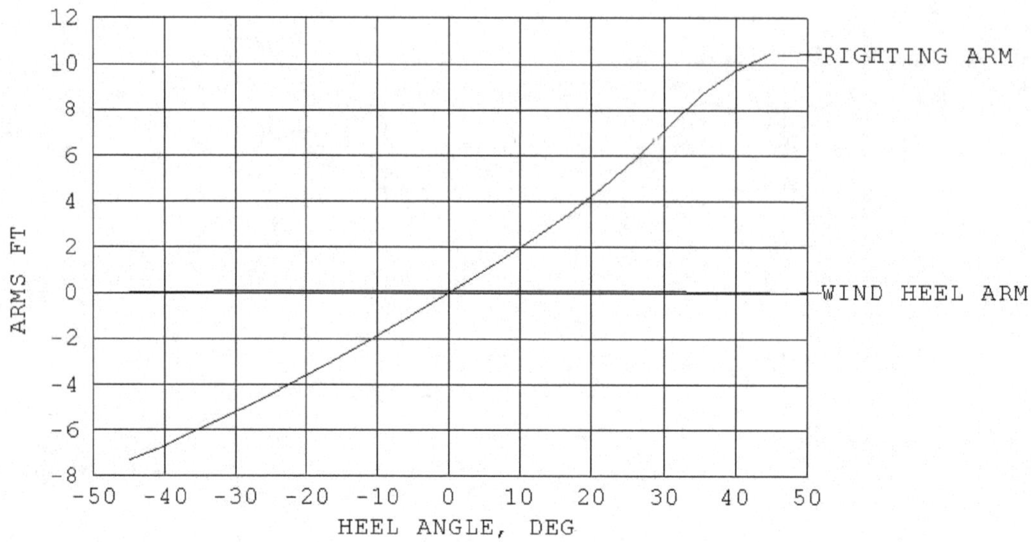

Figure VIII-7: Damaged Static Stability Curve

F. HULL STRUCTURE

The T-AO 201 class auxiliary ship is designed with sufficient strength to transport 26,000 long tons of fuel in the ship's midsection. Although the full load displacement of *Arsenal Ship* is approximately equal to the T-AO, the *Arsenal Ship*'s load is more evenly distributed than the T-AO. Additional structure is added to the interior of *Arsenal Ship* as well, which also aids in structural strength. A preliminary structural analysis was performed using ASSET to determine the number, sizes and weight of support structure (i.e., scantlings).

ASSET calculates the hogging and sagging bending moments based on the ship's weight distribution and section modulus. From these, the primary stresses at the keel and main deck are computed for hogging and sagging conditions. These stresses are then used to design the number and size of scantlings for various materials. Figure VIII-8 shows the midship section of the *Arsenal Ship* determined by ASSET. It should be noted that the double-hull and longitudinal bulkheads are not considered in this midship section. This is due to a limitation of the ASSET program, and results in an under-esimation of the ship's overall strength and over-design of the scantlings.

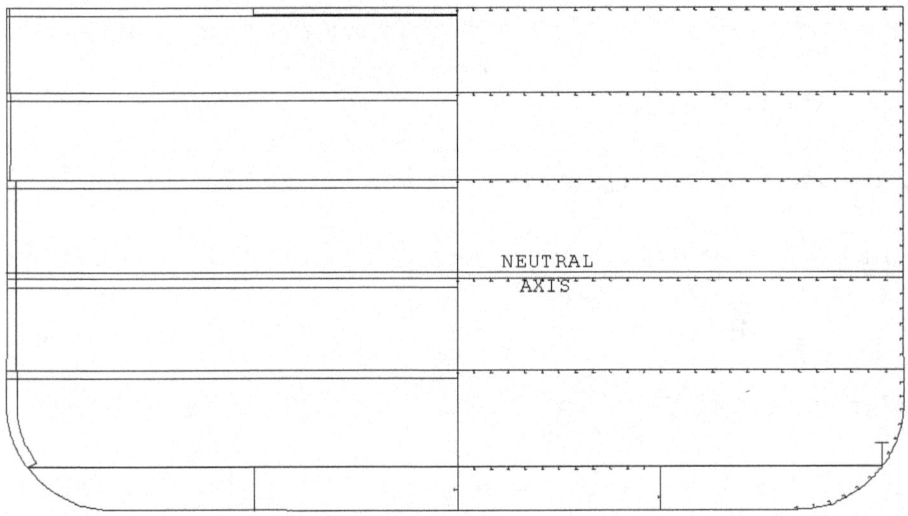

Figure VIII-8: *Arsenal Ship*'s Midship Section

IX. COST ESTIMATION

The modified-repeat design of the T-AO, together with the original construction of the T-AO using commercial standards, makes cost estimation a very difficult process. Cost figures for standard military construction of naval vessels are used in this study. The analysis spreadsheets are adjusted for the *Arsenal Ship* and T-AO because historical data for construction of this type of vessel is unavailable. The known unit sailaway cost of the T-AO is $167 million (1995) [42]. This price is scaled according to the *Arsenal Ship*'s design and construction techniques.

First, the material cost estimating relationships (CER) for ship's weightlist breakdown structure (SWBS) groups 100, 200 and 500 are reduced by 20 percent of the military specification (MIL-SPEC) values. Next, the material CERs for weight groups 320 and 340 are reduced by 25 percent. These adjustments are applied to the *Arsenal Ship* and the T-AO. Finally, since the *Arsenal Ship* is a modified-repeat design, the labor CERs are reduced by 15 percent, and the non-recurring engineering hours are significantly reduced (App. B).

The cost analyses are conducted using identical labor rates, learning curves, profit, and other costing variables for the *Arsenal Ship* and the T-AO. The weight breakdowns are obtained from the weight module of the Advanced Surface Ship Evaluation Tool (ASSET) computer model for each ship. The labor associated with installing weapons systems is accounted for in the *Arsenal Ship* computer run, however, the procurement costs of the systems are not included until after the construction costs are scaled. Inclusion of these prices in the analysis skews the data, giving the *Arsenal Ship* a less expensive than it realistically is. Appendix B contains the preliminary cost analyses performed on the T-AO and *Arsenal Ship*.

Table X-1 shows the scaled combat systems costs. Combining the ASSET model cost data, the combat systems costs, and a discount acquisition cost for the T-AO, the resultant unit sailaway cost for the *Arsenal Ship* is **$ 487 million**. This cost is well below the sailaway cost cap imposed by Defense Advanced Research Project Agency (DARPA).

512 MK-41 VLS Launchers	$ 215 M
CEC and ATWCS	$ 20 M
Communications Systems	$ 24 M
Self Defense Systems	$ 11 M
TOTAL	$ 270 M

Table IX-1. Combat Systems Estimated Costing Data

X. DISPOSAL OF WASTE AT SEA

Modern naval ships are expected to meet increasingly stringent environmental regulations. The *Arsenal Ship* is no exception to this, and to meet these challenges, the ship is designed to be an environmentally sound ship [58]. This designation requires that the ship must conduct operations, in port and at sea, in such a manner as to minimize or eliminate any adverse impact on the marine environment. An environmentally sound ship has the following attributes: [59]

- Environmental compliance
- No significant adverse environmental impact
- Wastes treated or destroyed on board to the maximum extent practicable
- No inappropriate dependence on shore facilities for waste off-load
- Minimal logistical costs for waste management
- Minimal use of hazardous materials on board.

The diplomatic, logistical, economical, environmental and public relations consequences of non-compliance are serious and require such a design.

Simply stated, the *Arsenal Ship* must be capable of operating world-wide with minimal potential for regulatory constraint or dependence on shore facilities for waste management support. To meet this challenge, the *Arsenal Ship* design incorporates several shipboard systems that destroy or appropriately treat waste generated on board. The following wastes and emissions are considered:

- Oily waste
- Non-oily wastewater
- Solid waste

Oily waste and waste oil are unavoidable byproducts of operating ships. There are three primary sources for oily waste: bilge, ballast and waste oil storage tanks. Navy policy limits the oil content of ship discharges to less than 20 ppm within 12 nm of land and to less than 100 ppm beyond 12 nm. To meet these requirements, the *Arsenal Ship* employs an Oil-water Separation (OWS) and an Oil Content Monitoring (OCM) systems. The U.S. Navy has two OWS designs, and one OCM design, in production that are suitable for this requirement. [60].

Non-oily wastewater includes blackwater (sewage) and graywater (discarded water from deck drains, heads, showers, galley, etc.). With regard to sewage and graywater, the Navy policy requires that ships employ marine sanitation devices (MSD) which prevent the discharge of untreated sewage within 3 nm from land. To meet this requirement, the *Arsenal Ship*, uses the existing collection, holding and transfer (CHT) system of the T-AO oiler. Sewage and graywater production are directly proportional to the number of personnel aboard, therefore with a minimally manned crew, the generation of such waste is expected to be significantly less than the capacity of the existing CHT system. An additional consideration precluding enhancement to the CHT system, is that the *Arsenal Ship* rarely operates within 3 nm from land.

The majority of shipboard solid waste is comprised of the following: paper, metal, glass, food waste, plastic waste and medical waste. These waste items are regulated and managed differently. For non-plastic waste, Navy policy prohibits discharge within 3 nm of any coastline, with a relaxation to this requirement for pulped garbage discharge beyond. To handle the disposal of non-plastic waste, the *Arsenal Ship* is equipped with a solid waste shredder (SWS) for processing of metal, glass and ceramic waste, and a trash pulper for the processing of paper, cardboard and food waste. Both items have completed operational testing on USS George Washington (CVN-73) [61] and are ready for installation on *Arsenal Ship*.

For plastic waste, the Navy continues to intimate a future policy of zero discharge. To achieve this on *Arsenal Ship*, the ship is outfitted with a plastics waste processor (PWP). This system heats, compacts and sanitizes plastic waste, reducing large volumes of plastic to easily handled dense disks. This system has also completed testing on USS George Washington (CVN-73) [59] and is also ready for installation.

Medical waste generation on board *Arsenal Ship* is expected to be minimal. The small crew size, coupled with the stringent medical screening requirements, reduces the demand for medical attention and thus medical waste processing. Most medical procedures will be performed at medical facilities off of the ship (battlegroup asset, ashore) therefore, the *Arsenal Ship* is equipped with only the minimum requirement for medical waste disposal.

By designing the *Arsenal Ship* as an environmentally sound ship, operations are not impaired by environmental regulatory issues. The installation of systems that reduce or destroy waste are integral to meeting this challenge.

XI. DESIGN EVALUATION

Overall, this design meets the objectives of the proposed *Arsenal Ship* concept. Although many objectives in the SCD and CONOPS are subjective for ship designers, there are several that are measurable and crucial for a successful design. These are listed in priority in the design philosophy. An attempt is made and documented where the design can be measured against an objective. Detailed explanations are provided to justify objectives that are subjective in nature. Overall, this design not only demonstrates the ability to meet mission needs, but is also straight- forward in its design reasoning.

First and foremost, the most important objective for the design is acquisition cost. This design meets that objective. Cost drives the entire design and is the reason for selection of a modified-repeat concept. The cost of the combat system suite is considered before HM&E systems. Quantifying exact costs for equipment designed around Advanced Concept Technology Demonstrator (ACTD) progress is difficult. A concerted attempt is made to quantify these costs based on estimates received from researchers. The figures may be debatable, but estimates produced by others are subject to the same margin of error. The HM&E costs are more tangible. The figures presented in this category are based upon extrapolation of known costs of similar designs. Hence, these have a much smaller margin of error. The combined-costs for the combat system suite and HM&E systems are credible and meet the $550 million cost ceiling objective.

Another important objective in the design is the ability of the *Arsenal Ship* to operate as a remote-firing battery. Technology for remote-firing capability is still in the research and development stage. There is sufficient confidence, however, that it will be cost-effective and ready for installation on the first *Arsenal Ship*. The system will be capable of remotely-firing SM-2, TLAM, and ATACMS from other targeting platforms, using the Advanced Tactical Weapon Control System on the *Arsenal Ship*. If this technology fails, the entire *Arsenal Ship* delivery schedule will be in jeopardy.

Arsenal Ship has a minimal amount of sensors, aside from CEC and standard communications. *Arsenal Ship's* self-defense capability is driven by reduction in costs and manning. The vessel's self-defense relies on EW systems (i.e., AIEWS, IR, RF decoys) and RAM. These systems, in conjunction with mass tonnage and inherent survivability features, meet the CONOPS objects.

Manning reduction is another important design goal. The objective is a maximum crew of 50 personnel. This *Arsenal Ship* is designed to have a crew of 44. Automated-systems are used wherever possible, with sufficient redundancy to reduce system failure. In addition, standard naval operating practices and procedures are reviewed and tailored for a minimum manning concept. Safety is looked at first in every case. Some special evolutions are removed all together. This leaves *Arsenal Ship* without the capability for conducting that evolution. However, in such cases, a full explanation for removal is presented in detail. The reduced-manning concept also places a large percentage of equipment repairs on shore-based, deployable if necessary, maintenance teams. Equipment is chosen with minimum maintenance requirements in mind.

As mentioned above, cost considerations are paramount in this design. Therefore, *Arsenal Ship* is designed from a modified-repeat T-AO 201 class auxiliary ship with double hull. Alternative hull forms are reviewed, however, a new hull design removes money that can be used to purchase valuable combat system equipment. Not only is this hull a proven design, but it also has the added double-hull benefit. The double-hull has greater survivability characteristics over conventional monohulls. Different engines are used to account for a higher sustained-speed. Taking into account these modifications, HM&E costs are still within an acceptable range for supporting the combat system package selection. The T-AO 201 class auxiliary ship with double-hull is believed to be the best and most cost-effective solution for the *Arsenal Ship* concept.

This report is only a "first turn around the design spiral." There is sufficient confidence that the detailed design stage will prove that this *Arsenal Ship* surpasses all CONOPS and SCD objectives. The *Arsenal Ship* is designed with state-of-the-art technology for the warfighter of the 21st century. It is a design which supports not only naval surface warriors, but also aviators, submariners, and infantrymen. It provides the first significant platform for joint-service missile targeting and sets the stage for follow-on joint military hardware.

In summary, this design is based on at-sea experience and thorough analysis of warfighters and engineers from the Navy, Coast Guard, and Marine Corps, and presents a realistic solution to the Navy's call for an *Arsenal Ship* of the 21st century.

LIST OF REFERENCES

1. U.S. Navy, Office of the Chief of Naval Operations and Commandant of the Marine Corps,
   , 1994
2. Commandant of the Marine Corps, *Operational Maneuver From the Sea*, Washington D.C., 1996
3. Office of the Chief of Naval Operations, *2020 Vision,* Washington, D.C., 1996
4. DARPA, *Arsenal Ship Concept of Operations*, 23 May 1996
5. DARPA, Arsenal *Ship Capability Document*, 23 May 1996
6. R. Ball, C. Calvano, "Establishing the Fundamentals of a Surface Ship Survivability Design Discipline," Naval Engineers Journal, Jan. 1994, p. 72
7. Raines, D., "Ship Affordability," Naval Engineer's Journal
8. NWP-14revE, Replenishment at Sea
9. Carvey, BM2, USS Stethem, Personal interview, Aug. 17, 1996.
10. NWP-42revJ chg1UC3, Shipboard Helicopter Operations
11. USS Stethem Deck Division PMS Schedules, 43PI, Aug 17, 1996
12. *Smart Ship Implementation Proposal FLEET-0022, Automated Division Officer's Notebook: COMPASS for Windows*
13. *Smart Ship Implementation Proposal FLEET-0023, Shift All Hands to Coveralls.*
14. Barker, R., CAPT, *Smart Ship Project*, Presentation, Core Installation slide, Jun. 11 1996.
15. Sperry Marine, *Vision 2100 Voyage Management System VT.* Sperry Marine Inc., U.S., 1994
16. Dahl, Ernest, A., *TWARSES - Two Wire Automatic Remote Sensing Evaluation System Training Manual.* Naval Surface Warfare Center, Port Hueneme Division, Code 4L03, Port Hueneme, CA.
17. *Smart Ship Implementation Proposal FLEET-0020, Standard Monitoring and Control System (SMCS).*
18. CAE Electronics LTD. Standard Monitoring Control System (SMCS). CAE Electronics LTD., Arlington, VA.

19. *Smart Ship Implementation Proposal FLEET-009, Integrated Condition Assessment System (ICAS).*
20. *Smart Ship Implementation Proposal FLEET-010, Damage Control System (DCS).*
21. CAE Electronics LTD, *Damage Control System (DCS).* CAE Electronics LTD., Arlington, VA.
22. *Smart Ship Implementation Proposal FLEET-042, Coordinated Physical Security System (COPS).*
23. Naval Surface Warfare Center, Port Hueneme Division, Combat Systems Department
24. Naval Surface Warfare Center, Port Hueneme Division, Verticle Launch System Department
25. Rothe, Michael, NAVSEA PEO (TAD) D111, Telephone interview, Aug. 15, 1996.
26. Naval Research Labs, Surface EW Division, *SLQ-32A(V) Short Course*, Presentation
27. PEO (TAD), *Rolling Airframe Missile (RAM)*, Program Decision Meeting, 5
28. PEO (TAD), *NULKA Information Briefing*, Presentation
29. Surface Force Tacnote, X20050-1-94.DARPA,
30. Center for Naval Analysis, NSFS COEA, Alexandria, VA, Dec 93, May 94, Aug 94,Oct 94
31. NAVSEA Fax, ONR FY 98 Navy Advanced Technology Demonstration VGAS - Vertical Guns for Advanced Ships
32. Dr. Randy Rohr HQ/DNA, VGAS - Vertical Guns for Advanced Ships Power Point Presentation, 28 Nov 95
33. Ange, Chris, *Naval Surface Fire Support 5-Inch MK 45 Gun Mount Modification*, Surface Warfare Magazine Vol.21 No.3, Alexandria, VA, May/June 1996
34. Lyons, Kenneth T., *Naval Surface Fire Support Extended Range Guided Munition*, Surface Warfare Magazine Vol.21 No.3, Alexandria, VA, May/June 1996
35. Morral, CAPT Dennis, *Naval Surface Fire Support Fire Support Missile Options*, Surface Warfare Magazine Vol.21 No.3, Alexandria, VA, May/June 1996
36. Liese (PMS 429C1), David and Morral (PMS 429), CAPT Dennis, *Naval Surface Fire Support Technology Overview*, Power Point Presentation, 31 July 1996
37. Blosser, O. Kelly, *Naval Surface Fire Support Mission Planning and*
38. *Joint Maritime Communications Strategy,* by Robert Taylor, Sept/Oct Surface Warfare, vol. 21, No 5,

39. *JCS Pub 1, Joint Warfare,* Joint Chiefs of Staff

40. *Command Arrangements for Peace Operations,* by Dr. David Alberts and Dr. Richard Hayes, NDU, May 1995

41. *Naval Shipboard Communications Systems,* by John Kim and Eugene Mueldorf, Prentice Hall PTR, Englewood Cliffs, NJ, 1995.

42. DeHaven, S. D., and LCDR Glen Darling, *Low Observable Multi-Function Stack*, Presentation, July, 1996.

43. DDR&E S&T Review Surface Ship Combatants Subarea Handout, 24 April, 1996.

44. J.C. Kim, and E.I. Muehldorf, *Naval Shipboard Communications Systems*, Englewood Cliffs, N.J: Prentice Hall 1995.

45. *Smart Ship Implementation Proposal FLEET-034, AN/URC-131A(V) High Frequency Radio Group (HFRG).*

46. Law, Preston E., *Shipboard Antennas*, 2nd ed.,., Dedham, Massachusetts: Artech House, Inc 1986.

47. Williamson, J., ed., *Janes Military Communications 1995-96*, Janes Information Group, Inc.: Alexandria, VA., 1996.

48. Hemming, Leland H., *Shipboard Antennas Systems*, MacDonnell Douglas Helicopter Systems Facsimile, Mesa, AZ., Nov. 6, 1996.

49. Surface Warfare Magazine, Sep/Oct 1996, vol. 21, no. 5, Alexandria, VA.

50. Applied Physics Laboratory (APL), *The Cooperative Engagement Capability*, John Hopkins APL Digest, vol. 16, no. 4, 1995.

51. Naval Surface Warfare Center, Port Hueneme Division, Underway Replenishment Department

52. T-AO costing data conversation with Contracting Officer, Supervisor of Shipbuilding, New Orleans, LA., 10/8/96

53. OPNAVINST 9070 Ser 09/8U501139 23Sep1988 Survivability Policy for Surface Ships of the U.S. Navy

54. Bush, R. LT, and Simea, G. LT, *Operational Analysis for the Arsenal Ship*, Aug 1996

55. Naval Surface Warfare Center, Carderock Division, *Vertical Launch System Survivability Report,* 1995

56. Principles of Naval Architecture, 2nd rev., vol. 1, Society of Naval Architects & Marine Engineers, 1988.

57. Gillmer, T.C., and Johnson, B., *Introduction to Naval Architecture*, Naval Institute Press, 1982.

58. U.S. Navy, Office of the Chief of Naval Operations, *OPNAVINST 5090.1B, Chapter 19, Environmental Compliance Afloat*, 01 Nov. 1994, pp. 1-35.

59. Koss, Larry, "Technology development for environmentally sound ships of the 21st century: an international perspective", J Mar Sci Technol, vol. 1, 1996, pp. 127-137.

60. Koss, Larry, "Environmentally sound ships of the 21st century", white paper, Office of the Chief of Naval Operations (N452), 1992.

61. "Navy Solves its Plastics Problem," Shipboard Environmental Update, vol. 1, no. 1, Sep 1995, pp. 3-4.

62. Myers, John, Foam-In Salvage System. Internet page of Battelle Corporation, http://www.battelle.org/oe/FIS.html, March, 1996.

APPENDIX A

Functional Analysis And Flow Diagrams

FUNCTION NAME AND NUMBER	FUNCTION PERFORMANCE AND REQUIREMENTS	TEAM
1.0 Inport/Moored Operations	Extended Inport/Moored Operations with extensive shore support	HOC
1.1 Exchange/Load Weapons		
1.1.1 Carry 500 VLS cells (64x8=512)	MK41 VLS vertical launch system (tradeoff with CCL - Concentric Canister Launcher)	C H
1.1.2 Determine Mission Requirement Loadout	Loadout determined by Area of Operation and JCWC requirements	O
1.1.3 Load 500 cells/Exchange out-of-date weapons for maintenance	•Ordnance Inventory tracked on ROLMS (Retail Ordnance Logistics Management System) SSP-FLEET-0037 •Missile Between Maintenance Storage Time: SM2 >90 days TLAM >90 days ATACMS >90 days SLAM >90 days	C
1.2 Maintain Charts	•Electronic Chart Updates (Shoreside Support) •Sperry VMS VT (Voyage Management System) Planning&Navigation Management Station •DNC (Digital Nautical Chart) for AN/SSN-6 NAVSSI (Navigation Sensor System Interface) SSP-FLEET-0025 •CIS (Chart Inventory System) SSP-FLEET-0031 integrates with IBS (Integrated Bridge System) SSP-FLEET-0013, tracks Notice to Mariners electronically through NIN (Navigation Information Network) and is a module of Digital Chart System of IBS (Integrated Bridge System)	O
1.3 Update Crew Records	(ID Card/Service Record/Medical Record/Dental Record/Security Clearance/ Security Access) •Electronic Record Updates (Shoreside Support) •MARC (Multi-technology Automated Reader Card) SSP-FLEET-0011	O
1.4 Medical/Dental Screening	Physicals / Dental Checks / Electronic Record Updates / PRT (Shoreside Support)	O
1.5 Intelligence Briefings	•Deployment Threat Brief / Mission Assignment (Shoreside Support) •Secure Teleconferencing Capability for ONI briefs	O
1.6 Maintain Consumables / Resale Operations	•Automatic re-order on issue of item •Computer request for delivery •Bar Coding for Non-SNAP Shipboard Applications SSP-FLEET-0016 •ROM EDI (Resale Operations Management Electronic Data Interchange) SSP-FLEET-0038	HOC
1.7 Physical Security	Protection of ship and crew	HOC
1.7.1 External	•Quarterdeck Video Surveillance System SSP-FLEET-0040 •MARC (Multi-technology Automated Reader Card) SSP-FLEET-0011 for crew ID/muster/status	
1.7.2 Internal	•MARC (Multi-technology Automated Reader Card) SSP-FLEET-0011 •COPS (Coordinated Physical Security System) SSP-FLEET-0042 for security access	
2.0 **Underway Preparations**	Preparations for Getting Underway	HOC

2.1 Tailor Weapons Loadout	Specific Theater / Mission Loadout	C
2.2 System Checks	Engineering System Checks Combat Systems System Checks	H C
2.2.1 Ship Control	•IBS (Integrated Bridge System) SSP-FLEET-0013 •Sperry VMS VT (Voyage Management System)	
2.2.2 Navigation	AN/SSN-6 NAVSSI (Navigation Sensor System Interface) SSP-FLEET-0025	
2.2.2.1 Navigation Receivers	•WRN-6 GPS Receiver ?antenna requirement? •SRN-25 (Omega,Transit,GPS) ?antenna requirement?	
2.2.2.2 Inertial Navigation	•WSN-5 INS •Ring Laser Gyro Navigator (RLGN)	
2.2.3 Engineering Systems	•IC*Arsenal Ship* (Integrated Condition Assessment System) SSP-FLEET-0009 •SMCS (Standard Monitoring Control System) SSP-FLEET-0020	
2.2.4 Ship Integrity	DCS (Damage Control System) SSP-FLEET-0010	
2.2.5 Combat Systems	•Input to SMCS (Standard Monitoring&Control System) SSP-FLEET-0020 • Input to ICAS (Integrated Condition Assessment System) SSP-FLEET-0009	
2.3 Refuel (Top-Off)	100% fuel on departure (Enough for 90 days)	H
2.4 Plant Lightoff	Main Engines / Generators up and online	H
2.5 Shift Services	Shift from shore to ship services	H
2.6 Initialize/test PNAV	Energize / Initialize Precision Navigation (pcode-GPS) Inertial Navigation Equipment	O
2.7 SOC Detachment	Embark Special Operations Capable (SOC) Detachment and/or Equipment	O
2.8 Shift Communications	•Communication Equipment Checks •Shift comm guard from shore to ship	C O
2.9 Load Mission Stores	Load Mission Specific Stores	HOC
3.0 Departure Operations	Sea&Anchor Detail to Underway	HOC
3.1 Sea and Anchor Detail (Special Evolution)	Watch Station Positions (minimum manning) Bridge (?) Engineering (?) Deck (?) CIC/Comms (?)	HOC
3.1.1 Linehandlers / Tugs	Manning (4) Linehandlers per station to work lines / receive tugs (OPS Special Evolutions Doc)	O
3.1.2 Anchor Detail	Manning (6) (OPS Special Evolution Doc)	O
3.1.3 Navigation Detail	Manning (?) / Electronic Assisted Piloting	O
4.0 Transit to theater		HOC
4.1 Mission Briefing	Crew Briefings by Closed Circuit TV / Scenario Drills	O
4.2 Disposal of Waste at Sea	Federal / International Environmental Regulations Store/Compact/Incinerate/Discharge Overboard Method and Equipment	O H
4.2.1 Solid Waste	•For Dry Waste - Strachan&Henshaw Waste Procesing Machine (*Navy International - JAN/FEB 1993)* •*Large Waste Pulper - SSP-FLEET-0019*	
4.2.2 Plastics	Plastic Waste Processor SSP-FLEET-0015	
4.3 Discharges Over the Side	Federal / International Regulations Required Intake/Discharge	O H

5.0 Theater Operations	90 day employment	HOC
5.1 On Station Presence	Continuous Availability / Conventional Deterrence	O
5.2 Combat Operations	Employ Weapon System	C
5.2.1 Position Ship	Position for defense of ship by AEGIS Control Platform Position ship for firing arcs/blast/smoke control	H O C
5.2.2 Weapon Inventory Status to Control Platform	Weapon inventory and status sent to Control Platform MK41 VLS LCU to ATWCS	C
5.2.2.1 Missiles	Missile Inventory and Status Indications MK 41 VLS LCU (Launch Control Unit)	C
5.2.2.2 Extended Range Gun System	•Reserve Space/Weight/Power to accomodate future fit •VGAS (Vertical Gun for Advanced Ships) 155mm Manning (0) Maintenance (1-3) •Trade-off with 5inch 54 Manning (10) Maintenance (3+)	H C
5.2.3 Process Remote Missile Selection	Processing Equipment: ATWCS	C
5.2.4 Initialize Selected Missile	Specific Requirements for each missile type:	
5.2.4.1 Standard Missile	Missile Initialization Requirements: ATWCS (Advanced Tomahawk Weapon Control System) SM-2 Blk IIIA/B: SM-2 Blk IVA: SM-2 LEAP: STRIKE SM:	C
5.2.4.2 ATACMS	Missile Initilization Requirements: ATWCS ATACMS: ATACMS-BAT: ATACMS-ER:	C
5.2.4.3 SLAM	Missile Initialization Requirements: ATWCS SLAM:	C
5.2.4.4 TLAM	Missile Initialization requirements: ATWCS TLAM: TLAM-BAT: TLAM-C/D: TLAM BLKIV:	C
5.2.5 Receive Remote Launch Order	Receive Method: Primary-Data Link / Secondary Receive Equipment: CEC / LINK 16	C
5.2.6 Process Remote Launch Order	Processing Equipment: (ATWCS) Advanced Tomahawk Weapons Control System	C
5.2.7 Launch Missile / Fire Gun	•Weapons Release Authority from CWC (analogy on/off safety switch)	C
5.2.7.1 Missile Misfire	Automatic Reselect Alternative Cell	C
5.2.8 "Missile Away" Message to Control Platform	Method: Primary-Data Link / Secondary /Tertiary Equipment: CEC / LINK 16/ Link 22	C

6.0 Return Transit		HOC
6.1 Ballast/Deballast	As required for stable transit	H
6.2 Post Mission Briefs	Data Disks/Voice Recording/Computer Printout	O
7.0 Prearrival Operations		HOC
7.1 Sea Detail		HOC
7.1.1 Linehandlers	Embark Linehandlers from tugs	O
7.1.2 Anchor Detail	Automatic Keel Anchor	O
8.0 Mooring Operations		HOC
8.1 Tugs		O
9.0 Post-Arrival Operations	Immediately Upon Arrival	HOC
9.1 Default Missile Loadout	Reload VLS with default loadout	C
9.2 Shift Services	Shift from Ship to Shore Connections	H
9.3 Refuel	Refuel Ship (DFM - F76)	H
9.4 Secure Nonessential Systems	Shut down non-essential equipment Computer Equipment to Monitor Status	HC
9.5 Crew Change	Crew Debrief and Changeout	O
9.6 Shift Comms	Ship to Shore	OC
9.7 Load Stores	Load pre-ordered stores - Use of ship storeage elevators and loadout conveyors	HC
10.0 Sustain the Crew		O
10.1 Messing	Crew and SOC Detachment	HO
10.1.1 Plan Meals		O
10.1.2 Prepare the Meals	Easy Prep Galley Operations SSP-FLEET-0024	HO
10.1.3 Serve the Meals	Self Help	O
10.1.4 Cleanup the Meal	Self Help	HO
10.1.5 Disposal of Food Waste	Large Waste Pulper SSP-FLEET-0024	HO
10.2 Provide Hotel Services		HO
10.2.1 Provide Hot Water Service		H
10.2.2 Provide Cold Water Service		H
10.2.3 Provide HVAC service		H
13.2.4 Provide Potable Water	Reverse Osmosis	
10.3 Berthing	Crew and SOC Detachment	HO
10.3.1 Provide Sanitary Head Facilities		H
10.3.2 Provide Sanitary Washroom Facilities		H
10.3.3 Provide Sanitary Linen Service		H

	•Consider Mast/Antenna Requirement	OC
11.1 Inport	Shore Guard	OC
11.2 Interior Communications	•Digital Pipes/Bells •Phone System •Computer LAN	
11.3 Underway Exterior Communications	•Certain Comms guarded by Control Ship •SACCS (Ship Automated Communications Control System)	OC
11.3.1 LF	Ship-Shore	
11.3.2 HF	•(HFRG) AN/URC-131A(V) High Frequency Radio Group SSP-FLEET-0034 (Ship-Ship / Ship-Submarine / Ship-Aircraft / Ship-Shore)	
11.3.3 UHF	•FLTSATCOM (Ship-Satellite) •Ship-Ship / Ship-Aircraft	
11.3.4 VHF	•Ship-Ship	

11.3.5 EHF millimeter wave	Milstar (Ship-Satellite)	
11.3.6 FLEET BROADCAST	(HSFB) AN/USQ-122A(V) High Speed Fleet Broadcast	
11.3.7 Data Link	OTH Satellite Link Tactical Data Link / Weapons Control Link	C
11.3.7.1 CEC (Cooperative Engagement Capability)	Primary	C
11.3.7.2 LINK 16	Secondary	C
11.3.7.3 LINK 22	Tertiary	
11.3.8 IFF	Transponder	C
11.3.9 TACAN	Use Control Ship	C
12.0 Maintainence	•Reliability Centered Maintenance Based Review of Planned Maintenance Requirements SSP-FLEET-0002 •Automated Tag Out System (TAGLINK) SSP-FLEET-0032	HOC
12.1 Inport		HOC
12.1.1 Equipment Monitoring and Evaluation		HOC
12.1.1.1 System Condition Monitoring		HOC
12.1.1.2 Logkeeping/Documentation		HOC
12.1.1.3 Trend Analysis		HOC
12.1.1.4 Diagnostics and Decision Making		HOC
12.1.2 System Repair		HOC
12.1.2.1 Troubleshooting	Video Teleconferencing / Remote System Analysis	HOC
12.1.2.1.1 Access to technical manuals		HOC
12.1.2.1.2 Access to off ship experts	Video Teleconferencing	HOC
12.1.2.2 Organizational level repair		HOC
12.1.2.3 Intermediate level repair		HOC
12.1.2.4 Depot level repair		HOC
12.1.3 Preventive Maintenance System and Documentation	SNAP III SSP-FLEET-0021	HOC
12.1.4 Integrated Logistic Support	SNAP III SSP-FLEET-0021	HOC
12.1.4.1 Maintenance History Reporting	SNAP III SSP-FLEET-0021	HOC
12.1.4.2 Parts Procurement	SNAP III SSP-FLEET-0021	HOC
12.1.4.2.1 Identification of Needed Parts	SNAP III SSP-FLEET-0021	HOC
12.1.4.2.2 Documentation/Failed Parts Reporting	SNAP III SSP-FLEET-0021	HOC
12.1.5 Preservation and Cleaning	Wet/Dry Vacuums	HOC
12.1.5.1 Painting	Minimize	HOC
12.1.5.1.1 Paint Locker	Paint Stick SSP-FLEET-0039	HOC
12.1.5.1.2 Paint Clean-up	Paint Stick SSP-FLEET-0039	HOC
12.1.5.1.3 Interior Spaces		HOC

12.1.5.2 Cleaning	•Easy Clean Surfaces •Wet/Dry Vacuums	HOC
12.1.5.2.1 Interior Spaces		HOC
12.1.5.2.2 Ventilation System		HOC
12.2 Maintain the Ship Underway		HOC
12.2.1 Equipment Monitoring /Evaluation	(SMCS) Standard Monitoring & Control System SSP-FLEET-0020	HOC
12.2.1.1 System Condition Monitoring	(ICAS) Integrated Condition Assessment System SSP-FLEET-0009	HOC
12.2.1.2 Logkeeping/Documentation		HOC
12.2.1.3 Trend Analysis		HOC
12.2.1.4 Diagnostics and Decision Making		HOC
12.2 Underway		HOC
12.2.1 Online System Checks	•(SMCS) Standard Monitoring & Control System SSP-FLEET-0020 •(ICAS) Integrated Condition Assessment System SSP-FLEET-0009 •Video Teleconferencing / Remote System Analysis	HOC
13.0 Survivability	Balanced Design	HOC
13.1 Damage Control	•Maximum automation •(DCS) Damage Control System SSP-FLEET-0010	HO
13.1.1 Firefighting	•Remote monitor/detection/response •Firefighting Equipment (P-100a&SCBA) SSP-FLEET-0014 •EZ Pup: Firehose Nozzle Handling Device SSP-FLEEY-0036	HO
13.1.2 Flooding	Remote monitor/detection/response	HO
13.2 CBR	Citadel	HO
13.3 Passive Defense Systems		H
13.3.1 Mass Tonnage		H
13.3.2 Multiple Hull / Box Girders		H
13.3.3 Signature Reduction		H
13.3.3.1 Low Radar Cross Section		H
13.3.3.2 Acoustic Quietening		H
13.4 Active Defense Systems		
13.4.1 Gun/Missile System		C
13.4.2 Chaff/Flares		C
13.4.3 Electronic Countermeasures		C
14.0 Training	Inport vs Underway	O
14.1 Man Overboard Drill	Ship, Swimmer, Ship Boat Recovery (?) Control Ship/Helo Recovery (?)	O
14.2 Abandon Ship Drill	How to (Liferafts, Ship boats, Helo (?))	O
14.3 Imbedded Training	Built into systems (Scenarios, Failures, Operation)	HC
14.3.1 Operator	Console/System Operator training	HC

19.0 ADMIN		
19.1 Admin/Record Keeping	Automated Division Officer's Notebook: COMPASS for Windows SSP-FLEET-0022	

Arsenal Ship
Top Level Functional Flow Diagram

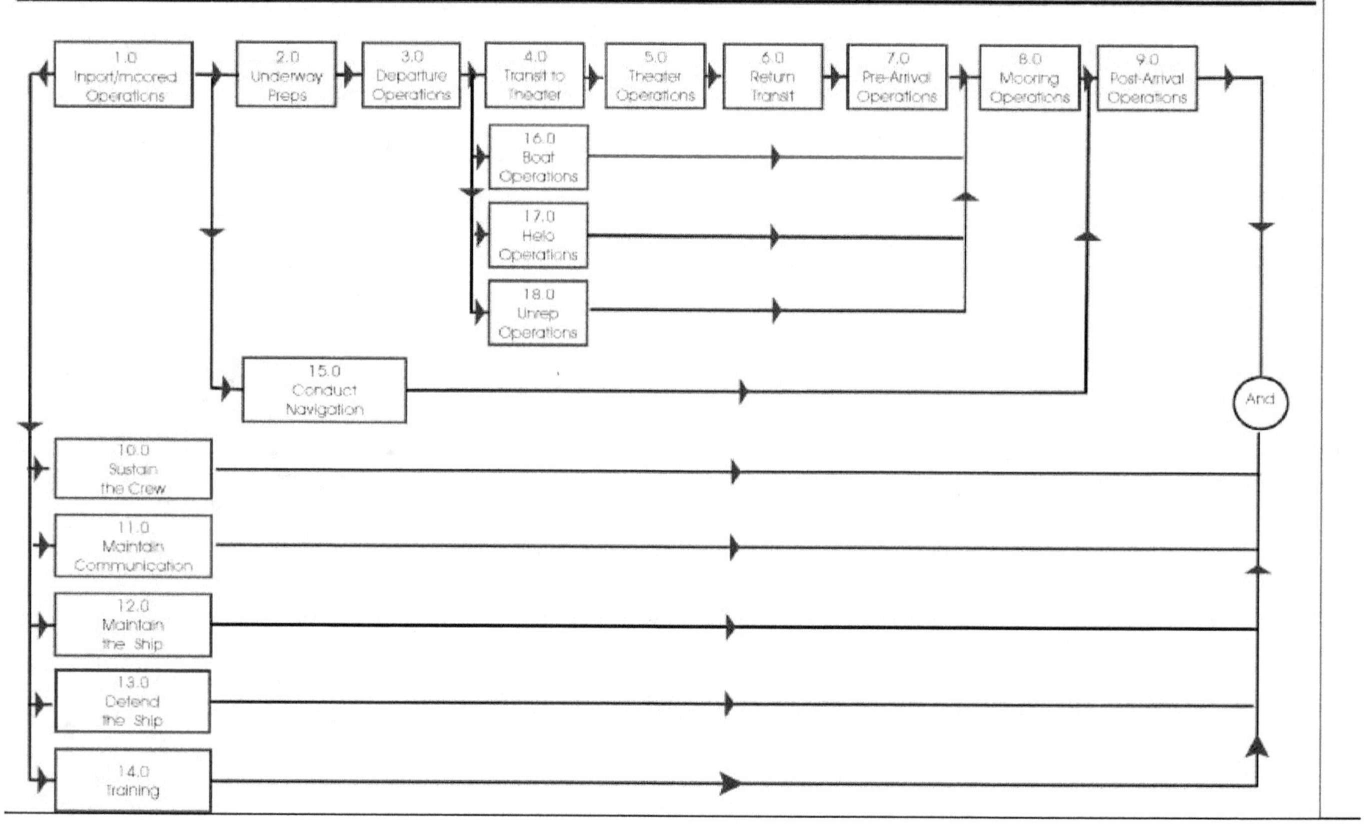

Arsenal Ship
Inport/Moored Operations (Ref. 1.0)

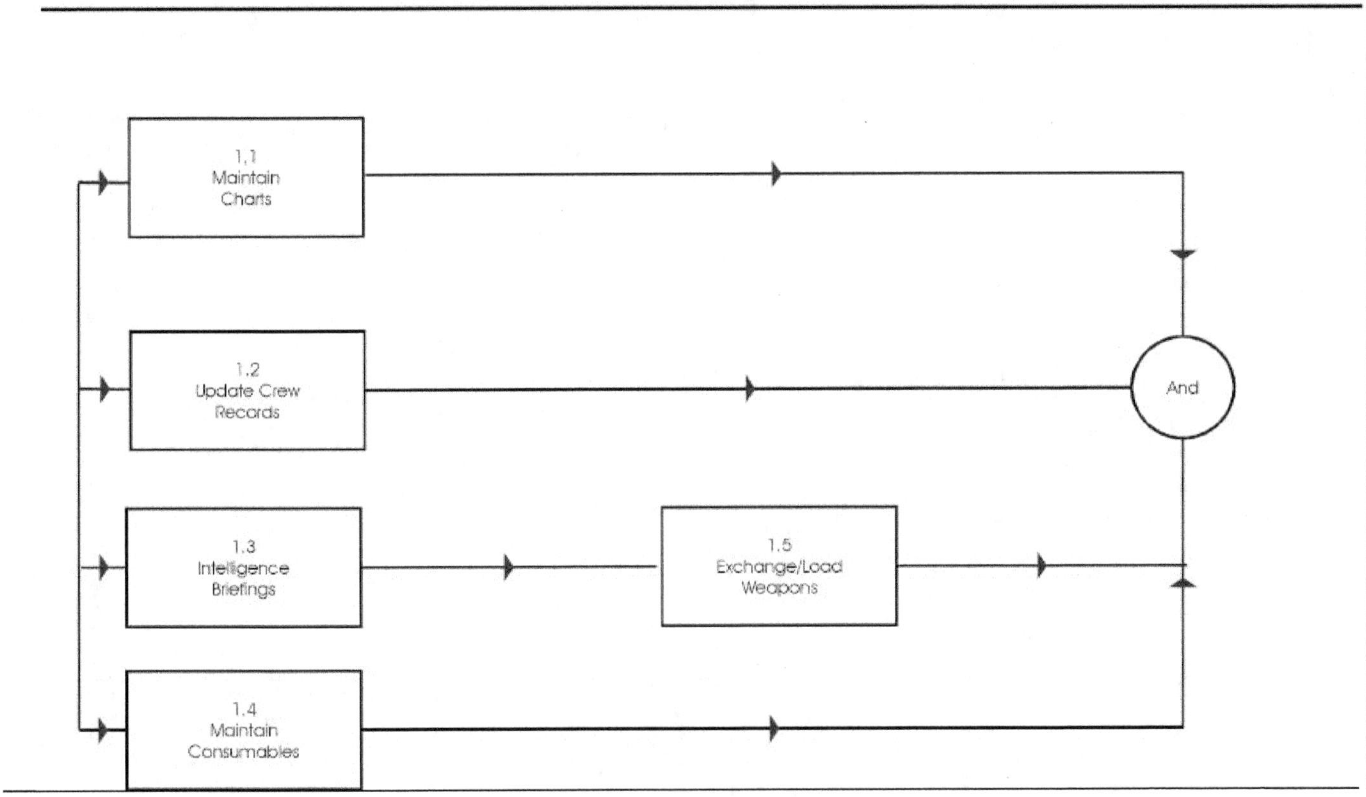

Arsenal Ship
Underway Preparations (Ref 2.0)

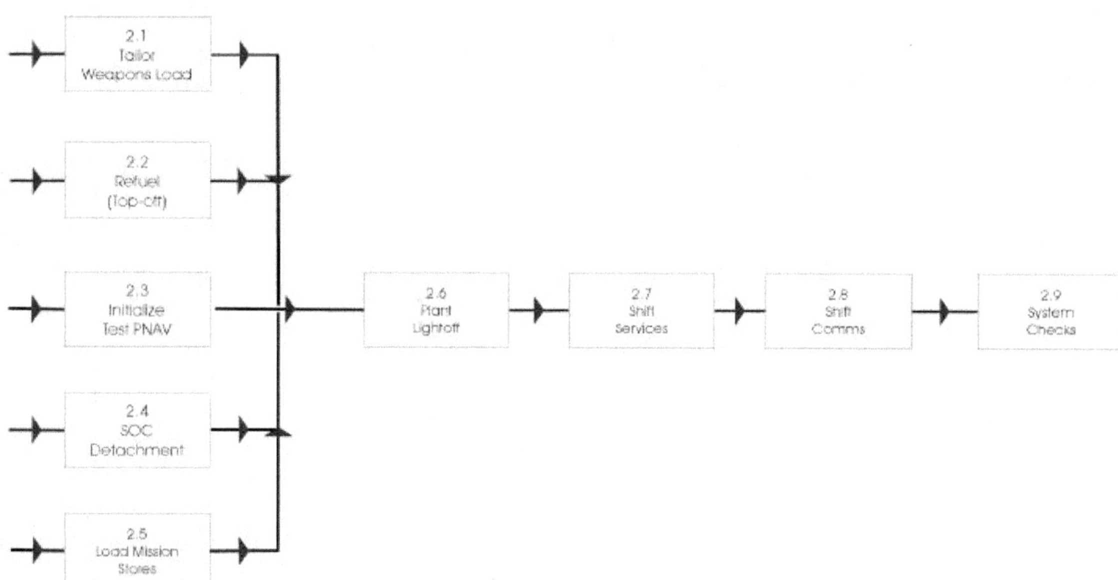

Arsenal Ship
Departure Operations (Ref 3.0)

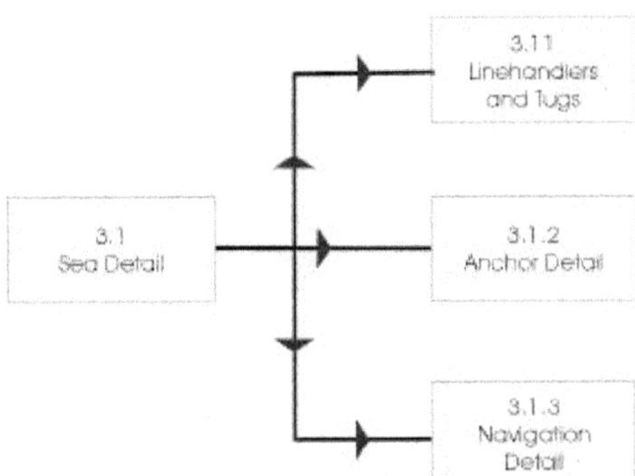

Arsenal Ship
Transit to Theater (Ref 4.0)

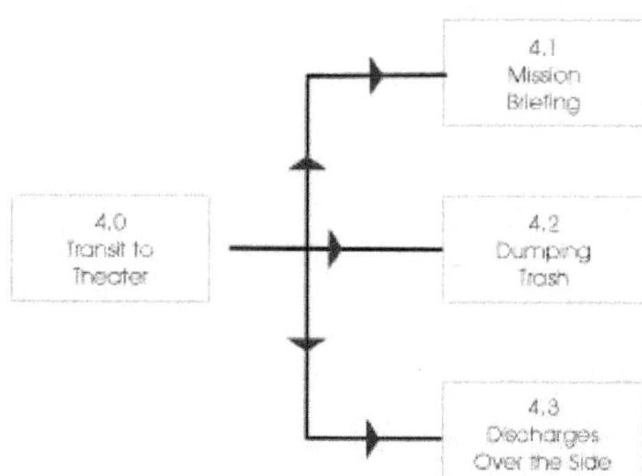

Arsenal Ship
Theater Operations (Ref 5.0)

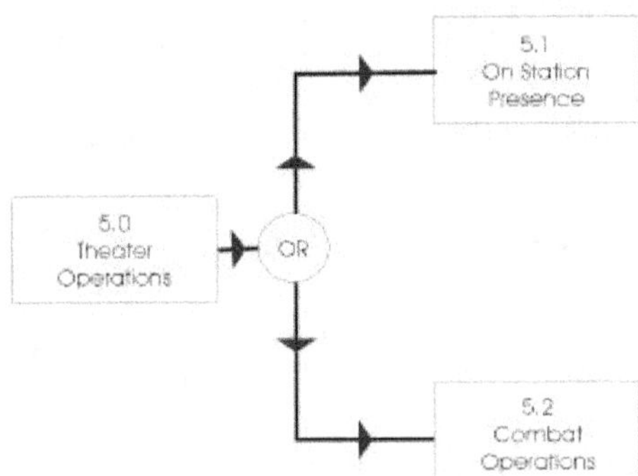

Arsenal Ship
Return Transit (Ref 6.0)

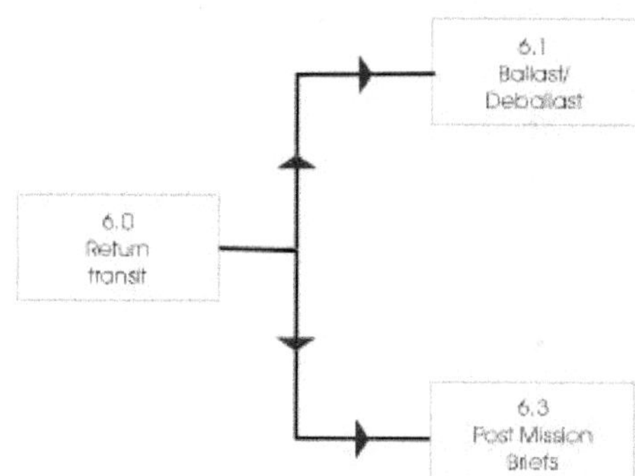

Arsenal Ship
PreArrival Operations (Ref 7.0)

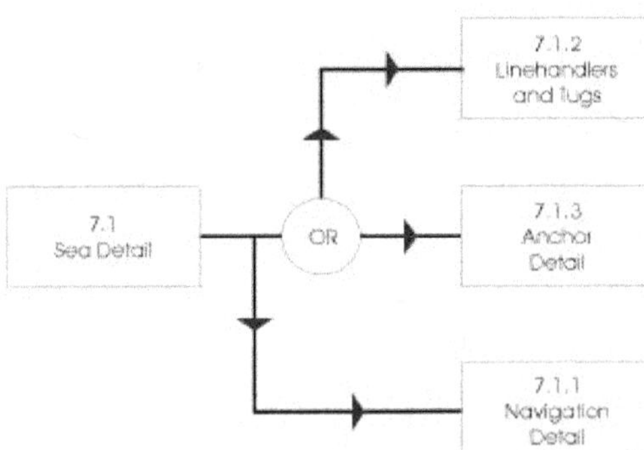

Arsenal Ship
Mooring Operations (Ref 8.0)

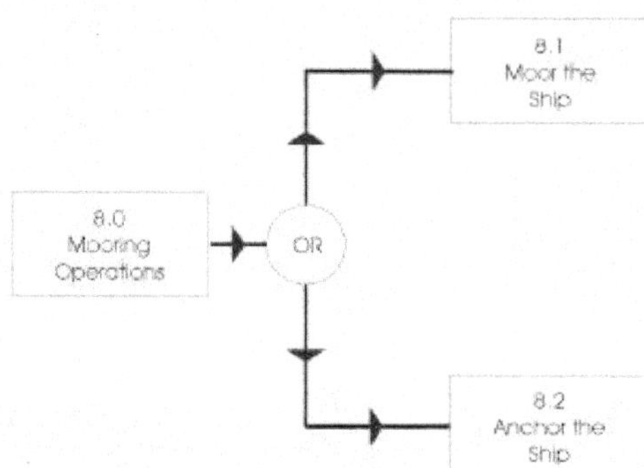

Arsenal Ship
Post Arrival Operations (Ref 9.0)

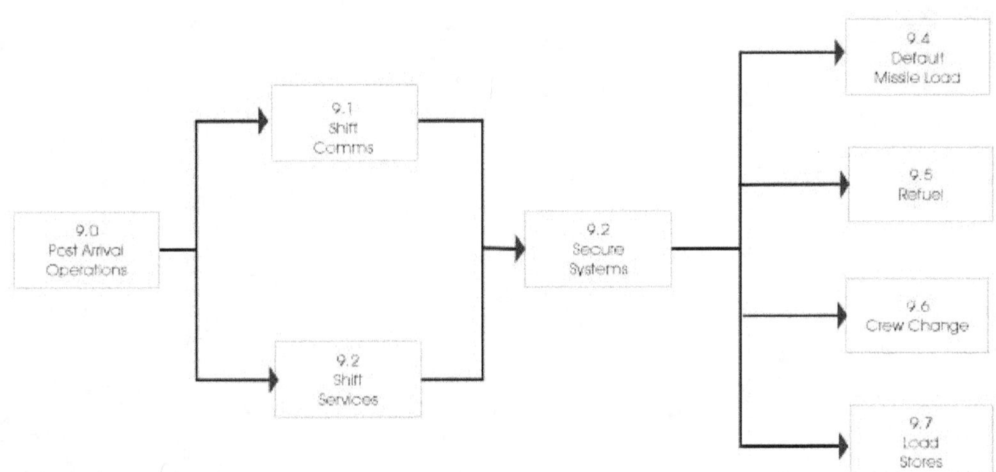

Arsenal Ship
Sustain the Crew (Ref 10.0)

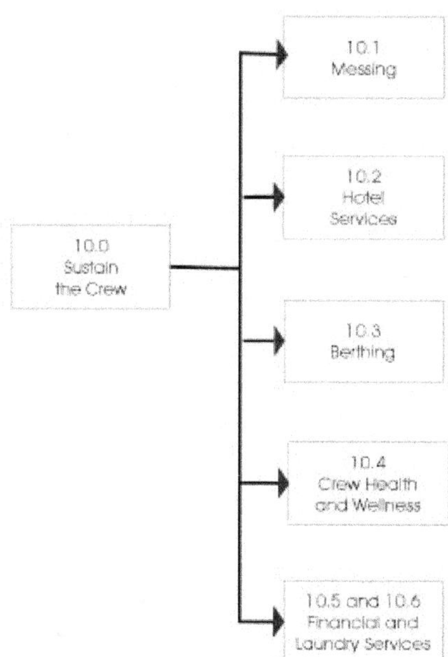

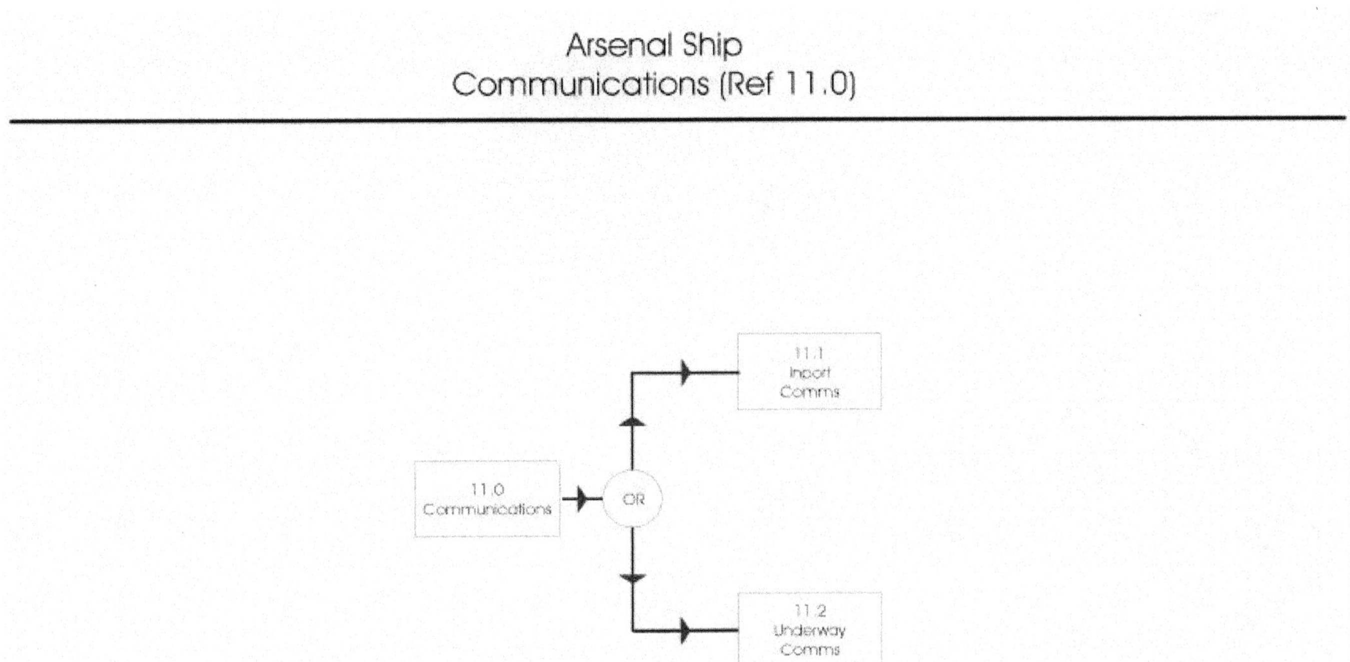
Arsenal Ship
Communications (Ref 11.0)
11.0
Communications
OR
11.1
Inport
Comms
11.2
Underway
Comms

Arsenal Ship
Maintenance (Ref 12.0)

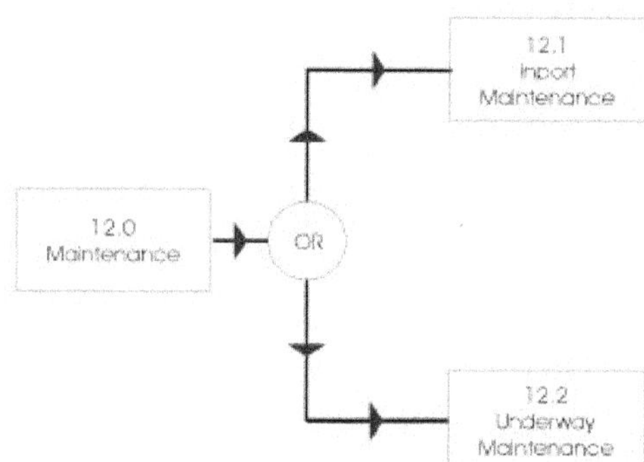

Arsenal Ship
Survivability (Ref 13.0)

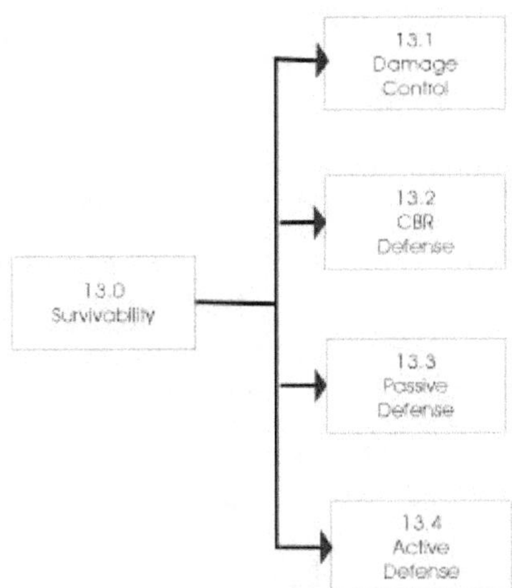

Arsenal Ship
Training (Ref 14.0)

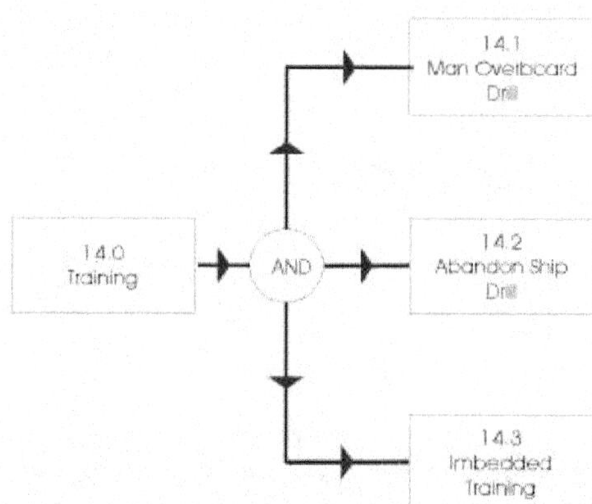

Arsenal Ship
Navigation (Ref 15.0)

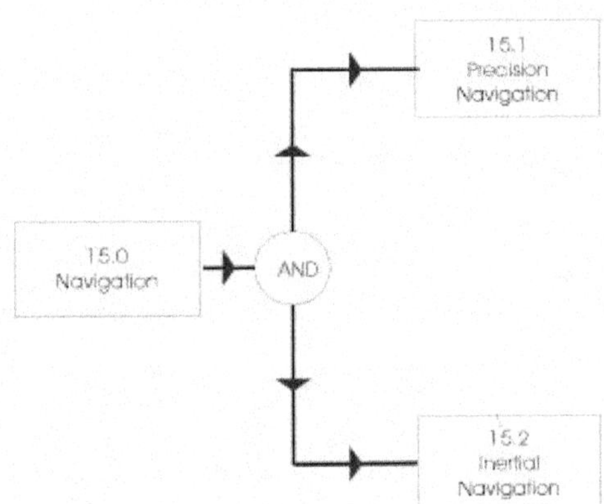

Arsenal Ship
Helo Operation (Ref 17.0)

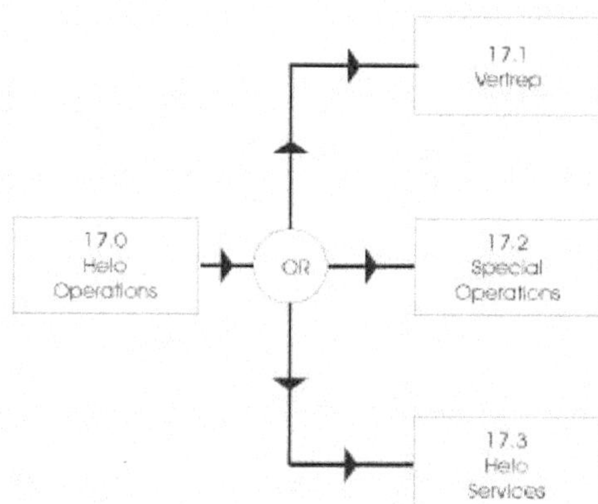

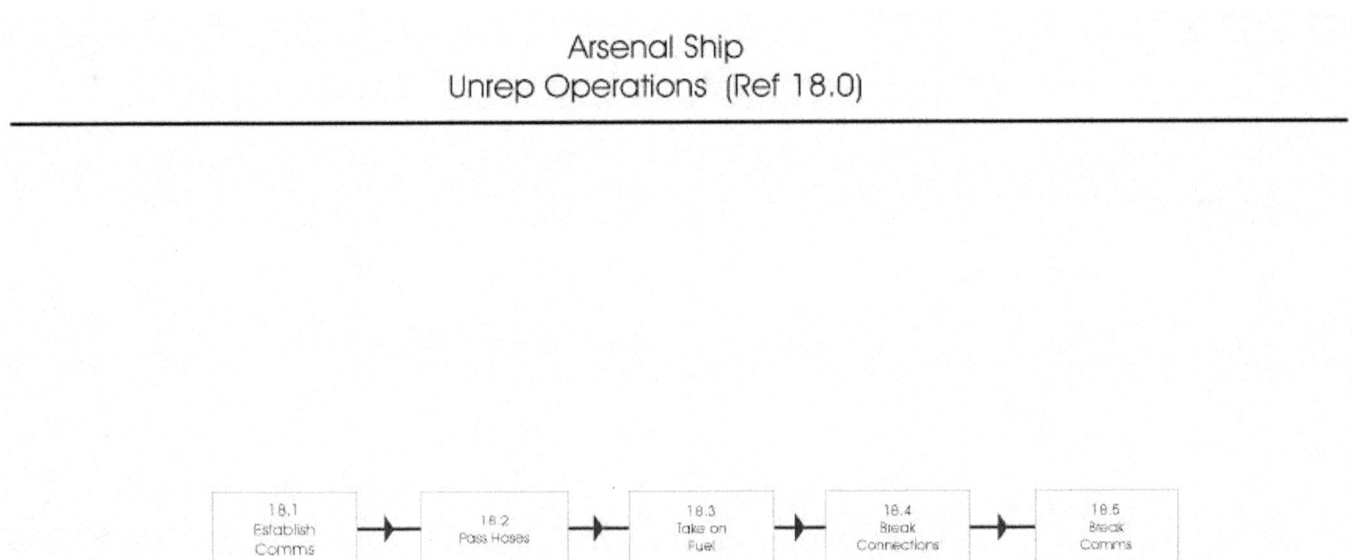
Arsenal Ship
Unrep Operations (Ref 18.0)
18.1
Establish
Comms
18.2
Pass Hoses
18.3
Take on
Fuel
18.4
Break
Connections
18.5
Break
Comms

APPENDIX B

Asset Naval Architecture Data

This Appendix has been deleted because of its length

APPENDIX C

Arsenal Ship Cost Analysis

SWBS	Description	Weight	Other	Material CER	Material Cost	Labor CER	Labor Cost
110	SHELL + SUPPORTS	4,177.0		945	3,945,347	269	1,121,946
120	HULL STRUCTURAL BULKHDS	3,679.0		945	3,474,966	269	988,183
130	HULL DECKS	2,470.0		945	2,333,016	269	663,444
140	HULL PLATFORMS/FLATS	1,169.0		945	1,104,168	269	313,995
150	DECK HOUSE STRUCTURE	181.2		822	149,019	588	106,546
160	SPECIAL STRUCTURES	315.0		13,052	4,111,380	213	67,095
164	BALLISTIC PLATING	3,909.0		1,200	4,690,800	100	390,900
170	MASTS+KINGPOSTS+SERV PLATFORM	4.0		4,946	19,782	139	556
180	FOUNDATIONS	251.0		822	206,422	305	76,555
190	SPECIAL PURPOSE SYSTEMS	566.0		3,806	2,154,422	343	194,138
210	ENERGY GEN SYS (NUCLEAR)						
220	ENERGY GENERATING SYSTEM (NONNUC)						
230	PROPULSION UNITS	731.0	57,750	115	6,612,375	178	130,118
240	TRANSMISSION+PROPULSOR SYSTEMS	414.0	57,750	50	2,887,500	138	72,105
243	SHAFTING	108.5	317	16002	5,069,749		
250	SUPPORT SYSTEMS	96.0	32,500		0	350	33,600
260	PROPUL SUP SYS- FUEL, LUBE OIL	116.0	32,500	29,533	3,425,847	1,200	139,200
290	SPECIAL PURPOSE SYSTEMS	55.0	32,500		0		0
310	ELECTRIC POWER GENERATION	170.0	7,500	650	4,875,000	3	21,733
320	POWER DISTRIBUTION SYS	372.0	3,000,000	73,747	10,636,094	1,100	409,200
330	LIGHTING SYSTEM	59.0		5,450	321,550	1,130	66,670
340	POWER GENERATION SUPPORT SYS	43.0	3,000,000	10,909	1,573,290	1,600	68,800
390	SPECIAL PURPOSE SYS	17.0		788	13,396	400	6,800
410	COMMAND+CONTROL SYS	10.0			0	200	2,000
420	NAVIGATION SYS	6.8			0	200	1,360
430	INTERIOR COMMUNICATIONS	17.0			0	200	3,400
440	EXTERIOR COMMUNICATIONS	13.0			0	200	2,600

450	SURF SURV SYS (RADAR)	1.2		0	200	240
460	UNDERWATER SURVEILLANCE SYSTEMS			0	200	0
470	COUNTERMEASURES	70.0		0	200	14,000
480	FIRE CONTROL SYS	0.0		0	200	0
490	SPECIAL PURPOSE SYS	4.6		0	200	920
510	CLIMATE CONTROL	189.0	26,294	4,969,566	420	79,380
520	SEA WATER SYSTEMS	257.0	40,564	10,424,948	577	148,289
530	FRESH WATER SYSTEMS	129.0	27,226	3,512,154	450	58,050
540	FUELS/LUBRICANTS,HANDLING+STORAGE	200.0	33,700	6,740,000	230	46,000
550	AIR,GAS+MISC FLUID SYSTEM	415.0	56,212	23,327,980	550	228,250
560	SHIP CNTL SYS	138.0	11,220	1,548,360	300	41,400
570	UNDERWAY REPLENISHMENT SYSTEMS	0.0	6,428	0	150	0
580	MECHANICAL HANDLING SYSTEMS	266.0	13,482	3,586,212	220	58,520
590	SPECIAL PURPOSE SYSTEMS	160.0	1,510	241,600	240	38,400
610	SHIP FITTINGS	24.4	55,033	1,342,805	750	18,300
620	HULL COMPARTMENTATION	455.7	11,160	5,085,612	630	287,091
630	PRESERVATIVES+COVERINGS	356.0	10,789	3,840,884	420	149,520
640	LIVING SPACES	36.4	29,677	1,080,243	1,050	38,220
650	SERVICE SPACES	16.7	26,174	437,106	115	1,921
660	WORKING SPACES	38.9	27,376	1,064,926	248	9,647
670	STOWAGE SPACES	22.0	86,901	1,911,822	10	220
690	SPECIAL PURPOSE SYSTEMS	0.4	35,511	14,204	590	236
710	GUNS+AMMUNITION				200	0
720	MISSLES+ROCKETS	906.0			200	181,200
730	MINES				200	0
740	DEPTH CHARGES				200	0
750	TORPEDOES				200	0
760	SMALL ARMS+PYROTECHNICS	0.8			200	160
770	CARGO MUNITIONS				200	0
780	AIRCRAFT RELATED WEAPONS				200	0
790	SPECIAL PURPOSE SYSTEMS	7.1			200	1,420
F10	SHIPS FORCE	5.5				
F20	MISSION RELATED EXPENDABLES+SYS	1,760.0				

F30	STORES	22.0
F40	LIQUIDS, PETROLEUM BASED	12,298.0
F50	LIQUIDS, NON-PETRO BASED	134.0

Total 1991 Material Cost:		126,732,545
Total 1996 Material Cost:		146,917,754
1996 Cost of 512 VLS Cells:		0
1996 Cost of Command & Control Systems:		0
Total Reference Labor Hours:		11,672,566
Learning Curve Exponent:	95.00%	
Total 1996 Lead Ship Labor Hours:		12,933,591
Total 1996 Second Ship Labor Hours:		12,286,911
Total 1996 Third Ship Labor Hours:		11,923,723
Total 1996 Fourth Ship Labor Hours:		11,672,566
Total 1996 Fifth Ship Labor Hours:		11,481,402
Total 1996 Sixth Ship Labor Hours:		11,327,536
1996 Labor Cost per Hour:	25.0	
Total 1996 Lead Ship Labor Cost:		323,339,768
Total 1996 Second Ship Labor Cost:		307,172,780
Total 1996 Third Ship Labor Cost:		298,093,065
Total 1996 Fourth Ship Labor Cost:		291,814,141
Total 1996 Fifth Ship Labor Cost:		287,035,053
Total 1996 Sixth Ship Labor Cost:		283,188,412
Non-recurring Engineering Hours:		650,000
Engineering Burdened Rate:		45.0
Non-recurring Engineering Cost:		29,250,000

```
'Ship Assembly and Support Labor

SASLabor = 0.478 * Labor

'Integration and Engineering Labor

IELabor = 0.186 * Labor

'Program Management Labor

ProgManLabor = 0.194 * Labor
```

Shipyard General & Administrative Overhead:	6.50%	
Shipyard Insurance:	1.00%	
Shipyard Contingency:	10.00%	
Shipyard Profit:	4.00%	
Total Fee + Overhead Rate:	22.98%	
Navy Program Cost Factor:	1.00%	
Total 1996 Acquisition Costs		
Non-recurring Engineering Cost:		29,542,500
Lead Ship:		584,105,928
Second Ship:		564,024,944
Third Ship:		552,747,049
Fourth Ship:		544,948,010
Fifth Ship:		539,011,914
Sixth Ship:		534,234,009

T-AO Cost Analysis

SWBS	Description	Weight	Other	Material CER	Material Cost	Labor CER	Labor Cost
110	SHELL + SUPPORTS	3,759.0		945	3,550,529	269	1,009,671
120	HULL STRUCTURAL BULKHDS	1,788.0		945	1,688,839	269	480,259
130	HULL DECKS	1,789.0		945	1,689,783	269	480,527
140	HULL PLATFORMS/FLATS	372.0		945	351,369	269	99,920
150	DECK HOUSE STRUCTURE	833.0		822	685,059	588	489,804
160	SPECIAL STRUCTURES	304.5		13,052	3,974,334	213	64,859
164	BALLISTIC PLATING	0.0		1,200	0	100	0
170	MASTS+KINGPOSTS+SERV PLATFORM	213.0		4,946	1,053,413	139	29,607
180	FOUNDATIONS	217.0		822	178,461	305	66,185
190	SPECIAL PURPOSE SYSTEMS	431.0		3,806	1,640,558	343	147,833
210	ENERGY GEN SYS (NUCLEAR)						
220	ENERGY GENERATING SYSTEM (NONNUC)						
230	PROPULSION UNITS	443.0	57,750	115	6,612,375	178	78,854
240	TRANSMISSION+PROPULSOR SYSTEMS	235.0	57,750	50	2,887,500	138	44,022
243	SHAFTING	84.0	245	16002	3,924,967		
250	SUPPORT SYSTEMS	125.0	32,500		0	350	43,750
260	PROPUL SUP SYS- FUEL, LUBE OIL	86.0	32,500	29,533	2,539,852	1,200	103,200
290	SPECIAL PURPOSE SYSTEMS	40.0	32,500		0		0
310	ELECTRIC POWER GENERATION	137.0	7,500	650	4,875,000	3	21,733
320	POWER DISTRIBUTION SYS	248.0	3,000,000	73,747	10,636,094	1,100	272,800
330	LIGHTING SYSTEM	59.0		5,450	321,550	1,130	66,670
340	POWER GENERATION SUPPORT SYS	43.0	3,000,000	10,909	1,573,290	1,600	68,800
390	SPECIAL PURPOSE SYS	17.0		788	13,396	400	6,800
410	COMMAND+CONTROL SYS	0.3			0	200	60
420	NAVIGATION SYS	6.8			0	200	1,360
430	INTERIOR COMMUNICATIONS	17.0			0	200	3,400
440	EXTERIOR COMMUNICATIONS	13.0			0	200	2,600
450	SURF SURV SYS (RADAR)	1.2			0	200	240
460	UNDERWATER SURVEILLANCE SYSTEMS				0	200	0
470	COUNTERMEASURES	75.0			0	200	15,000
480	FIRE CONTROL SYS	1.6			0	200	320

490	SPECIAL PURPOSE SYS	4.6		0	200	920
510	CLIMATE CONTROL	227.0	26,294	5,968,738	420	95,340
520	SEA WATER SYSTEMS	257.0	40,564	10,424,948	577	148,289
530	FRESH WATER SYSTEMS	132.0	27,226	3,593,832	450	59,400
540	FUELS/LUBRICANTS,HANDLING+STORAGE	386.0	33,700	13,008,200	230	88,780
550	AIR,GAS+MISC FLUID SYSTEM	255.0	56,212	14,334,060	550	140,250
560	SHIP CNTL SYS	139.0	11,220	1,559,580	300	41,700
570	UNDERWAY REPLENISHMENT SYSTEMS	398.0	6,428	2,558,344	150	59,700
580	MECHANICAL HANDLING SYSTEMS	266.0	13,482	3,586,212	220	58,520
590	SPECIAL PURPOSE SYSTEMS	160.0	1,510	241,600	240	38,400
610	SHIP FITTINGS	24.4	55,033	1,342,805	750	18,300
620	HULL COMPARTMENTATION	455.7	11,160	5,085,612	630	287,091
630	PRESERVATIVES+COVERINGS	356.0	10,789	3,840,884	420	149,520
640	LIVING SPACES	71.0	29,677	2,107,067	1,050	74,550
650	SERVICE SPACES	29.0	26,174	759,046	115	3,335
660	WORKING SPACES	43.0	27,376	1,177,168	248	10,664
670	STOWAGE SPACES	120.0	86,901	10,428,120	10	1,200
690	SPECIAL PURPOSE SYSTEMS	0.4	35,511	14,204	590	236
710	GUNS+AMMUNITION				200	0
720	MISSLES+ROCKETS	17.0			200	3,400
730	MINES				200	0
740	DEPTH CHARGES				200	0
750	TORPEDOES				200	0
760	SMALL ARMS+PYROTECHNICS	0.8			200	160
770	CARGO MUNITIONS				200	0
780	AIRCRAFT RELATED WEAPONS				200	0
790	SPECIAL PURPOSE SYSTEMS	2.1			200	420
F10	SHIPS FORCE	19.0				
F20	MISSION RELATED EXPENDABLES+SYS	7.2				
F30	STORES	61.6				
F40	LIQUIDS, PETROLEUM BASED	1,645.0				
F50	LIQUIDS, NON-PETRO BASED	68.0				

Total 1991 Material Cost:		128,226,790	
Total 1996 Material Cost:		163,653,487	
1996 Cost of 512 VLS Cells:		0	
1996 Cost of Tomahawk Weapon Control Sys:		0	
Total Reference Labor Hours:		9,064,157	'Ship Assembly and Support Labor
Learning Curve Exponent:	95.00%		SASLabor = 0.478 * Labor
Total 1996 Lead Ship Labor Hours:		10,043,387	
Total 1996 Second Ship Labor Hours:		9,541,218	'Integration and Engineering Labor
Total 1996 Third Ship Labor Hours:		9,259,189	IELabor = 0.186 * Labor
Total 1996 Fourth Ship Labor Hours:		9,064,157	
Total 1996 Fifth Ship Labor Hours:		8,915,712	'Program Management Labor
Total 1996 Sixth Ship Labor Hours:		8,796,230	ProgManLabor = 0.194 * Labor
1996 Labor Cost per Hour:	25.0		
Total 1996 Lead Ship Labor Cost:		251,084,681	
Total 1996 Second Ship Labor Cost:		238,530,447	
Total 1996 Third Ship Labor Cost:		231,479,730	
Total 1996 Fourth Ship Labor Cost:		226,603,925	
Total 1996 Fifth Ship Labor Cost:		222,892,795	
Total 1996 Sixth Ship Labor Cost:		219,905,743	
Non-recurring Engineering Hours:		2,000,000	
Engineering Burdened Rate:		45.0	
Non-recurring Engineering Cost:		90,000,000	
Shipyard General & Administrative Overhead:	6.50%		
Shipyard Insurance:	1.00%		
Shipyard Contingency:	10.00%		
Shipyard Profit:	4.00%		

Total Fee + Overhead Rate:	22.98%	
Navy Program Cost Factor:	10.00%	
Total 1996 Acquisition Costs		
Non-recurring Engineering Cost:		99,000,000
Lead Ship:		561,049,500
Second Ship:		544,066,383
Third Ship:		534,528,314
Fourth Ship:		527,932,422
Fifth Ship:		522,912,080
Sixth Ship:		518,871,256

BIBLIOGRAPHY

D-1. Rivera G. and Le T., "*Arsenal Ship* Tomahawk Design Options" NSWC Dahlgren, May, 1996.

D-2. Herrington, J. B. "Uniform Framework For IMU/GPS Integration Using Kalman Filtering," Master's Thesis, Department Of Aeronautics, Naval Post Graduate School, Monterey, CA, June, 1995.

D-3. Clynch, J., "A Global Positioning System (GPS) Users Seminar," Department of Oceanography, Naval Post Graduate School, Monterey, CA, March, 1996.

D-4. Barton, D. K., *MODERN RADAR SYSTEMS*, Artech House, Inc., 1988.

D-5. Gates, P. J., *SURFACE WARSHIPS Vol. III,* Brassey's Sea Power: Naval Weapons Systems and Technology Series, Volume 3, 1987

D-6. Gill, C., "AA4276 Design Project - A Guidance, Navigation, and Control System for the Boeing 747," Department Of Aeronautics, Naval Post Graduate School, Monterey, CA, December, 1995.

D-7. Harney, R. C., Class Notes for TS3003, U. S. Naval Post Graduate School, Monterey, CA, 1996

D-8. White, C. And Reiber, C. "Navy Technical Assessment of LTN-72 Inertial Navigation Software Program 72-9-20," Technical Report, Naval Air Test Center Patuxent River, Maryland, June 1985

A Proposed Littoral Dominant Battle Group Centered Around The Arsenal Ship

CCS 1997

Subject Area - Warfighting

EXECUTIVE SUMMARY

Title: A Proposed Littoral Dominant Battle Group Centered Around The Arsenal Ship

Author: Lieutenant Commander John P. Looney, United States Navy

Thesis: For the projected $500 million investment in a demonstrator Arsenal Ship, the Navy will likely get what it is asking for, a ship designed to help the aircraft carrier battle group make the transition from a blue-water dominant team to a littoral dominant team. Just as the Navy built the aircraft carrier battle group team and doctrine to dominate the blue-water for the past 50 plus years, the Navy now needs to build the team and the doctrine that can dominate the littoral battle space of the 21st century. The Arsenal Ship concept should center around having industry build a ship that could be the centerpiece of a littoral dominant battle group.

Background: The United States Navy has shifted its strategic focus from the blue-water to the littorals; therefore, it would make sense for the Navy to develop doctrine and force structure to dominate the littorals. The Navy is in the process of building a revolutionary new class of warship--Arsenal Ship. The Arsenal Ship's operational concept is focused on a ship that will enhance the fire power of existing aircraft carriers, land attack capable combatants and submarines. In this period of austere funding and downsizing, the Navy must look not only to new technologies but to new organizations and doctrine to effectively and efficiently meet its mandated missions. There are two areas where shortfalls exist to building a littoral dominant team: 1) the Navy does not have a staff that is organized and trained to plan and execute littoral dominance operations; and 2) the Navy has critical shortfalls in naval surface fire support assets.

Recommendations: The United States Navy should build a littoral dominant battle group centered around the Arsenal Ship: a typical aircraft carrier battle group with an Arsenal Ship in the place of the aircraft carrier, and an Amphibious Group (PHIBGRU) commander leading it instead of a Carrier Group (CARGRU) or Cruiser Destroyer Group (CRUDESGRU) commander. An Arsenal Ship Battle Group, led by an experienced littoral warfare expert, would be a viable forward presence and power projection entity; furthermore, it would be an effective building block for follow-on forces if a large-scale crisis were to develop.

Report Documentation Page	*Form Approved* *OMB No. 0704-0188*

1. REPORT DATE **1997**	2. REPORT TYPE
3. DATES COVERED	**00-00-1997 to 00-00-1997**
4. TITLE AND SUBTITLE	**A Proposed Littoral Dominant Battle Group Centered Around The Arsenal Ship**
5a. CONTRACT NUMBER	
5b. GRANT NUMBER	
5c. PROGRAM ELEMENT NUMBER	
6. AUTHOR(S)	
5d. PROJECT NUMBER	
5e. TASK NUMBER	
5f. WORK UNIT NUMBER	
7. PERFORMING ORGANIZATION NAME(S) AND ADDRESS(ES)	**Marine Corps War College,Marine Corps Combat Development Command,Quantico,VA,22134-5067**
8. PERFORMING ORGANIZATION REPORT NUMBER	
9. SPONSORING/MONITORING AGENCY NAME(S) AND ADDRESS(ES)	
10. SPONSOR/MONITOR'S ACRONYM(S)	
11. SPONSOR/MONITOR'S REPORT NUMBER(S)	
12. DISTRIBUTION/AVAILABILITY STATEMENT	**Approved for public release; distribution unlimited**
13. SUPPLEMENTARY NOTES	
14. ABSTRACT	
15. SUBJECT TERMS	

16. SECURITY CLASSIFICATION OF:			17. LIMITATION OF ABSTRACT	18. NUMBER OF PAGES	19a. NAME OF RESPONSIBLE PERSON
a. REPORT	b. ABSTRACT	c. THIS PAGE			
unclassified	**unclassified**	**unclassified**	**Same as Report (SAR)**	**46**	

Standard Form 298 (Rev. 8-98)
Prescribed by ANSI Std Z39-18

Table of Contents

A PROPOSED LITTORAL DOMINANT BATTLE GROUP CENTERED AROUND THE ARSENAL SHIP

THE ARSENAL SHIP'S IMPACT ON FUTURE NAVAL OPERATIONS

> We have entered a period of uncertainty where threats are indeterminate even as changes in technology accelerate. Rapid innovation--apparent in the impact of stealth and precision weaponry in the Gulf War--appears likely to continue. Yet the Armed Forces are not apt to receive anything close to the resources enjoyed during the Cold War. With less money and greater ambiguity on the nature of opponents and wars in the future, we must innovate.[1]

The United States Navy is investing scarce resources in the development of a revolutionary new class of capital ship--Arsenal Ship. The Arsenal Ship will capitalize on technological advances in ordnance, hull design and manufacturing capability, propulsion and damage control automation, and command and control systems. The Navy should capitalize on the Arsenal Ship by leveraging, through innovation, the opportunities that its myriad technological advances provide. Unfortunately, the Navy's operational concept for the Arsenal Ship states that it will be an affordable strike and naval fire support ship that will enhance the fire power of existing aircraft carriers, land attack capable combatants and submarines, in essence aircraft carrier battle groups.[2]

---

[1] Williamson Murray, "Innovation: Past and Future," *Joint Force Quarterly*, Summer 1996, p. 51.

[2] Department of the Navy, "Arsenal Ship Concept of Operations," 1996, p.1.

It is not a replacement for these or for land-based air. Instead, it is part of the whole-just as the Battleship was a part of the whole for nearly a century. Operating under the control and umbrella of regularly deployed Aegis combatants, arsenal ship will supply substantial firepower, early: giving unified Commanders-in-Chief (CinCs) the capability to halt or deter invasion and, if necessary, enable the build-up of coalition land-based air and ground forces to achieve favorable conflict resolution.[3]

During a recent hearing of a House subcommittee on military procurement, in response to a question concerning the "core need" for the Arsenal Ship, the Assistant Secretary of the Navy John Douglass said, "I don't think there is an incremental core need."[4] Additionally, the Deputy Chief of Naval Operations Vice Admiral Donald Pilling said essentially that the Arsenal Ship was "...a "land-attack platform," intended to influence events ashore from the sea. When asked whether the Arsenal Ship could attack a specific target that could not be struck by an existing weapon system, the admiral admitted there was not."[5] That prompted the observation that "...the Arsenal Ship seems like the ship of the future with everything. Everything except a mission."[6]

Thesis

For the projected $500 million investment in a demonstrator Arsenal Ship, the Navy will likely get what it is asking for, a ship designed to help the aircraft carrier battle group make the transition from a blue-water dominant team to a littoral dominant team. That operational concept is strikingly similar to the operational concept espoused by the battleship sailors of the 1930s who tried to keep battleships relevant as blue-water dominant ships by neatly fitting the aircraft carrier into a supporting role of providing aerial spotters for battleship battle lines.

---

3 Department of the Navy, "Arsenal Ship Concept of Operations," 1996, p.1.
4 C.P. Cavas, "Is Arsenal Ship's Mission Adrift?," *Navy Times*, 17 March 1997, p. 30.
5 Ibid., p. 30.
6 Ibid., p. 30.

The Navy has shifted its strategic focus from the blue-water to the littorals. Now, it should focus its intellectual power on being innovative because resources will continue to be scarce and force structure is likely to go through additional cuts. Just as the Navy built the aircraft carrier battle group team and doctrine to dominate the blue-water for the past 50 plus years, the Navy now needs to build the team and the doctrine that can dominate the littoral battle space of the 21st century. The Arsenal Ship concept should center around having industry build a ship that could be the centerpiece of a littoral dominant battle group.

Naval Power Projection Today

The Navy currently employs two types of forward deployed entities, the Aircraft Carrier Battle Group (CVBG) and the Amphibious Ready Group (ARG). The CVBG with its composite air wing (F-14, F/A-18, S-3, E-2C, EA-6, and SH-60), surface combatant escorts (cruisers, destroyers, and frigates), direct support attack submarines, and logistics ship is clearly a blue-water dominant battle group. The ARG consists of an amphibious assault ship (LHA, LHD, or LPH), a Dock Landing Ship (LSD), an amphibious transport dock (LPD), and a Special Operations Capable Marine Expeditionary Unit (MEU(SOC));
by itself, except in the most benign battlespace, the ARG is neither a blue-water dominant group nor a littoral dominant group. The only littoral dominant group is a combination of the ARG and elements of the CVBG.

There are two clear points in the Navy's strategic vision: (1) the Navy must retain its ability to dominate the blue-water regions of the world (i.e., keep sufficient Aircraft Carriers to accomplish that mission), and (2) the Navy must improve its ability to provide decisive crisis response in the littoral regions of the world. This thesis suggests an alternate and fiscally responsible way to improve decisive crisis response capability in the littoral battlespace. It will address the who and what of decisive operations in the littoral battlespace.

EMPLOYING A LITTORAL DOMINANT BATTLE GROUP

Background

The United States Naval Service--the Navy and Marine Corps--operates in the oceans and littoral regions of the world to "project power and influence of the nation across the seas to foreign waters and shores in both peace and war."[7] For most of the twentieth century, the United States Navy was mostly interested in blue-water operations. During the Cold War, the Navy's strategic focus was on achieving global maritime dominance. Due to the demise of the Soviet Union and with it the only global fleet seen as a potential adversary, the United States Navy's strategic focus shifted to littoral operations and power projection from the sea. That shift in strategic focus was articulated in the Department of the Navy's publications--"From the Sea" and "Forward...From the Sea."

Currently, the Navy contributes combat power to theater commanders in the form of Aircraft Carrier Battle Groups, Amphibious Ready Groups, and Independent Deployers (e.g., Maritime Patrol Aircraft Detachments and one or more surface combatants). Those forces are trained, organized, and equipped to conduct a wide range of tactical evolutions including peacetime engagements (e.g., port visits and exercises), maritime dominance (e.g., protection and denial of sea lines of communications), and power projection from the sea (e.g., amphibious demonstrations and forced entry operation). Because forward deployed Naval forces are available to respond quickly, require minimum support, and are less constrained by potential diplomatic restrictions that could be imposed on land-based forces by their all too necessary host

---

7 Department of the Navy, Forward...From the Sea, (Washington, D.C.: Department of the Navy), 1995, p. i.

nations, naval forces enjoy far greater employment options as compared to the Army and Air Force. This means that the naval services can respond to most crisis situations quicker than other services and with a wider range of tactical options.

The congressionally mandated roles of naval forces are maintaining maritime superiority, contributing to regional security, conducting operations from the sea, seizing or defending advanced naval bases, and conducting such land operations as may be essential to the prosecution of naval campaigns.[8] The management of those naval campaigns is the job of navy staffs. During the early days of a crisis situation the Navy has basically two staffs and their associated forces readily available to respond to the crisis: (1) the Aircraft Carrier Battle Group (CVBG) commanded by a Carrier Group Commander (CARGRU) or a Cruiser Destroyer Group Commander (CRUDESGRU) both Rear Admirals (O-7 or 8 grade officers), and (2) the Amphibious Ready Group (ARG) with its embarked Marine Expeditionary Unit (MEU) commanded by an Amphibious Squadron Commander (PHIBRON) and a MEU commander, both O-6 grade officers, respectively. If the crisis situation is too large in scope or has the potential for escalation, then those forward deployed staffs and assets become the "building blocks" for follow-on naval forces.

> Building on normally deployed forces, the Navy can mass, if the situation requires, multiple Aircraft Carrier Battle Groups into Carrier Battle Forces, Amphibious Ready Groups with embarked Marine Expeditionary Units, and as needed project those naval expeditionary forces ashore using the afloat Maritime Prepositioning Force. Such massing of naval units can be

---

8 Department of the Navy, Naval Doctrine Publication 1, *Naval Warfare*, (Washington, D.C.: Department of the Navy), 1994, pp. 15-16.

> complimented by the deployment of Army and Air Force units to provide a joint force capable of the full range of combat operations that may be required.[9]

Those follow-on naval forces of the littorally focused future will likely include a Marine Expeditionary Force Forward (MEF Forward), which is a Marine Air Ground Task Force (MAGTF) whose size is variable but somewhere between a MEU-sized force and an entire MEF. A MEF Forward will typically be commanded by a Marine Corps Brigadier General. It is safe to assume that if a MEF Forward sized force is dispatched overseas to respond to a crisis, then Maritime Preposition Force (MPF) ships will also be routed to that crisis area to provide sustainment for the MEF Forward. Therefore, per current doctrine, the Navy will send an Amphibious Group (PHIBGRU) commander (Rear Admiral) and staff to the area to coordinate the off-load and marrying-up of the MEF Forward and the MPF assets. Once in theater, that PHIBGRU commander and staff will most likely stay to command, as Amphibious Task Force Commander (CATF), any MEF Forward-sized amphibious operation, as was the case during the Gulf War. This is due in part to the limitations of a PHIBRON staff to effectively coordinate much more than an ARG sized Amphibious Task Force (ATF). However, there is one flaw with that practice which has to do with the organization and employment of the PHIBGRU staff.

Key Issue

A PHIBGRU staff is primarily organized for administrative functions (i.e., managing maintenance, basic training, and other non-operational functions such as personnel issues for the amphibious ships that are not on deployment or working-up for a deployment) rather than

---

[9] Department of the Navy, Forward...From the Sea, (Washington, D.C.: Department of the Navy), 1995, p. 5.

tactical operations. The PHIBGRUs, 2 in the Atlantic, 3 in the Pacific, and 1 in the Western Pacific, own all the amphibious ships and the nine PHIBRONs (four each assigned to PHIBGRU 2 and 3, and the remaining one assigned to PHIBGRU 1). Three ship ARGs are assigned to a deploying PHIBRON approximately six months prior to deployment. The PHIBRON oversees the operational employment and training of the ships throughout the work-up phase and the deployment while maintenance and administration are still the responsibility of the PHIBGRU staff.

If a PHIBGRU commander is called on to conduct tactical operations (normally large-scale operations), standard procedures call for the PHIBGRU commander to subsume one of his or her PHIBRON commanders as the chief of staff for the tactical operation while PHIBGRU's billeted chief of staff and other selected members stay behind to perform the PHIBGRU's administrative functions. In contrast, nothing like that happens in the CVBG structure: CARGRU, CRUDESGRU, and even Destroyer Squadron (DESRON--the equivalent of a PHIBRON) staffs retain all administrative and tactical duties at all times. The CARGRUs, CRUDESGRUs and DESRONs have, since August 1995, permanently assigned ships. Their staffs are organized and manned to enable them to manage simultaneously and over the life-cycle of the ships both the administrative and operational employment of their assigned ships

The subsumption of a PHIBRON commander also includes subsumption of the rest of the PHIBRON staff to help perform the tactical planning and execution functions of the PHIBGRU. The subsumption of a PHIBRON builds an ad hoc amphibious staff that has a great deal of individual operational experience but lacks operational experience and cohesiveness as a tactical staff. That potentially promotes unnecessary friction within the staff. It also promotes unnecessary external friction, as the ad hoc PHIBGRU staff inter-operates with other staffs and ships that have had little tactical experience working with and for the PHIBGRU staff. Those internal and external frictions are potentially magnified if an amphibious operation, often deemed the most complex type of military operation, is being employed.

The Navy, if truly committed to its littoral and maneuver warfare focus, must promote the understanding of the land battle from its inception well out to sea through its transition of command ashore. The littoral battlespace and maneuver warfare are far different from Alfred Thayer Mahan's massed fleet theory of seeking out and destroying the enemy's fleet to gain unimpeded use of the sea. Tomorrow's coastal defense threat (mobile, lethal, and survivable) will require the Navy to build and train forces to employ maneuver warfare from over the horizon. The Arsenal Ship, a strike and fire support platform, goes a long way towards building the required forces. What is needed are littoral warfare experts trained and experienced employing maneuver warfare.[10] They must be able to understand how the Navy can best maneuver and shape the enemy while optimizing force protection and sustainability of naval forces at sea and ashore. They must also be equipped with the best command, control, communications, computers, intelligence, surveillance, and reconnaissance (C4ISR) systems available in order to maximize their ability to operate on the increasingly fast-paced and widely dispersed battle field.

The Navy faces some challenges developing commanders and their staffs to be capable of planning and executing operational maneuver from the sea. Currently, the only navy staffs getting tactical experience in littoral warfare and ground operations in the overseas operating areas, where crises are likely to occur in the future, are the PHIBRONs: CARGRUs and

---

[10] LaPlante, J.B., VADM, USN, "The Future of Amphibious Warfare Takes Shape." *USNI Proceedings*, May 1994, p. 88.

CRUDESGRUs have little to do with ARG operations overseas, and PHIBGRUs do not deploy nor do large-scale amphibious exercises occur frequently enough, due to their high costs, to provide them with high quality at-sea training. To borrow a phrase from the Marine Corps, the MEF or MEF Forward commander is the "warfighter." Who is the Navy's littoral "warfighter?" The answer is not the CVBG staff due to experience and knowledge limitations in amphibious and ground combat operations. The answer is not the PHIBRON staff due to their size limitation and lack of strike warfare experience to coordinate the Navy's shaping assets. More than likely, the Navy will turn to the PHIBGRU commander who suffers from staff limitation foremost in readiness because of the staff's low operating tempo and manning that is not focused on tactical littoral operations. The PHIBGRU commander also suffers in the area of C4ISR capabilities. It is not very often that a PHIBGRU commander conducts amphibious warfare exercises from the decks of a Amphibious Command Ship (LCC), a ship designed specifically to serve as the command ship for the Commander, Amphibious Task Force (CATF) and the Commander, Landing Force (CLF). LCCs have become the home of the numbered fleet commanders. LHAs and LHDs have good C4ISR capabilities but they have not received the same priority for upgrades as the navy's premier capital ship, the aircraft carrier.

Proposed Littoral Dominant Staff

To rectify the problems that plague the Navy's ability to develop littoral warfare dominance, I propose the creation of deploying PHIBGRU staffs, two per coast, organized more along the lines of the CARGRU and CRUDESGRU staffs. Furthermore, the PHIBGRUs must redistribute custody of amphibious ships and have PHIBRONs assume the permanent ownership of their ARGs, a concept recently proven in the DESRON community. That will require some minor changes to the PHIBRON's organization and manning, and a reduction in the number of PHIBRONs from nine down to eight in order to keep one staff for each deploying ARG.

The reorganization of the PHIBGRU staff bears some additional comments. The flag support (N0x personnel that work for the Chief of Staff), administration (N1), and intelligence (N2) departments are adequately structured and do not need to be changed. Specifically, the PHIBGRU's intelligence department works out of the amphibious flagship's intelligence center, therefore, the intelligence center becomes the focal point for all-source intelligence fusion through the combined efforts of the flagship, PHIBGRU, and MEU(SOC) intelligence personnel.

The operations department should have the following divisions to support execution of maneuver warfare from the sea: surface, ground, air, and submarine. Heading up the operations department should be a navy captain or a marine colonel. The surface operations division should be led by a post-commander command navy surface warfare officer with strong amphibious warfare experience. The surface operations division should be staffed with an appropriate mix of supporting personnel, such as operations, combat systems, or deck experienced officers. The air operations division should be led by a post-commander command aviator with strong air assault experience. The air operations division should be staffed with appropriate supporting personnel, such as strike warfare specialists in tomahawk and attack aircraft communities, and assault support aviators. The ground operations division should be led by a field grade ground combat officer. The ground operations division should be staffed with an appropriate mix (infantry, artillery, or armor) of company grade or junior field grade ground combat officers as well as a marine logistician. The submarine division should be led by a post department head submarine officer. Also, the operations department requires a host of talented enlisted personnel to support the management and execution of the departments responsibilities.

The material department (N4) should be reorganized by decreasing the amount of navy engineering influence, due to the reduction in ships assigned, and emphasize more the logistics and sustainment at sea. The MAGTF forces that go ashore should look to the MEU(SOC) staff and PHIBGRU's ground and air operations divisions for logistics support. An operational planning group (N5) should be organized to support future planning. The personnel should come from the various departments and supporting or associated ships and staffs (e.g., MEU(SOC) and DESRON). The communications department (N6) should expand to take on a more robust command and control warfare posture. The training and readiness department (N8) should decrease in size, due to the reduction in ships assigned, and come under the control of the operations department or become the core of the operational planning group.

In order to achieve operational expertise in maneuver warfare and littoral operations, the PHIBGRU commander needs to work throughout the work-ups and deployment with the ARG and MEU. Therefore, the PHIBGRU's flagship should be the LHA or LHD, both very capable command and control platforms. But, what happens to the CVBG and its commander because clearly the three ship ARG is not such a significant command to warrant a flag officer and staff? It does not seem reasonable to assume that during a crisis situation the entire CVBG command could be efficiently and effectively commanded by a PHIBGRU staff embarked on an amphibious ship: it is a matter of the very close relationship between the carrier air wing staff and the CVBG staff.

The alternative that I propose is to have the Arsenal Ship take the place of an aircraft carrier as the centerpiece of what used to be called the CVBG. But can the Navy's mission still be accomplished effectively without the aircraft carrier forward deployed? The answer to that question is yes. During every recent Mediterranean deployment, the aircraft carrier has spent at least one quarter of its deployment out of the Mediterranean theater, and sometimes as much as half of the deployment. The ships that remained in the theater accomplished, in conjunction with other joint and combined forces, all of the required theater missions. If a crisis situation had developed in the theater, the remainder of the CVBG stood ready to provide TLAM strike support and escort missions for the ARG to allow it to perform its mission. Would that strike and escort support have been enough to accomplish maritime dominance in crisis area's littoral without the aircraft carrier? Would there have been enough strike assets to have a decisive or substantial shaping effect on the enemy? Clearly that is all situational dependent. This is where the Arsenal Ship can play a key role. But, what is an Arsenal Ship, and what are its capabilities?

ARSENAL SHIP--THE CENTERPIECE OF A LITTORAL DOMINANT BATTLE GROUP

> The arsenal ship concept is a direct outgrowth of the Navy's Shift in focus from the open ocean to the littoral. It is fully consistent with "Forward...from the Sea", and "Operational Maneuver from the Sea", and addresses current as well as anticipated future requirements for more decisive, responsive and varied naval support to the land battle. Through concentration of massive firepower, continuous availability and application of netted targeting and weapons assignment, the arsenal ship will increase dramatically the scope and relevance of surface strike and fire support.[11]

BACKGROUND

The concept of a modern arsenal ship is not a new idea. Vice Admiral Joseph Metcalf III, USN (retired), the former Deputy Chief of Naval Operations for Surface Warfare first introduced the concept in his 1988 article, "Revolution at Sea Initiative." In that article he foresaw the development of a new kind of "dreadnought" based on the principle of maximizing a ship's ability to deliver ordnance on target. He surmised that the a ship would be built, a strike cruiser, to exploit vertical launch and smart missile technologies.[12]

Arsenal Ship Requirements

Theater Commander in Chiefs have identified capability requirements based on existing shortfalls within their respective theaters. These requirements include:

- Conventional Deterrence against regional aggression inimical to United States interests,

---

[11] Department of the Navy, "Arsenal Ship Concept of Operations," 1996, p. 1.

[12] VADM Joseph Metcalf III, USN (retired), "Revolution at Sea Initiative," *USNI Proceedings*, January 1988, p. 36.

• Flexible response for demonstration of power independent of diplomatic limitations'

• Credible forward firepower support to joint and coalition land forces early in a regional contingency if deterrence fails. The forward theater arsenal ship weapons load out will be robust, flexible and tailorable to Commander in Chief requirements in order to expand Commander in Chief options for use of assigned joint forces.[13]

Based on the above requirements, the United States Navy has embarked on an acquisition program to develop an arsenal ship to satisfy those requirements. Additionally, the Navy is looking to satisfy some of its long-standing warfare deficiencies and predictable future shortfalls.

LONG-STANDING DEFICIENCIES

The missions of naval surface fire support include: (1) suppression of enemy artillery, (2) interdiction of enemy reinforcing elements, (3) area neutralization and denial, (4) close support fires, and (5) protection for evacuation operations. Those missions must be performed in all weather, day and night, and from over-the-horizon.[14] Since the early 1960s, the navy has seriously decreased its Naval Surface Fire Support (NSFS) capability by the decommissioning of its Battleships and Heavy Cruisers with their multiple 16 inch (406mm) and 8 inch (203mm) guns respectively. Although re-activated twice to truss up its surface fire support capability, once for the Vietnam conflict and again during the mid-1980s through the Gulf War, the Navy's Battleships were once again de-activated primarily because of their extremely high operating costs and the fact that, at the time of their de-activation, the Navy

---

13 Department of the Navy, "Arsenal Ship Concept of Operations," 1996, p. 3.

14 Jere G. Mackin, "Navy Warfighting Formula Needs Fire Support Boost," *National Defense*, May/June 1995, p. 34.

leadership had not shifted its focus from "blue-water" operations to power projection ashore from the sea.

Combine the decommissioning of Battleships and Heavy Cruisers with the post-Cold War era military drawdown, which is not yet complete, and the results are an all-time paucity of naval surface fire support in this century. In fact, the current inventory of United States. Navy gun weapon systems and ammunition is inadequate to meet naval surface fire support requirements.[15]

PREDICTED DEFICIENCies

A forward deployed aircraft carrier battle group has a finite number of vertical launch missile cells and large caliber naval guns. The load out of those cells and gun magazines is a question of theater requirements. Those requirements are driven by theater commanders and naval doctrine.

Vertical Launched Ordnance. Theater commanders recognize the flexibility provided by sea-based ordnance operating in international waters. They demand that a large percentage of the vertical launch missile cells available to a battle group commander are loaded with theater level weapons such as the Tomahawk Land Attack Cruise Missile.

Another potential theater level weapon carried by ships is the Theater Ballistic Missile Defense (TBMD) surface-to-air standard missile that is being developed by the Navy. It is

---

15 Jere G. Mackin, "Navy Warfighting Formula Needs Fire Support Boost," *National Defense*, May/June 1995, p. 34.

reasonable to assume that theater commanders will require additional battle group vertical launch cells be reserved for TBMD missiles.

The requirement for a certain number of those two types of theater weapons reduces the number of vertical cells available to a battle group for defensive ordnance such as surface-to-air missiles and vertical launched anti-submarine rockets. Additional ordnance that will be competing for space within the vertical launch missile cells are the self-defense surface to air Evolved Sparrow Vertical Launched Missile and the naval version of the Army Tactical Artillery Missile that is compatible with the current vertical launch weapons system.

Naval Gun System. Maneuver warfare from the sea requires shaping by lethal and non-lethal fires, and robust fire support capabilities. The navy which relies on a 5 inch 54 caliber (127 mm) gun system has severely limited shaping and fire support capabilities due to the range (maximum range of 13 nautical miles) and lethality of that weapon system's ordnance. Planned improvements to the existing inventory of 5 inch naval guns will not of itself meet NSFS requirements.

Although a 62 caliber upgrade to the 5 inch 54 caliber gun and an associated Extended Range Guided Munition (ERGM) are being programmed, for every 5 inch ERGM that is put into the magazine of a surface ship, two existing rounds are removed. That is significant because shipboard gun magazines have a standard 600 round capacity; approximately 40 percent of that capacity is taken up by rounds specifically designed for air defense, surface ship engagement, training, and special purpose ground support ordnance such as white phosphorus and illumination rounds. Thus, either the multi-purpose capability of the 5 inch gun system will be sacrificed by reducing the amount or types of ordnance within its magazine or the amount of ammunition available for NSFS missions will be reduced by 50 percent. Perhaps the accuracy of ERGM will off-set the reduction in the amount of NSFS available, and then again, less than 200 five inch ERGMs per shipboard magazine does not seem to be a lot of fire power. Conversely, the benefit of those gun systems on escort ships is the ability to re-arm them at sea.

There are other trade-offs associated with the improved range (approximately 63 nautical miles) offered by the 62 caliber and ERGM upgrades to the 5 inch gun, those include: less destructive power of ERGM due to ordnance weight restrictions that require the use of submunition technology; increased potential for fratricide based on the failure rate of submunition technology; decreased responsiveness due to rate of fire decreases associated with gun loading procedures; and extended times of flight due to a high-altitude fly-out phase of an ERGM.

SPECIFICATIONS

Arsenal ship is designed to be a "fly before buy" acquisition program. The Navy has provided industry with a Concept of Operations and a Ship Capabilities Document. Industry teams take the functions described in the ships capability document and treat them as goals when conducting trade studies against the cost threshold. The goal is to leverage commercial technology and manufacturing processes with existing Department of Defense investments to rapidly (i.e., in less than half of the time of a typical major acquisition program) produce a functional and cost effective prototype arsenal ship for operational evaluation and testing. This prototype will be capable of being converted into a fully mission capable ship.[16]

---

16 Department of the Navy, "Arsenal Ship Capabilities Document," 1996, p. 1.

The following is a synopsis of the relevant operational concepts and specifications that the Navy is asking industry to deliver in the Arsenal Ship:

- The Arsenal Ship's launching system should have approximately 500 vertical launch cells, and room for an extended range gun system. The ship must be capable of firing the following weapons in support of the land campaign: Tomahawk Land Attack Cruise Missiles using off-ship targeting and mission planning, Standard Missiles using Cooperative Engagement Capability, and a vertical launched naval surface fire support weapon using digital call for fire technology.[17]

- Optimize the Arsenal Ship's survivability in the littoral environment using passive means. If survivability evaluations dictate, then provide the Arsenal Ship with limited active self defense systems to augment passive defenses.

- Make Arsenal Ship a "remote magazine" that will receive all targeting, mission planning, and command and decision functions from off-ship through Cooperative Engagement Capability or an equivalent data link. Make the remote connectivity with other navy assets reliable, rapid, and secure over both satellite and line of sight media. Make it jointly interoperable with minimum impact. Successful employment of the Arsenal Ship will rest on effective command and control systems both internal and external.

- The Arsenal Ship will be designed for 35 years of service with an overall availability rate of 95 percent. It will be capable of underway refueling and vertical stores replenishment. Underway re-arming of vertical launch cells is not required. Provide it with the ability to store 90 days consumable and maintenance stores.

- Make Arsenal Ship capable of at least 22 knots and sufficient fuel to conduct a 90 day mission at the most economical fuel consumption speed. The ship shall have precise navigation capability independent of location, weather, or visibility.

Advantages

"The **protection of forces** will often be a **friendly center of gravity** during early entry operations. Therefore, early entry forces should deploy with sufficient organic and supporting capabilities to preserve their freedom of action and protect personnel and

[17] Department of the Navy, "Arsenal Ship Capabilities Document," 1996, pp. 1-2.

equipment...."[18] The Arsenal Ship enhances force protection because its vertical launch cells free up cells on cruisers and destroyers to be loaded with weapons that can be used to protect friendly forces. Also, Arsenal Ship's long-range firepower can weaken potential counterattacks against joint forces by disrupting enemy C4ISR capabilities with standoff unmanned strikes and provide substantial capabilities to suppress enemy defenses (e.g., air and coastal), damaging logistics infrastructure, and attrite some of the enemy's long-range offensive weapons.

The Arsenal Ship's striking power is increased as the ship moves closer to the shore where its NSFS ordnance can effect the enemy. If designed properly (signature reduction and armored), Arsenal Ship may not need to worry about anti-ship cruise missiles. The primary threats that it will face are mines, diesel-electric submarines, and manned attack aircraft.

With the exception of counter-mine warfare, the aircraft carrier airwing is a vital element of subsurface (anit-submarine) and air (anti-air) warfare. Therefore, if battle group strike can be taken over by the arsenal ship and tomahawk loaded escorts, then carrier aviation assets can focus on air superiority missions and maritime patrol. This would strengthen defense in depth around the battle group and minimize the exposure of manned aircraft to integrated enemy air defenses over the beach.

The Arsenal Ship will have substantial on-scene sustainment capabilities; much more robust than any other surface combatant. Her ordnance load should fulfill any current CINC's strike requirements with room to meet requirement growth. The remaining cells will improve

---

[18] Joint Chiefs of Staff, Joint Publication 3-0, *Doctrine for Joint Operations*, (Washington, D.C., Office of the Chairman), 1995, p. IV-5.

the Navy's ability to meet fire support ordnance requirements and augment air defense and TBMD missions.

"Maneuver warfare is a warfighting philosophy that seek to shatter the enemy's cohesion through a series of rapid, violent, and unexpected actions which create a turbulent and rapidly deteriorating situation with which he cannot cope."[19] Those actions must be viewed across both time and space.

From a maneuver warfare perspective, the implications of the Arsenal Ship seem very favorable: with its massive amount of unmanned, long-range precision firepower, effectively double the amount nominally available to a standard aircraft carrier battle group, critical enemy targets over a vast area can be attacked simultaneously or in a specific time sequence developed in the joint commander's campaign plan. Striking multiple critical Command, Control, Communications, Computer, Intelligence, Surveillance, and Reconnaissance (C4ISR) nodes in a rapid fashion will promote achieving dominant maneuver by impacting the enemy's ability to fight his forces in an organized and effective manner. The shock created by this kind of attack in theory would paralyze an enemy.

Likewise, the Arsenal Ship will enhance mobility because of the suppressive effects of naval surface fire support and the potential to decrease the criticality of getting field artillery and the logistics footprint that it requires ashore to support ground combat elements. Amphibious operations of the past were typically ushered in by massive naval gunfire and aviation dropped ordnance evolutions. Future amphibious operations, whether they are under-the-horizon linear assaults or over-the-horizon highly maneuver warfare oriented

---

[19] Department of the Navy, Fleet Marine Force Manual 1, *Warfighting*, (Washington D.C., Headquarters United States Marine Corps), 1989, p. 59.

assaults will likely require even greater naval gunfire support, especially in the over-the-horizon type assault.

Seemingly gone are the days when a host of massive naval guns (8 to 16 inches) would "soften-up" an enemy, prepare a landing area, and provide supporting arms for ground forces until their organic artillery and its logistics footprint were established ashore.

Arsenal Ship seems to be the only effective answer for the future because the Navy is not likely to commit to re-activating the battleships or to building a large caliber gun ship and the short-term gunfire deficiency solution that the Navy is pursuing to upgrade the current fleet inventory of escort ships with a 5 inch 62 caliber gun mount is not going to provide enough firepower to solve the problem. For example: using each two-gun escort (nominally 6) of two aircraft carrier battle groups in fire support areas, a total of 12 guns and 2,700 light-weight 127mm (5 inch) rounds would be simultaneously available to provide 120 rounds per minute for twenty minutes of landing area preparation and ground forces fire support. The Arsenal Ship with one twin 155mm Vertical Gun Advanced Ship (VGAS) systems (currently under development) would provide 1,200 light-weight 155mm rounds at a firing rate of 40 rounds per minute for thirty minutes. Thus, one Arsenal Ship with only one VGAS could provide roughly half of the explosive weight at a third of the firing rate of six dual gunned surface combatants. This is very significant because when a surface combatant is engaged in NSFS operations, the ship's capabilities in other warfare areas is seriously degraded. The development of digital NSFS command and control systems may mitigate the performance degradation in other warfare areas while a ship is engaged in NSFS.

The forward deployed Arsenal Ship will allow the vertical launch cells of battle group escorts to be loaded with more defensive weapons. That load will enable the surface combatants to better fulfill their primary role as an escort. It will also allow the escorts more versatility to be able to go to critical maritime areas of the battle field and more effectively influence that battlespace through presence, capabilities, and ordnance. For example: theater ballistic missile defense may require an Aegis ship to be stationed away from battle group operating areas. That escort will require sufficient self-defense missiles (e.g., Standard Missiles and Antisubmarine rockets) and TBMD missiles to accomplish the mission at acceptable risk.

Limitations

The limitations of the Navy's proposed requirements are technological and doctrine related. The Arsenal Ship will not have a high volume of inexpensive ordnance available to support shaping and NSFS operations. Nor will it have a large caliber weapons system capable of delivering two million pounds of ordnance to support the close battle space like the two battleships, USS MISSOURI and USS WISCONSIN, were able to provide during the Gulf War.[20] It would take more than 17 VGAS systems to deliver that amount of ordnance. The navy probably will never employ 17 VGAS systems simultaneously during a crisis, however, in theory the accuracy of the ERGM round should compensate for some the tonnage differential between the 16 inch guns and the VGAS. To compensate for the difference in destructive power of the 16 inch shells and the ERGM, the navy is conducting research in advanced ordnance systems that will have improved warheads for better destructive effects as

[20] Gourley, Scott, "Naval Surface Fire Support in Forcible Entry Operations." *Naval Forces,* January 1995, p. 62.

well as significant ordnance speed increases (up to mach eight) to enhance target penetration.[21]

Other limitations include the extensive mission planning that is required to deliver the over-the-horizon ordnance planned for the Arsenal Ship. Current versions of Tomahawk require detailed planning for both the over-water and over-land portions of the missile's flight path.

Planning the over-land flight path is a highly complex, labor intensive, slow process that requires a significant amount of information support. That portion of the missile's flight path is primarily planned by planning experts located within the continental United States. Although an Afloat Planning System (APS) does exist; that system only gives a commander a limited ability to build, or more likely modify, a tomahawk mission to enhance strike flexibility because of the highly complex, labor and information intensive, slow mission planning process. In either circumstance, however, the new missions are not likely to be responsive enough nor developed on a large enough scale to be a significant flexible fire power capability.

With future GPS equipped versions of the Tomahawk missile, the APS should provide significant speed and flexibility in planning the over-land portion of a missile's flight path. The drawback to this capability is the potential future vulnerability of GPS to Command and Control Warfare (C2W) and interdiction.

Planning the over-water portion of the flight path is critical for a variety of reasons. The issue of airspace deconfliction must be considered because of our own air defense capabilities and friendly air assets operating between the launcher and the target. The over-water portion of

21 Townes, John W. III, CAPT, USN, "Navy Surface Fire Support: On Target." *Surface Warfare*, January/February 1997, p. 26.

the Tomahawk missile's flight path is also planned to safely avoid surface contacts to prevent having the missile clobber into a ship or obstacle, and to minimize potential early warning and counter-targeting by third-parties and enemy surveillance assets. Also, the over-water portion of the Tomahawks flight path is critical to ensuring that the missile gets to the beginning point of its over-land flight path on time in order to ensure that the missile achieves its designated time-on-top of the target, if specified.

Because the Arsenal Ship will not be manned to conduct any mission planning, an extra burden is placed on the mission planners onboard other ships. As already witnessed, human and technological errors in the complex world of mission planning have caused one ship to fail to meet its assigned mission (USS MONTEREY's September 1995 failure to launch assigned Tomahawk missions against targets in Bosnia).[22] How much more complex is having another ship do its own jobs and also plan all of the missions for the Arsenal Ship? Combine that with the additional burden of ensuring command, control, and communications connectivity between the itself and the Arsenal Ship, and potential fragileness with the Arsenal Ship concept begins to surface. The solution to the command and control challenge of total off-ship mission planning and launch may be solvable through a high degree of redundancy in the Arsenal Ship's communications architecture, an increase in the number of communications paths that are capable of sending the required data to the Arsenal Ship, and by increasing the number of mission planners and "trigger pullers" that are able to launch the Arsenal Ship's weapons.

The theater level weapons planned for the Arsenal Ship are less time of flight critical than are the NSFS weapons. The NSFS weapons that are planned for the Arsenal Ship are assisted in

---

[22] Stearman, William L., "The Navy's Proposed Arsenal Ship - Con." *The Retired Officer Magazine,* November 1996, p. 41.

achieving their extended ranges by a boost phase which is marked by a steep vertical assent to high-altitude, followed by a fly-out phase in the thin upper atmosphere, ending with a GPS controlled descent phase onto the target. For targets requiring that type of flight profile, the ordnance's time of flight could be as long as five minutes. Against moving targets or time-sensitive targets, that may be unacceptable. The navy is working on developing a variety of next-generation ordnance that will have dramatic increases in speed for example: Fasthawk--a surfaced launched missile capable of a 700 mile range at mach four; and Scramshell--a hypersonic ballistic round capable of mach 5 or greater.[23]

"Outstanding weapons won't be effective without the ability to accurately designate targets and ensure fires are coordinated over an extended and fast-paced battlefield."[24] The Navy plans on providing responsive and accurate fires for the rear, close, and deep land battlefield through the Cooperative Engagement Capability (CEC). CEC already exists to support rear and to a large extent close area air defense including TBMD. CEC allows for the remote selection, initialization, and remote launch orders of TBMD and standard surface-to-air missiles. That capability allows an Aegis combatant that is equipped with the CEC upgrade to remotely launch and control surface-to-air missiles within the limits of its fire control director's terminal guidance range limitation.

The Advanced Tomahawk Weapons Control System (ATWCS) coupled with CEC communications data links should solve the deep battlefield strike mission planning, coordination, and command and decision functions. An Arsenal Ship's TLAM missions for

---

23 Townes, John W. III, CAPT, USN, "Navy Surface Fire Support: On Target." *Surface Warfare*, January/February 1997, p. 26.

24 O. Kelly Blosser, "Naval Surface Fires and the Land Battle," *Field Artillery*, September-October 1996, p. 44.

strategic and operational level targets will be issued to an Aegis combatant for over-water mission planning and launch control. Aegis ships with CEC and ATWCS to CEC upgrades would then remotely select, initialize, and launch TLAMs from an Arsenal Ship to execute ordered missions.

Close and deep area fire support target designation and weapons coordination is a critical requirement for all services. The Army is the lead agent for developing a joint artillery command and control system; that digital command and control system is the Advanced Field Artillery Tactical Data System (AFATDS). It is scheduled for fielding in early 1997, and has programmed upgrades to make it more interoperable with other services command and control systems. "AFATDS provides integrated, automatic support for mission planning, coordinating and controlling fire support assets...for close and deep operations."[25] With programmed upgrade that will link it to the Air Force's Contingency Theater Automated Planning System (CTAPS) and the Marine Corps Fire Support System (MCFSS), AFATDS has the potential to deconflict all air and ground battlefield coordination and clearance issues for supporting fires from the Arsenal Ship.[26]

The Navy, recognizing that it must move into the digitized fire support arena, has experimented with AFATDS in a major Joint Task Force Exercise. A lot was learned from that experiment but clearly the Navy is committed to overcoming its shortfalls in this area through common operating environment systems and communication architectures to tap into AFATDS from over the horizon.

---

[25] MGEN Randall L. Rigby, "Mapping the Future: FA State of the Branch 1996," *Field Artillery*, November-December 1996, p. 4.

[26] MGEN Randall L. Rigby, "Mapping the Future: FA State of the Branch 1996," *Field Artillery*, November-December 1996, p. 5.

The impact that the Arsenal Ship has on the future battlefield will be largely dependent on the command and control system that is developed for it. As a remote magazine for navy weapon systems, the command and control challenges presented by the Arsenal Ship do not seem insurmountable because of the Cooperative Engagement Capability (CEC). On the other hand, if all of the strike and NSFS ordnance is called for through one or a few designated "Arsenal Control Ships," then those control ships could very rapidly become overwhelmed trying to do their own potentially vital role while being responsible for command and control of the Arsenal Ship's vast array of long-range lethal firepower. Furthermore, using exclusively navy ships as the command and control link from sensor to shooter would be placing a middleman in the loop which only provides another critical link that could fail.

The command and control architecture appears to rely on digital over-the-horizon connectivity. The ability to secure the use of satellite channels during a crisis may prove untenable. If acquired, the ability of those satellite links to remain functioning are also in doubt thus jeopardizing reliability. No back-up connectivity systems have been proposed yet.

The overall responsiveness of the Arsenal Ship's ordnance to tactical calls for fire will depend largely on the ability to develop a robust information support system to perform airspace deconfliction and manage ground fire coordination and control measure in the dynamic and fast paced operational environment of the future battlespace.

In order to achieve its place among the capable warships to the United States Navy, the Arsenal Ship will have to do more than just communicate with the Navy: it will have to be accepted by the Marine Corps, Army, and Air Force. Theater CINCs will have to demand that their forces are capable of and trained in automated command and control of the ordnance stowed in the Arsenal Ship.

The Arsenal Ship, like all current VLS surface combatants, will not be capable of at-sea ordnance replenishment. The only way to reload it will be to take it out of the operation and send it to a suitable port. Due to the likelihood that it will be loaded with a mixture of ordnance, especially NSFS ordnance, the Arsenal Ship will probably be kept on station thereby precluding it from retiring to a port for reloading its depleted ordnance.

Airspace coordination issues resulting from tactical fires will get more complicated because the Arsenal Ship will be providing fire support to widely dispersed tactical elements. Although the steep boost and high altitude fly-out phases are less coordination intensive, the descent phase and the flatter trajectory ordnance (i.e., sea SLAM and ATACM) from a central location to dispersed targets causes significant airspace coordination versus on-target responsiveness challenges. Improvements in automated systems to coordinating airspace usage over the battle field and in the ability to get information concerning on-call fire missions into the cockpit of aircraft already over the battle field will help mitigate the potential for fratricide.

Arsenal ships may have an adverse effect on the operational mobility of naval forces do to their self-defense limitations. As described above, because of the requirement to defend the Arsenal Ship, the ability of escorts to project power and conduct needed missions elsewhere in the battlespace may be jeopardized. Conversely, maximizing the range of the Arsenal Ship's weapon systems, low-detectability, and survivability may prove sufficient to allow battle group escorts to perform other missions than direct escort for the Arsenal Ship.

Potential Operational Effects of the Arsenal Ship

Consider for a moment the potentially decisive effects that the presence of a battle group made up of an Arsenal Ship, two Cruisers, three strike destroyers, one frigate, two submarines, and the ARG, augmented by the forward stationed Arsenal Ship (three of the proposed six will be permanently operated overseas one each in the Mediterranean, Indian Ocean, and Western Pacific) could have on an adversary: the numbers of vertical launched missile cells alone would be close to 1500. It is reasonable to say that at least half of those would be armed with TLAM. How many sorties (at what sortie rate) would it take for an aircraft carrier to generate 50 percent of those 1500 potential strike cells? That kind of firepower could stand off an enemy's coast at least 500 miles and deliver wave after wave of shaping fires while follow-on forces were massing for decisive ground operations, if necessary. At the stand-off range provided by TLAMs, the battle group escorts could be very effective at force protection by combining speed, maneuver, and deception with defense in depth.

The disadvantage of relying on Arsenal Ships is that they might lead to a "use of lethal force" quandary. The presence of two Arsenal Ships, safely over the horizon at 150 miles, may not have any influence on a crisis situation. An aircraft carrier, on the other hand, operating safely over the horizon can project its power by non-lethal means: flying strike aircraft over a target without committing to firing ordnance at the enemy. To overcome that limitation, aircraft from the amphibious ships could be sent over enemy territory to demonstrate presence. The drawback to that approach is the lack of defense in depth that an aircraft carrier's air wing can provide for ships operating in the littorals. In certain circumstances, joint forces, such as United States. Air Force assets (AWACS and overseas based fighters), and/or planned multi-purpose organic fixed and rotary-wing aircraft (Joint Strike Fighter and SH-60R with an airborne early warning radar variant) could provide defense in-depth for maritime power demonstrations. The development and fielding of those organic aircraft, mentioned above, are scheduled prior to the planned completion of the six Arsenal Ships by 2007.

Dominant Maneuver Enhanced by Naval Surface Fire Support--Arsenal Ship

It must be remembered that, as long as TLAM are considered the theater CINC's weapon, the Arsenal Ship will be viewed first as a strike asset. The other significant role that the Arsenal Ship will perform is that of the Navy's premier Naval Surface Fire Support asset, with its array of weapons under development such as:

•Navy Tactical Missile System (NTACMS)--a supersonic, global positioning system guided, 150 nautical mile, submunition carrying missile. Successfully fired in its Army version from a ship. Successfully launched from a vertical launch cell located at a test sight.

•Sea SLAM--a subsonic, electro-optical guided, digital data link controlled, 75 nautical mile, liquid fueled, 500 pound warhead missile. An existing Navy weapons system requiring human-in-the-loop control usually from a airborne control platform (organic to the battle group).

•Land Attack Standard Missile--a supersonic, global positioning system guided, 100 nautical mile, fragmentary or possibly submunition armed missile. Standard Missile is a proven weapon, land attack variant is under development.

•Vertical Gun Advanced System--a twin 155 millimeter, smooth bored gun, automated magazine with 1200 or more ready service rounds. Using Extended Range Guided Munitions (ERGM) technology that incorporates global positioning system guidance and submunition technology currently under development. Technology demonstration is expected to yield a 10-15 round per minute system that delivers the 155mm round in excess of 100 nautical miles.

As the Navy produces the above systems, and fields upgrades to the Tomahawk missile system, those upgrades will make Tomahawk more of a navy asset than a CINCs weapon, the Navy will be taking a giant leap towards making Operational Maneuver From the Sea possible. But with those possibilities come further responsibilities, like the responsibility to effectively maneuver naval assets and shape the battlefield for success in follow-on land operations, and the responsibility to provide protection fires on demand to a wide range of units dispersed throughout the battlefield. And finally, the responsibility to help the ground forces capitalize on the sustainment benefits provided by having those supporting fires based at sea rather than having to be moved ashore in the early stages of an amphibious assault. Because of those responsibilities I proposed that a PHIBGRU staff, in conjunction with the MEU(SOC) command element, take the lead in becoming the operational planning and executing experts of decisive littoral operations with the Arsenal Ship Battle Group.

RECOMMENDATIONS

The Navy, like the other services, has been struggling with force structure issues since the end of the cold war. The 1993 "Bottom Up Review" defined the Navy's need for ten aircraft carrier battle groups to fight and win two nearly simultaneous major regional conflicts. It also validated the need to retain 12 aircraft carrier battle groups to meet the peacetime forward presence missions while still maintaining a satisfactory deployment rotation.[27] The Navy is standing firm on the issue of twelve aircraft carrier battle groups. The projected aircraft carrier force structure bears that out. Unfortunately, the budget reality is likely to only get more austere while the price of a nuclear powered Nimitz-class aircraft carrier does not get any cheaper (approximately six billion dollars a copy).

The serious issue that needs to be raised regarding the Arsenal Ship is the Navy's ability to afford building another power projection ship without realizing infrastructure savings

[27] Scott C. Truver, "Tomorrow's Fleet - Part I," USNI Proceedings, July 1996, p. 55.

elsewhere, specifically in the area of strike warfare. The Arsenal Ship concept should center around having industry build a ship that could be the centerpiece of a littoral dominant battle group. The Battleships Battle Groups (BBBG) of the 1980s were an integral part of the 600 ship navy and arguably a potent power projection force. Their primary mission was again focused on the blue-water battle as a surface action group designed to meet and defeat Soviet ships on the high seas with an additional mission of conducting long-range missile strikes or power projection against land targets.[28] The aside benefit of the BBBG was that the Navy knew where it would go to get the fire support required for amphibious operations.

Although there were some drawbacks to BBBGs, such as a lack of organic air power, the Navy had worked out a joint and combined solution to the problem of lack of layered defenses. "Tactical Air Support for Maritime Operations" (TASMO) was the title of NATO doctrine that was worked out between the air forces and the Navy to provide airborne assets to improve layered defenses for the BBBG while it was transiting through or operating in a threat area. The scheduling and management of the aircraft tasked to fly TASMO was the responsibility of the theater CINC's staff where the BBBG was operating.[29] Also, TASMO fostered better joint air defense training by forcing the Navy and the Air Force to work together.

An Arsenal Ship Battle Group's ability to start to maneuver and shape a battlespace from over 500 miles away could cause a potential aggressor to rethink his use of force options because at those stand-off distances, only a skilled blue-water adversary could pose a serious challenge. The mere fact the United States has shifted its focus away from the blue-water

---

[28] LCDR John D. Furness, USN *The Battleship Battle Group: Reversing Anti-air Warfare from a Liability to an Asset*, (Newport, RI: Naval War College), 1988, p. 2.

[29] Ibid., p. 11.

threat seems to justify the case that a blue-water battle group recapitalized with better littoral warfare assets (e.g., Arsenal Ship and F/A-18E/F) does not always need to be present to provide substantial power projection. If the adversary does have a significant blue-water navy and air force, then even with the combined power of the forward deployed CVBG and ARG, a theater CINC would probably be unwilling to commit to any early, aggressive naval campaign until a massed fleet could be assembled for a Mahanian style naval campaign. Adding one Arsenal Ship to the combined power of the CVBG and ARG would not likely change the theater CINC's options. Therefore, just as the traditional battle groups will be the building blocks for naval campaigns, so can a non-traditional battle group like an Arsenal Ship Battle Group be the building block for naval campaigns.

But building a littoral dominant battle group is only part of the equation: the doctrine necessary to effectively operate an Arsenal Ship Battle Group must be developed. Since the Navy will likely forward deploy at least three of its proposed six Arsenal Ships, it can be deduced that they will be predominantly loaded out with theater level weapons like the long-range Tomahawk missile and TBMD missiles. From this proposal, the remaining Arsenal Ships would deploy as the centerpiece of a forward presence battle group. So what should the deploying Arsenal Ship be loaded out with? The deploying Arsenal Ships need to address the Navy's critical shortfall in NSFS capabilities. They would clearly carry a significant number of TLAM missiles, perhaps as many as 150 to 200. The rest of the ordnance should consist of VGAS modules and a preponderance of vertical launched fire support rockets and missiles such as ATACMS. The deploying Arsenal Ship would likely need at least four VGAS systems unless vertical launched fire support rocket or missile technology realizes significant reductions in cost per round. Four VGAS systems would likely provide adequate redundancy and sufficient number of rounds to meet short term maneuver warfare fire support requirements. It must also be recognized that the Arsenal Ship will likely be the only ship in the short term that will be able to carry the VGAS system without major structural or weight distribution modifications to active ships.

That type of ordnance load would make the deploying Arsenal Ship clearly a navy asset to control and employ in the most advantageous way. Under the control of an experienced littoral warfare group commander, the Navy and Marine Corps team would be able to take littoral warfare to new levels of effectiveness through rapid and violent application of force if needed.

CONCLUSION

In conclusion, the Navy is bringing a very powerful new strike and fire support ship to the fleet inventory. More than ever before, the Navy will have the ability and responsibility to support the land battle effectively and efficiently. Supporting the land battle and amphibious operation will require a new level of expertise by the Navy. Building that expertise will not be easy because it requires a significant amount of change from a large bureaucratic system that has been extraordinarily successful for the past 55 years. But there exists a rare opportunity to make significant changes that will not only better prepare the Navy for future littoral operations but at the same time, allow it to achieve dramatic cost saving by substituting three 50-personnel, 500 missile, less than 500 million dollar a copy, Arsenal Ships for three 5,000 personnel, 80 aircraft, 6,000 million dollar a copy, aircraft carriers.

The Navy cannot afford to pass up the potential savings made possible by the Arsenal Ship. It must work hard now to develop doctrine, force structure, and organizational changes to complement Operational Maneuver from the Sea (OMFTS). 2007 is fast approaching; it will soon be time to marry-up the Amphibious Ready Groups (ARG) with a striking battle group centered around an Arsenal Ship. The savings alone from retiring just two aircraft carriers would easily pay for the command and control modifications necessary to allow the chosen LHD(s) and LHA(s) to have basically the same capability to run a battle group as that of an aircraft carrier. That would capitalize on the Arsenal Ship's primary mission as a strike warfare asset and secondary mission as a NSFS ship. The net result would be the optimization of the striking power of the Arsenal Ship by having it employed as a maneuver and shaping asset by an on-scene amphibious warfare (littoral) experienced admiral in close coordination with an on-scene land warfare commander (Commander of the Landing Force (CLF)) starting at ranges upwards of one thousand miles from the enemy's coast. This would also provide the theater CINCs more flexible options to respond to crises in their theaters than currently available with a traditional CVBG.

The Arsenal Ship Battle Group is a viable forward presence option. The PHIBGRU as a deploying battle group commander is a key improvement that the Navy must make if it has truly shifted its focus to the littoral battle. The best way to realize the potential of the Arsenal Ship is to take it out from under the umbrella of the aircraft carrier, just as aircraft carriers were taken out from under the umbrella of the battleship dominated Navy almost 60 years ago. The key to unlock the potential of the Arsenal Ship is to demand littoral and maneuver warfare expertise from the amphibious community.

BIBLIOGRAPHY

Balisle, Phil, CAPT, USN. "Joint Solutions: Air Defense in the 'Ultimate JEZ'." *Surface Warfare*, January/February 1997.

Balisle, Phil, CAPT, USN. "Theater Ballistic Missile Defense: Blunting the Attack." *Surface Warfare*, January/February 1997.

Bally, Jacques J., CAPT, (Ret.). "The Shape of Ships to Come." *Armada International*, January 1994, 28-32.

Bally, Jacques J., CAPT, (Ret.). "The Shape of Ships to Come II." *Armada International*, January 1996, 6-15.

Blazar, Ernest. "Future Shock/Arsenal Ship Will Have Small Crew & Big Punch." *Navy Times*, 29 July 1996.

Blazar, Ernest. "How the Arsenal Ship Will Fight." *Navy Times*, 29 July 1996.

Blazar, Ernest. "Navy's Future: Is it Boorda's of Owens'?" *Navy Times*, 26 December 1994.

Blazar, Ernest. "Ships With 2 Points of View." *Navy Times*, 4 September 1995.

Blosser, O. Kelly. "Naval Surface Fires and the Land Battle," *Field Artillery*, September-October 1996.

Bradshaw, Bruce, CAPT, USN. "Fleet Readiness: Right Sailor, Right Training, Right Ship." *Surface Warfare*, January/February 1997.

Brooks, Donovan. "Admiral: China no 'Blue-water Threat'." *Pacific Stars and Stripes,* 28 February 1997.

Buckley, Patrick E., LCDR, USN. *The Submarine Threat to Naval Operations...From the Sea.* Joint Military Operations Department Paper. Newport, RI: Naval War College, 1995.

Callahan, Timothy G., CAPT, USMC. "Automated Fire Support Coordination." *Marine Corps Gazette,* July 1996.

Cavas, C. P. "Is Arsenal Ship's Mission Adrift?" *Navy Times*, 17 March 1997, 30.

Cayce, David M., LT, USN. "Composite Maneuver Warfare Commander." *Marine Corps Gazette*, March 1995.

Conn, Daniel J., MAJ, USMC. "USMC Fire Support Conference 1994." *Field Artillery*, August 1994, 29-31.

Department of the Navy, "Arsenal Ship Capabilities Document," 1996.

Department of the Navy, "Arsenal Ship Concept of Operations," 1996.

Department of the Navy, *Forward...From the Sea*, Washington, DC: Department of the Navy, 1995.

Dziminowicz, John W., LCDR, USN. *Maritime Action Groups: The Expeditionary Building Block of the Future*. Department of Operations Essay. Newport, RI: Naval War College, 1993.

Fleet Marine Force Manual (FMFM) 1. *Warfighting*. Washington, DC: Department of the Navy, 1989.

Fleet Marine Force Manual (FMFM) 1-1, *Campaigning*, Washington, DC: Department of the Navy, 1990.

Furness, John D., LCDR, USN. *The Battleship Battle Group: Reversing Anti-air Warfare from a Liability to an Asset.* Department of Maritime Operations Paper. Newport, RI: Naval War College, 1988.

Gordon, John, IV, MAJ, USA. "Naval Fire Support and the Force Projection Army." *Field Artillery,* October 1993.

Gourley, Scott. "Arsenal Ship." *Popular Mechanics*, June 1996.

Gourley, Scott. "Naval Surface Fire Support in Forcible Entry Operations." *Naval Forces,* January 1995.

Gourley, Scott. "U.S. Naval Surface Fire Support: The Gun Factor." *Military Technology*, August 1992, 33-36.

Halloran, Joe. "AFATDS: 21st Century C^2 for the Total Force." *Military Technology*, May 1992.

Hilliard, Jay, MAJ, USA. "ATACMS Block II: Killing Armored Targets Deep." *Field Artillery*, January-February 1996.

Holzer, Robert. "Doubling the Sortie Rate." *Navy Times*, 10 March 1997.

Holzer, Robert. "Sharing Arsenal Ship Assets." *Navy Times*, 1 July 1996.

Holzer, Robert. "The Future of Naval Gunfire Support." *Navy Times*, 22 April 1996.

Ingersoll, Alvah E., III, MAJ, USMC. *A Naval Expeditionary task Group in Operations Other Than War*. Department of Joint Military Operations Paper. Newport, RI: Naval War College, 1996.

Joint Chiefs of Staff, Joint Publication 3-0, *Doctrine for Joint Operations*, Washington, DC: Office of the Chairman of the Joint Chiefs of Staff, 1995.

Joint Chiefs of Staff, "Joint Vision 2010: America's Military - Preparing for Tomorrow." *Joint Force Quarterly*, Summer 1996.

Joint Chiefs of Staff, *National Military Strategy of the United States of America, 1995*, Washington, DC: GPO, 1995.

Jones, Davy, JO2, USN. "Retired Surface Warrior Details New Arsenal." *Campus News: The Navy's University*, Public Affairs Office, Naval Postgraduate School Monterey, California, 21 October 1996.

Kelly, Patrick M., ENS, USN. "The U.S. Navy Must Re-Evaluate Its Doctrine." *USNI Proceedings*, July 1996, 68-69.

Lane, Lawrence B., MAJ, USMC. "The Future of Fire Support--The Future of the Marine Corps." *Marine Corps Gazette*, September 1994.

LaPlante, J.B., VADM, USN. "The Future of Amphibious Warfare Takes Shape." *USNI Proceedings*, May 1994, 86-88.

Lasswell, James A., COL, USMC. "Why the Arsenal Ship Concept is Gaining Momentum." *Marine Corps Gazette*. January 1996, 31-32.

Loire, Rene. *The Striker - A Warship for the 21st Century*. Houston, TX: A. Ghosh, 1996.

Mackin, Jere G. "Navy Warfighting Formula Needs Fire Support Boost," *National Defense*, May/June 1995.

Martin, John R., LTC, USA and Gunzinger, Mark, LTC, USAF. "Fight Fast First," *USNI Proceedings*, July 1996.

McAuliffe, Amy. "COTS to Frame New Surface Warship Architecture." *Military & Aerospace Electronics*, November 1995.

McConnell, Kevin M., MAJ, USMC. "1996 MAGTF Fire Support Conference." *Field Artillery*, September-October 1996.

Metcalf III, Joseph, VADM, USN (Ret.). "Revolution at Sea Initiative," *USNI Proceedings*, January 1988.

Morrison, Archibald, IV, MAJ, USMC. *Sea Dragon at the Operational Level.* Department of Operations Paper. Newport, RI: Naval War College, 1996.

Murray, Williamson. "Innovation: Past and Future." *Joint Force Quarterly*, Summer 1996.

Naval Doctrine Publication (NDP) 1, *Naval Warfare*, Washington, DC: Department of the Navy, 1994.

Naval Studies Board Commission on Physical Sciences, Mathematics, and Applications National Research Council. *Navy-21 Update: Implications of Advanced Technology for Naval Operations in the Twenty-first Century*, National academy Press, Washington, DC: 1993.

Naval Surface Warfare Center. *Arsenal Ship Combat Systems.* Study. NSWCDD/MP-95/194, 1995.

Patton, Jim, CAPT, USN(ret). "Keeping Up With the Revolution." *USNI Proceedings*, July 1996.

Perry, Tony. "Military Poised for Era of Super Surveillance." *Los Angeles Times*, 17 March 1997.

President of the United States, *A National Security Strategy of Engagement and Enlargement*, Washington, DC: GPO, February 1995.

Polmar, Norman. "The Navy's Proposed Arsenal Ship - Pro." *The Retired Officer Magazine,* November 1996, 36+.

Quist, Burton C., COL, USMC. "Naval Expeditionary Warfare Update." *Marine Corps Gazette*, March 1995.

Quist, Burton C., COL, USMC. "Operationalizing the NEF Concept." *Marine Corps Gazette*, June 1995.

Rhea, John. "DARPA Backing COTS for Arsenal Ship." *Military & Aerospace Electronics*, May 1996

Rigby, Randall L., MGEN, USA. "Mapping the Future: FA State of the Branch 1996," *Field Artillery*, November-December 1996.

Rigby, Randall L., MGEN, USA. "Report Out: 1996 Senior Fire Support Conference--Focusing Fires for Force XXI." *Field Artillery*, November-December 1996.

Scott, J.T., LTGEN, USA. "Fire Support for Power Projection: A Special Operations Perspective." *Field Artillery*, April 1995.

Smith, Carole J., LCDR, USN. *Small Fleet--Big Risk.* Department of Operations Paper. Newport, RI: Naval War College, 1995.

Snyder, Jim, CAPT, USN. "Undersea Warfare: The Battle Below." *Surface Warfare*, January/February 1997.

Soderburg, Bruce W., CAPT, USMC. "Tearing Down the Stovepipes in the Marine Corps Fire Support." *Marine Corps Gazette*, July 1996.

Stearman, William L. "A Look At...The Pentagon's Misguided Priorities: A Misguided Missile Ship." *The Washington Post,* 7 July 1996, C03.

Stearman, William L. "The Navy's Proposed Arsenal Ship - Con." *The Retired Officer Magazine,* November 1996, 37+.

Stover, Dawn. "Floating Arsenal." *Military Technology,* February 1996.

Townes, John W. III, CAPT, USN. "Navy Surface Fire Support: On Target." *Surface Warfare*, January/February 1997.

Townes, John W. III, CAPT, USN. "Surface Strike: The Powerful Punch of Deterrence." *Surface Warfare*, January/February 1997.

Truver, Scott C. "Tomorrow's Fleet - Part I," USNI Proceedings, July 1996.

Truver, Scott C. "Tomorrow's Fleet - Part II," USNI Proceedings, August 1996.

Van Buskirk, James H. "Synchronizing Fire Support and Joint SOF Operations." *Field Artillery*. February 1993.

Wilson, John G.R., LCDR, USN. *An Examination of Naval Fires in Support of Future Amphibious Operations.* Fort Leavenworth, KS: School of Advanced Military Studies United States Army Command and General Staff College, 1992.

Wilson, John G.R., LCDR, USN. *An Examination of the United States Navy's Ability to Conduct Operational Fires*, Fort Leavenworth, KS: School of Advanced Military Studies United States Army Command and General Staff College, 1992.

Woeltjen, Donald A. CAPT, USMC. "Operational Maneuver from the Sea and the Marine FIST." *Marine Corps Gazette*, April 1996.

Wright, Richard, CAPT, USN. “Arsenal Ship: Potent and Punishing,” *Surface Warfare*, January/February 1997.

Wright, Richard, CAPT, USN. “Coastal Defense: Suppressing the Enemy.” *Surface Warfare*, January/February 1997.

Wright, Richard, CAPT, USN. “Information Superiority: Increasing the Warfighter's Advantage.” *Surface Warfare*, January/February 1997.

Wright, Richard, CAPT, USN. “Killing Machines in the 19th Century: Wooden Warships and Surface Fires.” *Surface Warfare*, January/February 1997.

Wright, Richard, CAPT, USN. “The 21st Century Surface Navy.” *Surface Warfare*, January/February 1997.

Wright, Richard, CAPT, USN. “War of Attrition: Kamikaze Attacks at Okinawa.” *Surface Warfare*, January/February 1997.

NAVAL POSTGRADUATE SCHOOL
MONTEREY, CALIFORNIA

THESIS

THE ARSENAL SHIP AND THE U.S. NAVY: A REVOLUTION IN MILITARY AFFAIRS PERSPECTIVE

by

Dawn H. Driesbach

December, 1996

Thesis Advisor: Jan S. Breemer
Second Reader: W. M. Dunaway

Approved for public release; distribution is unlimited.

THE ARSENAL SHIP AND THE U.S. NAVY: A REVOLUTION IN MILITARY AFFAIRS PERSPECTIVE

Dawn H. Driesbach
Lieutenant, United States Navy
B.S., United States Naval Academy, May 1987

Submitted in partial fulfillment
of the requirements for the degree of

MASTER OF ARTS IN (NATIONAL SECURITY AFFAIRS)

from the

NAVAL POSTGRADUATE SCHOOL
December 1996

Author: Dawn H. Driesbach

Approved by: Jan S. Breemer, Thesis Advisor

W. M. Dunaway, Second Reader

Frank Petho, Chairman
Department of National Security Affairs

ABSTRACT

This thesis examines one of the U.S. Navy's major capability initiatives aimed at addressing the changed strategic and technological environment since the end of the Cold War. This initiative is known as the Arsenal Ship. This thesis considers operational concepts for use of the Arsenal Ship; it examines some of the cost and technical issues that have been raised, and it reviews possible alternatives.

The thesis concludes that the Arsenal Ship is a viable platform for meeting the U.S. Navy's requirement for littoral power projection to meet the strategic and technological requirements of the 21st century. While there are limitations to the Arsenal Ship, the program is conceptually sound and the Arsenal Ship demonstrator program should be pursued.

TABLE OF CONTENTS

X

EXECUTIVE SUMMARY

This thesis examines one of the U.S. Navy's major capability initiatives aimed at addressing the changed strategic and technological environment since the end of the Cold War. This initiative is known as the Arsenal Ship. This thesis considers operational concepts for use of the Arsenal Ship; it examines some of the cost and technical issues that have been raised, and it reviews possible alternatives.

This thesis addresses these questions in seven chapters. Chapter I is the introduction. Chapter II discusses the fundamental changes in U.S. naval strategy following the end of the Cold War. It points out that the Navy's basic preoccupation with security on the high seas has shifted toward a strategy of joint operations conducted from the sea. Basic national guidance for planning and programming Navy operations from the sea is premised on the requirement that U.S. military forces be able to fight and win two near-simultaneous major regional conflicts (MRCs).

Chapter III discusses the so-called Revolution in Military Affairs (RMA). It specifically focuses on what some military planners believe will be a vastly enhanced U.S. capability to engage in "precision warfare." The chapter addresses innovative technologies which the Navy can capitalize on in attempts to maintain its dominant international position.

Chapter IV ties together the discussions in Chapters II and III. It argues that the strategic changes in the wake of the collapse of the Soviet Union, when combined with impending technological changes, amount to a revolution in naval affairs. It goes on to examine elements required to provide a link between the current Navy and the Navy intended for the 21st century.

Chapter V discusses the concept of the Arsenal Ship, including principal features such as a planned 500-cell Vertical Launch System, substantial automation, greatly reduced crew requirements, and the incorporation of the Cooperative Engagement Capability. This chapter also addresses costs, survivability concerns and options proposed in lieu of the Arsenal Ship.

Chapter VI discusses some of the operational applications and deployment concepts of the Arsenal Ship. It is concerned in particular with how this capability may be fully integrated into a new Navy capability to directly and operationally affect the conduct of war on land.

The Arsenal Ship has its share of proponents and detractors. Supporters believe that the ship is critical to the Navy's requirement to meet the demands of both a littoral strategy and the current rate of technological progress. Opponents argue that it is imprudent to invest time and money in a new platform during an era of reduced global threat and while existing platforms are capable of meeting current and anticipated security requirements. This thesis concludes that, on balance, the Arsenal Ship meets the Navy's requirement for a platform capable of executing a naval power projection strategy in the 21st century. The Arsenal Ship may not be the perfect ship - important questions still remain. However, given the lengthy period of time associated with the development of new weapons systems, timely alternatives appear unlikely.

I. INTRODUCTION

The variety and design of ships in the Navy have been subject to very few large scale changes during the past 45 years. When they have occurred, they were typically the result of technological developments, new operational concepts or a combination of both. The result was commonly an incremental improvement upon existing ship designs. As the 21st century looms on the horizon, U.S. strategy is changing while technological innovation accelerates. The combination of these events raises the question whether or not improvements to existing ships will suffice for the future Navy or if there is a need for an entirely new type of platform.

The Navy has evidently decided upon the latter. On 18 March, 1995 the Navy signed an agreement with the Defense Advanced Research Projects Agency (DARPA) which formalized a cooperative effort toward the research and development of a radically different seagoing platform termed the "Arsenal Ship." The platform, when completed, will be vastly different from anything the Navy has in its inventory.

The Arsenal Ship is conceptually designed to carry a substantially higher ordnance payload than any current ship in the fleet. It is intended to provide massive firepower in the initial stages of a conflict while operating under an integrated C4I (command, control, communications, computers and intelligence) system consistent with the joint focus of the 21st century U.S. military.

This thesis addresses the strategic and technological changes that will have an impact on the future of the Navy. Initially, the work addresses the fundamental changes which have occurred in U.S. strategic thinking following the end of the Cold War. Chapter II analyzes those changes and the factors which have influenced them. Additionally, it examines the increasing emphasis placed on joint strategy in order to assess recent changes and determine whether or not those changes provide a catalyst for revolutionary measures associated with new platform designs.

Chapter III addresses the so-called Revolution in Military Affairs (RMA.) It reviews emerging innovative technologies and their potential impact on the way U.S. forces will operate in peacetime and, when required, during combat.

Chapter IV addresses the combined impact of strategic and technological change which, this thesis proposes, has set the stage for a naval revolution. The chapter focuses on one dimension of this revolution in particular, i.e. the concept of "precision strike." This concept is central to the Arsenal Ship. The Arsenal Ship, when mated with precision strike capabilities and innovative operational strategies, has the potential of changing the Navy's institutional make-up in a manner as dramatic and far-reaching as occurred with the creation of nuclear-powered submarines, the emergence of the aircraft carrier as the capital ship, and the development of amphibious warfare techniques during in World War II.

Chapter V considers the Arsenal Ship's principal features. It also addresses the main concerns that have been raised as well as some of the alternatives which have been proposed.

In previous eras revolutions have resulted in institutional changes for the Navy. The transition of the Navy from a service identified with the battleship to one identified with the aircraft carrier was significant and had a far-reaching organizational and professional impact throughout the service.[1] Fifty years later the possibility of such a change occurring again has arisen with the concept of the Arsenal Ship. Chapter VI analyzes the operational potential of the Arsenal Ship in the current geopolitical and security environment. It assesses whether or not the Arsenal Ship might be an appropriate choice for the Navy for the next 25 years.

In summary, this thesis assesses strategic and technical issues which will potentially effect the future security environment in which the Navy will be operating. The primary question of this thesis is whether or not changes associated with the shift in the Navy's strategic focus have created a gap in current U.S. naval ship capabilities. If so, the next question is whether there is a need for a new ship within the Navy and whether the proposed Arsenal ship is the appropriate solution.

---

[1] References include: Richard Hough, *Death of the Battleship*, (New York: The MacMillan Company, 1963) and Clark G. Reynolds, *The Fast Carriers: The Forging of An Air Navy*, (New York: McGraw-Hill, 1968).

4

II. THE STRATEGIC REVOLUTION

A. LITTORAL FOCUS

Recent changes in strategic direction and doctrine of U.S. naval forces have shifted the focus of naval roles and missions in such a dramatic fashion that some might view them as revolutionary. One of the questions this thesis attempts to answer is whether, as a result of revolutionary changes in U.S. naval strategy, an innovative new ship concept is necessary or appropriate for the strategic environment facing the U.S. military in the 21st century.

During the five years or so preceding the collapse of the Berlin Wall, Department of the Navy strategic thinking was exemplified by the Maritime Strategy.[2] This strategic concept focused on the use of Navy and Marine Corps forces in a global war with the Soviet Union. Specifically, the Maritime Strategy was centered around the concept of "blue water" or open-ocean naval warfare against a powerful Soviet surface, submarine

---

[2] Admiral James D. Watkins and General P.X. Kelley, "The Maritime Strategy" and "Amphibious Warfare Strategy", as published in U.S. Naval Institute Proceedings, 112, no. 1 (January, 1986): 2-29. This was the official statement of policy during the 1980's for U.S. naval forces, combined with other U.S. services and allied forces, on how preparations would be made for fighting and terminating war on favorable terms. (henceforth cited as the Maritime Strategy.)

and air fleet.[3]

The end of the Cold War and collapse of the Soviet Union, spelled the demise of the Maritime Strategy. As a result, the Navy announced a "fundamental shift away from open-ocean warfighting on the sea toward joint operations conducted from the sea."[4] This shift served to re-focus U.S. naval capabilities on the need for joint and combined operations, forward presence and power projection, strategic deterrence, on-scene crisis response; and sea-lift.[5] Collectively, these and related missions are expected to be carried out primarily in littoral waters adjacent to the land mass.

From the Sea and Forward...From the Sea are not the sole sources for this "fundamental shift." The need for U.S. naval forces to be able to influence events on land was highlighted in the results of the Department of Defense Bottom-Up Review (BUR).[6] This document, signed by then - Defense Secretary Les Aspin and published in 1993, announced the restructuring of U.S. military forces to cope with the new dangers of the post-Cold War era, specifically aggression by regional powers requiring that U.S. forces

---

[3] Ibid.

[4] Department of the Navy, From the Sea: Preparing the Naval Service for the 21st Century, (henceforth noted as ...From the Sea), (Washington, D.C.: September 1992), 3. This document is the foundation for the U.S. Navy's strategy and doctrinal shift from planning associated with war at sea toward joint operations on land supported by naval forces "from the sea". The strategic concept was expanded in the 1994 publication of Forward... From the Sea, (Washington, D.C.: Undated).

[5] Ibid, 2.

[6] Secretary of Defense Les Aspin, Report on the Bottom-Up Review, (henceforth cited as BUR), (Washington, D.C.: October 1993).

henceforth be designed to fight and win two nearly simultaneous major regional conflicts (MRCs).[7]

The BUR's two-MRC scenario posed a formidable doctrinal challenge for the Navy. The BUR made clear that the principal adversaries likely to be encountered would be land-based, and their forces would be predominantly comprised of tanks, armored personnel carriers, land-based aircraft and ballistic missiles, etc.[8] The question before Navy strategists was how the service, whose capabilities and operational concepts had historically been focused on engagement on the high seas, could best ensure its operational relevance against a land-based adversary.

According to the BUR, the U.S. response to a major regional conflict would entail four main phases: (1) halt the enemy invasion, (2) build up U.S. combat power, (3) decisively defeat the enemy, and (4) provide for post-war stability.[9] The ability of U.S. forces to halt an enemy assault quickly and before the opponent reached key strategic objectives was believed central to U.S. success. Concentrated firepower early in the conflict was believed to hold the key to this success.[10] The Arsenal Ship appears to be the Navy's solution.

---

[7] Ibid.

[8] Ibid, 13.

[9] Ibid, 15 - 16.

[10] Ibid, 15.

B. JOINT WARFARE

The strategic necessity for naval forces to re-direct their capabilities and operational concepts away from open-ocean warfighting toward operations on and over land, has re-enforced the necessity for joint operations with land-based Army and Air Forces. The Chairman of the Joint Chiefs of Staff's vision of how American military forces are to fight and win their next "joint" war, was articulated in a document entitled Joint Vision 2010.

1. Joint Focus on Strategy

Joint Vision 2010 provides a template for the evolution of the U.S. Armed Forces into the next century.[11] Four broad operational concepts are highlighted in Joint Vision 2010: (1) dominant maneuver, (2) precision engagement, (3) focused logistics, and (4) full-dimensional protection. Collectively, they define "full-spectrum dominance." Its spearhead in a MRC will be the Navy. Joint Vision 2010 indicates naval forces will substantially contribute to the initial means by which joint operations will commence in the event of a crisis. Forward deployed naval expeditionary forces are extremely mobile

[11] General John M. Shalikashvili, Chairman, Joint Chiefs of Staff, Joint Vision 2010, (Washington, D.C.: Government Printing Office, as published in *Joint Force Quarterly*, Summer 1996), 38. References for policy on joint activities include: Joint Warfare of the U.S. Armed Forces, Joint Pub 1, (11 November 1991), and Unified Action Armed Forces, Joint Pub 0-2, (24 February 1995).

on global sea lanes. They are largely free from foreign basing and overflight constraints that can hinder land-based forces. Additionally, they can be assembled out of range of enemy forces, yet ready for action in the event a decision is made to commit forces. If the political-military situation changes, they can be withdrawn without entanglement ashore.[12]

It appears that the Navy is striving to re-invent itself in concert with the changing international security environment so as to ensure its decisive contribution to future operations. The change from the Maritime Strategy to littoral warfare necessitates a reassessment of the structure of the fleet. Changes are discussed further in Chapter IV.

---

[12] Office of the Chief of Naval Operations, Naval Operational Concept, (draft copy dated 14 June 1996), 4.

10

III. REVOLUTION IN MILITARY AFFAIRS

As described in the document Joint Vision 2010, "full-spectrum dominance" is essentially the concept of dominating the full range of military operations from humanitarian assistance to full-fledged combat through a synergistic utilization of dominant maneuver, precision engagement, full dimensional protection, and focused logistics. In order for U.S. forces to ensure the capability to achieve this goal, they must modernize. According to Andrew Krepinevich, "a military whose modernization can be sustained over the long term is in the best position to contribute to an overall national security strategy designed to dissuade would-be proliferators and great power challengers, while maintaining sufficient capability for the regional rogue states and peace operations that confront it in the near term."[13]

Modernization of the U.S. military will most likely take advantage of what many commentators believe is an impending RMA. The latter has been variously defined, but a most useful description has been offered by Dan Goure'. He explained the RMA as "complex changes in military organization, operations, and capabilities that are linked to

---

[13] "The Coming Defense Train Wreck and What to Do About It," *The Washington Quarterly*, (Winter 1996), 107. Comments made by Andrew F. Krepinevich who is the Director of the Center for Strategic and Budgetary Assessment, formerly known as the Defense Budget Project.

new technologies with greatly advanced capabilities."[14]

A. CONCEPTS OF THE RMA

As the idea of an RMA has gained increased acceptance, it is essential to understand the concept. An RMA is commonly triggered by new technologies which lead to new military equipment that, in turn, prompt changes in operational concepts, as well as doctrine and organization.[15] As a result, the conduct of war is fundamentally altered. Two leading interpreters of the contemporary RMA, Andrew Krepinevich and Andrew Marshall, have explained the process as follows: RMAs, claims Krepinevich,

> ... often are stimulated by significant changes in technology that facilitate a discontinuous leap in military effectiveness over a relatively short period of time, as occurred, for example, between the world wars. They are typically characterized by the rise of new military organizations which are created to fill new roles and to execute new missions.[16]

---

[14] Dan Goure, "Is There a Military-Technical Revolution in America's Future?," *The Washington Quarterly*, (Autumn 1993), 16, no. 4, 179.

[15] James R. Fitzsimonds and Jan M. Van Tol, "Revolutions in Military Affairs", *Joint Forces Quarterly*, (Spring 1994), 25 - 26.

[16] Andrew F. Krepinevich Jr., The Military Revolution: Restructuring Defense for the 21st Century, Testimony Presented to The Senate Armed Services Subcommittee on Acquisition and Technology, (5 May 1995), 2.

Marshall has offered this clarification:

> ...innovations in technology make a military revolution possible, but the revolution itself takes place only when new concepts of operation develop and, in many cases, new military organizations are created.[17]

1. Components of the RMA

The currently proposed RMA is characterized by stealth, global mobility, long-range precision strike, information warfare, and the effective use of space.[18] It embodies the idea of a change in warfare that is so drastic, the results completely alter the way future wars will be fought. Historical examples include the aircraft which gave commanders a view of the entire battlefield and spawned aerial bombing; the submarine which presented a hidden and deadly threat to surface ships; and nuclear weapons which brought massive indiscriminate destructive capability to warfare. Each changed the way in which war was to be prepared for and fought.

What are the components of a new RMA in the next century? According to Martin Libicki, "the twin roots of the RMA emerged in the late 1970's. First came the

---

[17]Andrew W. Marshall, Testimony Before the Senate Armed Services Committee, Subcommittee on Acquisition and Technology, (5 May 1995), 1.

[18] Andrew F. Krepinevich, Jr., "Keeping Pace with the Military-Technological Revolution," *Issues in Science and Technology*, (Summer 1994), 24.

development and proliferation of short-range precision-guided munitions (PGMs)."[19] The second root of the RMA, writes Libicki, is the "rapid and sustained improvements in information technology."[20]

With improvements in weapons technology long-range PGMs increasingly appear to be the weapon of choice for future strike operations.[21] The development of computers, satellites, and imagery has been occurring at an astounding rate, and there is no indication this will slow down in the foreseeable future. The inference is that the U.S. military will expand the ability to collect, evaluate and disseminate information relevant to the battlefield at a rate far greater than most current military opponents are capable of countering.

The key to the RMA then appears to be precision warfare. Precision warfare is the utilization of information technology in determining essential targets, combined with PGMs tasked to strike those targets. Precision warfare should increase the target kill rate while reducing the number of munitions and platforms necessary to achieve an

---

[19] "The Coming Defense Train Wreck and What to Do About It," 110. Comments by Martin Libicki.

[20] Libicki, "The Coming Defense Train Wreck...," 110.

[21] Andrew F. Krepinevich, Jr., "A New Navy for a New Era," (Washington D.C.: Center For Strategic and Budgetary Assessments, May 1996), i.

operation's goal.[22] According to Libicki, future precision strike capabilities will mean that, "to be seen on the battlefield is to be killed."[23] The life span of equipment currently in inventory is rapidly decreasing. It therefore seems that many of the principal weapons systems of the next quarter century will likely be designed and built around the turn of the century.[24]

Acknowledging the rapid change in military affairs is crucial in order to assimilate the capabilities required of weapons or platforms that will contribute to winning a war in twenty-five years. Coupled with this is the critical need for ensuring the development of concepts of operations that incorporate the new technologies, and of organizations to permit effective exploitation of the new capabilities.[25]

B. TECHNICAL REVOLUTION

1. Information and Communications Warfare

---

[22] Captain Edward A. Smith, Jr., a senior member of the CNO's Executive Panel, this information was gathered during an interview conducted 20 May 1996.

[23] Libicki, "The Coming Defense Train Wreck...," 110.

[24] Fitzsimonds, "Revolutions in Military Affairs", 28.

[25] Ibid, 29.

A most prominent aspect of the RMA is the revolution in information technology. Increasing focus is being paid to the development of technologies which enhance the ability to locate, identify, and track targets in greater numbers as well as over an extended period of time. Innovation associated with satellites and fixed platforms is rapidly improving these efforts. Future advances in sensor technology may open up unused portions of the electromagnetic spectrum that, when matched with improved computational capabilities and deployment in space, offer the prospect for a truly transparent battlefield. Meanwhile, improvements in the area of electronic systems and propulsion systems may be redesigned in a manner which will make platforms virtually undetectable.[26]

Future battlefields may be dominated by a struggle for "information superiority." Much, if not most, information gathering, transmission and dissemination will be done automatically, without human interference.

Such automation, though necessary, also creates new vulnerabilities. The most obvious is that associated with effective electronic countermeasures. It has been and will remain a primary military goal to render the opponent deaf, dumb and blind. Therefore, while the U.S. strives to improve its information and communication components, it must remain alert to associated vulnerabilities and strive to overcome them.

---

[26] Goure', "Is There as Military-Technical Revolution in America's Future?," 180.

2. Weapons Systems and Platforms

Information warfare will probably be a central focus of future wars. But it alone cannot deter or win wars. Essential elements which must be present to exploit the benefits of information warfare are appropriate and sufficient weapons and platforms, i.e. precision strike weapons and launch platforms. Long-range precision strike weapons mated with stealthy platforms will allow mission accomplishment while minimizing platform vulnerability to enemy countermeasures. The best known examples are the U.S. Air Force's F-117A and B-2.

3. Cooperative Engagement Capability

Another component of the RMA is the Navy's concept known as the Cooperative Engagement Capability (CEC). It entails a concept for the control of naval combat forces that "fuzes sensor data from multiple platforms into coherent, so-called composite tracks of fire control quality, effectively transforming a naval battle group into a single distributed weapons system."[27] If the CEC proves effective it will improve the timeline

---

[27] Joris Janssen Lok, "Three Thrusts for Naval Air Defense," *Janes Defence Weekly,* (3 April 1996), 25.

of target data collection, analysis, dissemination, and weapon assignment while enhancing redundancy against jamming or environmentally-caused interference of targeted weapons.[28] More interesting, CEC will allow a ground force commander to task a Tomahawk cruise missile aboard a destroyer to strike a target using targeting data gathered from an Air Force surveillance aircraft. The potential of this integration is encouraging, however, caution is necessary pending resolution of certain critical command and control issues discussed in Chapter VI.

[28] Ibid.

IV. REVOLUTION AT SEA

Rapid change in both U.S. military technologies and strategies has compelled the Navy to radically revise some of its operational concepts. This chapter considers the shift in the Navy's operational focus as reflected in recent key documents.

A. CHANGE IN OPERATIONAL FOCUS

In response to the demise of the Soviet threat, then Chief of Naval Operations, Carlisle Trost, modified the Maritime Strategy with the issuance of an article entitled "Maritime Strategy for the 1990's."[29] After noting the end of the Cold War, the document expressed concern for the dangers posed by military applications of rapid technological advancements throughout the world. It noted that the "most worrisome aspect of the increasing diffusion of global political and military power is the accompanying spread of high-technology weaponry."[30]

The trend suggested in "Maritime Strategy for the 1990's" was supported by the Navy's 1991 document "The Way Ahead."[31] The dissipation of a bipolar world was

[29] Admiral Carlisle A. H. Trost, Chief of Naval Operations, "Maritime Strategy for the 1990's", (as published in the Naval Institute Proceeding/Naval Review, 1990.)

[30] "Maritime Strategy for the 1990's", 94.

[31] "The Way Ahead" was written by Secretary of the Navy H. Lawrence Garrett III with Chief of Naval Operations, Admiral Frank B. Kelso II and Commandant of the Marine Corps, General A. M. Gray. Citations are from this document as published in The Naval

acknowledged as well as the increasing regional threat which accompanied it. This prompted the Navy to declare a change in ground rules and assumptions, leading to a restructured naval force with revisions in strategy, tactics and operating patterns.[32] Additionally, political and fiscal pressures associated with reducing the national debt were recognized as a concern which would unquestionably affect the level of resources available for defense.

...From the Sea and its successor, Forward...From the Sea, established the basic direction for the Navy today. The Navy's most recent attempt at spelling out the operational implications of ...From the Sea and Forward...From the Sea, is a document known until the fall of 1996 as 2020 Vision.[33] Though unpublished as of this writing, the ideas expressed in that document are suggestive of how Navy planners envision the exploitation of long-range precision strike capabilities. The ideas are formulated in a three-tier structure consisting of: (1) strategic options to break the enemy's will using precision operations to exploit critical enemy vulnerabilities; (2) operational options to attack the enemy's infrastructure which will require joint forces to identify specific key nodes in order to abort the enemy's means of fighting effectively; and (3) tactical options requiring direct combat with the support of information dominance.[34]

Institute Proceedings, (April 1991).

[32] ...From the Sea, 2.

[33] 2020 Vision: A Navy for the 21st Century, (Office of the Chief of Naval Operations, Washington, D.C.: draft copy 15 August 1996.) Cited with permission of Captain Smith of the CNO's Executive Panel, (henceforth noted as 2020 Vision.)

[34] Ibid, 6-7.

B. NAVAL EXPEDITIONARY WARFARE

With the Navy's principal Cold War mission - to secure command of the sea against the Soviet threat - having effectively been obviated, the Navy reoriented its strategy toward influencing events on land. This entails a significant change in focus for the Navy, but adoption of a policy which concentrates on operations associated with the littoral is a logical and necessary strategic development.

The use of naval forces in littoral waters against the land mass is not a new concept per se'. According to then Admiral William Owens:

> ...Affecting events on land is not a new concern of naval forces. But in concert with the Marines (or mobile Army units) the navy's new operational concept goes far beyond the traditional notion of power projection to a broader concept, better understood as battlefield dominance.[35]

Naval expeditionary war, i.e. amphibious warfare, is one "traditional" way for naval forces to project power ashore. Carrier-based air strikes are another. Yet, neither are sufficient if the Navy is to make a decisive impact on the "halt-the-invasion" phase of a future MRC. The early drafts of 2020 Vision proposed the Navy place reliance on precision operations, smart targeting, and massed fires from the sea.[36] One new naval platform, the Arsenal Ship, is being advocated as the means to bring early, massive, and

---

[35] William A. Owens, High Seas: The Naval Passage to an Uncharted World. (Annapolis, MD: Naval Institute Press, 1995), 78-79.

[36] 2020 Vision, 9.

sustained firepower to bear against an aggressor. This ship is emerging as the Navy's answer to the question of how sea-based forces might have a decisive impact in a future MRC.

V. THE ARSENAL SHIP: A SYNTHESIS OF REVOLUTIONS

In order to operationalize the strategic concepts laid out in the draft 2020 Vision paper, and specifically, make a decisive difference in the early phases of an MRC, the Navy must acquire the capability to bring large volumes of combat power to bear. Moreover, it must do so discriminately, that is to say, against targets that count and with little collateral damage. Air Force bombers can deliver large amounts of ordnance with relatively high precision. However, forward operating bases necessary for a high tempo of strike operations may not be available. Forward deployed Arsenal Ships are the Navy's solution.

According to Rear Admiral Daniel J. Murphy, Director of the Surface Warfare Division, Office of the Chief of Naval Operations (N86), the Arsenal Ship is "best defined as providing multifunctional support to the land battle in its role as the 'battleship of the 21st century'."[37] Norman Polmar, a leading naval strategist, noted that "this is the first completely new warship concept by the Navy since the 1950s when it developed the fleet ballistic missile submarine...it's an opportunity for Navy admirals to show they're not fighting the battle of Midway, but taking advantage of the newest technologies."[38]

Naval planners have proposed that the Arsenal Ship can provide massive

---

[37] Scott C. Truver, "Floating Arsenal to be 21st Century Battleship," International Defense Review, 29, no. 7, (1 July 1996), 44.

[38] Norman Polmar, as cited by John Mintz in "Navy Developing Remote-Control Ship," Washington Post, (23 June 1996), 1. Mr. Polmar is a naval analyst, historian and former advisor to three secretaries of the Navy.

firepower in the initial stages of a conflict to halt an aggressor's advancement.[39] The ship is envisioned to be a highly automated, missile-laden forward-deployed platform which will be capable of providing not only fire support for ground forces ashore, but also theater air defense, and theater ballistic missile defense.[40] Although it is a naval platform, as such, the Arsenal Ship advocates insist it will be a de facto joint system due to the variety of capabilities it is intended to have.[41]

A. CONCEPT OF OPERATIONS

On 11 April 1996 the Navy promulgated The Arsenal Ship Concept of Operations.[42] It contends that the Arsenal Ship is compatible with ...From the Sea and Forward...From the Sea, and that it is designed to meet the challenges of current and anticipated future naval requirements in support of land battles. Given budget constraints, the ship is to be designed, built, and manned in an affordable manner, while providing a much needed enhancement to the existing force of carriers and land attack-

---

[39] Captain Smith, interview conducted 20 May 1996. Commander R. Cameron Ingram, Surface Warfare Division, Plans Office, Office of the Chief of Naval Operations (N863). Information from Commander Ingram was gathered during an interview conducted 23 May 1996.

[40] Ibid.

[41] Captain Smith, interview conducted 20 May 1996.

[42] Rear Admiral Daniel J. Murphy, Memorandum for Distribution, Promulgation of The Arsenal Ship Concept of Operation, (11 April 1996).

combatants and submarines.[43]

1. Weapons Capability

The Arsenal Ship is being designed with approximately 500 Vertical Launch System (VLS) cells and is intended to have space for future extended range gun systems.[44] This is done so that the ship will be able to launch a diverse array of both current and future weapons. According to Captain Smith weapons under consideration for the Arsenal Ship program include:

- a 155mm gun firing extended-range guided munitions out to 100nm (185km)

- Tomahawk Land Attack Missile (TLAM) with a range of approximately 1,500 miles and a speed of 548mi/hr (885km/hr)

- Fast Hawk, a variant of the Hawk surface-to-air missile system in development, with an anticipated range of approximately 1,000 miles and a terminal velocity of Mach 3.5 to Mach 5

- a land attack variant of the Navy's Standard Missile, known as the "Strike Standard"

- SM-2 Block IV/IVA Standard Missile with extended range and a speed of over Mach 2.5 for theater air-defense

---

[43] Ibid, 1.

[44] Commander Jan Van Tol, Office of Net Assessment, Office of the Secretary of Defense. This information was gathered during an interview conducted on 20 May 1996.

- a naval variant of the current 24-inch diameter battlefield missile Army Tactical Missile System (ATACMS) with a range of 60-160nm (110-295km)[45]

- Sea SLAM, a variant of the air-launched Standoff Land-Attack Missile (SLAM) which is a modification of the Harpoon and can deliver a 500lb warhead over 50 miles (80km)

- Evolved Sea Sparrow Missile (ESSM) for self-defense.[46]

The Arsenal Ship will not necessarily be deployed with all of these weapons, nor is it intended to be limited to them. They simply illustrate the range of weapons capabilities the Navy has in mind.

2. Interoperability

The Arsenal Ship Concept of Operations states that the ship will employ the CEC discussed in Chapter III. According to Captain Smith, Navy planners view the ship as a "remote magazine" with remote missile selection, onboard missile initialization and remote launch capabilities. Initially the concept intends the "remote magazine" to be controlled by an accompanying Aegis weapons system; however, future plans are for this

---

[45] Test firing from a surface ship was conducted in February according to "The American Scud," *Navy News & Undersea Technology*, 12, No. 41, (23 October 1995).

[46] This list of weapons was compiled from interviews conducted with Captain Smith, 20 May 1996, Commander Ingram, 23 May 1996, Commander Joseph Bouchard, Strategy and Concepts Division, Office of the Chief of Naval Operations (N513), 21 May 1996 and from "Arsenal Ship...21st Century Battleship," a brief prepared for Admiral Hogg, USN (Retired) by OPNAV (N86), (23 May 1996), 3.

control to be extended across the joint spectrum, including the Air Force AWACS and JSTARS aircraft. There is even an employment concept that may eventually permit weapons' designation and launch by an individual ground force unit.[47]

To illustrate how this might work the following example is provided: (1) a satellite, reconnaissance plane, shipboard radar or ground based unit detects a battery of SCUD missiles; (2) networking capabilities of the CEC allows all U.S. units in the area of responsibility (AOR) to instantly share the data; (3) a command to launch weapons from the Arsenal Ship deployed in the AOR is given by a remotely located commander (possibly on another ship, with ground forces, or on an AWACS aircraft); (4) resulting in weapons from the Arsenal Ship being remotely controlled to strike a designated target. Naval officials are quoted as saying, "the stockpile of missiles deployed aboard the Navy's future Arsenal Ship could be controlled by Army or Air Force officers depending on the mission."[48]

If this capability is realized it would potentially provide the Joint Task Force (JTF) commander with increased flexibility in determining the most appropriately suited weapon and platform for utilization in a strike. If this proves successful, then the Arsenal Ship would be one of the first truly joint platforms in the U.S. military inventory.

---

[47] Captain Smith, interview conducted 20 May 1996.

[48] Robert Holzer, "Commanders May Share Arsenal Ship Assets," Defense News, (17 June 1996 - 23 June 1996), 10.

3. Manning

The Arsenal Ship Concept of Operations says little about manning other than that the crew will be limited to no more than 50 personnel. It does, however, allow for some additional berthing space to accommodate special evolution detachments. Further review of this program indicates that advances in information technologies, stealth and weapons guidance should allow the Navy to operate the ship with a greatly reduced crew.[49] Automation in engineering, as well as weapons and damage control systems could reduce the need for large crews. This, in turn, could contribute to an increase in space available for VLS cells. In addition, a reduction in crew size could allow the ship to remain at sea for extended periods of time. This would be made possible by the rotation of crews via helicopters or ferrying them by transport boats. Merchant tankers have been crewed in this fashion for years. However, the concern has been expressed that small crews could seriously degrade damage control capabilities. Immediate damage control measures are often essential to the survival of a combat vessel in the event it sustains damage during combat. With a minimal crew on board the Arsenal Ship, questions remain unanswered as to its capability to effectively control combat damage or fire and flooding.

[49] Robert Holzer with Pat Cooper, "Warships May Use Leaner Crews; Report Recommends Additional Firepower for U.S. Navy Vessels," Defense News, (29 January - 4 February 1996), 4.

4. Ship Design

The Arsenal Ship Concept of Operations states the design is to be based on commercial practices and incorporate previously mentioned automation. It is expected to have: (1) a double hull, which has not been used on a U.S. warship before; (2) a hull which will ride low in the water and incorporate additional stealthy features to degrade detection by enemy radar and sonar, such as those which have been tested on the Sea Shadow experimental ship or developed in the B-2 and F-117A programs; and (3) have minimum structural features exposed on the deck and above with navigation being directed from a small deck-level conning station.[50] A possible submersion capability has also been discussed.[51]

Prior to selection of the Arsenal Ship, there were four alternatives the Navy considered. The first would have been a modification of Ohio-class SSBNs which the Navy is removing from strategic service. The second was a modified Arleigh Burke-class destroyer (DDG-51.) A third possibility considered the use of a modified commercial oiler. And the last was a new hull designed from the keel up, i.e. the Arsenal Ship itself.[52]

According to Captain Smith studies were conducted to assess the feasibility of

---

[50] Robert Holzer, "U.S. Navy Eyes Options as Arsenal Ship Takes Shape," Defense News, (5 February - 11 February 1996). 20, and Holzer, " U.S. Navy's New Arsenal Ship Takes Shape," Defense News, (8 April - 14 April 1996), 4.

[51] Commander Van Tol, interview conducted on 20 May 1996.

[52] Ibid.

each option and the first three options were losing support for the following reasons.[53] A modified SSBN, though very stealthy and able to accommodate a reasonably large number of VLS tubes, would be far too expensive to justify. The modified DDG-51 concept would limit space available for VLS cells to approximately 150; therefore, the associated cost per launcher would have been steep, thus making it infeasible. The third option, the modified oiler, seemed to be fairly reasonable considering survivability of oilers struck by Iraqi and Iranian missiles during the "Tanker War" in the early 1980s. Obstacles presented with this option were the limited availability of U.S. oilers, limited lifetime coinciding with those that might be available, slow speeds and overall reduced capabilities.

The fourth option of an entirely new hull has gained acceptance. The Navy, in an agreement with the Defense Advanced Research Project Agency (DARPA), has awarded contracts to five shipbuilding/systems integration consortia "to develop a new low-observable surface ship to provide firepower for long-range strike, naval surface fire support and theater air defense."[54] According to the agreement the five consortia will have six months to conduct studies and develop designs leading to the selection of two teams for a 12 month period of refining concepts and performance specifications into a functional design. Ultimately, one team will be selected to produce a detailed design, build a demonstrator, and submit an irrevocable offer for construction of five additional

---

[53] Captain Smith, interview conducted on 20 May 1996.

[54] "Arsenal Ship Programme Launched," *Jane's Navy International*, 101, No. 7, Richard Scott, Editor (1 September 1996), 5.

vessels as well as the conversion of the demonstrator into an operational unit.[55] The goal for the demonstrator to go to sea for trials is fiscal year 2000.

B. COSTS

With budgetary restrictions playing such an important role in the post Cold War era, the Arsenal Ship may be well suited to filling the littoral gap in a cost efficient manner. The following three areas are viewed as primary cost saving segments of the Arsenal Ship program: (1) research, development, test and evaluation (RDT&E), (2) weapons capacity, and (3) personnel costs.

1. Research, Development, Test and Evaluation

On 18 March 1996, the Navy and DARPA signed an agreement to jointly develop the Arsenal Ship. The cost of development, production of the demonstrator and performance testing and fleet evaluation was agreed not to exceed $520 million with the Navy and DARPA funding $350 million and $170 million respectively.[56] Compared to

---

[55] Ibid.

[56] Larry Lynn, John W. Douglass and J. M. Boorda, "Arsenal Ship Program," Joint Memorandum from the office of the Chief of Naval Operations, Assistant Secretary of the Navy (Research, Development, and Acquisition), for Commander, Naval Sea Systems Command, Chief of Naval Research, (18 March 1996.) Mr. Lynn is Director, Defense Advanced Research Project Agency, and Mr. Douglass is Assistant Secretary of the Navy (Research, Development and Acquisition). Admiral Boorda was Chief of Naval

an estimated building cost of $4.5 billion for a new Nimitz-class aircraft carrier, $1.9 billion for a new Ohio-class submarine, or $800 million for a new Arleigh Burke-class destroyer, the project funding goal appeared modest and considerate of the constraints of current budgeting issues. The Navy and DARPA also agreed that only off-the-shelf systems were to be used. Any new systems proposed for development would require the specific approval of the Assistant Secretary of the Navy (Research, Development, and Acquisition.)

Extended deployment cycles are planned for this platform. The automation of shipboard systems is expected to permit a substantial reduction in crew. With an anticipated crew of 50 or less, rotations will not require the Arsenal Ship to pull into port. Meanwhile, the capability of housing a large number of weapons potentially allows the Arsenal Ship to remain at sea for extended periods of time during combat.

2. Weapons Capacity

The cost of weapons intended for the Arsenal Ship poses a concern. With a capability of housing 500 PGMs, the Arsenal Ship will potentially carry approximately $500 million worth of munitions.

The trade off for the benefit of advanced technology incorporated in PGMs is the more expensive cost per round than that associated with conventional munitions. At

Operations.

$800,000 each, the Tomahawk cruise missile is the most expensive missile expected to be placed on the ship.[57]

Nevertheless the technology provided by precision strike might also contribute to an overall reduction in cost. If the Arsenal Ship's stealth features, coupled with the defenses of the Aegis system, are as effective as Navy planners anticipate, then this platform would seem to have an increased chance of delivering its weapons without significant interference from enemy defenses. Therefore, the PGMs potentially will have an increased rate of direct hits on their designated targets. This combination could reduce the number of weapons required to be employed to destroy a target. Meanwhile, logistical requirements for an operation could be substantially reduced.

Additionally, the precise load-out of an Arsenal Ship may vary based on predetermined threats and objectives. Accordingly, the numbers of any given weapons system would likely vary for each deployment.[58] Therefore, the exact cost associated with weapons housed on board an Arsenal Ship seems difficult to determine in advance. It is prudent to say, however, that the cost associated with the weapons on an Arsenal Ship, in both the type of weapons and the numbers which it can carry, will be substantial.

3. Personnel

---

[57] Andrew F. Krepinevich, "Letters To The Editor," *Washington Post*, (27 July 1996).

[58] Captain Smith, interview conducted 20 May 1996.

Emerging RMAs and their products are typically underwritten by advances in technology. In today's environment such advances are predominantly associated with computers and related information technologies. With the improvements in computer technology made during the past several decades, far more can be accomplished with fewer people and in much less time, speaking in general terms. From this arises the question to what extent would the technologies incorporated into the Arsenal Ship impact manpower requirements? Little information is currently available on this issue, but it appears as though Navy officials are assuming increased automation will equate to decreased personnel.

Recalling that the Navy intends to incorporate innovative stealthy features on the Arsenal Ship, stealth itself might assist the Navy in its efforts to reduce personnel requirements. If, in fact, stealth platforms do have greatly increased survivability due to their ability to evade detection, it seems to logically follow that fewer platforms will be required. Having only a few platforms, coupled with extensive automation, would concurrently reduce the number of operators.

Staging requirements could potentially be reduced as well. With limited numbers of Arsenal Ships having extended deployment periods - currently no more than six are planned - there would not be an extensive need for shipyard support. However, since maintenance personnel might require additional training to support the needs of advanced systems associated with this ship, the initial indication is that these manpower numbers might be increased. However, considering that fewer platforms will be required, these manpower needs could possibly be reduced as well. As a result it appears as though the cost savings, associated with a limited number of personnel required to support the Arsenal Ship, could be substantial.[59] Should all the pieces of the Arsenal Ship fall into place, the Navy may very well reap benefits from the platform's reduced personnel requirements and their associated costs. However, questions still exist regarding the previously mentioned damage control issue and until those are satisfactorily answered, as well as questions which follow in the next section, more investigation into actual costs and benefits from drastically reduced manning will be needed.

C. QUESTIONS AND CONCERNS

Like any new program the Arsenal Ship has its share of critics. There are those who believe the Navy does not need any additional ships in light of the lack of a dominant threat to U.S. national security. Others acknowledge the threats posed by regional conflicts, but do not feel that the Arsenal Ship presents a substantial improvement to current capabilities.[60]

---

[59] Ibid.

[60] Critical assessments include Greg Pickell's article"Arsenal Ship Fails to Hit the Mark", Defense News, (16 October 1995 - 22 October 1995), 55; William Lloyd Stearman's article "A Misguided Missile Ship; Old Battleships Would Do a Better Job Than a Pricey New Boat" *Washington Post*, (7 July 1996), C03; and Robert C. Penison's article "Gunfire to Get Them Off the Beach", *Washington Post*, (14 August 1996), A20.

1. Weapons Cost

The cost of weapons seems to be one of the largest points of contention between the program's supporters and its antagonists. With the large volume of weapons intended to be housed on an Arsenal Ship and the intention of those weapons to be mostly technically advanced systems, adversaries of the program contend that the ship itself may not have a high price tag, but the associated costs of its weapons will be excessive.

The array of missiles being considered for the Arsenal Ship will cost $500,000 to $1.5 million per missile, according to the Navy Times (July 29, 1996.) If these costs are accurate then an Arsenal Ship could potentially carry an estimated $2 billion worth of weapons. Although these estimates seem excessive, with the cost of a Tomahawk being the most expensive of the intended weapons at $800,000, a completely loaded ship would still have a very high weapons cost. Skeptics of the program cite Admiral Boorda's own concern that "it is nice occasionally if the target costs more than the bullet you shoot," as a summation of their argument.[61]

An opposing view to this concern is that the U.S. military will have a vast inventory of weapons anyway, so why not have them forward deployed and immediately available to a theater Commander-in-Chief (CINC) instead of stockpiled in the United States? The capability of the Arsenal Ship to help alleviate the stockpiling problem, while housing the weapons in a manner which makes them ready for immediate use, is

[61] David Evans, "The Navy's Blues," *New York Times*, (8 June 1996,) 15.

one of the goals presented in a brief on 29 August by the Navy's Surface Warfare Division.[62]

2. Survivability

Loading up one ship with such an expensive cache of weapons raises the issue of platform survivability. If a CINC concentrates a majority of theater assets in one platform, as opposed to the Cold War strategy of dispersion, that platform would logically become the primary target of adversaries in the region. Therefore, the protection of an Arsenal Ship would be essential.

Current designs for the Arsenal Ship incorporate few self-defense measures. The most touted measure is that of a double hull, yet, it does not in itself seem sufficient. Additionally, there have been discussions of a small complement of last ditch self-defense systems such as the Phalanx Close-in-Weapons-System but a final decision is yet to be made.[63] The vulnerability of the platform has been acknowledged by naval planners, however, they never intended to jeopardize this platform by deploying it without an escort.[64] The Arsenal Ship is being designed in affiliation with the Aegis system as noted in the operational concept specifying, "the Arsenal Ship will operate in

---

62 "The Arsenal Ship," brief prepared by OPNAV (N86), (29 August 1995), 3.

63 Commander Van Tol, interview conducted 20 May 1996.

64 Captain Smith , interview conducted 20 May 1996.

any threat environment under the protective umbrella of battle force combatants, however, it must have self-defense measures which reinforce its survivability against 21st century anti-ship missiles, torpedoes and mines."[65]

Passive defense is the predominant self-defense measure that the ship will rely on. "Passive defense will allow the Arsenal Ship to capitalize on the benefits of mass tonnage, innovative applications of multiple hull integrity, and signature reduction."[66] Stealth technology is intended to be utilized in the design of a hull with a low radar cross section in order to evade enemy radar. Navy planners intend for the Arsenal Ship to ride very low in the water and have radar-suppressing surfaces and shapes to assist in reducing its radar cross section.[67] Another stealth feature to be incorporated is that of technology associated with wake reduction and acoustic propagation.[68] A number of stealth related technologies such as these have been successfully tested on the Navy prototype ship know as the "Sea Shadow".

The second major selling point for survivability is that of a double hull. This is a design which would enhance the hull integrity against mines, missiles and torpedoes. Never before has a U.S. warship used such a design; yet, the additional protection provided by this structure will increase the probability of survival. Strikes against current ships in the U.S. fleet with single hulls have demonstrated respectable survivability as

---

[65] Rear Admiral Murphy, "Promulgation of the Arsenal Ship Concept of Operations," 4.

[66] Ibid.

[67] John Mintz, "Navy Developing Remote-Control Ship," A22.

[68] Commander Ingram, interview conducted 23 May 1996.

recently as the Persian Gulf War, therefore, doubling the hull on the Arsenal Ship is a strong selling point for the Navy.[69] The problem with this approach is that most mines which have been struck have been very small and of old design.

Skeptics do not seem to believe these measures will be sufficient. In a recent article, William Stearman questions "how can it be a 'stealth ship' when it requires an easily located escort fleet for protection?"[70] Logic implies that if the Arsenal Ship is deployed without an escort fleet, its stealth features, while likely to reduce its possibility of detection, would not be able to guarantee it remained undetected. In the event the ship were to be detected, by perhaps a diesel submarine, a double hull could prove insufficient to guarantee survivability of the platform.

If this program cannot provide more assurances for the survival of an Arsenal Ship, then some skeptics of the program seem justified with their concerns. The apparent vulnerability of an Arsenal Ship may present difficulties for JTF commanders who may be risking a substantial amount of their theater assets if Arsenal Ships are fully loaded yet haven't been able to ensure enhanced survivability.

---

[69] Captain Smith, interview conducted 20 May 1996.

[70] William L. Stearman, "The Navy's Proposed Arsenal Ship," *The Retired Officer Magazine,* 102, No. 11, (November 1996), 39. Mr. Stearman is a former Navy officer and member of the White House National Security Council staff.

3. Battleships

Besides his concerns over survivability, Stearman claims that the Iowa-class battleships could provide substantial firepower more cost-effectively. He insisted that, missiles are "fine for distant, high-value, fixed-targets, but for supporting troops locked in combat, 16-inch rounds are not only infinitely cheaper, but have substantially greater penetration and blast effects and disperse submunitions as well, if not better."[71] He claimed that with current technology the range of these rounds can be extended from 30 to 100 miles.

Although there are valid points in arguments such as these, other issues seem to be either ignored or misinterpreted. An example is the vulnerability of 50 year old battleships. According to Andrew Krepinevich we need only to be reminded that "a single shell from the German warship Bismarck sunk the British battleship HMS Hood, ... or of the Japanese torpedo attack on Pearl Harbor's 'Battleship Row.'"[72] In light of weapons advancements in the past 50 years, the vulnerability of these ships appears to be substantially increased.[73]

Additionally, the spectrum of capabilities planned for the Arsenal Ship is far

---

[71] Stearman, "A Misguided Missile Ship...".

[72] Krepinevich, *Washington Post.*

[73] Ibid.

greater than that of a battleship.[74] Arsenal Ships are intended to carry anti-ballistic and anti-tactical missiles while also having a 155mm gun that will provide ground support firepower. Unmanned aerial vehicles may also be carried. "The Arsenal Ship's flexibility is as varied as the weapons we choose to put aboard," says Captain Smith.[75] Meanwhile, many more personnel are required to operate a battleship than an Arsenal Ship and the need for logistical support for a battleship is substantially greater.

D. EARLY SUPPORT

In January 1988 then Vice Admiral Joseph Metcalf III published thoughts on a "Revolution at Sea."[76] He took note of the enormous expansion over the past decade in the warship's battlespace especially as the result of the 1,200-mile Tomahawk missile."[77]

Metcalf's article discussed issues such as redesigning ship superstructures for a modern era with little more than a conning station visible on the deck. He also proposed that the warfighting policy of a revolutionary Navy would be to maximize a warship's ability to deliver ordnance on target. With this tasking, he conceptualized a ship designed

---

[74] Edward A. Smith, "Naval Firepower for the 21st Century," *Washington Post*, Op-Ed Section, (27 July 1996).

[75] Ibid.

[76] Joseph Metcalf III, Vice Admiral, USN (Ret), "Revolutions at Sea," U.S. Naval Institute Proceedings, 114, no. 1019, (January 1988), 34-39.

[77] Ibid, 36.

with an extensive volume of weapons, most likely VLS cells for cruise missiles.[78]

An emerging concern of the RMA is a need to increase the number of individually targetable missiles in theater to counter future threats. The Arsenal Ship conceptually provides up to 500 VLS cells which could substantially improve capabilities to counter these threats.

Issues such as increased missile capacity, automation, reduced manpower requirements, and a stealthy ship design espoused by Metcalf are being incorporated into the Arsenal Ship.[79] Additionally, efforts by the Navy to examine potential RMA issues have resulted in the utilization of the Arsenal Ship in a number of recent wargames.

Metcalf concluded that technology alone could not lead the way of the future Navy. He stressed the need for the Navy and its leadership to "step up to the mark," as revolutions occurring across the spectrum require innovation within the naval ranks. Supporters of the Arsenal Ship assert the Navy has acknowledged this necessity and through the Arsenal Ship program the Navy is pursuing innovative methods to ensure its viability in the next century.

---

[78] Ibid, 38.

[79] Captain Smith, interview conducted 20 May 1996.

VI. OPERATIONAL USE OF THE ARSENAL SHIP

Having reviewed the strategic and technological changes currently impacting the U.S. military and examining the physical attributes of the Arsenal Ship and its associated systems, the next step is to address the actual employment of the ship. In doing so the ship needs to be assessed as to whether it will provide the United States with a joint capability that fills the strategic gaps of the post Cold War and presents innovative ways, consistent with the RMA, of dealing with the future. This includes a review of its anticipated effectiveness in carrying out its military missions, discussions regarding command and control issues, and its potential for joint interoperability.

A. MISSION

The missions of the battleship, aircraft carrier and Arsenal Ship share a number of similarities as well as differences. The similarities revolve around their primary goal of power projection, while their differences are defined by the technology associated with their respective eras of development. With respect to the battleship, Richard Hough has written how, "like most weapons of destruction created in times of peace, the first function of the battleship was to instill fear in the hearts of men. For almost a century the threat of the armored battleship fleet was the first instrument of power diplomacy.[80] It

[80] Richard Hough, *Death of the Battleship*, 4.

was specifically designed for destroying similar targets, i.e. other battleships.

With the modernization of the U.S. fleet, through the incorporation of aircraft carriers during World War II, continuity of the battleship mission of power projection and forward presence was maintained. However, the carrier redefined this mission through the use of sustained long-range strike. With further modernization in strategy and technology it appears as though Navy planners intend for the Arsenal Ship to play a similar role in redefining the Navy's mission through massive firepower and precision warfare.

The Air Force has long had the ability to extend its range of strike through the use of intercontinental bombers and intercontinental ballistic missiles. These options have arguably been favorable to carrier based aircraft due to a reduced risk of collateral damage. According to Commander Ingram, the Arsenal Ship would risk relatively little collateral damage; yet it could provide a JTF with a substantial strike capability that is more timely than the Air Force alternatives.

Technology has added PGMs to the equation. Cruise missiles, such as the TLAM, have gained utility as a primary means of conducting strikes. The combination of extended range and precision guidance system enhances the chances of commanders reducing the risk to their platforms and personnel. This is accomplished by launching weapons from far greater distances than was possible in previous eras. At the same time PGMs provide theater platforms with an immediate response capability due to their increased delivery speeds over more conventional weapons.

As Krepinevich stated in his recent publication "*A New Navy for a New Era*," these systems and munitions have increasingly been encroaching on operations "turf" that was once the sole preserve of the carriers.[81] If such changes continue to occur, might it be possible that the role of the aircraft carrier could decrease while other platforms such as the Arsenal Ship take on a greater naval role? This is not to say the aircraft carrier will no longer be needed. In the foreseeable future there remain a number of missions which will still be more aptly suited to the abilities of the carrier than a platform such as the Arsenal Ship, specifically that of sustained strike operations.[82]

A closer look at the potential effects of the Arsenal Ship shows a variety of operational capabilities. Some are redundant, yet this platform is intended to provide an enhanced capability tailored toward massive strike during the initial stages of a conflict to meet the demands of... From the Sea and Forward...From the Sea. Currently, there is no U.S. national asset which can accomplish this task as rapidly and decisively as a permanently deployed ship containing 500 VLS cells.

1. Forward Operations

The Bottom-Up Review focused attention on the U.S. military's need to prepare for two MRCs. Accordingly, Forward ... From the Sea states the Navy shall maintain a

---

[81] Andrew Krepinevich, Jr., "A New Navy for A New Era," 6.

[82] Richard Hough, *Death of the Battleship*, 13.

posture of forward operations.[83] This is visualized as a means to provide peacetime presence and crisis response in the event of a regional conflict. The current means by which the Navy accomplishes this mission is through the deployment of carrier battlegroups (CVBGs) and Amphibious Ready Groups (ARGs). However, budgetary reductions in the 1990's make continual overseas presence of CVBGs increasingly difficult to sustain.

A number of options for sustaining the Navy's forward presence have been proposed. Among those proposals, the Arsenal Ship is relatively cheap to build, much less expensive to operate than a carrier and could provide a theater CINC with capabilities currently unavailable.

One of the primary requirements for the Navy's forward deployed forces is to "satisfy joint naval expeditionary force warfighting requirements early in regional conflicts."[84] The Arsenal Ship is designed to accomplish this via three avenues: "(1) massive firepower, (2) exploiting technology, and (3) providing resources to meet the tasks."[85] According to Captain Smith and Commander Ingram, the first could be provided by the overwhelming firepower of the Arsenal Ship. The littoral focus of this ship is a shift from traditional naval concerns of fighting open-ocean battles. Instead, this

---

[83] Forward...From the Sea, 2.

[84] "The Arsenal Ship", N86 brief of 29 Aug 95, 4.

[85] Ibid, 5.

ship is intended to provide massive firepower in the initial stages of a conflict effectively halting an adversary's aggression.[86]

The second is proposed to be accomplished through an array of technical advancements associated with the Arsenal Ship, such as the integration of PGMs and the Cooperative Engagement Capability. Effective utilization of the CEC would conceivably allow enhanced distribution of targeting data among theater assets. This, in turn, should provide a CINC with the capability of selectively targeting vital enemy assets. The combination of such targeting and the use of PGMs in attacking those targets is intended to enhance the probability of hitting vital targets with fewer rounds.[87]

The third avenue is intended to be accomplished through the diversity of weapons available in an Arsenal Ship. Navy planners insist that when a CINC determines he needs ground fire support, the Arsenal Ship will have the capability to deliver it. If the CINC determines theater ballistic missile defense (TBMD) is the primary threat, the ship will be able to deploy weapons in support of this. The Arsenal Ship is also intended to provide theater air defense (TAD). In theory, the ship should contribute to the joint concept of "full-spectrum dominance" through its ability to provide support to allied assets throughout the theater and in numerous missions. The Arsenal Ship's weapons capability and mission requirements are not to be limited to traditional naval operations. Instead it is proposed to combine the strategic requirements of ...From the Sea with those of the

---

[86] Commander Ingram, interview conducted 23 May 1996.

[87] Captain Smith, interview conducted 20 May 1996.

National Military Strategy and Joint Vision 2010 resulting in a platform that can potentially fill very diverse mission requirements across the joint spectrum.

It has also been proposed that the Arsenal Ship can provide a solution to the problem of vulnerable ammunition stockpiles. During Desert Storm, if the SCUD missile that struck near a pier stockpiled with U.S. ammunition at Jubayl, Saudi Arabia on 16 February 1991, had done so closer-in, vast amounts of ordnance could have been lost. In order to alleviate vulnerabilities such as this in the future, the Arsenal Ship is envisioned as having the means to house a vast amount of ammunition while also maintaining it in such a manner that it is combat ready.

Another concern of Navy planners is the "force architecture which identified vertical launch system shortfalls early in a major regional conflict."[88] This architecture identifies less than 1000 VLS cells available during the first eight days of a conflict under current surface combatant capabilities.[89] With a regional presence of two Arsenal Ships, which is the intention of Navy planners, the number of VLS cells available during these first eight days would be more than doubled. Such a presence indicates a vast increase in naval firepower support in the early stages of an MRC.[90] With the complement of weapons projected for an Arsenal Ship, its diversity should allow the ship to provide

---

[88] "Arsenal Ship...21st Century Battleship", 3.

[89] Ibid.

[90] Ibid.

massive firepower ranging from strike to surface fire support to theater air defense.

2. Strike

The actual strike role of the Arsenal Ship seems to have a number of possibilities. The weapons and the design of the ship are proposed to afford the Navy, as well as the other branches of the armed forces, an opportunity to create innovative strike packages. According to Commander Bouchard, this is not to say that current strike packages are obsolete. On the contrary in many cases they may prove to be more than adequate, but if innovative packaging can provide more effective results with less risk to U.S. personnel then the Arsenal Ship may put a new face on the way strike packages are viewed.[91]

The DOD specifically required the incorporation of long-range strike capabilities in addition to expeditionary ground support and fleet support capabilities into the Arsenal Ship. The increasing technological development in the area of PGMs should enhance the Arsenal Ship's capability of destroying strategic targets. It should also provide increased means of suppressing enemy air defenses as well as demolishing enemy military infrastructures.[92] These long-range strike measures could be coupled with the ability to provide early intervention against hundreds of targets.

According to Captain Smith the benefits of successful effects from early

---

[91] Commander Bouchard, interview conducted on 21 May 1996.

[92] Commander Ingram, interview conducted 23 May 1996.

firepower provided by the Arsenal Ship can be substantial. First it could ensure the continued exploitation of freedom of the seas by the deterrence and power projection role it is envisioned to have. This could be accomplished by respecting the sovereignty of partners due to the range and maneuverability of the Arsenal Ship. This exploitation could allow minimizing the use of early land-based units, that have recently lost access to a number of overseas staging facilities through base closures. Secondly, massive firepower is intended to punish the enemy from the initial stages of combat operations. According to the recent OPNAV brief "Arsenal Ship...21st Century Battleship," the intent is to have 1,000 VLS cells available in less than a week. This capability is intended to enhance the credibility of those forces that are forward deployed. Next, the Arsenal Ship could free-up early airlift tasked with transporting munitions into the theater. This capability will potentially make airlift available to provide support across a much larger spectrum. Finally, it will increase the time tactical air components have for preparations of traditional roles such as combat air support, combat air patrol, and strategic strike.

3. Interoperability

Perhaps the most dramatic change from traditional strike operations that the Arsenal Ship would incorporate is its operation under the CEC. As previously indicated the Arsenal Ship "allows for remote missile selection, on-board missile initialization and remote launch orders, and provides remote 'missile away' messages to the control platform."[93] This "program seeks to combine all the combat systems and major sensors on ships into a single, integrated architecture for intelligence, surveillance, reconnaissance, and C4I."[94] The CEC has the potential to provide commanders with "a dramatically improved picture of the extended battle area and greatly improve their ability to engage targets successfully at extended ranges employing long-range precision strikes."[95] As a fully integrated component of the Arsenal Ship, CEC could potentially utilize the weapons capability of this ship to enhance the Navy's long-range strike assets in a way never before attainable.

The benefits the Navy could reap from such an asset seem worth the investment. "The Navy's strike capabilities are not likely to be sustained at today's level of effectiveness unless the Navy can meet the challenge of increasingly capable Third World military systems."[96] To do this the Navy will have to "increase its emphasis on conducting long-range precision strikes, and on defending the fleet from the threat of submarines, mines, and anti-ship missiles."[97] According to Krepinevich's "*A New Navy For A New Era*", the Arsenal Ship will be able to outrange carriers by virtue of its

---

93 "Promulgation of the Arsenal Ship Concept of Operations," 2.

94 Andrew F. Krepinevich, Jr., "A New Navy For A New Era," 40-41.

95 Ibid, 41.

96 Ibid.

97 Ibid.

missiles and the missiles' reduced vulnerability. The latter, claims Krepinevich, will allow Arsenal Ships to operate more closely to shore than carriers. The Arsenal Ship is touted as having the ability to undertake offensive mining and countermine operations, and still operate with less detection - and less protection - in the littoral areas than carriers, however, the skepticism presented in the previous chapter regarding vulnerability raises questions as to the accuracy of these points. Meanwhile, Krepinevich's point that "an emphasis on a more balanced, more distributed fleet that exploits advancing technologies under the CEC, could provide future fleet commanders as well as CINCs with a variety of options for conducting strike and presence operations," seems well founded.[98]

B. COMMAND AND CONTROL

The implementation of CEC, though promising in many respects, raises fundamental questions about command and control. With the control of Arsenal Ship weapons being located elsewhere in the theater, the Arsenal Ship's commanding officer will have relatively little responsibility for actually fighting the ship. This would present a major paradigm shift from the historical role of a ship's commanding officer and would represent a major institutional change for the Navy.

Aside from institutional concerns, who decides what goes into the tubes and who will actually control the launching are two primary command and control questions.

[98] Ibid.

According to Captain Smith, the theater CINC would determine the load-out of an Arsenal Ship prior to deployment based on the theater threat assessment. The decision for control of targeting and shooting is still being refined; however, the best estimate is that control would be handled in a manner similar to that currently associated with the Tomahawk cruise missile.

Questions regarding who decides the actual location of the Arsenal Ship, control of the multipurpose radars, and which mission will be assigned for multiple-mission weapons are also yet to be resolved, though discussions seem to be leaning toward the JTF commander.[99] Whether or not the JTF commander will delegate any or all of these roles, or will have the legal option to do so, remains unclear.

One additional concern is that of too much access leading to confusion. This concern is meant to address the issue surrounding future projections for the CEC. If the CEC ultimately includes a ground force unit with the capability of launching a weapon from the Arsenal Ship, while additional units such as AWACS, JSTARS, Aegis system and F/A-18 simultaneously have that same capability, command and control issues could become tangled. There would have to be a control point to ensure different units were not tasking the same weapon and that conflicting use of weapons does not occur.

---

[99] Commander Bouchard, interview conducted 21 May 1996.

C. JOINT WARFIGHTING CAPABILITIES

The Arsenal Ship has been conceived and espoused as a fully integrated joint platform.[100] First, it is intended to have strategic mobility and protection capabilities with the ability to operate in forward positions. With fewer personnel on board and the protective umbrella of the Aegis system, the Arsenal Ship should be able to remain forward deployed for extended periods of time. Therefore, it should be able to provide continuous force presence and a forward magazine capacity.[101]

Secondly, as previously mentioned, it is intended to have an excellent strike capability. With the massive volume of weapons an Arsenal Ship is anticipated to have, the Navy proposes the Arsenal Ship will be able to provide sustained and coordinated long-range strikes.[102]

The third capability intended to support joint warfighting is direct support to the land campaign. The Arsenal Ship is perceived to have the flexibility of positioning in such a manner as to effectively interdict advancing armor and to deliver battlefield operational fires in support of U.S. ground forces in the region.[103]

---

[100] Captain Smith, interview 20 May 1996.

[101] "The Arsenal Ship," 7.

[102] Commander Ingram, interview conducted 23 May 1996.

[103] Ibid.

A fourth joint capability would seem to be that of contributing to air superiority and countering Weapons of Mass Destruction (WMD). This is envisioned through the combination of theater air defense and theater ballistic missile defense weapons planned for the Arsenal Ship. Having a permanent availability of these weapons coupled with the remote magazine concept would considerably enhance air superiority and efforts to counter WMD.

Finally, there is the area of command and control. The Arsenal Ship is intended to be fully joint interoperable and capable of remotely firing weapons.[104] The combination of these capabilities appears to be the Navy's proposal for its contribution to affecting events on or over land in the event of an MRC.

[104] Ibid.

VII. CONCLUSION

This thesis asked the question whether changes that brought about the shift in the Navy's strategic focus from open-ocean to littoral warfare have created a shortfall in U.S. naval land attack capabilities; and if so, whether the Navy's proposed Arsenal Ship is the right choice for rectifying the situation? To analyze these questions, four areas were examined: the strategic revolution (Chapter II), the Revolution in Military Affairs (Chapter III), the resultant naval revolution (Chapter IV), the Arsenal Ship as a synthesis of revolutions (Chapter V), and the operational concepts of the Arsenal Ship (Chapter VI).

Neither the strategic revolution in the wake of the end of the Cold War, nor the Revolution in Military Affairs alone appear to be sufficient justifications for a new ship capability. However, when the two are combined, implications for future warfare indicate a need for a new ship capability. The combination of a new strategy which focuses on operating in the littoral regions of the world so as to affect events on land, plus an evolving era of technological advancements, suggests that current Navy ships would lack the capabilities necessary to ensure the Navy's ability to support dominant battlespace superiority.

A. STRATEGIC CHANGES DRIVING THE 21ST CENTURY NAVY

The Navy no longer bases its strategic thinking on the concept of open-ocean naval warfare. This strategy was the focal point of the Cold War; however, in the post Cold War era and for the foreseeable future this strategic approach is no longer warranted. With the publication of ...From the Sea, the Navy made a fundamental shift away from open-ocean warfighting on the sea toward a strategy of joint operations conducted from the sea. How and against what "kinds" of opponents the Navy should prepare to fight was spelled out in the Bottom-Up Review.

The question Navy planners have found themselves confronted with is how their service can effectively contribute to operations during a regional conflict that will predominantly occur on or over land. This must be done in a manner which enables the Navy to retain enough warfighting capabilities in reserve to thwart a second regional conflict should one arise. In order to support joint operations required to accomplish these goals, the Navy is focusing on increasing its naval expeditionary warfare capabilities. Naval planners will have to concentrate efforts on training in the littoral; however, current ship capabilities limit the extent to which this can be done. A new ship designed to operate in the littoral might provide the Navy with an effective means to accomplish these goals.

B. RMA IMPLICATIONS

The extensive advancements in technology in today's military environment indicate the necessity for modernization of U.S. capabilities if the United States is to maintain its dominant role in international security and protect U.S. interests overseas.

Stealth, global technology, long-range precision strike, information warfare, and the effective use of space characterize the current technological movement. The rate at which battlefield information can be collected, analyzed and disseminated is rapidly approaching a rate far greater than most current capabilities can counter. The key to the RMA appears to be precision warfare. The utilization of precision guided munitions tasked to strike targets which are determined essential, through the use of information warfare technology, will alleviate unnecessary expenditure of weapons and should vastly increase the target-kill ratio. In addition to the increasing emphasis on PGMs, the concept of a Cooperative Engagement Capability which fuzes sensor data from multiple platforms into composite tracks of fire control quality could effectively transform a naval battlegroup into a distributed weapons system. Innovative technologies and capabilities such as these will most likely be the norm in the 21st century; therefore, the Navy must continue to modernize in concert with advancements in technology and in military strategy.

C. THE ARSENAL SHIP

Under the Arsenal Ship's operational concept, each ship will have 500 VLS cells capable of carrying an array of current and future weapons. These weapons will range in mission capability from land attack to TAD and TBMD. The Arsenal Ship is to be highly automated, requiring a crew of no more than 50 personnel, and it is to be operated in conjunction with the CEC for remote targeting.

1. Design

The actual design of the Arsenal Ship is still in development; however, Navy planners expect it will have a double hull which will ride low in the water, have stealthy features so as to minimize the probability of detection by enemy radar and sonar, and have minimal structural features above the water line. After reviewing a number of platform options, the Navy, working with DARPA, has decided to design an entirely new ship. A demonstrator ship is scheduled for sea trials in fiscal year 2000.

2. Cost

An obvious concern is cost. The RDT&E cost of the Arsenal Ship was agreed by the Navy and DARPA not to exceed $520 million which is far less than the building cost of an Arleigh Burke-class destroyer. Meanwhile, the associated cost of weapons to be carried on the Arsenal Ship is estimated to be rather high. Nevertheless the configuration of weapons on the Arsenal Ship will be determined by the theater CINC based on his threat assessment. As a result, the associated cost will fluctuate. Additionally, having these weapons housed on the Arsenal Ship ready for use, rather than stockpiled in warehouses or on piers, make the ship an attractive option.

Personnel related costs have been presented with the potential of substantial savings. The incorporation of extensive automation indicates the potential for dramatic reductions in crew size. Consequently fewer lives will be put at risk. With crew changes via helicopter or other capabilities, ship deployments can be greatly extended. Subsequently, this indicates the manpower requirements associated with the Arsenal Ship could be dramatically reduced, resulting in a potential cost benefit for the Navy.

3. Detractors

Survivability of the Arsenal Ship poses some concern. With few self-defense measures intended to be incorporated other than a double hull and some stealth features, it would seem that the Arsenal Ship could be a "sitting duck." The Navy has considered this point and intends to deploy the Arsenal Ship in conjunction with an Aegis equipped platform and perhaps additional ships from a battlegroup.

Command and control issues pose another concern. If the Arsenal Ship does incorporate the CEC which eventually expands to permit targeting by an array of forces, this structure could become disorganized. The potential for trouble associated with dual tasking is disturbing unless a specific command and control blueprint is established.

4. Mission

The Arsenal Ship is intended to operate forward in a manner which provides peacetime presence and crisis response in the event of a regional conflict. With massive firepower, Navy planners hope to create the ability to halt a regional aggressor quickly, allowing time for a counter-build up of friendly forces. The Arsenal Ship will attempt to exploit technology through the integration of PGMs and the CEC, enhancing distribution of targeting data among theater assets. Additionally, it is intended to provide resources to meet required tasks through the diversity of weapons which it is proposed to carry.

These capabilities suggest the Arsenal Ship could provide the Navy with an effective means of affecting events on land and contributing to the joint concept of "full-spectrum dominance."

LIST OF REFERENCES

Books

Hough, Richard. Death of the Battleship. New York: MacMillan Company, 1968.

Owens, William A. High Seas: The Naval Passage to an Uncharted World. Annapolis: Naval Institute Press, 1995.

Reynolds, Clark G. The Fast Carriers: The Forging of An Air Navy. New York: McGraw-Hill, 1968.

Articles

Evans, David. "The Navy's Blues." The New York Times, 8 June 1996, 15.

Fitzsimonds, James R. and Jan M. Van Tol. "Revolutions in Military Affairs." Joint Force Quarterly, 4 (Spring 1994): 25-26.

Goure, Daniel. "Is There a Military-Technical Revolution in America's Future?" The Washington Quarterly, 16, no. 4 (Autumn 1993): 179.

Krepinevich, Andrew F. Jr. "Keeping Pace with the Military-Technological Revolution." Issues in Science and Technology, (Summer 1994): 24.

Krepinevich, Andrew F. Jr. "Letters To The Editor." The Washington Post, 27 July 1996.

Krepinevich, Andrew F. Jr., and Martin Libicki, commentary. "The Coming Defense Train Wreck and What to Do About It." The Washington Quarterly, (Winter 1996): 107.

Holzer, Robert. "Commanders May Share Arsenal Ship Assets." Defense News, (17 June - 23 June 1996): 10.

Holzer, Robert with Pat Cooper. "Warships May Use Leaner Crews: Reprt Recommends Additional Firepower for U.S. Navy Vessels." Defense News, (29 January - 4 February 1996): 4.

Holzer, Robert. "U.S. Navy Eyes Options as Arsenal ship Takes Shape." Defense News, (5 February - 11 February 1996): 20.

Holzer, Robert. "U.S. Navy's New Arsenal Ship Takes Shape." Defense News, (8 April - 14 April 1996): 4.

Lok, Joris Janssen. "Three Thrusts for Naval Air Defense." Janes Defense Weekly, 3 April 1996: 25.

Lok, Joris Janssen. "Arsenal Ship Will Pilot Future USN Combatants." Janes Defense Weekly, 17 April 1996: 3.

Metcalf, Joseph III. "Revolutions at Sea." U.S. Naval Institute Proceedings, 114, no. 1019 (January 1988): 34-39.

Mintz, John. "Navy Developing Remote-Control Ship." The Washington Post, 23 June 1996: 1.

Penison, Robert C. "Gunfire to Get Them off the Beach." The Washington Post, 14 August 1996: A20.

Pickell, Greg. "Arsenal Ship fails to Hit the Mark." Defense News, (16 October 1995 - 22 October 1995): 55.

Scott, Richard, ed. "Arsenal Ship Programme Launched." Jane's Navy International, 101, no. 7 (1 September 1996): 5.

Smith, Edward A. "Naval Firepower for the 21st Century." The Washington Post, 27 July 1996.

Stearman, William L. "The Navy Proposes Arsenal Ship." The Retired Officer Magazine, 102, no. 11 (November 1996): 39.

Stearman, William L. "A Misguided Missile Ship: Old Battleships Would Do a Better Job Than a Pricey New Boat." The Washington Post, (7 July 1996): C03.

-------, "The American Scud." Navy News & Undersea Technology, 12, no. 41 (23 October 1995).

Truver, Scott C. "Floating Arsenal to be 21st Century Battleship." International Defense Review, 29, no. 7 (1 July 1996): 44.

Interviews

Bouchard, Joseph. Strategy and Concepts Division, Office of the Chief of Naval Operations (N513). Washington, D.C.: 21 May 1996.

Ingram, R. Cameron. Surface Warfare Division, Plans Office, Office of the Chief of Naval Operations (N863). Washington, D.C.: 23 May 1996.

Smith, Edward A. CNO Executive Panel. Washington, D.C.: 20 May 1996.

Van Tol, Jan. Office of Net Assessment, Office of the Secretary of Defense. Washington, D.C.: 20 May 1996.

U.S. Government Publications

U.S. Department of Defense. Arsenal Ship...21st Century Battleship. Brief prepared by OPNAV (N86). Washington, D.C.: 23 May 1996.

U.S. Department of Defense. Arsenal Ship Program. Joint memorandum signed by Larry Lynn, John W. Douglass and J. M. Boorda. Washington, D.C.: 18 March 1996.

U.S. Department of Defense. Joint Vision 2010, by John M. Shalikashvili. Washington, D.C.: GPO, 1996.

U.S. Department of Defense. Joint Warfare of the U.S. Armed Forces, Joint Pub 1. Washington, D.C.: GPO, 11 November 1991.

U.S. Department of Defense. National Military Strategy of the United States of America: a Strategy of Selective Engagement, by John M. Shalikashvili. Washington, D.C.: GPO, February 1995.

U.S. Department of Defense. Maritime Strategy for the 1990's. As published in the Naval Institute Proceedings/Naval Review, 1990.

U.S. Department of Defense. Mobility Requirements Study, Bottom-Up Review Update, by Captain J.J. Langer and Lieutenant Commander J. Browne, eds. Newport.

U.S. Department of Defense. Report on the Bottom-Up Review, by Les Aspin. Washington, D.C.: GPO, October 1993.

U.S. Department of Defense. Promulgation of The Arsenal Ship Concept of Operations. Memorandum for Distribution by Daniel J. Murphy. Washington, D.C.: 11 April 1996.

U.S. Department of Defense. The Arsenal Ship. Brief prepared by OPNAV (N86). Washington, D.C.: 29 August 1996.

U.S. Department of Defense. The Military Revolution: Restructuring Defense for the

21st Century. Testimony presented by Andrew F. Krepinevich, Jr. to the Senate Armed Services Subcommittee on Acquisition and Technology. Washington, D.C.: 5 May 1995

U.S. Department of Defense. Testimony presented by Andrew W. Marshall to the Senate Armed Services Subcommittee on Acquisition and Technology. Washington, D.C.: 5 May 1995.

U.S. Department of Defense. The Way Ahead, by H. Lawrence Garrett III with Frank B. Kelso II and A. M. Gray. As published in The Naval Institute Proceedings. (April 1991).

U.S. Department of Defense. The Maritime Strategy/Amphibious Warfare Strategy, by James D. Watkins and P. X. Kelley. Washington, D.C.: GPO.

U.S. Department of Defense. Unified Action Armed Forces, Joint Pub 0-2. Washington, D.C.: GPO, 24 February 1995.

U.S. Department of the Navy. Forward...From the Sea. Washington, D.C.: GPO, Undated.

U.S. Department of the Navy. From the Sea: Preparing the Naval Service for the 21st Century. Washington, D.C.: GPO, September 1992.

U.S. Department of the Navy. Naval Operational Concept. Office of the Chief of Naval Operations. Washington, D.C., draft copy: 14 June 1996.

U.S. Department of the Navy. 2020 Vision. Office of the Chief of Naval Operations. Washington, D.C., draft copy: April 1996.

General Publication

Krepinevich, Andrew F. Jr. A New Navy for a New Era. Washington, D.C.: Center For Strategic and Budgetary Assessments, May 1996.

INITIAL DISTRIBUTION LIST

		No. Copies
1.	Defense Technical Information Center 8725 John J. Kingman Road., Ste 0944 Ft. Belvoir, VA 22060-6218	2
2.	Dudley Knox Library Naval Postgraduate School 411 Dyer Rd. Monterey, CA 93943-5101	2
3.	Captain Frank Petho, USN National Security Affairs (NS/PE) Naval Postgraduate School Monterey, CA 93943	1
4.	Professor Jan S. Breemer National Security Affairs (NS/BE) Naval Postgraduate School Monterey, CA 93943	1
5.	Captain W. M. Dunaway, USN National Security Affairs (NS/DU) Naval Postgraduate School Monterey, CA 93943	1
6.	Ambassador Rodney Minott National Security Affairs (NS/MI) Naval Postgraduate School Monterey, CA 93943	1
7.	Professor Dana P. Eyre National Security Affairs (NS/EY) Naval Postgraduate School Monterey, CA 93943	1
8.	Major General Alain Rickenbacher Plantay 23 CH - 1163 Etoy, Switzerland	1

9. Charles N. Calvano (ME/Ca) 1
Mechanical Engineering
Naval Postgraduate School
Monterey, CA 93943

10. Lieutenant Dawn Driesbach, USN 1
Naval International Program Office
1111 Jeferson-Davis Highway, Suite 701
Arlington, VA 22202-1111

11. Captain Edward A. Smith, Jr., USN 1
CNO Executive Panel
4401 Ford Avenue
Alexandria, VA 22302

12. Commander R. Cameron Ingram, USN 1
The Pentagon Room 4D529
Washington, D.C. 20350-2000

13. Commander Jan Van Tol, USN 1
Office of Net Assessment
The Pentagon Room 3A930
Washington, D.C. 22301-2950

14. Commander Joseph Bouchard, USN 1
The Pentagon Room 4E577
Washington, D.C. 22301-2950

15. Andrew F. Krepinevich 1
Center for Strategic and Budgetary Assessments
1730 Rhode Island Ave., NW
Suite 912
Washington, D.C. 20036

16. LtCol and Mrs. M.G. Driesbach 1
1390 Independence Ave.
Melbourne, FL 32940

CAN THE ARSENAL SHIP REPLACE THE BATTLESHIP?

A MONOGRAPH
BY

Major Joseph M. Lance III
United States Marine Corps

School of Advanced Military Studies
United States Army Command and General Staff College
Fort Leavenworth, Kansas

First Term AY 96-97

REPORT DOCUMENTATION PAGE

1. AGENCY USE ONLY (Leave blank)

2. REPORT DATE: 9 Dec 1996

3. REPORT TYPE AND DATES COVERED: Monograph

4. TITLE AND SUBTITLE

Can the Arsenal Ship Replace the Battleship? (U)

5. FUNDING NUMBERS

6. AUTHOR(S)

Major Joseph M. Lance III, USMC

7. PERFORMING ORGANIZATION NAME(S) AND ADDRESS(ES)

SCHOOL OF ADVANCED MILITARY STUDIES
ATTN: ATZL-SWV
Fort Leavenworth, Kansas 66027-6900
COM (913) 758-3300 DSN 552-3300

8. PERFORMING ORGANIZATION REPORT NUMBER

9. SPONSORING/MONITORING AGENCY NAME(S) AND ADDRESS(ES)

10. SPONSORING/MONITORING AGENCY REPORT NUMBER

11. SUPPLEMENTARY NOTES

12a. DISTRIBUTION/AVAILABILITY STATEMENT

Approved for public release, distribution is unlimited.

12b. DISTRIBUTION CODE

13. ABSTRACT (Maximum 200 words)

This monograph examines the U.S. Navy's proposed arsenal ship and its potential for support to land forces operating in littoral regions. The undisputed platform of choice for naval surface fire support was the now deactivated Iowa class battleships. The arsenal ship will attempt to replace the battleships while adding a new deep strike capability. The paper addresses the genesis of the arsenal ship program by tracing the history of its predecessors. From the "long ship" of the Phoenicians, through the various ships-of-the-line, and finally to the Iowas. It examines the improvements in armor and armament along the way, and it culminates with the operational history of the Iowas. The monograph then introduces the arsenal ship program as the logical descendant of the battleship. It highlights the arsenal ship's required capabilities and concept of operations, and it attempts to compare battleship and arsenal ship capabilities. This monograph concludes with the decision that comparing the battleship and the arsenal ship is difficult at best. Their purposes are too different, and the intervening sixty years make objective assessments questionable. The arsenal ship is clearly the future of naval surface fire support, just as the battleship is clearly the past.

14. SUBJECT TERMS

Arsenal Ship, Battleship, Naval Fire Support

15. NUMBER OF PAGES: 58

17. SECURITY CLASSIFICATION OF REPORT	18. SECURITY CLASSIFICATION OF THIS PAGE	19. SECURITY CLASSIFICATION OF ABSTRACT	20. LIMITATION OF ABSTRACT
UNCLASSIFIED	UNCLASSIFIED	UNCLASSIFIED	UNLIMITED

SCHOOL OF ADVANCED MILITARY STUDIES

MONOGRAPH APPROVAL

Major Joseph M. Lance III

Title of Monograph: Can the Arsenal Ship Replace the Battleship?

Approved by:

COL Mary Goodwin, MA — Monograph Director

COL Danny M. Davis, MA, MMAS — Director, School of Advanced Military Studies

Philip J. Brookes, Ph.D. — Director, Graduate Degree Programs

Accepted this 9th Day of December 1996

Can the Arsenal Ship Replace the Battleship?

A Monograph
by
Major Joseph M. Lance III
United States Marine Corps

School of Advanced Military Studies
United States Army Command and General Staff College
Fort Leavenworth, Kansas

First Term

ABSTRACT

CAN THE ARSENAL SHIP REPLACE THE BATTLESHIP? by Major Joseph M. Lance III, USMC, 58 pages.

This monograph examines the U.S. Navy's proposed arsenal ship and its potential for support to land forces operating in littoral regions. The undisputed platform of choice for naval surface fire support was the now deactivated Iowa class battleships. The arsenal ship will attempt to replace the battleships, while adding a new deep strike capability.

The paper addresses the genesis of the arsenal ship program by tracing the history of its predecessors. From the "long ship" of the Phoenicians, through the various ships-of-the-line, and finally to the Iowas. It examines the improvements in armor and armament along the way, and it culminates with the operational history of the Iowas. The monograph then introduces the arsenal ship program as the logical descendant of the battleship. It highlights the arsenal ship's required capabilities and concept of operations, and it attempts to compare battleship and arsenal ship capabilities.

This monograph concludes with the decision that comparing the battleship and the arsenal ship is difficult at best. Their purposes are too different, and the intervening sixty years make objective assessments questionable. The arsenal ship is clearly the future of naval surface fire support, just as the battleship is clearly the past.

TABLE OF CONTENTS

CHAPTER ONE

INTRODUCTION

> Since men live upon the land and not upon the sea, great issues between nations at war have always been decided--except in the rarest cases--either by what your army can do against your enemy's territory and national life or by the fear of what the fleet makes it possible for your army to do.
> Sir Julian Corbett

If Corbett's point is still valid, naval ships should contribute to what an army can do against an enemy's territory. The U.S. Navy's proposed arsenal ship may directly support an army as a weapons delivery platform. The former Chief of Naval Operations (CNO) Admiral Mike Boorda emphatically stated, "...it's a fire support ship. It's all about providing fire support to troops ashore."[1] If the CNO's vision comes true, the arsenal ship will put more sea-based firepower at a land commander's disposal than ever before. This concept makes the arsenal ship a revolutionary vessel, and it does not fit neatly into any traditional category. Definitely not a frigate, destroyer, cruiser or aircraft carrier, the arsenal ship will not resemble any current surface combatant. It will not even resemble its closest relative, the battleship, but it must attempt to replicate the battleship's ability to provide fire support to land forces. This paper attempts to answer the basic question, can the arsenal ship replace the battleship?

Battleships have always been synonymous with big guns. When the British executed the first modern amphibious assault at Gallipoli, the battleship *Albion's* big guns pounded coastal fortifications and set an enduring precedent.[2] The American battleship experience culminated during World War II with the Iowa class ships. Congress authorized six Iowas as a counter to the German *Bismarck* and the Japanese *Yamato*. Although only four Iowas saw service, the U.S. Navy's battleship fleet provided invaluable gunfire support to landing forces and to land forces operating along or near a coastline. Most of the battleship fleet went out of commission after World War II, but the Iowas returned to service whenever the nation called. They fired during both the Korean and Vietnam conflicts and they returned again in 1983 as the growing Navy recognized a continuing need for their unique capabilities. Finally in 1991, the Navy decommissioned all four ships for what was intended to be the last time.[3] The Iowas left the active inventory, but the need for their capabilities did not diminish.

With Desert Storm completed and the Soviet Union in the process of dismantling itself, the Navy could take the time to assess the future and revise its course and speed. *...From the Sea* was the Navy's 1992 approach to an uncertain future and a rapidly changing world situation. Without an immediate blue water open ocean challenge, the Navy redirected itself toward the brown water littorals. *...From the Sea* emphasized the Navy's contributions to forces coming from the sea onto the land projecting power. The new approach acknowledged the landing forces required surface fire support, but there was no credible Navy response as to how to provide that support. Forced by fiscal reality to reduce its size, the Navy believed it could not afford to man or support the battleships any longer. Minus the Iowas, the only remaining naval guns were the short range 5-inch and 3-inch guns of the traditional surface combatants--the frigates, cruisers and destroyers. The short range guns were less than ideal because they could barely range into the landing force's area of operations. The Navy also needed to keep the surface combatants in the carrier battlegroups to protect the carriers. So, for many reasons, all of them important to the Navy, the naval gunfire mission went begging.

In 1994, the Navy followed *...From the Sea* with *Forward...From the Sea*. It reaffirmed the primacy of littoral operations, but it failed to address the obvious discrepancy between the focus on the littorals and the lack of gunfire support necessary to succeed there. Traditional carrier-based, land-based and expeditionary air support could attempt to fill the gunfire gap, but there will never be enough airplanes to ensure continuous close support for committed ground forces. Battleships, or something with a battleship's capabilities were still necessary components of littoral warfare. After a year's debate and amid sharp questioning from Congress, the Navy began to respond to its critics. The arsenal ship concept emerged throughout 1995, and in May of 1996 the Navy released the concept of operations and ship capabilities documents.[4] The proposed arsenal ship will be an impressive delivery platform with high technology missiles and stealthy design. It appears to be the Navy's preferred response to the obvious shortfall in surface fire support for troops ashore, but can the arsenal ship fill the void left by the Iowa class battleships?

To evaluate the arsenal ship, this paper begins by examining the history and legacy of the battleships. The story of the battleship is long, but it is crucial to any study of the arsenal ship. To understand the debate surrounding the arsenal ship, even the most casual reader must be familiar with the history of the battleship. After tracing the battleship story, then the development of the arsenal ship can begin. The arsenal ship's genesis, its required capabilities, the proposed concept of operations, and its postulated performance help frame the basic question and provide a basis for comparison. The paper's conclusion summarizes the findings and the current debate while answering the practical question, what does this study tell us and why is it worth knowing?

CHAPTER TWO

THE BATTLESHIP

> The battleship became the one accepted symbol of naval power throughout the world; and, because of both strategic and diplomatic implications inherent in a ship of such aggressive potential, it became the most controversial and publicly discussed type of ship in the United States. Brayton Harris

Maritime nations have always employed sea power to protect their commerce and project their power. Sea power required nations to invest in fighting ships.

> "The fighting ships evolved into fleets. These have always needed heavyweight champions, powerful warships in the front line of battle that can hit the hardest and take the most punishment--for war demands the ability to take as well as to give."[5]

The epitome of the fighting ship was the battleship. A seafaring nation identified with and was identified by its ships-of-the-line. American esteem and historical significance revolved around the early "Frigate Navy," the *Monitor*, the *Maine* , the Great White Fleet, Battleship Row at Pearl Harbor, and the culmination of the battleship line with the Iowa class ships. These proud warriors had their beginnings in the galleys of ancient Phoenicia.[6]

The Phoenicians controlled the eastern half of the Mediterranean from 1100 to 800 BC. To extend their economic prosperity, the Phoenicians built two distinct classes of ships. The "round ships" carried the bulk of the trade, and the "long ships" fought when necessary. The earliest battleship was easily recognizable by its narrow, menacing appearance capped by a long ram. The embarked archers and spearmen provided the "long ship" with the ability to engage enemies from a distance if ramming was not feasible. The "long ship" eventually gave way to faster, more maneuverable biremes and triremes. By 500 BC, the trireme, a galley with three tiers of rowers on each side of the centerline, ruled the Mediterranean Sea. Seaworthy but fragile, it set the standards for naval warfare for the next 2000 years.[7]

Ship design steadily evolved, but naval tactics languished. Even the addition of rudimentary cannons in the 15th Century did not change the basic tactical maneuvers. However, the 16th Century saw a great interest in maritime affairs as the major European powers rushed to expand their empires. The lure of untapped riches provided the impetus to expand and improve the navies of the day. The galleys grew into galleons, and soon ocean-going vessels bristled with cannons.[8] Spain, Portugal, England, France and Holland competed for glory, conquest and wealth.

Great Britain seized the command of the sea from Spain in 1588 by defeating the Spanish Armada. The English admirals did not close with the Spanish galleys and fight at close quarters. Never intending to ram or board, the British employed their heavier guns at great ranges and nibbled at the fringes of the Spanish Armada. The Spaniards valiantly fought terrible weather and the more sea-worthy English galleons, but Spain's seemingly invincible Armada went down in defeat. The British became the uncontested masters of the sea, and the galleon replaced the galley as the latest battleship.[9] The English Navy continued to evolve as it expanded the Empire to the furthest reaches of the globe. Bigger ships-of-the-line with more sails, more guns, and more space for supplies replaced the galleons.

The American experience with battleships began with the first 74-gun ship-of-the-line built in North America appropriately named *America*.[10] However, in an alarming precedent, the *America* was given to France in 1782 to replace a French ship lost in Boston Harbor.[11] The United States allowed itself to be dependent upon France for its protection. America's early leaders, though conscious of their fledgling nation's naval helplessness, did not feel they could raise enough revenue to justify investing in naval protection.

France's benevolence lasted until the period 1785 to 1795, when the British and French Navies, privateers, and Mediterranean corsairs preyed upon defenseless American ships.[12] Although still thrift conscious, the U.S. government could no longer ignore the threats to her commerce, and a modest "Frigate Navy" evolved around the famous *Constitution* and *Constellation*.[13]

As the immediate threat waned, so did the Navy. No new construction followed the "Frigate Navy" until the War of 1812 jolted the U.S. government awake. Massively outgunned and outmaneuvered by the British Fleet, the U.S. Navy responded with construction of the first true American ships-of-the-line in 1813. The Independence class warships employed 74 guns. The U.S. Navy had finally convinced Congress to authorize true battleships, but they could not be built and commissioned in time to serve during the war. The new battleships saw service in the Mediterranean and their impressive results convinced Congress to authorize a newer series of battleships, the North Carolina class. Many of the North Carolina class continued in service until and during the Civil War.[14]

Between the War of 1812 and the Civil War, naval innovations proceeded apace the evolution of technology in the private sector. Steam power, iron plating, increasingly strong metal alloys, and screw propulsion were all experimental technologies which emerged during this period. However, the commencement of the Civil War provided the impetus to bring the new technologies to maturation. Both the Union and the Confederacy rushed to launch a new battleship which would rule the continent.

The South gained an early advantage by recovering the scuttled *U.S.S. Merrimac* from the Gosport Navy Yard at Portsmouth, Virginia.[15] The *Merrimac's* formerly beautiful masts and superstructure were reduced to the waterline, and a new iron-plated gundeck completed the design that favored function over form.[16] The new *Merrimac*, rechristened the *C.S.S. Virginia* but known forever as the *Merrimac*, boasted two 7-inch pivot guns, two 6.40-inch guns and six smoothbore 9-inch Dahlgrens. Well-armed and well-armored, the *Merrimac* had no equal on the Union side.[17]

The Union scrambled to catch up to the Confederate lead. A Navy Board on Ironclads approved three designs for immediate construction, the *New Ironsides*, the *Galena*, and the *Monitor*.[18] A Swedish inventor, John Ericsson, promised the Board his "floating battery" would be more than a match for the *Merrimac*.[19] The *Monitor*, the first ship of her kind, combined steam power, iron-plated armor, screw propulsion, a rotating turret capable of mounting the biggest guns of the day, engines below deck level, forced air ventilation and several other patentable devices into the most potent vessel the world had ever seen.

The first readily recognizable ancestor of the modern battleship held up well in her baptism of fire. Earning no less than a draw on the tactical level, the *Monitor* clearly scored a decisive victory at the operational and strategic levels. The *Merrimac* left the *Monitor* in possession of the battlefield, and the Southern behemoth never again ravaged the Union's wooden ships. The Battle of Hampton Roads between the *Monitor* and the *Merrimac* attracted the world's attention.[20] All naval construction programs required reevaluation based on the results of the first battle between ironclads.[21] Thirty of the *Monitor*'s direct descendants remained in service or under construction through 1883.[22]

When the Civil War ended, the United States adopted its typical course and speed regarding naval development. The seven hundred ship fleet was pared to merely forty-eight active gun-mounting vessels by 1880. The U.S. Navy stood twelfth in the world, trailing Denmark, Chile and China.[23] As America turned inward to develop the country's interior, the Navy could only watch as it almost withered away. During the period 1883-1889, the Navy finally received funding to build some new ships. The new cruisers *Texas* and

Maine, and some smaller gunboats began the naval restoration process.[24]

By 1890, Congress finally funded three true battleships. The *Oregon*, *Indiana*, and *Massachusetts* lead the program which carried the Navy to her new second place status among the world's navies by 1910.[25] The resurgent Navy and the conquest of all available continental frontiers made expansion-minded Americans look overseas for new challenges. The sinking of the *Maine* lead the U.S. into war with Spain in 1898. The Navy soundly defeated the Spanish, and the American public embraced their Navy as heroes.[26] Congress quickly authorized enough funding to build a new Navy with a legitimate battleship fleet. The U.S. Navy rose in size, stature and capability, but the Navy enjoyed its position for only a short time. The launch of the *H.M.S. Dreadnought* made all other ships obsolete in 1906.[27] Obsolete or not, President Roosevelt directed the voyage of the "Great White Fleet" around the world, in December 1907, to demonstrate American interest and resolve in free trade issues. Amid increasing European tensions, America chose to continue its naval build-up at a moderate pace of one or two battleships per year. When World War I broke, the U.S. Navy--even with its many battleships--was still inadequately prepared for war. President Wilson's administration embarked on an aggressive program to build a fleet "equal to the most powerful maintained by any other nation in the world."[28]

World War I did not provide any opportunities for the U.S. Navy's battleships to test their mettle, but the U.S. Fleet did yeoman service providing convoy escorts to ward off submarine attacks. The U.S. continued the naval construction program begun during the war, but in 1921 the Washington Conference convened to place limits on naval development. The United States invited Great Britain, France, Italy and Japan initially to discuss and agree upon naval ratios, tonnages and armaments. The U.S. and Great Britain emerged with the highest ratio, but Japan remained strongest in the Far East. France and Italy were insulted by their inferior, third-tier positions. The U.S. scrapped or stopped construction on eleven ships to achieve compliance. The treaty also prohibited battleship construction for a ten-year period, and it signaled the end of the battleship as the primary ship-of-the-line. Nations devoted their resources to building aircraft carriers and submarines. The London Conference of 1930 met to extend the limitations. Cruisers, destroyers and submarines came under review, but the results were less in favor of the U.S. and England. The Japanese delegation successfully argued for more favorable ratios. The signatories extended the construction moratorium until 1936, but by that time, the Japanese Empire was expanding, and the Imperial Navy needed larger fleets to support the expansion. The Japanese Navy would not agree to any restrictions that might thwart its plans to enlarge Japan's sphere of influence. Only France, England and the U.S. agreed on the 1936 limits.[29] By 1938, it was obvious that Japan was building new battleships. The U.S. began designing and programming new ships of her own, but it continued to deploy her old battleships. Thus, many of the WW I vintage ships were lined up on "Battleship Row" when the Japanese struck Pearl Harbor in 1941.[30]

The effect of the Japanese bombing of Pearl Harbor sobered the country and embarrassed the Navy. *Arizona* blew up and sank, *Oklahoma* capsized, *West Virginia* and *California* sank, *Nevada* was beached, and *Maryland*, *Tennessee*, and *Pennsylvania* suffered light to moderate damage. Air power had apparently triumphed over the battleships, but the preliminary results did not tell the whole story. The old battleships had been built to take a beating and survive. *Arizona* and *Oklahoma* were beyond rescue, but the Navy repaired and modernized the other six battleships. Air power advocates pointed to Pearl Harbor as proof of the airplane's ascendancy, but the attack against unprepared, docked vessels accounted for the only U.S. battleship losses of the war.[31] The battleship was not dead.

While acknowledging the rise of the aircraft carrier, the Navy continued to build and improve its battleship fleet. Ten new battleships joined the fleet between 1941 and 1944. The new battleships were fast, powerful and incredibly durable. The combination of 16-inch guns, 17 inches of armor, and 33 knots maximum speed made the battleships the most formidable warships ever built. They performed admirably in many roles during World War II, but their biggest contribution came in the naval gunfire role. Battleships, old and new, pounded the beaches prior to, during and after the Marines' and Army's major amphibious assaults. Attu & Kiska, Makin, Tarawa, Kwajalein, Peleliu, Normandy, Iwo Jima, and Okinawa all received thunderous gunfire support. All the battleships received the appropriate recognition when the U.S. chose the *Missouri* to be the platform for the Japanese surrender.[32]

The advent of the atomic age, and the culmination of WW II signaled the demise of the battleship. The rise of the aircraft carrier and the submarine, and the promise of long-range missiles tipped with nuclear warheads seemed to negate the armor and firepower offered by even the Iowa class battleships. The Navy could not afford to continue manning and arming massive wartime fleets with the country and the Congress anxious to return to peacetime appropriations. A rapid battleship decommissioning program rid the Navy of its underappreciated veterans. Museums, drydocks and scrapheaps claimed all the battleships save the *Missouri* by 1950.[33]

When North Korea attacked South Korea, the *Missouri* responded. Her devastating firepower destroyed railroad bridges, marshaling yards, tank and troop concentrations and rolling stock. Congress quickly authorized the Navy to recommission the *Iowa*, *New Jersey*, and *Wisconsin*. The Iowa class ships wreaked destruction up and down the coasts of the Korean peninsula. The *Missouri* was the first battleship on the scene, and she alternated on and off station until April 1953. The *Iowa* fired mainly around the Wonsan area until 1952, and then she returned to the United States. The *New Jersey* also fired around the Wonsan area, but she remained in theater until November 1953. She and her crew won a Korean Presidential Unit Citation for their service. The *Wisconsin* returned from mothballs in 1951, and she served on the Korean gunline from November 1951 until March 1953.[34] The Iowas performed so effectively during the Korean Conflict that they earned a respite from retirement until 1957. Without any easily discernible threat, the Iowa class battleships again headed for mothballs and retirement.

By 1958, the U.S. Navy was without a true battleship for the first time since 1895.[35]

During the 1960s, the world's militaries continued their pursuit of faster jet aircraft and guided missiles. The battleships were viewed as resource intensive anachronisms. Although many planners designed modifications to convert the Iowas to guided-missile ships, mini-carriers, and amphibious assault ships, the Navy never found the will nor the funds, during peacetime, to reactivate and convert the battleships.[36] Experiences in Vietnam spurred renewed interest in long-range, all-weather fire support delivered from naval platforms. Cruisers and destroyers were unable to neutralize lucrative targets inland because of the range and lethality limitations of 5-inch guns. The *New Jersey* steamed out of drydock again to take her place on the gunline along the Vietnamese coast in 1968. She pounded North Vietnamese troop concentrations north and south of the border, in all weather conditions, and she did it without jeopardizing any aircraft or pilots. In only 120 days of firing, she fired 10,000 rounds. However, her reactivation was short-lived. In 1969, the Secretary of Defense announced the impending decommissioning of the *New Jersey* because of funding restrictions.[37]

As the world's last active battleship returned to mothballs in 1969, few observers pictured any further service for the Iowas. The North Vietnamese resurrected the possibility of again reactivating the Iowas when they attacked into South Vietnam with conventional forces in 1972. President Nixon questioned the Navy about an express, 30-day reactivation for the *New Jersey*. Reluctantly responding, the Navy was unable to support the President's request. The heavy cruiser *Newport News* answered the call, but the Seventh Fleet Commander had wanted and needed the *New Jersey*. Some naval experts had recommended the Navy completely divest itself of the battleships in 1973 and papers to authorize disposal circulated, but common sense prevailed. The Navy decided to continue the minimal preservation efforts in case the Iowas were ever needed again.[38]

Between 1974 and 1980, naval analysts conducted new studies and updated old studies concerning revisions and upgrades to the Iowa class battleships. In 1980 the Navy requested funds to reactivate the *New Jersey*, but funding restrictions prohibited the reactivation. When President Reagan's 1981 supplemental budget took effect, the *New Jersey*'s funding came through. The Falkland Islands conflict of 1982 raised the interest in the Iowas' reactivation. Modern naval ships had shown their vulnerability to bombs and missiles. The Iowas' armor, designed to take hits from 16 and 18-inch guns, would protect it from Exocet missiles. Secretary of the Navy John Lehman authorized the *New Jersey*'s return to the fleet with some major modifications. The 16-inch main gun battery would remain, but Tomahawk and Harpoon missiles would join the arsenal. As the *New Jersey*'s improvements and modifications continued, Congress approved additional funding to reactivate the *Iowa*. By 1988, all four of the Iowas once more roamed the seas in fighting shape.[39] The renovations were costly and time consuming, but the capability and power projected by the Iowas more than outweighed the time and money invested.

When the United Nations challenged Saddam Hussein's invasion of Iraq, the *Missouri* was part of the early Desert Shield buildup. When it came time to launch Desert Storm's initial strikes against Iraq's air defense network, the *Missouri* was ready and willing, but it was not the traditional main battery doing the firing. The *Missouri* launched twenty-eight Tomahawk cruise missiles in the opening days of the air campaign. Soon the *Wisconsin* also appeared in the Persian Gulf and both battleships pounded Iraqi positions along the coast and well inland. The combination of the 16-inch guns delivering fires and a remotely piloted vehicle (RPV) providing observation proved devastating. Iraqi soldiers surrendered to the RPV rather than face more shelling from the battleships.[40] Both battleships performed well during the Persian Gulf War, but they could not be joined by the *Iowa* and *New Jersey*. Those two veterans were already on their way back into mothballs, and the *Missouri* and *Wisconsin* would soon leave active duty also. The most powerful ships ever built once again went quiet because of funding and manning shortfalls.

The history of the American battleship experience is an interesting but sad tale. America has designed and built some of the finest ships in the world, but the Navy and the Congress have repeatedly removed them from service at the earliest opportunity. From the *America* in 1782 to the final deactivation of the Iowas in 1992, perfectly good ships-of-the-line have consistently been given away or retired well before the expiration of their effective service life. The Iowas, planned and programmed before the attack on Pearl Harbor, are a testimony to the men (and women) who designed, engineered and built them. It seems ironic that they were deactivated again as the Navy was embracing a new focus on littoral warfare.[41] What would be a better weapons delivery platform for a possible slugfest in the littorals than the most heavily armored and most heavily armed warships of all time? Torpedoes, Exocet missiles and shore batteries would have a difficult time sinking an Iowa class battleship. These facts were not lost on Congress. The Navy knew it had to come up with a replacement for the Iowas, or it would continue responding to Congressional inquiries about the Iowas.[42] Can the proposed arsenal ship be the successor to the battleship?

CHAPTER THREE

ARSENAL SHIP

> --this thing could shoot ATACMS, Tomahawk and it's various varieties, standard missiles, or it could be a theater ballistic missile defense ammunition locker, and more.
> Admiral J.M. Boorda

"The arsenal ship concept is a direct outgrowth of the Navy's shift in focus from the open ocean to the littoral."[43] The arsenal ship appeared to burst upon the scene in 1996, but its genesis lies in the Navy's acknowledgment of the rise in priority of littoral operations.

In 1992, the U.S. Navy publicly announced its "landmark" shift, but in an ironic twist it also deactivated the most capable littoral warfare platforms ever built, the Iowa class battleships. As the Soviet Navy retreated to its harbors and dismantled itself, the U.S. Navy searched for relevancy in a confusing new world. Life was simpler when the Soviet Bear prowled the oceans. Without a clearly defined threat, America's Navy had to prepare for anything and everything. ...*From the Sea* represented the Navy's recognition of "the changing strategic landscape--away from having to deal with a global maritime threat and toward projecting power and influence across the seas in response to regional challenges."[44]

...*From the Sea* also recognized a change in Navy-Marine Corps relationships. The Navy realized Marines were necessary to project power ashore in littoral regions. To make the new naval doctrine plausible, the Marines and amphibious warfare would have to be elevated to a higher status. The Marines relished their importance, and they began to petition the Navy to address the most significant amphibious warfare challenges.[45] Amphibious ships were in short supply, countermine operations had been neglected for decades, a replacement helicopter for the aging CH-46 was still not in the budget, a replacement amphibious tractor for the AAVP-7 was necessary to speed up the landing process, and naval surface fire support had atrophied to the point of despair.[46] The Marines knew they had better capitalize on the current littoral attention before another attraction lured the Navy back to its traditional blue water focus.

While preparing for regional littoral challenges, the Navy examined a 1993 RAND Corporation study prepared for the Air Force, The New Calculus: Analyzing Airpower's Changing Role in Joint Theater Campaigns. The Navy gleaned that a major regional conflict (MRC) could present a threat target array consisting of up to 15,000 armor or mechanized infantry fighting vehicles. The Navy further deduced a coalition force would have to neutralize one-third of the array to successfully halt an invasion force.[47] U.S. forces had no capability to stop a force of that dimension without allowing for a significant time delay to deploy forces. Realizing a potential foe may not provide the U.S. the time to build up combat power--as in Desert Shield--the Navy began its search for options to address the obvious shortfall.

A search for a new capability required a new platform. Aircraft carriers could launch strike missions against an enemy mechanized force, but a carrier air wing, or even two or three carrier air wings, could not produce enough attack sorties to blunt a 15,000 vehicle invasion. Vertical launch system (VLS) configured cruisers and destroyers could not launch enough Tomahawk cruise missiles to stop an array that large either. The cruisers and destroyers 5-inch guns did not have the range or lethality to seriously affect a substantial armor force.[48] Even applying carrier air, Tomahawks, and naval gunfire together would not produce the desired capability.

A solution to the problem would demand a new weapon system, but a new system would be difficult to acquire in the face of declining budgets, personnel cutbacks, and competing demands among the various branches of the Navy. The surface, the submarine, and the aircraft carrier forces all had designs on procurement dollars. Even within the surface branch, the anti-air warfare community, the amphibious warfare community, the logistics community, the anti-submarine community, and the mine warfare community all had competing priorities. To build a consensus on any program, the Navy would have to try to satisfy as many communities as possible. To see the light of day, any new program must demonstrate more capability than any existing ship or combination of ships. The arsenal ship faces a difficult task. It will have to please everyone, to include the Marines, as well as Congress.

By 1993, even the Congressional Budget Office (CBO) recognized the Navy was woefully deficient in naval surface fire support assets. The CBO proposed, "Reactivating battleships may be the only option that could be carried out during the next few years to improve naval fire support."[49] The Navy knew it had to come up with a battleship replacement design, or it would face continued pressure to once again activate the Iowas. In 1994, rather than siphon off precious resources into designs from the pre-WW II era, the Navy searched for alternatives. One program looked at extending the range of existing 5-inch guns to allow the destroyers and cruisers to support ground forces on shore.[50] Another program examined converting Trident submarines to

Tomahawk shooters. These "stealth battleships" could be configured from submarines no longer needed in the nuclear triad concept.[51] A third study came up with the arsenal ship idea.[52] It would be the obvious successor to the battleship, but it would not be solely dedicated to replacing the battleship's 16-inch guns. The arsenal ship would be more.

The arsenal ship idea received a high priority from the Chief of Naval Operations Admiral J.M. Boorda. In an August 1995 interview, Admiral Boorda laid out the arsenal ship concept and explained his vision. Although reluctant to provide hard figures, Admiral Boorda established the year 2000 as a tentative delivery date for the first ship. He also admitted his role as the driving force behind the program, to include remarking he had heard the rumors that "this is Boorda's legacy."[53] Given the CNO's obvious interest, it was no surprise when the "Arsenal Ship Combat Systems Study" appeared in November 1995.

> Conceived as a Fire Support Ship, the Arsenal Ship will serve primarily as a sea based, long-range strike and invasion stopping platform, providing direct maritime support of the land battle. On station, it will provide massive firepower early with little or no strategic warning. Capable of launching advanced long-range strike missiles equipped with submunitions [sensor-fuzed weapons (SFW), brilliant antitank (BAT) weapons, or wide-area mines (WAM)] this ship can deter mechanized invasions from outside the range of enemy land-based defenses.

> Once friendly forces have gained a foothold ashore, both naval expeditionary and joint ground forces will require sustained on-call tactical strike and direct fire support during counter offensives. The Arsenal Ship, employing large numbers of extended range tactical missiles, can provide responsive support to forces ashore.[54]

The arsenal ship program began as an extremely ambitious and at least mildly controversial concept. The original 1995 study listed the proposed arsenal ship roles in sequence: 1) long range strike and invasion stopping; 2) tactical strike and fire support; 3) battlespace dominance support - theater air defense (TAD) to include tactical ballistic missile defense.[55] To build one platform with the required capabilities would be an impressive feat, but potential contractors face other challenges also. The Navy also desires reduced manning, reduced acquisition and life-cycle costs, and reduced onboard target acquisition sensors.[56]

In March 1996, the Navy and the Defense Advanced Research Projects Agency (DARPA) signed a joint agreement to execute the arsenal ship technology and demonstration program.[57] The Navy briefed Congress on the program and requested funding to match the milestones. By May 1996, the Joint Program Office had briefed industry, and had published the necessary documents to begin the proposal process.

The Arsenal Ship Capabilities Document (SCD) tasks industry to provide a technology demonstrator to show the program is feasible within the established price thresholds. The demonstrator must have at least the basic capabilities, but does not have to possess the full capability of an arsenal ship. It must be able to demonstrate:

1) Performance of the mission for 90 days.

2) Required architecture, communications, and essential data links to support the Concept of Operations.

3) The capability for remote launch of strike, area Anti-Air Warfare (AAW) and fire support weapons.

4) That passive survivability will be sufficient for the expected operating scenarios.

5) The demonstrator must be capable of being converted to full mission capabilities and configuration and used as a fleet asset.[58]

The SCD lists the additional capabilities the arsenal ship must possess. It must include the ability to fire weapons in support of a land campaign to fulfill the main roles analyzed in the Arsenal Ship Combat Systems Study, but the wording and emphasis is now refined. The primary role is still Long Range Strike followed by Invasion Stopping. The secondary role is Fire Support to Joint Ground Forces, and the tertiary role is now Tactical Ballistic Missile Defense and Air Superiority. The ship must have about 500 vertical launch cells, and it must be capable of launching all current and planned vertical launch weapons. Finally, in a sop to ground forces, the ship must also have capacity reservations for future installation of an extended-range gun system.[59]

To allow the arsenal ship to fire its impressive array of weapons it must be able to communicate, and it must be able to survive. To employ the Cooperative Engagement Capability (CEC) 'remote magazine' launch concept, the arsenal ship must possess a robust suite of communication data links, to include an over-the-horizon (OTH) satellite link. It must be able to communicate with other ships, aircraft, satellites, shore stations, joint force headquarters, and supported ground forces. To enhance its survivability, the arsenal ship will not carry its own targeting sensors. Without target acquisition radar, spotting personnel, or unmanned aerial vehicles (UAVs), the arsenal ship is blind. The arsenal ship is also dumb. All the requisite target acquisition capability, launch control and missile tracking data must come from another source. The arsenal ship will rely upon passive survivability techniques under the Aegis control umbrella.[60] The hull design should incorporate "stealth" technology to make it hard to find and difficult to hit. The ship must be "virtually unsinkable," and the design life of the ship must be at least 35 years.[61]

To be in position to fire its weapons, the arsenal ship must be mobile. It must be able to make a sustained speed of at least 22 knots using only 80% of its power. It must be capable of operating for up to 90 days without refueling, and it must demonstrate precise, continuous navigational control under any and all conditions--to include withstanding 100 knot winds and accompanying seas. It must also prepare to suffer moderate underwater damage with sufficient reserve buoyancy to continue with its mission. The arsenal ship's biggest challenge will be to perform all its required missions with a crew of no more than 50 people.[62]

The SCD lists an impressive array of desired capabilities, but it stops short of telling industry exactly what to design and build. The requirements are challenging but straightforward, and the necessary technologies are either readily available, under development, or within reach in the near future. The SCD tells the technical side of the arsenal ship story, and the Concept of Operations (CONOPS) section covers the tactical employment options.

The Overview section of the CONOPS document contains some key verbiage that should attract contractor and Congressional attention. The Navy used that section to detail the arsenal ship concept, and to restate its position as the forward-deployed force of choice.

> In the face of steadily decreasing overseas basing and a shrinking military budget, the United States must maintain the ability, in concert with allies, to execute timely combat operations across the spectrum of conflict. Naval forces, sustaining forward presence, will be key to successful introduction as well as early employment of ground forces. Arsenal ship represents an affordable and much needed enhancement to our existing force of carriers and land attack capable combatants and submarines. It is not a replacement for these or for land-based air. Instead, it is part of the whole--just as the Battleship was a part of the whole for nearly a century. . . . With a force totaling about six, arsenal ships will be stationed continuously forward, always available for rapid movement upon receipt of even ambiguous or limited strategic warning. Much like our maritime prepositioning force, they will remain on station in support of a Unified CinC for indefinite periods without dependence on host nation support or permission.[63]

In one succinct paragraph, the Navy has said a mouthful. It acknowledges fiscal reality and the world's growing reluctance to host American armed forces. It pats itself on the back as an enabling force, graciously allowing the ground forces to arrive at their leisure. It reiterates the arsenal ship's affordability while avoiding a direct challenge of the air force. It attempts to retain the link to the battleship while also making reference to the highly successful maritime prepositioning program. The paragraph closes with the tie to the obvious warfighters, the Unified Commanders in Chief (CinCs) of the regional commands.

By July 1996, the Joint Program Office announced the interested Phase I contractors.[64] In spite of the ambitious requirements, five industry teams have formed to bid on the arsenal ship program. They have six months to develop their designs. Two teams will receive funding to continue development before a final decision is made to choose one team to build a demonstrator. The arsenal ship program is clearly on the fast track. It is an innovative concept, but if it tries to be all things to all people, will it be able to do anything well? The CNO's vision of the arsenal ship "providing fire support to troops ashore, or to make it possible to get troops ashore"[65] will be difficult to balance against its other requirements.

CHAPTER FOUR

THE INEVITABLE COMPARISON

> The Marine Corps is looking for Naval surface fires aboard as many ships as possible. We believe we need to break the paradigm that there is a particular ship that provides Naval surface fires.
>
> Lieutenant General Paul K. Van Riper

The arsenal ship is clearly a controversial program. Critics and supporters do battle on a daily basis, in myriad forums. The obvious battle is for funding. In an already constrained fiscal environment, future programs must expect intense scrutiny from within the organization and from outside sources like Congress, the General Accounting Office and civilian budget watchdog groups. Comparisons between the arsenal ship and the Iowa class battleships are inevitable.[66] To achieve validity, the comparisons must be objectively grounded, and they must avoid emotional appeals. The key areas for evaluation are fire support capability, availability, and survivability.

An objective comparison between the fire support capability of an arsenal ship and an Iowa class battleship is extremely difficult to produce.[67] Although both platforms could contain guns and missiles, the weapons mixes are totally different and they reflect different outlooks.

The Iowas, in their current configuration, employ a main battery of nine 16-inch guns, twelve 5-inch guns in the secondary battery, thirty-two Tomahawk missiles, and sixteen Harpoon missiles.[68] This weapons mix provides a balanced attack with a primary focus upon great suppression or neutralization capability out to the range of the 16-inch guns. To engage targets with its main battery, an Iowa will need to deploy close to the shoreline, exposing it to coastal defenses. The Tomahawks extend the battleship's reach out to a much greater range, but the inability to reload while underway limits the amount of long-range striking power a battleship can generate.

The arsenal ship will employ a storehouse of up to 500 missiles, to possibly include advanced Tomahawk variants, Sea Launched Army Tactical Missiles (SLATACMS), and Stand-off Land Attack Missiles Expanded Response(SLAM-ER).[69] The arsenal ship may carry extended-range 5-inch guns, but their number will be limited due to space constraints. The arsenal ship's missile stockpile will allow it to strike deep with unprecedented power and accuracy.[70]

The critical support criteria for this study is the vessels' ability to assist a landing force in a beach assault. A threat array makes evaluating the different ships' capabilities easier. Figures 1 and 2 depict two possible scenarios. Both scenarios depict former Soviet units attempting a coastal defense. Both scenarios also demonstrate a defender's ability to deploy in depth, using prepared fighting positions. If a landing force attempted an amphibious assault against a defending enemy in a similar array, it would require tremendous amounts of firepower to suppress or neutralize the defenses. Many battleship advocates maintain that only a high-volume gun system can provide the type of fire needed by landing forces.[71] Clearly, battleships have a decided edge over an arsenal ship when it comes to gun size and number of guns.

An arsenal ship could attempt to compensate for its lack of gunfire support by firing its missiles against the prepared defenses, but its effectiveness would be debatable. Current Tomahawk anti-ship missiles deliver a single thousand pound high-explosive warhead suitable for fixed-location point targets.[72] Future versions will deploy multiple armor-seeking submunitions.[73] Neither version would be effective against infantry in prepared positions. SLATACMS could suppress a target, but to continually suppress an entire target set, SLATACMS missiles would be prohibitively expensive, hence the seemingly tacked-on requirement for an extended-range 5-inch gun. Over the course of an amphibious assault, not even an arsenal ship could carry enough missiles to answer every call for fire with a missile. While an arsenal ship is not the ideal system for beach assault support, its value increases markedly as landing forces move inland.

Beyond the beach defenses lurks an extremely dangerous threat to a landing force, an armor heavy counterattack force. Here the advantage shifts to the arsenal ship. Improved Tomahawks, SLATACMS with Brilliant Anti-Tank (BAT) munitions and SLAM-ERs are capable of disrupting or neutralizing an armor unit before it can engage a landing force. A battleship cannot carry the number of Tomahawk missiles necessary to counteract an armor brigade or larger force. Under ideal conditions, 16-inch guns could temporarily disrupt an armor force, but guns are relatively short-range weapons when compared to the 100+ kilometer ranges for missiles.

Assessing the availability criteria is relatively easy. Only two battleships remain easily accessible for reactivation.[74] Theoretically, other battleships could be reacquired from their resting places as floating monuments, but a realistic assessment discounts this possibility. If the Navy decided to reactivate the battleships, it would require another trip to the drydocks for refurbishing. Past reactivations have taken a year or two per ship to resurrect and enhance the battleship's capabilities. The further the Navy tries to stretch the battleships, the more difficult it is to bring them up to current capabilities. Historically, improvement processes have been frustrated by the antiquated ship design and construction materials. To radically modify the battleships as suggested by some advocates' plans, would be exceptionally difficult, expensive and questionable in their effectiveness. Besides the cost of reactivation and the continued cost of operating a fifty-five year-old vessel, the most expensive part of the reactivation equation is the manpower. Manning the battleships with the required crew of 1,500 would be very difficult for the Navy to manage.[75] For much of the Navy, it is difficult to fathom why they should divert resources to an obviously antiquated platform. When Ticonderoga class cruisers and Arleigh Burke class destroyers only need around 300 sailors, it is hard to justify 1,500 men and women on one battleship.[76] One of the arsenal ship's key requirements is to man it with 50 or fewer sailors. This is an attractive feature, but some observers question the feasibility of the arsenal ship's stated manning level.

One of the arsenal ship's biggest critics, William Stearman, questions many of the program's promises. He doubts a crew of 50 can perform the necessary maintenance to keep an arsenal ship afloat and operable.[77] The arsenal ship study found the goal achievable by offloading the requirement to acquire targets, select weapons and compute firing solutions. The remote launch concept could make the manning goal possible and keep the arsenal ship lean, but industry must demonstrate how lean the arsenal ship can be while still successfully performing its mission. Whatever the final manning level becomes, the arsenal ship program promises more vessels than there are remaining battleships. The ideal ship mix would be six total vessels. Three would be continually at sea and three would rotate off-station for rest and maintenance. The Navy identified the key stations as "European Command (Mediterranean Sea), Central Command (Persian Gulf), and Pacific Command (Western Pacific)."[78] The availability criteria clearly favors the arsenal ship, but availability can be affected by survivability.

To deliver its fire support, a vessel must be capable of surviving the many threats it faces. Torpedoes, mines, anti-ship missiles, coastal artillery and dumb bombs are all possible threats for a naval surface fire support platform. The Iowas have proven their ability to take punishment and continue the mission. The arsenal ship will not be able to rely on the battleship's seventeen inches of armor. The arsenal ship eschews thick armor protection in favor of a radar evading "stealthy" design to avoid detection. However, firing up to 500 missiles defeats passive techniques and "stealthy" design, so the arsenal ship must prepare to receive some hostile fire.[79] The SCD requires the arsenal ship to be "virtually unsinkable," but what does that mean? The *Titanic* was thought to be unsinkable, but it never completed its maiden voyage. Trying to design and build an unsinkable arsenal ship featuring up to 500 missiles may be impossible. Industry will have to demonstrate just how survivable an arsenal ship can be,[80] but even the best designs will probably not approach the battleship's impressive ability to absorb punishment and continue the mission.

It becomes clear that arguing battleship against arsenal ship is similar to comparing apples and oranges. They should not be viewed as competitors. Although worlds apart in age, purpose and capability, they are clearly related. In an initial amphibious assault, the battleship is superior at providing massive amounts of suppressive fire. However, when the landing force attempts to consolidate its beachhead and push inland, it could face an extremely dangerous armor counterattack. The arsenal ship would be the platform of choice for disrupting the counterattack. Both vessels in support of a landing force would be the ideal, however, constrained budgets prohibit this option.

The Iowa class battleships cannot be replaced in kind, but the future of naval surface fire support is clearly the arsenal ship. However, the future should not be the arsenal ship alone. All the Navy's 5-inch guns should be upgraded to the extended-range version, and the arsenal ship should continue to receive high-profile interest and high-priority support from Naval and Congressional leaders. Even hinting at reactivating the battleships creates a strong emotional response from proponents and opponents, but the issue is now almost a moot point. The Navy shows little interest in reactivating the battleships, and with the arsenal ship program underway, Congress appears content to let the naval surface fire support issue rest.

The arsenal ships may not be the perfect platform to support troops in an amphibious assault against a determined enemy, but they will provide CinCs and landing forces with an unprecedented amount of firepower from a naval surface vessel. Air support, naval gunfire from many ships, and an arsenal ship's missiles could be the key to effective 21st-Century operations in the littoral regions of the world.

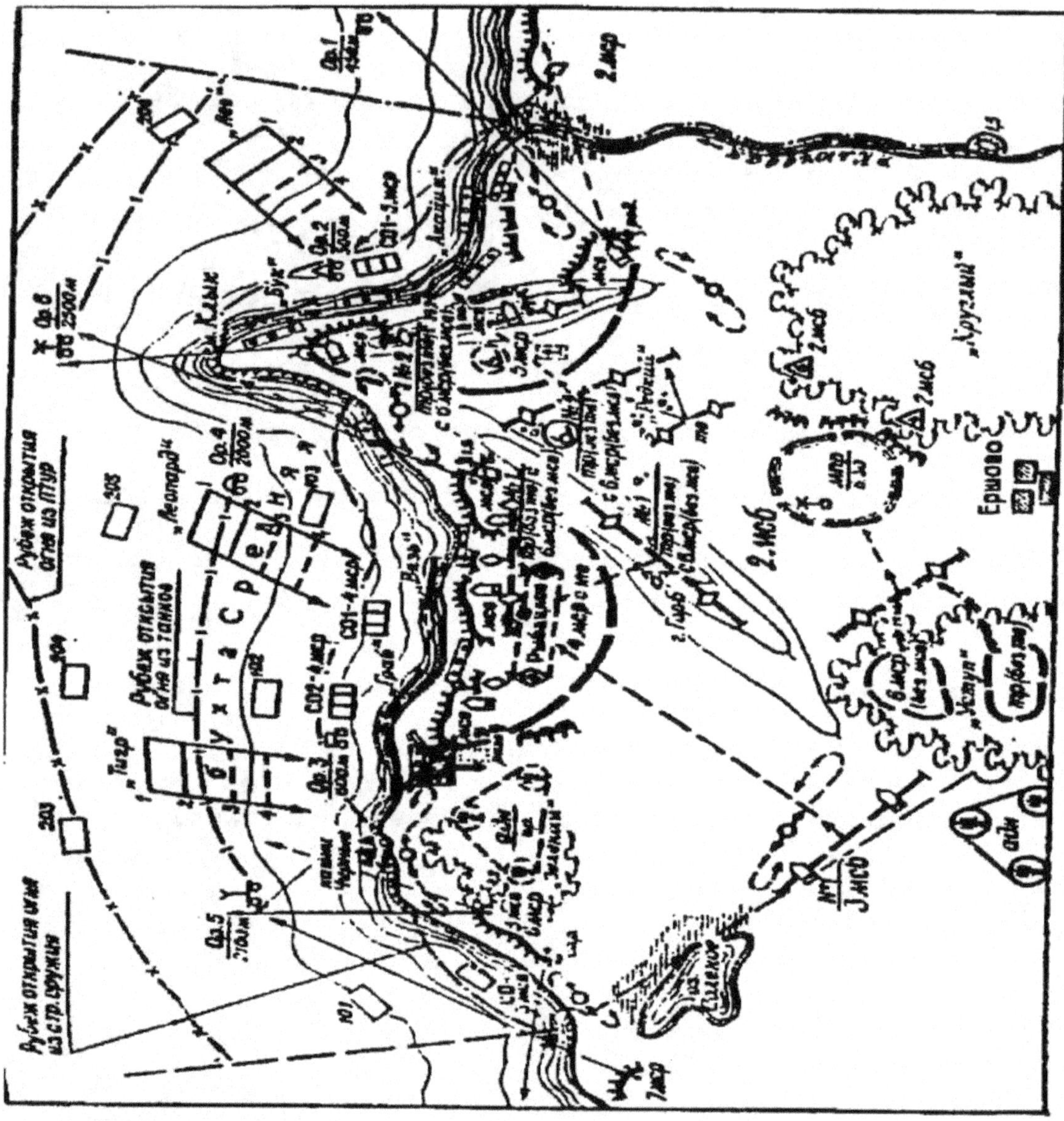

Fig. 1. A Battalion in a Coastal Defense. Armor counterattack force is positioned in the lower right corner of the diagram, directly in front of the artillery battalion. This figure only depicts one battalion in the defense. If a brigade or division established a coastal defense, it would present a target rich environment that would be difficult to neutralize with missiles alone. Reprinted, by permission, from Soviet Military Herald, Jan 1984.

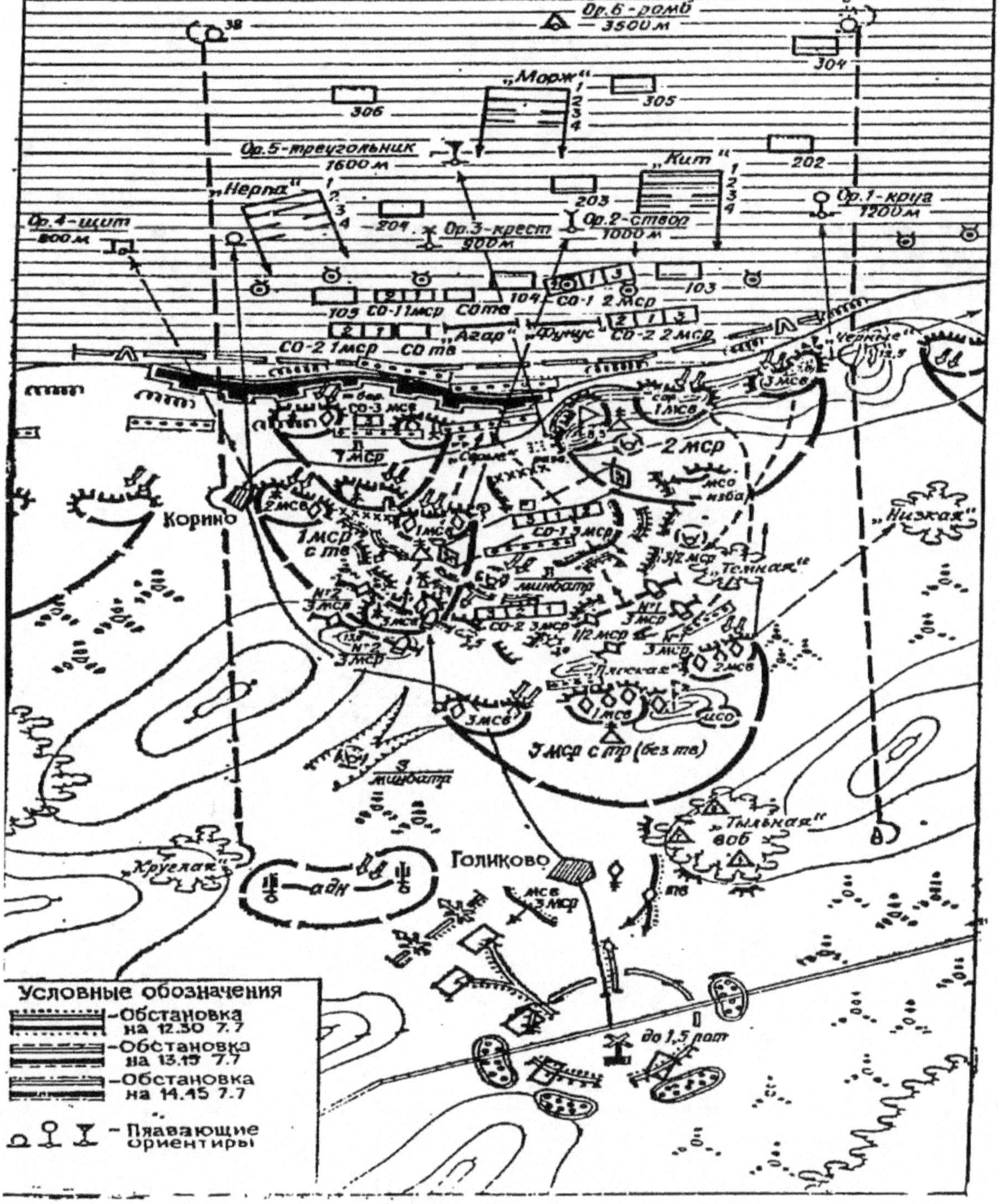

Fig. 2. A battalion defending a coastline along a narrow frontage. A concentrated defense presents another target rich environment. Missiles dispensing anti-personnel munitions could be effective against this defense. Reprinted, by permission from Soviet Military Review September 1989.

ENDNOTES

[1] Mike Boorda quoted by Jon P. Walman in "The New Generation of Combatants," Surface Warfare, 20, no. 6, (November/December 1995): 7.

[2] Alan Morehead, Gallipoli, (New York: Harper & Row, Publishers, Inc., 1956), 141.

[3] Ernest Blazaar, "Battleships Aren't Dead Yet," Navy Times, 44, no. 33, (22 May 1995): 25.

[4] Defense Advanced Research Projects Agency, "Arsenal Ship - Main Page," <http://www.darpa.mil/asjpo/>; http://www.darpa.mil/asjpo/documnts.html; http://www.darpa.mil/asjpo/solicitation/docs/pc_att1.zip CONOPS; http://www.darpa.mil/asjpo/solicitation/docs/pc_att1.zip SCD, November 1996.

[5] U.S. Navy, The Battleship in the United States Navy, (Washington: U.S. Naval History Division, 1970), 3.

[6] Richard Humble, Naval Warfare: An Illustrated History, (New York: St. Martin's Press, 1983), 9-12.

[7] E.B. Potter, Sea Power, (Englewood Cliffs, NJ: Prentice-Hall, Inc., 1960), 3-5.

[8] Humble, Naval Warfare: An Illustrated History, 38-42.

[9] Potter, Sea Power, 27-30.

[10] U.S. Navy, The Battleship in the United States Navy, 4.

[11] U.S. Navy, The Battleship in the United States Navy, 4.

[12] U.S. Navy, The Battleship in the United States Navy, 5.

[13] Carroll Storrs Alden and Allen Westcott, The United States Navy: A History, (Chicago: J.B. Lippincott

Company, 1943), 41-61 passim; U.S. Navy, The Battleship in the United States Navy, 5.

[14] U.S. Navy, The Battleship in the United States Navy, 5-7.

[15] William Chapman White and Ruth White, Tin Can on a Shingle, (New York: E.P. Dutton & Company, Inc., 1957), 10-11.

[16] Alexander A. Hoehling, Thunder at Hampton Roads, (New York: Da Capo Press, 1993), 32-35, 38-41; White and White, Tin Can on a Shingle, 16.

[17] James Phinney Baxter, The Introduction of the Ironclad Warship, (Cambridge, MA: Archon Books, 1968), 229; White and White, Tin Can on a Shingle, 16; Hoehling, Thunder at Hampton Roads, 56.

[18] White and White, Tin Can on a Shingle, 22-23.

[19] White and White, Tin Can on a Shingle, 46.

[20] Alden and Westcott, The United States Navy: A History, 171.

[21] White and White, Tin Can on a Shingle, 152.

[22] White and White, Tin Can on a Shingle, 153; Hoehling, Thunder at Hampton Roads, 194.

[23] Alden and Westcott, The United States Navy: A History, 281.

[24] Brayton Harris, The Age of the Battleship, 1890-1922, (New York: Franklin Watts, Inc., 1965), 6-8.

[25] Potter, Sea Power, 344.

[26] Harris, The Age of the Battleship, 1890-1922, 103.

[27] Potter, Sea Power, 388.

[28] Harris, The Age of the Battleship, 1890-1922, 153.

[29] Alden and Westcott, The United States Navy: A History, 370-375.

[30] Harris, The Age of the Battleship, 1890-1922, 394-395; U.S. Navy, The Battleship in the United States Navy, 22.

[31] U.S. Navy, The Battleship in the United States Navy, 22.

[32] Potter, Sea Power, 835.

[33] U.S. Navy, The Battleship in the United States Navy, 37.

[34] William H. Garzke, Jr. and Robert O. Dulin, Jr., Battleships: United States Battleships, 1935-1992, (Annapolis, MD: Naval Institute Press, 1995), 120, 128, 134, 203, 276.

[35] Garzke and Dulin, Battleships: United States Battleships, 1935-1992, 134.

[36] Henry Wyatt, "Retain the Iowas...As They Are," United States Naval Institute Proceedings, 121, no. 11, (November 1995): 75.

[37] Garzke and Dulin, Battleships: United States Battleships, 1935-1992, 128-129; U.S. Navy, The Battleship in the United States Navy, 39-40.

[38] Garzke and Dulin, Battleships: United States Battleships, 1935-1992, 212-213.

[39] Ronald M. Spector, Operational Experience of Fast Battleships: World War II, Korea, Vietnam, (Washington: U.S. Navy Historical Center, 1989), iii.

[40] Garzke and Dulin, Battleships: United States Battleships, 1935-1992, 239, 242.

[41] U.S. Navy, Naval Doctrine Publication 1: Naval Warfare, (Washington: Department of the Navy, 1994), 16.

[42] Ernest Blazar, "Battleships Aren't Dead Yet," Navy Times, 44, no. 33, (22 May 1995): 25.

[43] DARPA, "Concept of Operations (CONOPS), attachment 1 to Arsenal Ship Program Solicitation 23 May 1996," http://www.darpa.mil/asjpo/solicitation/docs/ pc_att1.zip, November 1996.

[44] U.S. Navy, "Forward ...From the Sea," Navy Public Affairs Library (NAVPALIB), http://www.navpalib@opnav-emh.navy.mil, November 1996.

[45] Congressional Budget Office, CBO Staff Memorandum: Selected Options for Enhancing Naval Capability in Regional Conflicts, (Washington: CBO, 1993), iv.

[46] David Silverstein, "As Military is Cut, America Still Needs the Marines," The Heritage Foundation Backgrounder, no. 871, (13 January 1992): 11-12.

[47] U.S. Navy, Arsenal Ship Combat Systems Study, (Dahlgren, VA: Naval Surface Warfare Center, 1995), 1.

[48] CBO, CBO Staff Memorandum: Selected Options for Enhancing Naval Capability in Regional Conflicts, viii.

[49] CBO, Selected Options for Enhancing Naval Capability in Regional Conflicts, viii.

[50] Silverstein, "As Military is Cut, America Still Needs Marines," Backgrounder, no. 871, 12.

[51] Ernest Blazar, "Mother of Invention: Do With Less," Navy Times, 45, no. 13, (1 January 1996): 28.

[52] U.S. Navy, Arsenal Ship Combat Systems Study, 1.

[53] U.S. Navy, "Transcript of an Interview with Admiral J.M. Boorda Chief of Naval Operations," <http://www.ncts.navy.mil/navpalib/ships/arsenal/boor0824.txt>, October 1996.

[54] U.S. Navy, Arsenal Ship Combat Systems Study, 7.

[55] U.S. Navy, Arsenal Ship Combat Systems Study, 8.

[56] U.S. Navy, Arsenal Ship Combat Systems Study, 9.

[57] Defense Daily, "Navy, DARPA Ink Agreement for Arsenal Ship," <www.phillips.com.defense/dd/dd032101.html>, October 1996.

[58] DARPA, "Arsenal Ship Capabilities Document (SCD)," http://www.darpa.mil/asjpo/solicitation/docs/pc_att1.zip SCD, November 1996, 1.2.

[59] DARPA, "Arsenal Ship Capabilities Document (SCD)," 2.0-2.2.

[60] DARPA, "Arsenal Ship Capabilities Document (SCD)," 2.3; DARPA, "Arsenal Ship Concept of Operations," Overview.

[61] DARPA, "Arsenal Ship Capabilities Document (SCD)," 2.3.

[62] DARPA, "Arsenal Ship Capabilities Document (SCD)," 2.4-3.2.

[63] DARPA, "Arsenal Ship Concept of Operations," Overview.

[64] DefenseLink News, "Arsenal Ship Program Selects Phase I Contractors," <www.dtic.dla.mil/defenselink/news/Jul96/b071196_bt422-96.html>, October 1996.

[65] U.S. Navy, "Transcript of an Interview with Admiral J.M. Boorda Chief of Naval Operations," Answer# 20, <http://www.ncts.navy.mil/navpalib/ships/arsenal/boor0824.txt>, October 1996.

[66] Ernest Blazar, "Battleships Aren't Dead Yet," Navy Times, 44, no. 33, (22 May 1995): 25.

[67] Edward A. Smith, "Naval Firepower for the 21st Century," Washington Post, 27 July 1996, 22.

[68] Garzke and Dulin, Battleships: United States Battleships, 1935-1992, 217, 252, 277.

[69] Jon P. Walman "The New Generation of Combatants," Surface Warfare, 20, no. 6, (November/December 1995): 7.

[70] Davy Jones quoting Joseph Metcalf in "Vessel Could Mean New Direction For Surface Warfare," The Campus News, May 1996, 1.

[71] Robert C. Peniston and Robert W. Selle, "Gunfire to Get Them Off the Beach," The Washington Post, 14 August 1996, A20.

[72] Garzke and Dulin, Battleships: United States Battleships, 1935-1992, 252.

[73] Edmund R. Anderson et. al., "TSTAR: Tomahawk Stops the Attacking Regiments," Surface Warfare, 21, no. 2, (March/April 1996): 13.

[74] Norman Polmar, "The Navy's Proposed Arsenal Ship: Pro," The Retired Officer Magazine, 52, no. 11, (November 1996): 40.

[75] Blazar, "Battleships Aren't Dead Yet," 25.

[76] Ernest Blazar, "Mother of Invention: Do With Less," Navy Times, 45, no. 13, (1 January 1996): 28.

[77] William L. Stearman, "The Navy's Proposed Arsenal Ship: Con," The Retired Officer Magazine, 52, no. 11, (November 1996): 39.

[78] DARPA, "Concept of Operations (CONOPS)," Operational Assumptions.

[79] Greg Pickell, "Fire Support: View From the Foxhole," The Washington Post, 11 July 1996, A24.

[80] U.S. Navy, "Ship Capabilities Document (SCD)," 2.4.

SELECTED BIBLIOGRAPHY

Articles

Andrienko, Vasily. "Battalion In Coastal Defence." Soviet Military Review. (January 1989): 26,46.

Ange, Chris. "Naval Surface Fire Support: 5-Inch Mk 45 Gun Mount Modifications." Surface Warfare. 21, no. 3. (May\June 1996): 2-6.

Blazar, Ernest . "Battleships Aren't Dead Yet." Navy Times. 44, no. 33. (May 22, 1995): 25.

________. "Future Shock." Navy Times. 45, no. 38. (July 29, 1996): 12-14.

________. "Mother of Invention: Do With Less." Navy Times. 45, no. 13. (January 1, 1996): 28.

________. "Ships With Two Points of View." Navy Times. 44, no. 48. (September 4, 1995): 22.

Boorda, Jeremy Mike. "An Interview with Adm. Mike Boorda." Surface Warfare. 21, no. 2. (March\April 1996): 2-4.

Gourley, Scott. "Arsenal Ship." Popular Mechanics. (June 1996): 58-60.

Jane's Navy International. "Budget Squeeze Blurs the Long-Range Vision." (June 1996): 1,2, 31-34.

Jones, Davy. "Retired Surface Warrior Details New Arsenal: Vessel Could Mean New Direction For Surface Warfare." The Campus News. (Monterey,CA: Naval Postgraduate School). (May 1996): 1.

Kokhanov, V. "Antilanding Defence." Soviet Military Review. (August 1972): 14-17.

Luddy, John. "Charting a Course for the Navy in the 21st Century." Heritage Foundation Backgrounder. No. 979, (March 9, 1994): 1-14.

Lyons, Kenneth T. "Extended Range Guided Munitions." Surface Warfare. 21, no. 3. (May\June 1996): 8-11.

Mackin, J. G. et. al. "Systems Impact of Extended Range Guns for NSFS." Paper presented at ASNE Symposium-1995, Mississippi Coast Coliseum & Convention Center, (16-17 February 1995).

Morral, Dennis. "Naval Surface Fire Support: Fire Support Missile Options." Surface Warfare. 21, no. 3. (May\June 1996): 12-13.

Norton, Scott. "Naval Surface Fire Support Advanced Technologies." Surface Warfare. 21, no. 3. (May\June 1996): 14-21.

Phillips, Mark D. and Kevin M. Norman. "BB's Big Guns Blast Away." Surface Warfare. 16, no. 2. (March\April 1991): 16-17.

________. "Tomahawk Strike." Surface Warfare. 16, no. 2. (March\April 1991): 8-13.

Pickell, Greg. "Fire Support: The View From the Foxhole." The Washington Post. 14 August 1996, A20.

Polmar, Norman and William L. Stearman. "The Navy's Proposed Arsenal Ship: Pro and Con." The Retired Officer. 52, no. 11. (November 1996): 36-41.

Sabalos Jr., Nicholas, ed. "An Interview With Lt.Gen Paul K. Van Riper Commanding General, Marine Corps Combat Development Center." Surface Warfare. 21, no. 3. (May\June 1996): 30-31.

Samigulin, K. "Coastal Defence," Soviet Military Review. (September 1977): 22-23.

Selle, Robert W. "Gunfire to Get Them Off the Beach." The Washington Post. 14 August 1996, A20.

Serig, Howard W., Jr. "The *Iowa* Class: Needed Once Again." U.S. Naval Institute Proceedings. 108. (May 1982): 134-149.

Silverstein, David. "As Military is Cut, America Still Needs the Marines." The Heritage Foundation Backgrounder. Number 871. (13 January 1992): 1-14.

Smirnov, V. "A Battalion in a Coastal Defense." Soviet Military Review. (Jauary 1984): 30-33.

Stearman, William Lloyd. "A Misguided Missile Ship". The Washington Post. July 7, 1996, 18.

Trainor, Bernard. "Anchors Awry: How to Fix the Sinking Navy." The Washington Post. May 26, 1996, C01.

Walman, Jon P. "The New Generation of Combatants: SC-21, Arsenal Ship Reflect 21st-Century Vision." Surface Warfare. 20, no. 6. (Nov\Dec 1995): 6-8.

Books

Alden, Carroll Storrs and Alan Westcott. The United States Navy: A History. Chicago: J.B. Lippincott and Company, 1943.

Baxter, James Phinney. The Introduction of the Ironclad Warship. Cambridge, MA: Archon Books, 1968.

Beach, Edward L. The United States Navy: 200 Years. New York: Henry Holt and Company, 1986.

Bowie, Christopher et al. The New Calculus: Analyzing Airpower's Changing Role in Joint Theater Campaigns. Santa Monica, CA: RAND, 1993.

Garzke, William H Jr. and Robert O. Dulin. Battleships: United States Battleships, 1935-1992. Annapolis, MD: Naval Institute Press, 1995.

Gregson, Barry. Amphibious Operations. London: Blandford Press, 1988.

Grove, Eric. The Future of Sea Power. Annapolis, MD: Naval Institute Press, 1990.

Harris, Brayton. The Age of the Battleship 1890-1922. New York: Franklin Watts, Inc., 1965.

Hastings, Max and Simon Jenkins. The Battle for the Falklands. New York: W.W. Norton and Company, 1983.

Hoehling, A.A. Thunder at Hampton Roads. New York: Da Capo Press, 1976.

Hough, Richard. Death of the Battleship. New York: The Macmillan Company, 1963.

Humble, Richard. Naval Warfare: An Illustrated History. New York: St. Martin's Press, 1983.

Keaney, Thomas A. and Eliot A. Cohen. Gulf War Air Power Survey Summary Report. Washington: U. S. Government Printing Office, 1993.

Keegan, John. The Price of Admiralty. New York: Penguin Books, 1988.

Legg, Stuart. Jutland: An Eyewitness Account of a Great Battle. New York: The John Day Company, 1967.

Moorehead, Alan. Gallipoli. New York: Harper & Row, 1956.

Morrison, Samuel Eliot. Victory in the Pacific 1945. Boston: Little, Brown and Company, 1975.

Potter, E.B. Sea Power - A Naval History. Englewood Cliffs, NJ: Prentice-Hall, Inc., 1960.

Reilly Jr., John C. Operational Experience of Fast Battleships; World War II, Korea, Vietnam. Washington: Naval Historical Center, 1989.

White, W.C. and Ruth. Tin Can on a Shingle. New York: E.P. Dutton Company, 1957.

Government Documents

Committee on Armed Services House of Representatives. Reactivation of Battleships. Washington: Government Printing Office, 1983.

Congressional Budget Office. CBO Staff Memorandum: Selected Options For Enhancing Naval Capability in Regional Conflicts. Washington: Congressional Budget Office, 1993.

Department of Defense. Conduct of the Persian Gulf War: Final Report to Congress. Washington: Government Printing Office, 1992.

U.S. Army. FM 6-20 Fire Support in the AirLand Battle. Washington: Department of the Army, 1988.

U.S. Navy. The Battleship in the United States Navy. Washington: U.S. Naval History Division, 1970.

________. Force 2001: A Program Guide to the U.S. Navy. Washington: Deputy Chief of Naval Operations Resources, Warfare Requirements and Assessments (N8), 1995.

________. Forward...From The Sea. Washington: Department of the Navy, 1994.

________. ...From The Sea: Preparing the Naval Service for the 21st Century. Washington: Department of the Navy, 1992.

________. Naval Doctrine Publication 1: Naval Warfare. Washington: Department of the Navy, 1994.

________. NWP 22-2 (Rev. C) Supporting Arms in Amphibious Operations. Norfolk, VA: Naval Doctrine Command, 1993.

________. NWP 3-09.11M Supporting Arms in Amphibious Operations. Norfolk, VA: Naval Doctrine Command, 1995.

Unpublished Materials

Defense Advanced Research Projects Agency and U.S. Navy. "Arsenal Ship Program - Main Page." <http://www.darpa.mil/asjpo/>. Joint Program Office, September 1996.

________. "Arsenal Ship Program Documents." <http://www.darpa.mil/documents.html>. Joint Program Office, September 1996.

________. "Arsenal Ship Program Introduction." <http://www.darpa.mil/asjpo/docs/ pcmay_p1.zip>. Joint Program Office, September 1996.

________. "Arsenal Ship Program Acquisition Overview." <http://www.darpa.mil/asjpo /docs/pcmay_p2.zip>. Joint Program Office, September 1996.

________. "Arsenal Ship Program Concept of Operations." <http://www.darpa.mil/ asjpo/docs/pcmay_p3.zip>. Joint Program Office, September 1996.

________. "Arsenal Ship Program Description." <http://www.darpa.mil/asjpo/docs/ pcmay_p4.zip>. Joint Program Office, September 1996.

________. "Arsenal Ship Program Phase I Kickoff Plans." <http://www.darpa.mil/ asjpo/docs/pcmay_p5.zip>. Joint Program Office, September 1996.

________. "Arsenal Ship Program Section 845 Review." <http://www.darpa.mil/ asjpo/docs/pcmay_p6.zip>. Joint Program Office, September 1996.

Defense Technical Institute Center. "Arsenal Ship Program Selects Phase I Contractors." <http.//www.dtic.dls. mil/defenselink/news/Jul96/b071196_bt422-96.html>. Department of Defense, August 1996.

Jordan, Michael W. "Soviet Antilanding Doctrine: Does it Matter?" Monterey, CA: Naval Postgraduate School, 1991.

Meeker, Thomas J. "The Tomahawk Land Attack Cruise Missile and the Missions of the United States Navy." Monterey, CA: Naval Postgraduate School, 1982.

Palermo, Anthony J. "Exploiting the Soviet Antilanding Defense With Amphibious Maneuver." Newport, RI: Naval War College, 1987.

NAVAL WAR COLLEGE
Newport, R.I.

THE ARSENAL SHIP MEASURES UP TO JOINT VISION 2010

by

Alan J. Abramson

Lieutenant Commander, United States Navy

A paper submitted to the faculty of the Naval War College in partial satisfaction of the requirements of the Department of Joint Military Operations.

The contents of this paper reflect my own personal views and are not necessarily endorsed by the Naval War College or the Department of the Navy.

Signature: ____________

7 November 1997

Paper directed by
CAPT George W. Jackson, USN
Chairman, Department of Joint Military Operations

Dennis M. Lunghofer 6FEB97
Faculty Advisor Date
CAPT Dennis M. Lunghofer, USN
Department of Joint Military Operations

UNCLASSIFIED
Security Classification This Page

REPORT DOCUMENTATION PAGE

1. Report Security Classification: UNCLASSIFIED	
2. Security Classification Authority:	
3. Declassification/Downgrading Schedule:	
4. Distribution/Availability of Report: DISTRIBUTION STATEMENT A: APPROVED FOR PUBLIC RELEASE; DISTRIBUTION IS UNLIMITED.	
5. Name of Performing Organization: JOINT MILITARY OPERATIONS DEPARTMENT	
6. Office Symbol: C	**7. Address:** NAVAL WAR COLLEGE 686 CUSHING ROAD NEWPORT, RI 02841-1207
8. Title (Include Security Classification): THE ARSENAL SHIP MEASURES UP TO JOINT VISION 2010 (U)	
9. Personal Authors: LCDR Alan J. Abramson, USN	
10. Type of Report: FINAL	**11. Date of Report:** 7 February 1997
12. Page Count: 22	
13. Supplementary Notation: A paper submitted to the Faculty of the NWC in partial satisfaction of the requirements of the JMO Department. The contents of this paper reflect my own personal views and are not necessarily endorsed by the NWC or the Department of the Navy.	
14. Ten key words that relate to your paper: arsenal, ship, joint, doctrine, precision, maneuver, protection, logistics, operations.	

15. Abstract: The arsenal ship concept combines new and emerging technologies into a new type of naval platform capable of massive offensive firepower. Concentrating hundreds of advanced weapons in one platform will provide the Joint Force Commander with a potent yet simple instrument for mass, maneuver and precision fires, as well as a credible deterrence prior to the onset of armed conflict.

When measured against Joint Vision 2010, the Navy's Arsenal Ship Concept of Operations supports an effective joint fighting force for the early 21st century in most areas. More development is needed, however, in the areas of full-dimensional protection and command structure for the ship in the joint force. The arsenal ship's potential contribution to joint warfighting compels the most rigorous efforts to make the most efficient and effective use of this platform and protect it from the threats of future enemies.

16. Distribution / Availability of Abstract:	Unclassified	Same As Rpt	DTIC Users
	X		

17. Abstract Security Classification: UNCLASSIFIED	
18. Name of Responsible Individual: CHAIRMAN, JOINT MILITARY OPERATIONS DEPARTMENT	
19. Telephone: 841-6461	**20. Office Symbol:** C

Security Classification of This Page UNCLASSIFIED

Abstract of

THE ARSENAL SHIP MEASURES UP TO JOINT VISION 2010

The arsenal ship concept combines new and emerging technologies into a new type of naval platform capable of massive offensive firepower. Concentrating hundreds of advanced weapons in one platform will provide the Joint Force Commander with a potent yet simple instrument for mass, maneuver and precision fires, as well as a credible deterrence prior to the onset of armed conflict.

When measured against Joint Vision 2010, the Navy's Arsenal Ship Concept of Operations supports an effective joint fighting force for the early 21st century in most areas. More development is needed, however, in the areas of full-dimensional protection and command structure for the ship in the joint force. The arsenal ship's potential contribution to joint warfighting compels the most rigorous efforts to make the most efficient and effective use of this platform and protect it from the threats of future enemies.

THE ARSENAL SHIP MEASURES UP TO JOINT VISION 2010

The Revolution in Military Affairs which has touched nearly every aspect of military operations can perhaps best be illustrated by the development of the arsenal ship. This innovative ship design combines existing and emerging technologies with a new concept of weapons employment; the early and massive use of firepower to seize an immediate offensive advantage at the onset of armed conflict.[1] Not a naval combatant in the traditional sense, the arsenal ship will perform no command and control functions and will possess only very modest active defensive systems.[2] The primary roles for the ship will be deterrence through sustained forward deployment and overwhelming firepower for deep strike and interdiction available at the onset of war. Subsequent to the initial strike, the arsenal ship will be a ready source of precision fires for the joint force commander as well as an additional supply of theater air defense weapons.

As history teaches us, a successful military innovation lies not merely in technology, but in the marriage of

---

[1] "3-R Strategy" for Rapid, Robust Response as delineated by Ernest Blazar, "How the arsenal ship will fight," Navy Times, 29 July 1996, 14:2.
[2] U.S. Navy Dept., Arsenal Ship Concept of Operations (Washington: 1996), 4.

technology and sound operational doctrine.[3] Just as earlier doctrinal advances assured the success of the aircraft carrier during World War II, thorough development and critical analysis of operating doctrine is essential if the arsenal ship is to succeed in the joint warfighting arena. Current plans envision the first demonstrator, or prototype, ship conducting sea trials in the year 2000,[4] and the initial operational deployment as soon as the year 2004.[5] Because the final design and full capabilities of the ship will not be known until early 1998, the Navy must continue to refine existing operating doctrine so that joint staffs will be prepared to include the new ship in their operational plans well before it appears in their force.

As detailed by the Navy, the arsenal ship concept fulfills the naval service's vision as expressed in Forward . . . From the Sea and Operational Maneuver From the Sea. The arsenal ship represents a fulfillment of the stated goals of using forward deployed sea-based platforms to influence events ashore in peacetime, and to provide swift, massive response

[3] Stephen Peter Rosen, "New Ways of War: Understanding Military Innovation," International Security, Summer 1988, 134.
[4] Ernest Blazar, "Future Shock: Arsenal ship will have small crew & big punch," Navy Times, 29 July 1996, 13:1.
[5] Telephone conversation with CDR Cam Ingram, USN, Office of Chief of Naval Operations, Surface Warfare Plans, Programs and Requirements, Pentagon, Washington, DC, 9 January 1997.

and flexible support of ground forces in war.[6] Some key weaknesses appear in the Navy's Concept of Operations (CONOPS), however, when measured against Joint Vision 2010, particularly in the areas of protection and unity of command. The Navy should more fully develop the full-dimensional protection concept and the command structure for the ship in joint operations. The Navy CONOPS should reflect continuing refinements and must always fully support joint doctrine if the arsenal ship is to achieve its potential of operational success by providing a key element to the joint force of the future.

BACKGROUND

The idea of a stealthy ship composed almost entirely of offensive weapons was first given public credence by VADM Joseph Metcalf III, USN (Ret.) in his "Revolution at Sea" of the mid-eighties. His overriding philosophy in conceptualizing a future strike cruiser was maximizing the delivery of "ordnance on target".[7] Though it took some time for VADM Metcalf's vision to enjoy wide acceptance, the idea

---

[6] U.S. Navy Dept., Forward . . . From the Sea (Pentagon, Washington: 1994), 2. and U.S. Marine Corps, Operational Maneuver From the Sea (Pentagon, Washington: 1996), 12.

[7] VADM Joseph Metcalf III, USN (Ret.), "Revolution at Sea," U.S. Naval Institute Proceedings, January 1988, 38.

was delivered nearly intact to the Navy's design concept for the arsenal ship.

Because of the innovative procurement process being pursued by the Navy, in collaboration with the Defense Advanced Research Projects Agency (DARPA), the final design for the arsenal ship will not be made public until January 1998, only about two years prior to initial sea trials of the prototype ship. Until then, the most complete description of the ship lies in the Navy's CONOPS which outlines the following requirements:

- crew size of no more than 50,

- about 500 missile cells, capable of launching Tomahawk Land Attack Missiles (TLAM), a navalized version of the Army Tactical Missile System (NTACMS), the latest and emerging family of Standard Missiles (SM-2 Block IVA and LEAP), and possibly a surface-launched variant of the Standoff Land Attack Missile (SLAM),

- advanced gun system firing long-range guided munitions,

- Cooperative Engagement Capability (CEC) remote targeting and launch system,

- and an "affordable balance" of active and passive defense features against 21st century missiles, torpedoes and mines.

CURRENT EMPLOYMENT CONCEPT

The Navy's CONOPS envisions a force of "about six" arsenal ships forward deployed to the Southwest Asia/Persian Gulf, Western Pacific and Mediterranean theaters to provide the joint force commander with conventional deterrence, flexible response, and credible forward firepower, if deterrence fails, in support of joint and coalition land forces. Supportable roles include long-range strike, invasion stopping, fire support, tactical ballistic missile defense and air superiority. Peacetime Operational Control (OPCON) will be exercised by a numbered fleet commander, and joint force OPCON will be given to the Joint Force Maritime Component Commander (JFMCC). Tactical Control (TACON) will normally be exercised by a naval commander. AEGIS cruisers and destroyers will perform weapons control, mission planning, targeting and joint connectivity functions for the arsenal ships.[8]

A detailed analysis of the arsenal ship's possible utility in war was conducted during a Joint Multiwarfare

[8] Ibid., 1-4.

Analytical Game (JMAG), and approved by a senior review board of retired general and flag officers from all four services. Key findings include:

- "arsenal ship missions, doctrine, and basing policy should place deterrence first in all considerations.",

- long-range weapons (TLAM) are most effectively used early in conflict, and shorter-range weapons (NTACMS, SLAM, Strike-SM, naval gunfire) are most effectively used later in conflict, once sea and air control is achieved,

- the optimum weapons mix is theater dependent,

- and the forward presence of two arsenal ships can replace two to three days of aircraft strike assets, freeing some critical airlift for other needs just prior to and early in conflict.[9]

Joint Vision 2010

Focused on creating a more effective fighting force for the early 21st century, Joint Vision 2010 was devised to foster innovative ways to employ new and emerging technologies in support of four operational concepts; dominant maneuver, precision engagement, full dimensional protection, and focused

---

[9] Global Associates, Ltd., Technology Services Group, Joint Multiwarfare Analytical Game, Arsenal Ship Analysis, Executive Summary (Arlington, VA: 1996), 4-132.

logistics. Above all else, Joint Vision 2010 emphasizes the "imperative of jointness" in all future military operations.[10] The arsenal ship measures up well with this plan for the future, espoused by the Chairman of the Joint Chiefs of Staff. It brings innovative solutions to many of the current problems faced by today's forces; however, it also brings some new problems to be resolved by future commanders. The following paragraphs highlight the positive and negative contributions of the arsenal ship to the future joint force, as measured against the four operational concepts of Joint Vision 2010.

Dominant Maneuver: ". . . the multidimensional application of information, engagement, and mobility capabilities to position and employ widely dispersed joint air, land, sea, and space forces to accomplish the assigned operational tasks."[11] This concept matures the process of maneuver warfare, emphasizing the need for forces capable of "conducting sustained and synchronized operations from dispersed locations."[12]

The arsenal ship provides the Joint Force Commander (JFC) with a powerful tool to support the land force's scheme of

[10] U.S. Joint Chiefs of Staff, Joint Vision 2010 (Pentagon, Washington: 1996), 1-8.
[11] Ibid, 20.
[12] Ibid, 20.

dominant maneuver through close support and interdiction weapons. It can be particularly effective as an instrument of asymmetric leverage against an enemy with little or no maritime capability. While Joint Vision 2010 de-emphasizes the traditional methods of massing forces, the arsenal ship's large number of weapons allows the JFC to mass fires through rapid targeting and near simultaneous engagements with "massive" amounts of "ordnance on target." The huge initial punch provided by the arsenal ship will allow the JFC to gain an immediate offensive advantage and continue to exercise the initiative through maneuver and synchronization. The preponderance of weapons contained in the arsenal ship will also free other naval combatants to conduct other vital missions within the force, making the most efficient use of available forces.

Dispersal of forces is a key consideration for dominant maneuver, according to Joint Vision 2010. In the sense that the arsenal ship consolidates many weapons into one platform, it opposes the movement toward dispersal. This will reduce flexibility somewhat and make the arsenal ship a primary and perhaps vulnerable target for enemy forces. However, in the sense that the arsenal ship provides an additional platform for large scale precision fires, interdiction, and close support, missions normally conducted by land and air forces, the ship may actually enhance dispersal of those forces. In order to overcome the inherent bulk of the platform, the JFC should employ methods to disperse the arsenal ship within the theater on a continual basis. Maneuver, stealth, and synchronization with land and air forces can make the arsenal ship a true "hit and run" platform, well suited to the fast-paced action prescribed in Joint Vision 2010.

Precision Engagement: ". . . a system of systems that enables our forces to locate the objective or target, provide responsive command and control, generate the desired effect, assess our level of success, and retain the flexibility to re-engage with precision when required."[13] This concept seeks to take full advantage of the current and projected U.S. technological edge in weapons accuracy, precision, range, and platform stealth to extend the battlespace and better protect U.S. forces.

Incorporating the latest technology in precision weapons, the arsenal ship will be a key element in the "system of systems." The ship possesses no targeting, command and

[13] Ibid, 21.

control (C2), or battle damage assessment capabilities, so maximum use of CEC to join other elements of the system is essential to successful employment. The Navy's CONOPS states that an AEGIS combatant will provide the C2 link to the joint force, performing all targeting, remote firing, and other C2 functions for the arsenal ship.[14]

While this may be necessary for the initial deployment period of the arsenal ship, the full development and joint use of CEC must be pursued. CEC will allow air and land force commanders to target and launch arsenal ship weapons directly, eliminating the "middle man" platform.[15] Using an AEGIS combatant as a full time C2 escort is not only an inefficient use of a multi-mission platform, but could also negate the most effective survivability feature of the arsenal ship, stealth, if conducted from close range. C2 functions performed off-ship should be undertaken from as great a distance as feasible. The targeting and launching functions are also best performed by the commander using the weapons· JFACC for strike and JFLCC for interdiction and close support. The AEGIS combatant is best suited for control of the arsenal

[14] U.S. Navy Dept., Arsenal Ship CONOPS, 4.
[15] CAPT Richard L. Wright, USN, "Arsenal Ship: Potent and Punishing," Surface Warfare, January/February 1997, 22.

ship for theater missile defense only. Less reaction time and direct control by the force requiring support will improve efficiency and effectiveness. The goal should be to place as few steps between the customer and the provider as possible.

Unity and efficiency of command is a principal issue for successful joint operations, particularly with regard to the arsenal ship, since it will support multiple component and functional commanders. The Navy CONOPS states that, within a joint force, OPCON will reside with the JFMCC and that TACON will reside with a naval commander. To effectively control the arsenal ship's movements and primary tasking, the JFC should retain OPCON, eliminating the inevitable competition for resources between the component and functional commanders.

A possible solution for the tactical level of control of the arsenal ship lies in the apportionment of the weapons to the component and functional commanders. The Area Air Defense Commander (AADC) would own air and theater missile defense missiles, the JFACC would own deep strike weapons, and the JFLCC would own interdiction and close support weapons. This division of resources would allow the JFC to direct primary tasking, flexible enough for the rapid pace of future conflicts, while also providing a measure of decentralized execution necessary for efficient mission accomplishment.

Full-Dimensional Protection: ". . . control of the battlespace to ensure our forces can maintain freedom of action during deployment, maneuver and engagement, while providing multi-layered defenses for our forces and facilities at all levels."[16] This concept emphasizes proactive measures, built on information superiority, to degrade the enemy's opportunity for offensive action.

The arsenal ship can bring a large number of defensive weapons to a theater to enhance the protective posture of the joint force and shore facilities. While not contributing directly to information superiority, it can provide additional resources for the AADC, JFMCC, and JFLCC to execute their defensive plans and also to pursue the destruction of enemy command and control targets.

The arsenal ship brings with it an added consideration for the JFC: how to protect such a valuable platform. The greatest weakness of the arsenal ship concept is the lack of robust active defense systems for such a large concentration of firepower. Although designed to enhance passive defense

[16] U.S. Joint Chiefs of Staff, Joint Vision 2010, 22.

measures (stealth, double hull construction, etc.), the arsenal ship's value to the joint force, as well as its value as a primary target for enemy forces, will necessitate a high level of attention to its protection. The Navy CONOPS states that an AEGIS combatant will provide a defensive escort for the arsenal ship. In light of the Navy's continued reliance on Aircraft Carrier Battle Groups and Amphibious Ready Groups for power projection,[17] and the projected reduced number of combatant ships available,[18] assigning a full time AEGIS escort for the arsenal ship may not always be possible. The JFC must pursue additional defensive options. Submarines could be used for surface and subsurface protection when an AEGIS ship with an embarked helicopter is not available, and aircraft (Navy or Air Force) could provide additional air defense. Innovative methods of operational deception, high levels of maneuverability, best use of the ship's reduced signatures, and the inherent toughness of the platform must be continually assessed and combined to maximize protection. The arsenal ship will be one of the most valuable platforms in the joint

[17] U.S. Navy Dept., Forward . . . From the Sea, 4.

[18] U.S. Navy Dept. Director of Surface Warfare, Surface Warfare Roadmap (Washington: 1996), 23.

force, and its protection must be a high priority at all levels of command.

Focused Logistics: ". . . the fusion of information, logistics, and transportation technologies to provide rapid crisis response, to track and shift assets even while enroute, and to deliver tailored logistics packages and sustainment directly at the strategic, operational, and tactical level[s] of operations."[19]

The arsenal ship could reduce the logistics requirements during a regional contingency conflict. As detailed in the JMAG report, weapons of deployed arsenal ships could replace 2-3 days of aircraft deep strike and interdiction assets early in such a conflict. The release of an equivalent amount of airlift required to transport the associated aviation support could be used for other vital needs. While realistically, the JFC would probably not forego an equivalent amount of aircraft firepower because of the presence of an arsenal sh   or two, some aircraft could be directed to other targets or delayed in transport, lightening the early airlift requirements to some lesser degree. Later in the conflict, the arsenal ship moves closer to the land mass as air and sea control are achieved,

---

[19] U.S. Joint Chiefs of Staff, Joint Vision 2010, 24.

and its close support and theater missile defense weapons could replace land-based weapons systems (ATACMS, Patriot, etc.), further lightening the logistics load for their support. The key for the realization of any logistics savings will be in loading of the correct mix of weapons in the arsenal ship prior to deployment. Long term theater needs must be considered if the weapons brought to the theater will continue to meet the JFC's needs throughout the arsenal ship's deployment. The arsenal ship will also reduce the need for naval ammunition replenishment by traditional methods, because it will be able to transport, position, and fire additional weapons on the command of combatant forces who need them most. In fact, "the most efficient replenishment ship yet designed" could be the most accurate assessment of the arsenal ship's primary contribution to the joint force of the future.

CONCLUSIONS AND RECOMMENDATIONS

The arsenal ship concept is a powerful and versatile instrument for the Joint Force Commander to support nearly all aspects of joint military operations. Focused on primary missions, yet flexible in responsive options, this innovative platform can be used to create an immediate advantage and to seize the initiative from the start of armed conflict.

The operational concepts of dominant maneuver, precision engagement, full-dimensional protection, and focused logistics, defined by Joint Vision 2010, can be realized more fully with the arsenal ship in the fleet, although serious consideration must be given to resolving command and control issues and to providing full-dimensional protection for the arsenal ship. The Navy Concept of Operations should be revised to meet the needs as follows:

- Place the arsenal ship under the OPCON of the JFC in all joint operations. Maneuver decisions, primary tasking and protection responsibilities must be made from the operational or strategic perspective, with no single-service bias, and with irrefutable authority.

- Rely less on a continual escort by an AEGIS combatant for protection and joint connectivity. Develop the CEC concept to its full joint potential, allocating weapons to the arsenal ship's customers (AADC, JFACC and JFLCC) to exercise weapons release directly, cutting out the middle man, streamlining command lines and cutting reaction times.

- Develop additional options for surface combatant, submarine and aircraft protection for the arsenal ship. Reliance on an AEGIS combatant for full-time close protection will not always be the most feasible, or even the most effective option.

With continued discussion and doctrinal development, the arsenal ship can provide a much needed force multiplier and powerful joint asset in the strategic, operational and tactical realms. As the final ship design takes shape, more refined solutions to doctrinal needs can be added, and further discussions of the best roles and missions can be pursued.

BIBLIOGRAPHY

Anderson, LCOL Charles A., USA and COL Richard G. Kurtz, USA (Ret.). "Air and Missile Defense: Who's in Charge?" ADA, July-August 1996, 2-8.

Beaumont, LCDR W., USN. "Counterair Is Still Dis Jointed." U.S. Naval Institute Proceedings, January 1997, 35-38.

Bingham, LCOL Price T., USAF (Ret.). "It's the Best Thing Since Gunpowder." U.S. Naval Institute Proceedings, January 1997, 45-49.

Blazar, Ernest. "Future Shock: Arsenal ship will have small crew & big punch." Navy Times, 29 July 1996, 12:1.

________. "How the Arsenal Ship Will Fight/Massive Precision Fire in a Hurry." Navy Times, 29 July 1996, 14:2.

Carey, Merrick and Loren Thompson. "Submarines and the Future of Seapower." Strategic Review, Fall Supplement 19 October 1996, 6-15.

Cate, LCDR David, USN. "Arsenal Ship, The Distributed Force, Force Planning and Innovation." Unpublished Research Paper, U.S. Naval War College, Newport, RI: 1996.

Concian, COL Mark, USMCR. "The Revolution is Incomplete." U.S. Naval Institute Proceedings, September 1996, 70-73.

Cushman, LGEN John H., USA (Ret.). "Joint Vision 2010: Can It Happen?" U.S. Naval Institute Proceedings, January 1997, 39-43.

Friedman, Norman. "Stealth and Survivability." Jane's Navy International, July/August 1996, 10-14.

Global Associates, Ltd. Technology Services Group. Joint Multiwarfare Analytical Game, Arsenal Ship Analysis Executive Summary. Arlington, VA: 1996.

Hayes, CAPT Bradd C., USN and CDR Douglas V. Smith, USN, ed. The Politics of Naval Innovation. Research Report 4-94. Newport, RI: U.S. Naval War College. Strategic Research Department, 1994.

Kre inevich, Andrew F., Jr. "Competing for the Future: Searching for Major Ellis." Marine Corps Gazette, November 1996, 28-37.

________. "Transforming the Navy's Warfighting Capabilities." Issues in Science and Technology, Fall 1996, 28-32.

Metcalf, VADM Joseph, III, USN (Ret.). "Revolution at Sea." U.S. Naval Institute Proceedings, January 1988, 34-39.

Moeller, CAPT R. T., USN. "JSTARS Works for Navy, Too." U.S. Naval Institute Proceedings, January 1997, 44.

Myers, Chuck. "The Arsenal Ship--She's Long Overdue." U.S. Naval Institute Proceedings, September 1996, 81-82.

Owens, ADM William A., USN. "The Emerging System of Systems." U.S. Naval Institute Proceedings, May 1995, 35-39.

Patton, CAPT Jim, USN (Ret.). "Keeping up with the Revolution." U.S. Naval Institute Proceedings, July 1996, 39-42.

Rosen, Stephen Peter. "New Ways of War: Understanding Military Innovation." International Security, Summer 1988, 134-168.

Scott, Richard, Joris Janssen Lok, Barbara Starr, E. R. Hooton, and Vincent Grimes. "Arsenal Ship Takes Shape." Jane's Navy International, June 1996, 47.

Smith, LT Neil A., USN. "Can the Arsenal Ship Survive?" U.S. Naval Institute Proceedings, January 1997, 80-81.

Telephone conversation with CDR Cam Ingram, USN, Office of Chief of Naval Operations. Surface Warfare Plans, Programs and Requirements Branch, Pentagon, Washington. 9 January 1997.

Trainor, Bernard E. "Anchors Awry: How to Fix the Sinking Navy." The Washington Post, 26 May 1996, C01.

Truver, Scott C. "Budget Squeeze Blurs the Long-range Vision." Jane's Navy International, June 1996, 28-39.

U.S. Dept. of Defense. Director of Net Assessment. RMA Update. Memorandum for the Record. Washington: 1994.

________. ________. Some Thoughts on Military Revolutions - Second Version. Memorandum for the Record. Washington: 1993.

U.S. Joint Chiefs of Staff. Doctrine for Joint Operations. Joint Pub 3-0. Washington: 1995.

________. Doctrine for Joint Theater Missile Defense. Joint Pub 3-01.5. Washington: 1996.

________. Joint Vision 2010. Washington: 1996.

U.S. Marine Corps. Operational Maneuver From the Sea. Washington: 1996.

U.S. Navy Dept. Arsenal Ship Concept of Operations. Washington: 1996.

________. Forward . . . From the Sea. Washington: 1994.

________. . . . From the Sea. Washington: 1992.

Wright, CAPT Richard L., USN. "Arsenal Ship: Potent and Punishing." Surface Warfare, January/February 1997, 21-23.

NAVAL POSTGRADUATE SCHOOL
Monterey, California

THESIS

TRADEOFF ANALYSIS MODEL FOR ARSENAL SHIP SURVIVABILITY AND SUSTAINABILITY

by

Ronald S. Bush
Arthur E. Cimiluca, Jr.

September, 1996

Thesis Advisor: Wayne P. Hughes, Jr.

Approved for public release; distribution is unlimited.

REPORT DOCUMENTATION PAGE			Form Approved OMB No. 0704-0188
Public reporting burden for this collection of information is estimated to average 1 hour per response, including the time for reviewing instruction, searching existing data sources, gathering and maintaining the data needed, and completing and reviewing the collection of information. Send comments regarding this burden estimate or any other aspect of this collection of information, including suggestions for reducing this burden, to Washington Headquarters Services, Directorate for Information Operations and Reports, 1215 Jefferson Davis Highway, Suite 1204, Arlington, VA 22202-4302, and to the Office of Management and Budget, Paperwork Reduction Project (0704-0188) Washington DC 20503.			
1. AGENCY USE ONLY *(Leave blank)*	2. REPORT DATE September 1996.	3. REPORT TYPE AND DATES COVERED Master's Thesis	
4. TRADEOFF ANALYSIS MODEL FOR ARSENAL SHIP SURVIVABILITY AND SUSTAINABILITY			5. FUNDING NUMBERS
6. AUTHORS: LT Ronald S. Bush, LT Arthur E. Cimiluca, Jr.			
7. PERFORMING ORGANIZATION NAME(S) AND ADDRESS(ES) Naval Postgraduate School Monterey CA 93943-5000			8. PERFORMING ORGANIZATION REPORT NUMBER
9. SPONSORING/MONITORING AGENCY NAME(S) AND ADDRESS(ES)			10. SPONSORING/MONITORING AGENCY REPORT NUMBER
11. SUPPLEMENTARY NOTES The views expressed in this thesis are those of the author and do not reflect the official policy or position of the Department of Defense or the U.S. Government.			
12a. DISTRIBUTION/AVAILABILITY STATEMENT Approved for public release; distribution is unlimited.			12b. DISTRIBUTION CODE

13. ABSTRACT *(maximum 200 words)*

The arsenal ship program is unique and requires examining the possible features of a paradigm shift in ship design. This thesis presents a user-friendly model with which a decision maker can perform tradeoff analyses between adding specific systems and technologies to the arsenal ship or adding the escort services of combatant ships. The goal of the model is to produce configuration alternatives with high arsenal ship survivability subject to a budget constraint. The model also examines operational logistics by predicting the sustainability of forces with specified arsenal ship configurations. As some inputs are necessarily speculative at this stage, the model is formatted parametrically to facilitate easy updating. A balanced arsenal ship design incorporating point defense, stealth, and hardening is the most attractive choice for littoral operations when life cycle costs are considered. The naval component must also be balanced, reinforcing the notion that stealth and staying power are important in an arsenal ship task force containing DDG-51s and SC-21s.

14. SUBJECT TERMS Arsenal ship, Ship Vulnerability, Operational logistics, Ship sustainability, Ship survivability, Tradeoff analysis, Missile Allocation			15. NUMBER OF PAGES 94
			16. PRICE CODE
17. SECURITY CLASSIFICATION OF REPORT Unclassified	18. SECURITY CLASSIFICATION OF THIS PAGE Unclassified	19. SECURITY CLASSIFICATION OF ABSTRACT Unclassified	20. LIMITATION OF ABSTRACT UL

NSN 7540-01-280-5500

Standard Form 298 (Rev. 2-89)
Prescribed by ANSI Std. 239-18 298-102

TRADEOFF ANALYSIS MODEL FOR ARSENAL SHIP SURVIVABILITY AND SUSTAINABILITY

Ronald S. Bush
Lieutenant, United States Navy
B.S., Iowa State University, 1988

Arthur E. Cimiluca, Jr.
Lieutenant, United States Navy
B.S., United States Naval Academy, 1987

Submitted in partial fulfillment of the requirements for the degree of

MASTER OF SCIENCE IN OPERATIONS RESEARCH
from the
NAVAL POSTGRADUATE SCHOOL
September 1996

Authors: Ronald S. Bush

Arthur E. Cimiluca, Jr.

Approved by: Wayne P. Hughes, Jr., Thesis Advisor

Charles N. Calvano, Second Reader

Frank Petho, Chairman, Department of Operations Research

Preceding Page Blank

ABSTRACT

The arsenal ship program is unique and requires examining the possible features of a paradigm shift in ship design. This thesis presents a user-friendly model with which a decision maker can perform tradeoff analyses between adding specific systems and technologies to the arsenal ship or adding the escort services of combatant ships. The goal of the model is to produce configuration alternatives with high arsenal ship survivability subject to a budget constraint. The model also examines operational logistics by predicting the sustainability of forces with specified arsenal ship configurations. As some inputs are necessarily speculative at this stage, the model is formatted parametrically to facilitate easy updating. A balanced arsenal ship design incorporating point defense, stealth, and hardening is the most attractive choice for littoral operations when life cycle costs are considered. The naval component must also be balanced, reinforcing the notion that stealth and staying power are important in an arsenal ship task force containing DDG-51s and SC-21s.

DISCLAIMER

This thesis contains a computer model which incorporates spread sheets, macros, Turbo Pascal source code, and executable files. While the authors have made every possible attempt to validate input data and exercise the model for all conditions, the computer model has not been validated by an independent source. The user is cautioned that some data included is perishible as costs will change. Additionally, the data incorporated in the model is unclassified; classified data may be required for the best results.

TABLE OF CONTENTS

Preceding Page Blank

EXECUTIVE SUMMARY

The development of an arsenal ship reaffirms the Navy's resolve to meet changing post-Cold War threats. *Forward . . . From the Sea* outlines the Navy's response for the current world order. Emphasis is on projecting power from littoral regions with naval expeditionary forces to support joint operations. In the littoral, the ship has little time to defend against advanced cruise missiles fired from land.

In designing the ship, a determination must be made on what configuration (of passive systems, active systems, and protection provided by other warships) will give the most survivability for a given price. The unique mission of the arsenal ship and the new contracting procedures of the arsenal ship program require that we examine the possibility of a paradigm shift in ship design. The ship will carry minimal active self-defense systems, so studies involving the incorporation of stealth and hardening design features (such as selective armoring) are warranted.

The Arsenal Ship Tradeoff Analysis Model consists of a spreadsheet interface for data input, result output, and Pascal program execution. Since by definition the design alternatives are speculative at this stage, the model is formatted parametrically to facilitate easy updating. Because the procurement, operating, and support costs of an arsenal ship are as yet unknown, we consider the incremental life cycle costs of adding systems over the base ship in our study. A baseline ship is an arsenal ship without any defensive systems. We believe that we can more accurately estimate these incremental costs than the cost of a conceptually new baseline ship.

The Arsenal Ship Tradeoff Analysis Survivability Sub-model is designed to track configurations, costs, and survivability. Our objective is to present a user-friendly model with which a decision maker can perform tradeoff analyses between adding specific systems and technologies to the arsenal ship and adding the protective services of other warships. The Survivability sub-model produces configuration alternatives with high arsenal ship survivability subject to a budget constraint. The model allows us to gain insight into and draw conclusions about the value of hardening and stealth features for naval ship designs in general.

For a wide range of circumstances and measures of effectiveness, the analysis shows that building survivability into the arsenal ship is almost always preferred to assigning escorts, even when only a small fraction (18.6%) of the surface combatant's life cycle cost is charged against the escort role. Our recommendation for arsenal ship survivability features is to incorporate stealth, point self-defense systems, and hardening into the design.

Stealth characteristics and hard kill systems are, by the analysis, the features appearing in preferred designs most often. Sensitivity analysis strongly supports a moderate investment in stealth.

Even though hard kill systems were more commonly seen in the analytically preferred results than soft kill systems, we believe soft kill systems are essential for littoral operations. First, soft kill is synergistic with stealth, and the effectiveness of soft kill is amplified. Second, soft kill measures have been highly effective in actual combat, but hard kill systems remain largely unproven Operating in littoral waters with current rules of engagement (weapons not free) makes the arsenal ship vulnerable to initial and sudden attacks without ample time to respond with hard kill.

Our model shows that when surprise attacks occur, hardening is a very attractive feature of ship design. The additional staying power to remain mission capable after at least one hit will help ensure that the arsenal ship's 500-plus missiles are not rendered useless by a lucky or cheap shot.

We have gone to some pains to show that stealth, ship hardening, and defensive short range, hard and soft kill systems are complementary. For example, stealth adds nothing when the arsenal ship is firing a large missile volley, radiating, subject to air attack with bombs, surface gunfire attack, or a submarine launched torpedo attack. But, hardening retains its effectiveness in all these circumstances.

Our overall conclusion is that stealth, ship hardening, and some set of modern point defense (hard and soft) are, in view of their modest cost in construction and operating personnel, well worth the modest cost on the margin because the arsenal ship's concept of operation requires that it be exposed to major attack.

Operational logistics is examined by exploring the sustainability of different naval component forces in the Sustainability Sub-Model. The arsenal ship program would make little sense if the ship could not improve the naval component's time on station. A highly survivable ship contributes to a force's sustainability. Sustainability is not only measured in days on station, but also in terms of incoming missiles that can be countered, since the solution is scenario-dependent. This key aspect of logistical robustness has not received the attention it deserves.

The arsenal ship alleviates the requirement to frequently cycle surface combatants with fewer VLS cells from the scene of action off station to a replenishment site and back again into the action. By its presence, the arsenal ship will sustain other surface combatants on station. It can remain on station longer than any other ship for a given missile delivery rate, but when its weapons are expended the reload problem will take it off station for a considerable amount of time.

A balanced design will seek to maximize the arsenal ship's net delivered firepower over the combat life of the ship. Incorporating *all features* listed is, according to the analysis, tantamount to over-designing arsenal ship survival features, when its survival *with those features* is compared to the survival of the accompanying present-day surface combatants. Since the DDG-51 and CG-47 cost roughly the same to procure as an arsenal ship-plus-missile load-out, and they will be less survivable, this may appear out of balance. However, in the future as more Arleigh Burke destroyers and SC-21 type warships enter the fleet, matching stealth and superior hardness will provide operational balance.

I. INTRODUCTION

A. BACKGROUND

On May 23, 1996, the Defense Advanced Research Projects Agency (DARPA) released the Arsenal Ship Program Solicitation. The arsenal ship will be designed commercially and benefit from streamlined contracting procedures with early contractor involvement. The program is managed by the Arsenal Ship Joint Program Office, under the cognizance of DARPA, and will follow a five stage process. In July, five industry teams were awarded Phase One contracts to develop preliminary designs. Phase One will last for six months, after which the top two teams will be awarded Phase Two (one year) contracts to develop more detailed, functional designs. In Phase Three, one team will be selected to construct a prototype or demonstrator. Phases Four and Five constitute operational evaluation and production [Refs. 1 and 2].

1. Requirements

The Naval Postgraduate School's Total Ship System Engineering (TSSE) team has been tasked to develop an arsenal ship design which might assist in exploring options being considered by industry teams. Before designing the ship, a determination must be made on what configuration (of passive systems, active systems, and protection from escort ships) will give the most survivability for a given price. Maximizing survivability of the arsenal ship subject to cost constraints will produce the numbers and types of system subcomponents necessary to meet a required level of survivability. This thesis focuses on the survivability of the arsenal ship and will provide a tool for the TSSE team. Validating the draft capabilities defined for the arsenal ship is not an objective.

The authors' recent participation in the Future Navy Game II provided insight for the model as well as another avenue for employing the model. We were invited to participate because of our familiarity with the arsenal ship program, and the game considered force calculi with differing numbers of arsenal ships. The seminar war game is part of a series of games sponsored by the Office of the Secretary of Defense Director of Net Assessment with the goal of examining the "operational and organizational implications of a Navy precision-maneuver strategy against a large peer competitor in the 2020 timeframe." Results from the game series will be used in the Navy Long Range Planning Process and the PR99 assessment cycle. [Ref. 3] The participants are given a limited budget to buy advanced systems, and then play a scenario with their chosen systems. The game is evolving and becoming more analytical; recently small Excel spreadsheet models have been incorporated to aid in estimating mine warfare capabilities and strategic airlift scheduling. Operational logistics and the effects of attrition on naval forces are largely neglected. The sustainability portion of our model could prove useful for this and other future war games.

2. Threats and Employment Concept

The development of an arsenal ship reaffirms the Navy's resolve to meet changing post-Cold War threats. *Forward . . . From the Sea* outlines the Navy's response for the current world order. Emphasis is on projecting power from littoral regions with naval expeditionary forces to support joint operations. Surface ships with theater ballistic missile defense capabilities presumably foster conventional deterrence by discouraging the proliferation of ballistic missiles by our allies and adversaries. Such forces provide theater Commanders-in-Chief with rapid response, measures to control escalation, and a transition force able to act before land-based troops arrive in theater in the event of a major regional contingency. [Ref. 4]

According to the Capabilities Document, the arsenal ship will have two main missions: its primary mission is long-range strike and invasion stopping, and its secondary mission consists of tactical strike, fire support, battlefield interdiction, and battlespace dominance support (theater air defense, anti-air warfare, tactical ballistic missile defense) [Ref. 1]. A total force of five or six forward-deployed arsenal ships is envisioned. Arsenal ships will bolster the essential surge response capabilities and provide the theater Commander-in-Chief with a new operational maneuver element. These ships will carry an array of vertically launched weapons into littoral regions to support a land campaign. They will operate under the control and protection of multi-mission combatants, including Aegis cruisers and destroyers. Table 1 delineates representative target sets the arsenal ship might engage, and Table 2 lists weapons the ship should be able to carry to counter each threat.

	Halt Invasion	Long Range Strike	Battlespace Dominance	Surface Fire Support
Complex Adaptive Armed Forces	Air Land Maneuve Battle Groups (e.g., OMGs)	National/Regional C4I Space Control	Manned A/C TBMs, UAVs Cruise Missiles SAM/AAA	Long-Range Artillery TBMs Logistics Assets
Armored Mech. Armed Forces	Armor-Heavy Comb. Arms Formations Divisions/BDEs	National and Regional C4I	Manned A/C TBMs SAM/AAA	Long-Range Artillery
Infantry Based Armed Forces	Armor/Mech. "Pure" units (BDEs/BNs)	Military Region District C4I	Manned A/C SAM/AAA	Medium-Range Artillery Logistics Assets
Internal Security Light Force	Transportation Railroads Trucking, Light Vehicles	National CMD Authority Military Concentrations	OP Bases Light A/C Coastal Patrol Craft	Logistics Assets Economic Asset Local Forces

Table 1. Target sets to be countered by arsenal ship from Ref. [A].

	Halt Invasion	Long Range Strike	Battlespace Dominance	Surface Fire Support
Complex Adaptive Armed Forces	SM-2/ATACMS-BAT SLAM TLAM-BAT TLAM-C	TLAM	ATACMS TLAM-C/D	ATACMS, SLAM TLAM-C/D Naval Gunfire (VGAS/SCRAM)
Armored Mech. Armed Forces	SM-2/ATACMS-BAT TLAM-BAT, SLAM Strike-SM	SM-2/ ATACMS-BAT TLAM-BAT SLAM Strike-SM	ATACMS TLAM-C/D	ATACMS, SLAM TLAM-C/D Naval Gunfire (VGAS/SCRAM)
Infantry Based Armed Forces	ATACMS SLAM Strike-SM	TLAM-D ATACMS-ER	ATACMS	ATACMS, SLAM TLAM-C/D Naval Gunfire (VGAS/SCRAM)
Internal Security Light Force	Naval Gunfire (VGAS/SCRAM)	TLAM-C	ATACMS Naval Gunfire (VGAS/SCRAM)	ATACMS Naval Gunfire (VGAS/SCRAM)

Table 2. Weapons to counter target sets from Ref. [1].

While the arsenal ship will add new capabilities, it will also face new threats. The littoral affords less space for operations, and the interface between sea and shore is where chaos is most prevalent. Fighting in the littoral subjects the ship to attacks from shore with more advanced missiles. These factors result in less time to react to an incoming missile attack.

3. Design Philosophy

Conceptually, the arsenal ship is a force multiplier and should function as a cost-effective, remote missile magazine. The target unit price is $450 million or less with a ceiling of $550 million in FY 98 dollars The design life is thirty five years. The ship may be unmanned or manned with 50 or fewer people. Projections for the arsenal ship include approximately 500 vertical launch cells; targeting will be accomplished by other ships with Cooperative Engagement Capability (CEC) or similar links and an over-the-horizon (OTH) satellite link. The ship must carry sufficient fuel, consumables, and repair parts to support a ninety-day mission. It must also be capable of underway refueling and accommodating SH-60, V-22, and CH-46 aircraft with landing area and limited services [Ref. 1].

Regarding arsenal ship survivability, OPNAV N8 states:

> Though (the) arsenal ship will operate in any threat environment under the protective umbrella of battle force combatants, it must be survivable against 21st century anti-ship missiles, torpedoes, and mines. An affordable balance of active and passive on-board self defense features is necessary. Active self defense should be roughly equivalent to that of a combat logistics force ship. Passive defense should capitalize on the benefits of mass (tonnage) and innovative applications of multiple hull integrity and signature reduction [Ref. 1].

B. PREVIOUS STUDIES

Numerous studies have assessed the staying power of ships or proposed means to extend surface ships' ability to take hits and continue fighting. Incorporating stealth technology reduces susceptibility. Many sources, including the *Joint Munitions Effectiveness Manuals* cite the effectiveness of combat systems on susceptibility or survivability. A thesis by John Schulte [Ref. 5] reminds us that ships can be hit by ASCMs even with layered defenses. Further, empirical evidence and predictive equations presented by Schulte and Humphrey [Ref. 6] show that current ships (with the exception of aircraft carriers) are not built to the Timex standard. They cannot "take a licking and keep on ticking." The Naval Surface Warfare Center (NSWC), Carderock Division analyzed the effectiveness of anti-vulnerability features in ship design and reports that the judicious application of hardening can greatly improve a ship's survivability without becoming cost prohibitive or losing much maneuverability. [Ref. 7]. In the Military Worth of Staying Power, Hughes states that through staying power, the military worth of a vessel is sustained and can be measured by the "maximum accurately delivered ordnance over the combat life of a warship" [Ref. 8].

A great deal of valuable work has been accomplished to date. Yet, no study consolidates the above issues in a cost effectiveness study of an arsenal ship's survivability and the task force's sustainabability.

C. MOTIVATION AND PROBLEM DEFINITION

The intent of the arsenal ship program is to provide a relatively low-cost platform to operate in the littoral regions. The unique mission of the arsenal ship and the new contracting procedures of the arsenal ship program require that we examine the possibility of a paradigm shift in ship design. The ship will carry minimal active self-defense systems, so studies involving the incorporation of stealth and hardening design features (such as selective armoring) are warranted. While not written specifically about the arsenal ship, the excerpt below from an article in the *Naval Engineers Journal* summarizes the problem well.

> There has been significant change in national policy and naval strategy over the last decade. We are now focused to provide forward presence in peacetime and make a critical contribution to power projection during the transition from crisis to conflict. Our recent experience indicates that these operations require force structures higher than we can sustain with our current investment accounts. The prospect for additional funding is dim, so we must become more efficient in the utilization of our resources. One of the keys is to determine the proper balance of stealth, EW and hard kill systems for our next generation of surface combatant.

> . . . By these means we can significantly improve ship staying power and overall mission effectiveness. . . . To accomplish these benefits, the Navy will need to develop new concepts for littoral operations and conduct tradeoffs where stealth, EW, and hard kill are balanced and not treated as mutually exclusive. The result will be force multiplying with a higher probability of mission success and survival. [Ref. 9]

Our objective is to develop a user-friendly model with which a decision maker can perform tradeoff analyses between adding specific systems and technologies to the arsenal ship and adding the protective services of other warships. The immediate goal of the model is to produce configuration alternatives with high arsenal ship survivability subject to a budget constraint. Since by definition the design alternatives are speculative at this stage, the model is formatted parametrically to facilitate easy updating. The model will also allow us to gain insight into and draw conclusions about the value of hardening and stealth features for naval ship designs in general.

Operational logistics is examined by exploring the sustainability of different naval component forces. The arsenal ship program would make little sense if the ship could not improve the naval component's time on station. A highly survivable ship contributes to a force's sustainability. Sustainability is not only measured in days on station, but also in terms of incoming missiles that can be countered, since the solution is scenario-dependent. This key aspect of logistical robustness has not received the attention it deserves.

Related to this issue is the projection in the *Surface Combatant Force Level Study* that around the year 2010 "The requirements for in-theater surface combatants exceeded the numbers needed to escort CVBGs and ARGs by a considerable portion" [Ref. 10]. The situation is true now more than ever. The surface fleet has down-sized while "Surface combatants have evolved into major combatants in their own right, able to make significant contributions to the joint campaign. . ." [Ref. 10].

Since VLS tubes are not currently replenished at sea, an arsenal ship task force may prove invaluable in the opening days of a conflict. The new platform may alleviate the need of removing warships from their tasks to rearm, which would require rotating replacement ships into the theater. Comparisons will be made between forces with and without an arsenal ship, and an arsenal ship missile load-out that maximizes the naval component's sustainability for a major regional contingency (MRC) will be offered.

This thesis focuses on the threat to an arsenal ship from anti-ship cruise missiles (ASCMs), which are presumed the most serious and likely threat in the littoral arena. Since 1970, cruise missiles have caused more damage to warships and shipping than all other threats combined. Obviously ASCMs are not the only threat. Mines and torpedoes are nemeses that must be dealt with. While a ship can be hardened to withstand damage from contact torpedoes, under hull torpedo or mine explosions are threats not easily designed against. Other weapons, including gunfire and bomb raids on the ship can be approximated in the model by changing the input ASCM explosive weight. The methodology of the model can be applied to mine and torpedo threats, but even if effective countermeasures could be traded-off, the authors believe that synergistic effects of vastly different threats cannot be modeled responsibly or reliably. A pragmatic approach to coping with the torpedo threat is to assume that ASW assets are required and deduct their cost from the top of the available budget. The mine threat can be minimized by collecting intelligence observations, avoiding areas known to be mined, and maintaining current surveys of underwater characteristes for areas of interest (Q-route surveys).

II. MODEL DEVELOPMENT AND PHILOSOPHY

A. SURVIVABILITY EQUATION DEVELOPMENT

This section defines requisite terms, states assumptions, and derives our survivability equation. The survivability equation is the base measure of effectiveness applied in our iterative survivability model which can then be compared against cost. Detailed derivations of individual probabilities appearing as multiplicative factors in said equation appear later.

A naval ship's survivability is "the capacity of a surface ship to avoid and/or withstand a man made hostile environment while performing its mission" [Ref. 11]. Hence survivability depends on susceptibility to attack and vulnerability once stricken.

Susceptibility is defined as the probability of being hit (P_{hit}) and represents "the inability of a ship to avoid the sensors, weapons, and weapons effects of that man-made hostile environment" [Ref. 11]. The susceptibility of the arsenal ship is a function of factors internal and external to the ship. If escorted by surface combatants, the number and types of ships in formation are important. Escorted or not, active and passive defensive systems on the arsenal ship affect susceptibility. For example, the susceptibility of an escorted arsenal ship under missile attack depends on the effectiveness of the defending escort's weapons. If missiles leak through the escort's protective umbrella (called leakers), the arsenal ship's point defense systems and degree of stealth factor into its susceptibility.

Vulnerability is the conditional probability that a ship is killed given that it is hit ($P_{kill|Hit}$) by a bomb, missile , torpedo, or mine [Ref. 11]. Values for ($P_{kill|Hit}$) depend on physical characteristics of the ship and incoming weapon employed. Historically, studies of ship vulnerability consider one or both of two degrees of vulnerability — the probability that a ship is rendered not mission capable or the probability that a ship is sunk. **This thesis only considers mission kills.** Hence, vulnerability is defined more narrowly as the conditional probability that the arsenal ship is rendered not mission capable by an incoming projectile.

Preceding Page Blank

The arsenal ship is unique in that it requires input from CEC capable ships or satellites to accomplish its mission. Consequently, the arsenal ship can suffer a mission kill if it remains intact but has no platform nearby to control its weapons. The effects of escort attrition are considered significant, are estimated in the model, and will be addressed later.

In the purest sense, vulnerability is assessed by the ratio of hits the ship sustains to its staying power — the number of hits from a specified weapon required to affect a mission kill (N_A). Applying the results of Richard Humphrey's *Warship Damage Rules for Naval Wargaming*, staying power can be calculated as: $\mathbf{N_A = C(D/H_E)^{1/2}}$,where C is a fitted constant based on an comprehensive study of ships damaged in World War II; D is the displacement of the arsenal ship in tons, and H_E is the explosive weight of the projectile in pounds [Ref. 6].

A more recent thesis by John Schulte derived an alternative formula for staying power [Ref. 5]. As shown Table 3 below, both methods produce similar results. Schulte's thesis considered Exocet missile and equivalents, so, for comparison, we used the Exocet's explosive weight in Humphrey's equation. We incorporated the Humphrey equation into our model because it considered larger ships. The Schulte thesis studied 222 cruise missile firings, but no targeted ship exceeded 7000 tons.

Ship Type	Displacement (tons)	Humphrey N_A	Schulte N_A
FFG-7	5500	1.51	1.74
DDG-51	8300	1.86	1.97
DD-963	9100	1.95	2.02
CG-47	9600	2.00	2.06
Arsenal Ship	20000	2.89	2.52
Arsenal Ship	40000	4.08	3.02

Table 3. Comparison of staying power calculation results using an Exocet equivalent net explosive weight.

The goal is to assess an arsenal ship's survivability. We define survivability (S_A) as the arithmetic complement of the probability of a mission kill (P_{kill}). The probability of a mission kill is the product of the arsenal ship's susceptibility and vulnerability. Therefore, arsenal ship survivability against a missile is: $\mathbf{S_A = 1\text{-}(P_{hit})(P_{kill|Hit}) = (1\text{-}P_{kill})}$.

Expressed in terms of the number of notional missiles fired and the resulting number of leakers, arsenal ship survivability can be calculated as: $\mathbf{S_A = 1\text{-}(MQ/N_A)}$, where M is the number of well-aimed missiles; Q is the probability of leakers; and N_A is the staying power in hits. Hardness features can be incorporated into the ship design that will increase the staying power by a factor of H. Assuming missile firings are independent, identically distributed events with **P(leaker) = Q**, then leakers are distributed binomially with parameters M and Q. Therefore, the expected number of leakers is: **E[leakers] = MQ** and the expected damage to the arsenal ship is: **E[damage] = MQ/(N_AH)**.

Q can be decomposed into probabilities for individual events as shown in Table 4. For a leaker to strike the arsenal ship the following must all occur.

EVENT	ASSOCIATED PROBABILITY
Defending escort does not kill incoming missile	Q_{AE}
Missile not affected by arsenal ship stealth	Q_{AS}
Arsenal ship soft kill measures fail	Q_{AP}
Arsenal ship point defense systems are ineffective	Q_{AA}
Note: Associated probabilities are derived below.	

Table 4. Probabilities Contributing to a Leaker Damaging the Arsenal Ship

The resultant equation for survivability is: $\mathbf{S_A = 1\text{-}(MQ_{AE}Q_{AS}Q_{AP}Q_{AA}/(N_AH))}$

B. KILL PROBABILITY DERIVATIONS

In this section the parameters for determining the survivability of the Arsenal ship are given.

<u>INDICES</u>

i Soft kill systems

j Hard kill systems

t Escort types (ship classes)

<u>UNIT QUANTITY</u>

n_t Number of units of type t

SOFT KILL SYSTEMS

These systems conduct soft kills on a shot. Systems like ESM and Chaff/Torch divert a shot after the arsenal ship has been targeted by the ASCM.

HARD KILL SYSTEMS

Point defense systems physically kill leakers. The Mark 15 Vulcan Phalanx Close-in-Weapon-System CIWS, rolling airframe missile (RAM), 5"54 gun, and Evolved Sea Sparrow (ESS) are examples.

ESCORT TYPES

Cruisers (CG-47 Class - Ticonderoga), Destroyers (DDG-51 Class - Arleigh Burke), (DD-963 Class - Spruance), Frigates (FFG-7 Class - Oliver Hazard Perry).

PROBABILITIES OF KILL

P_{AE} Probability that escorts can kill an incoming shot.

Note: P_{AE} is a representation of the aggregate probabilities. The value varies in the simulation. Details of the iterative approach appear in Chapter III.

P_i Probability that a system of type i can effect a soft kill of an incoming shot.

P_j Probability that a system of type j can kill an incoming shot.

P_m Probability that stealth precludes an incoming shot.

MISSILE LEAKER PROBABILITIES

Q_{AE} Probability that the arsenal ship is not protected by the defending escort.

$$= \left[(1 - P_{AE}) \right]$$

Q_{AS} Probability that stealth does not preclude an incoming shot.

$$= \left[(1 - P_m) \right]$$

Q_{AP} Probability that the arsenal ship does not soft kill an incoming shot.

$$= \left[\prod_i (1 - P_i) \right]$$

Q_{AA} Probability that arsenal ship point defenses <u>do not</u> kill an incoming shot.

$$=\left[\prod_{j}(1-P_j)\right]$$

C. COSTING ASSUMPTIONS

Since the procurement, operating, and support costs of an arsenal ship are as yet unknown, we consider the incremental cost of adding systems over the base ship in our study. A base ship is an arsenal ship without any defensive systems. We believe that we can more accurately estimate these incremental costs than the cost of a conceptually new ship itself.

By conducting the tradeoff analysis using incremental costs, we also avoid some difficult questions of arsenal ship O&S costs. These will depend on several unresolved issues, the two most significant are:

1. The ship's manning level is unknown, and a rotation scheme such as blue-and-gold crews may be used.

2. While desired capabilities have been published, how the ship will operate has not been established.

While most calculations in the model do not require the base cost of the arsenal ship, our estimates for the price of adding stealth and hardening do. In these instances, we use the target cost of $450 million reported in the Arsenal Ship Program Solicitation. This figure, which may be high, still affords $100 million for procuring defensive systems based on the ceiling established in Ref A. Details of cost calculations are provided in Chapter IV.

III. MODEL FLOW AND FUNCTION

The Arsenal Ship Tradeoff Analysis Model is designed to track configurations, costs, survivability and sustainability. It is not intended to be a combat simulation. Only those elements required to predict expectations of the desired output are incorporated in the model.

The Arsenal Tradeoff Analysis Model overview in Figure 1 is actually comprised of two smaller sub-models. The Survivability sub-model aids decision makers in determining the best combination of arsenal ship self defense features and escorts. The Sustainability sub-model tests a specifically designed arsenal ship deployed within a task force to determine the force's sustainability. Both sub-models are described below.

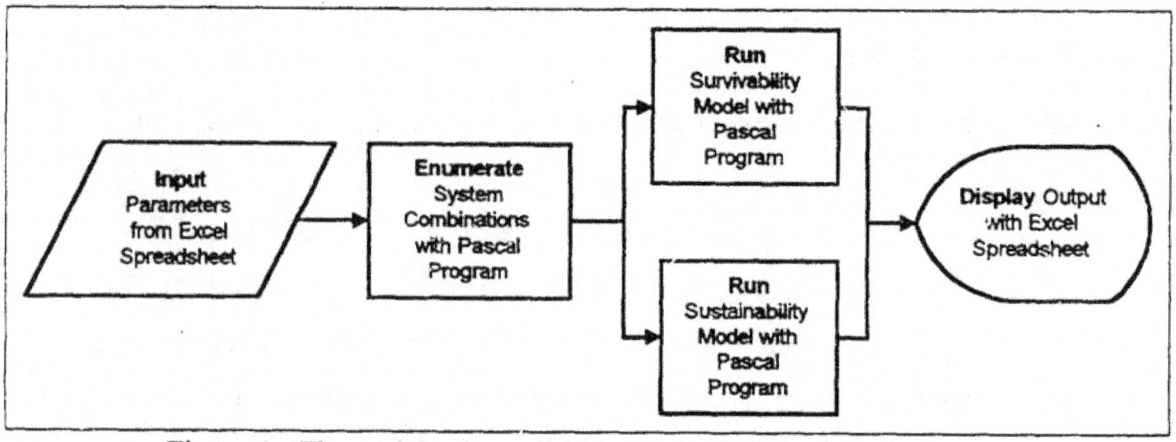

Figure 1: Flow of the Arsenal Ship Trade-Off Analysis Program.

A. SURVIVABILITY SUB-MODEL

1. Input Section

The Survivability sub-model consists of a spreadsheet input section, two Pascal programs, and a spreadsheet output display. In the input section, the user enters parameters, scenario, and limiting factors for consideration in generating arsenal ship configurations. After all inputs are made, the user is prompted to save the parameters to a data file which is formatted for use by other programs in the model. Chapter IV will cover the inputs in greater detail, and the User's Guide (Appendix C) describes the input procedures.

Preceding Page Blank

2. Enumeration Program

This second step in the survivability model uses a Pascal program to enumerate all possible combinations of arsenal ship self-defense features and escorts with the parameters from the input section. It stores them in a file formatted for reading by the Survivability program.

3. Survivability Program

The third step is to run the Survivability program -- also coded in Pascal. The model calculates the survivability and cost of an arsenal ship based on the defense features and escort variables input by the user. In addition, the model determines the task force's effectiveness in defending the arsenal ship and the number of SAMs remaining on the arsenal ship. Each major component in Figure 2 will be discussed, highlighting the assumptions and methodology. The enumeration file is opened, and the first system combination is selected. Using the Excel data file, each combination of systems is run through the survivability program outlined in the following pages. The output is a file listing the systems which meet the limiting criteria.

a. Designate Defender

If there are escorts, the most capable AAW platform will be chosen as the task force defender (see Figure 3). If a defender has already been designated in a previous run, it is re-confirmed as a mission capable defender. A mission capable defender is one that has additional staying power and SAMs.

Our model assumes that the defender will always be positioned between the threat and task force, hence, attack geometry is optimal. In reality, CEC and stationing tactics promote coordinated fire between defenders. This reduces multiple ship engagements, overkill and wasted ordnance. Hence our assumption that the best positioned defender is selected to engage the incoming missile is reasonable.

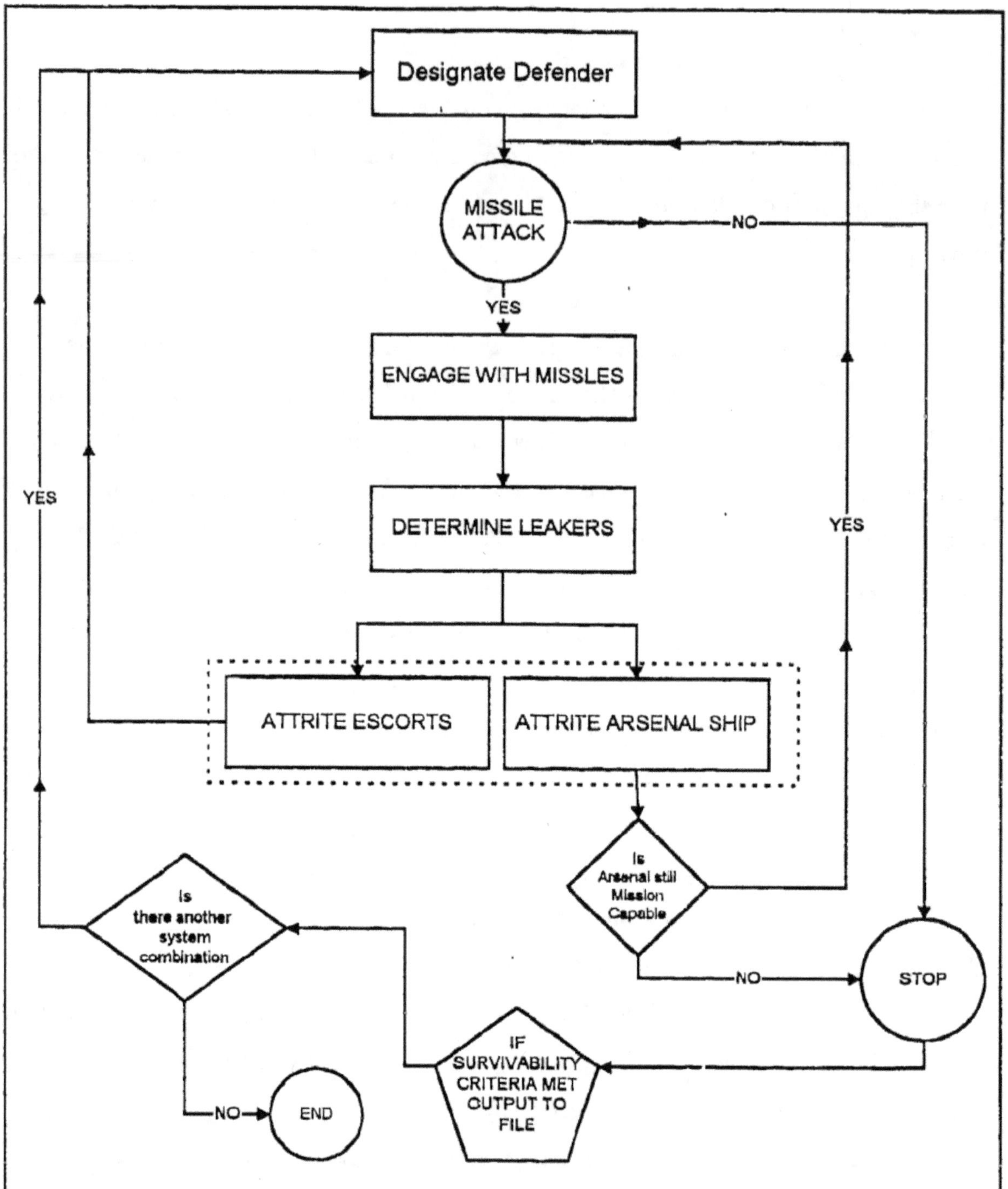

Figure 2: Major components and flow of the Arsenal Ship Survivability Model

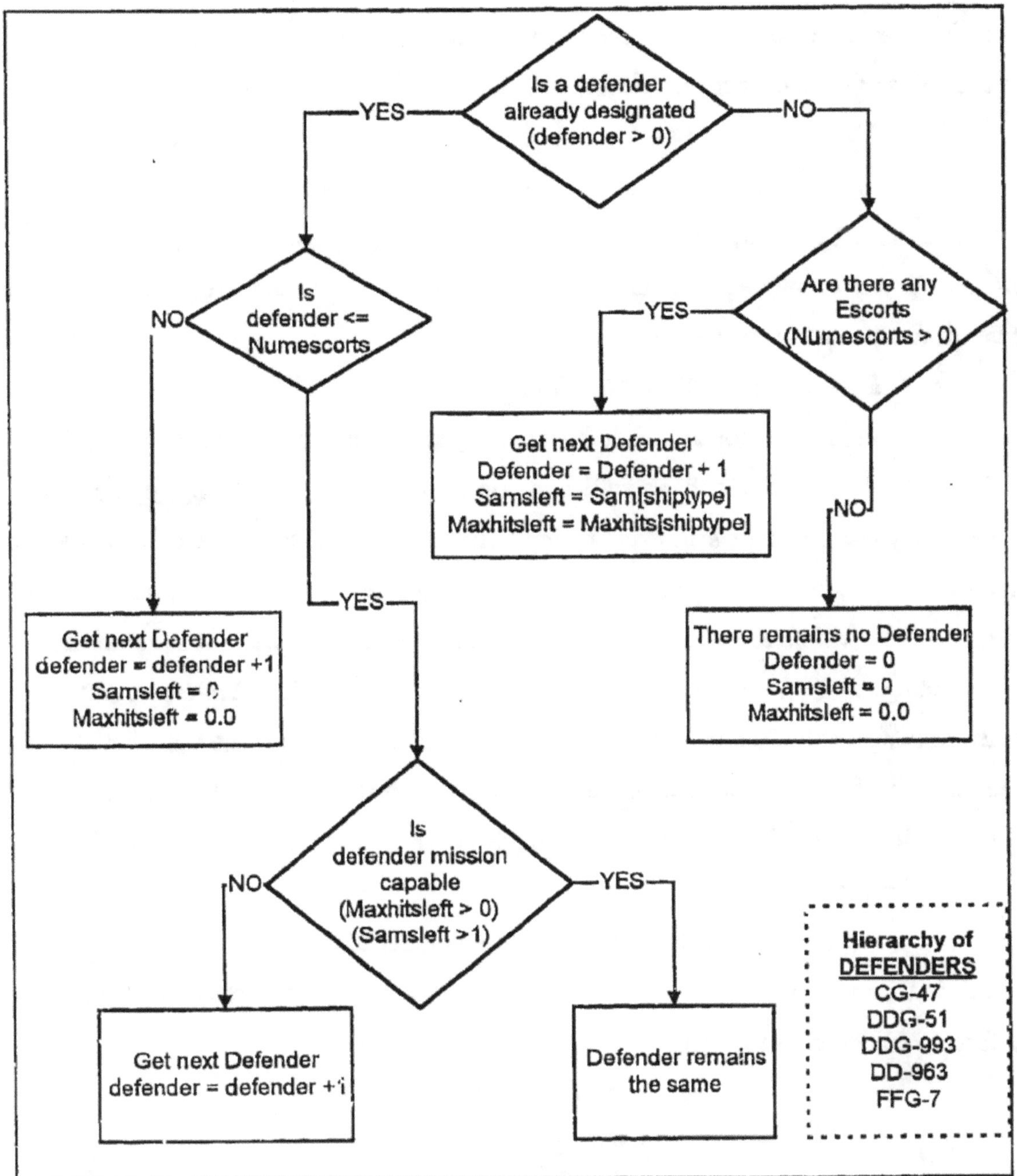

Figure 3: Model flow of Designating Defender

After each missile shot, the defender's cumulative damage is checked. If the accumulated damage does not exceed its staying power and it has SM's left, then the escort remains the defender for the next missile attack.

b. Missile Attack

Missile attacks are discrete independent occurances with the time between events such that they do not saturate a defender's ability to engage incomings. The user inputs the number of missiles to be engaged by defining a scenario that identifies the number of missiles per wave per day.

c. Engage With Missiles

In this component the defender engages the incoming missiles with SAMs (see Figure 4). If the defender has a sufficient number of missiles, it will engage the inbound using a shoot-shoot-shift firing policy. This policy is chosen based on a littoral scenario where the defender only has time for one engagement. This means every incoming missile will be engaged with two SAMs and the defender will not shoot at the same incoming missile even if it survives the initial salvo. An Aegis defender without missiles will engage the incoming missile with arsenal ship missiles via the CEC link. When the arsenal ship and Aegis defender are empty, a new defender is chosen. A non-Aegis defender (not CEC capable) without missiles is considered not mission capable, and a new defender is chosen.

After the inbounds are engaged, the number of leakers is determined. The leaker formulation is given below. The next chance to defeat the leakers happens with the task force's point defense systems.

$$\text{Leakers} = (1-PK_{SAM})(1-PK_{SAM})$$

d. Determine Leaker Allocations

This component allocates leakers among the task force by radar cross section (RCS) as shown below and in Figure 5. Simply stated, a unit's RCS percentage of the whole task force's RCS determines the portion of leaker to allocate to that unit. Two

$$\text{ArsenalAllocation}[AA] = \frac{\text{ArsenalRCS}}{\sum_i \text{EscortRCS}_i + \text{ArsenalRCS}}$$

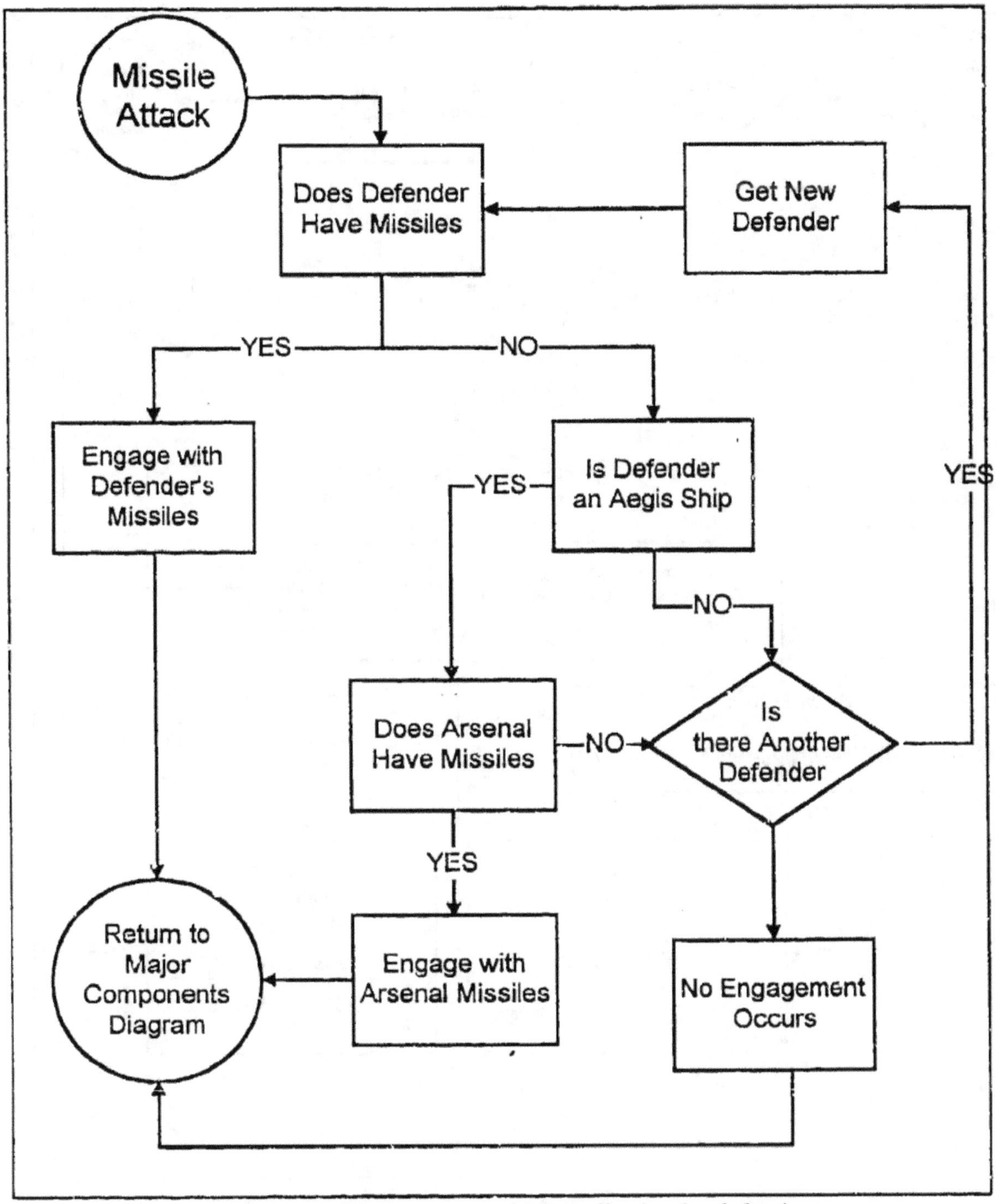

Figure 4: Engagement with Missiles by the defender

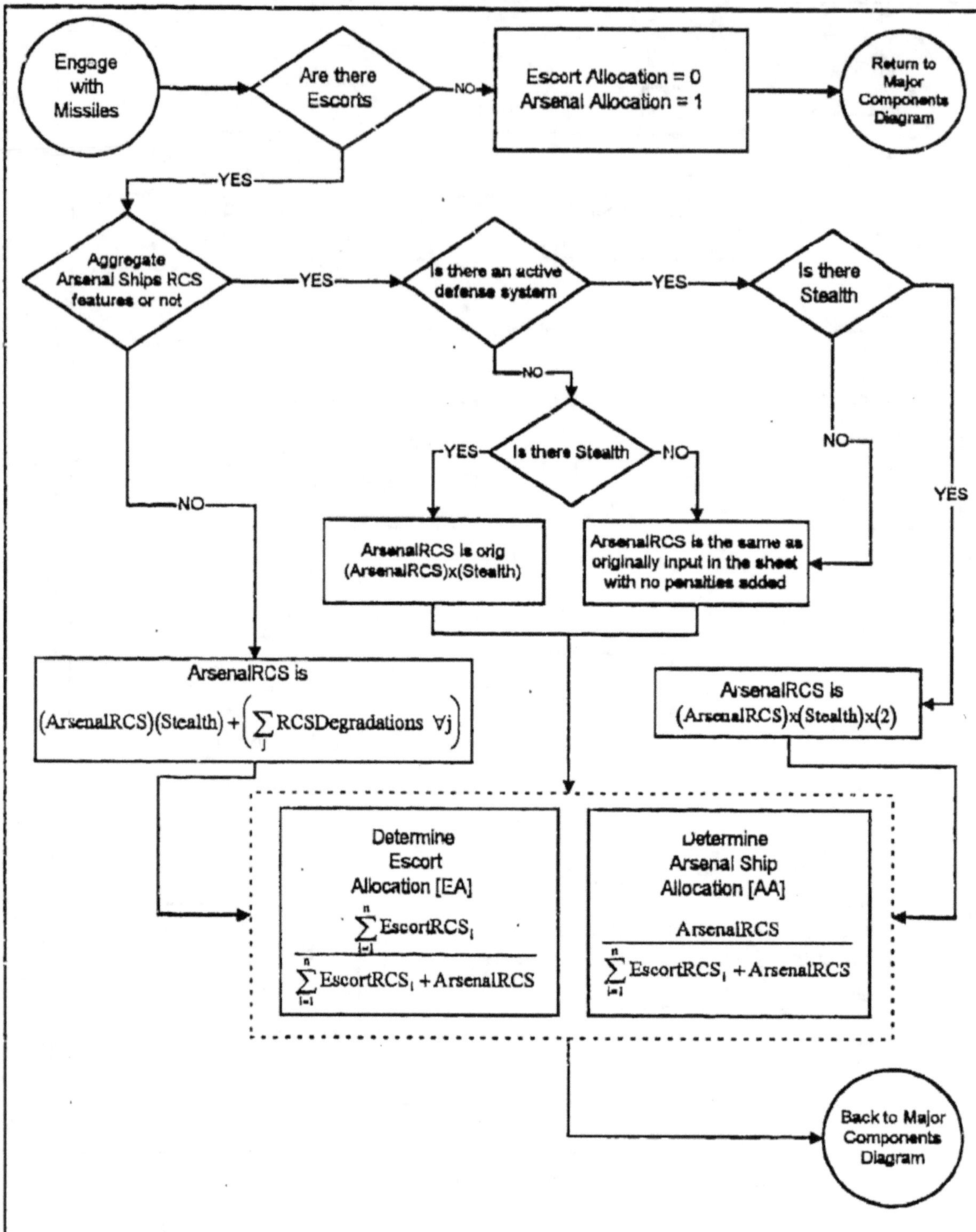

Figure 5: Model component Determine Leakers for escorts and arsenal ship

21

$$\text{Escort Allocation}[EA] = \frac{\sum_i \text{EscortRCS}_i}{\sum_i \text{EscortRCS}_i + \text{ArsenalRCS}}$$

methods of determining radar cross section are available. They are represented in Figure 5 and described in the input section of Chapter IV.

e. Attrite Escorts

Referring to Figure 6, the model attrites on a fair share basis. The EA is sub-allocated among the escorts, see formulation below. Each escort then engages its share of leakers with their point defense systems. The damage incurred from leakers on all available escorts is applied to the current defender. After a defender sustains damage,

$$\text{Escort's Fair Share of EA} = (EA)\frac{\text{EscortRCS}}{\sum_i \text{EscortRCS}_i}$$

its cumulative damage is compared to its staying power. If the damage exceeds the staying power, another defender is chosen (if there is another).

f. Attrite Arsenal Ship

Like the escorts, the AA is subjected to the arsenal ship's point defenses. See Figure 7. The escorts have a single aggregate PK factor, whereas the arsenal ship's PK depends on the system combination being considered. Arsenal ship damage is that amount of AA which defeats the hard kill, soft kill and anti-vulnerability features of the ship. After the arsenal ship sustains damage its cumulative damage is compared to its staying power. If the damage exceeds the staying power, the arsenal ship is considered a mission kill.

B. SUSTAINABILITY SUB-MODEL

1. Input Section

This model evaluates the long-run effects of an arsenal ship's survivability on the battle force's sustainability. The input section consists of designating a specific arsenal ship configuration, task force composition and arsenal ship's limiting (desired) survivability level.

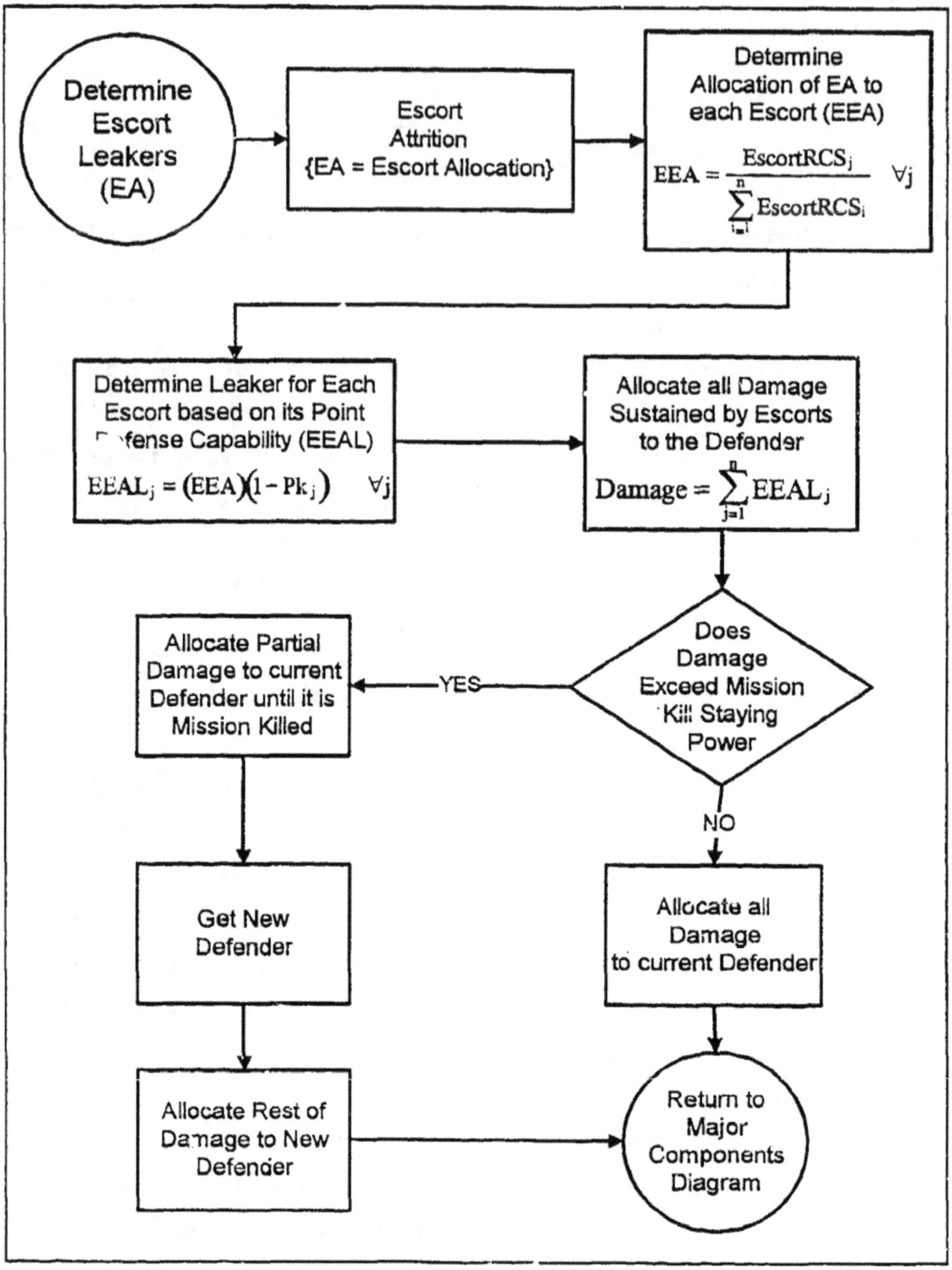

Figure 6: Model component Attrite Escorts

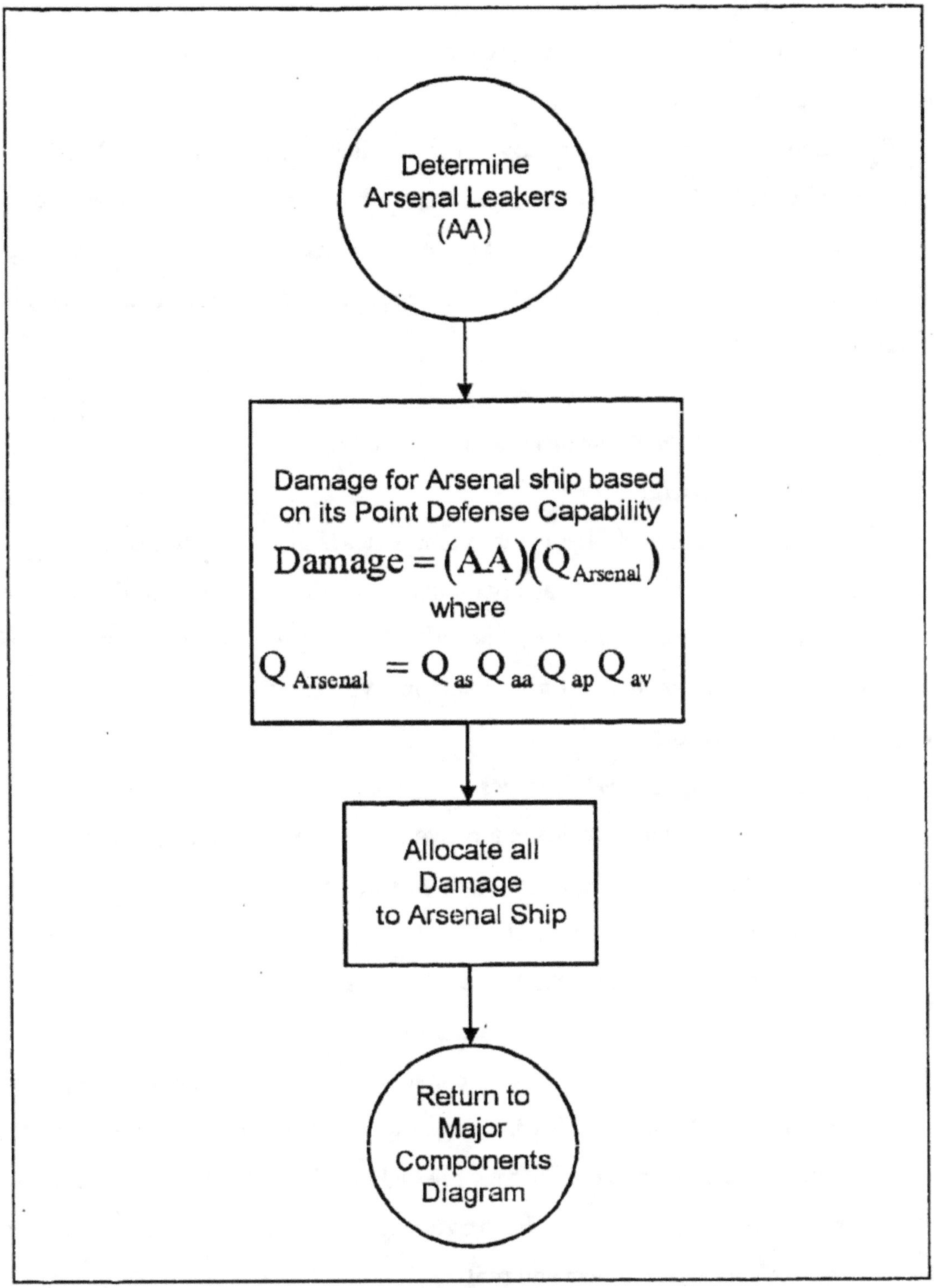

Figure 7: Model component Attrite Arsenal Ship

2. Enumeration Section

This program is identical to the Survivability model's enumeration program.

3. Sust~inability Program

The major components of this program are similar to those in the Survivability program (see Figure 2). The difference is that the arsenal ship's configuration is now constant and the number of missiles the task force can withstand before the arsenal ship reaches its limiting survivability is variable. Differences from the Survivability program are outlined as follows.

a. Designate Defender

This component remains the same. See Figure 3.

b. Missile Attack

In this component, the number of missiles launched at the tas. orce becomes variable, unlike in the Survivability model where the features were the variables and the maximum number of incoming missiles was fixed. The number of missiles a task force can endure is representative of time on station. Missile attacks continue so long as there is a ship in the task force not OOA.

c. Engage with Missiles

This component remains relatively the same as in Figure 4, except for code which records the status of the defender when it becomes OOA.

d. Leaker Determination

Same as in the Survivability model. See Figure 5.

e. Escort and Arsenal Attrition

Missile attacks and attrition occur so long as there are mission capable ships in the task force. The survivability limiting criteria entered by the user apply to the arsenal ship **only**. The escorts fight until they are OOA and the arsenal ship fights until it reaches the limit of its staying power. The program runs until the task force is OOA. The model results are displayed on screen and written to a file.

IV. MODEL INPUTS

Flexibility and power are demonstrated in the input section. The user must input susceptibility and vulnerability features of the escorts and arsenal ship. Each type of input will be discussed and then followed with the values used to exercise the model to investigate the relative values of survival attributes. Results will be presented in Chapter V.

A. ESCORT SHIPS

The user may consider the defensive effectiveness of escorts currently or potentially in the Fleet and their impact on the arsenal ship's survivability. Since this model focuses on the ASCM threat, the escorts must have the ability to defend against ASCMs. Ticonderoga Class (CG 47), Arliegh Burke Class (DDG 51), Kidd Class (DDG 993) and Oliver Hazard Perry Class (FFG 7) ships are used. SC-21 variants or other ships can be used if the user has characteristic data. The following inputs are required of each escort type.

1. Quantity

A minimum and maximum number of each candidate escort type is required. An input of '0' for minimum and '2' for maximum means that the model will enumerate all permutations with zero, one and two of this escort type. To fix or 'hardwire' the number of an escort type, set the minimum and maximum numbers the same. For example, changing the Ticonderoga minimum and maximum to '1' and '1' will force the model to consider one CG-47 in each permutation. This thesis examines the quantity inputs shown in Table 5 for our run of the model.

2. Single Salvo Kill Probability

AAW single salvo kill probability (PSSK) is the chance an escort will destroy an incoming ASM with one SAM salvo. This probability accounts for incomings with direct (the escort itself is the target) or crossing (another ship in the task force is the target) flight paths. As explained in Chapter III, this model uses a shoot - shoot - shift SAM firing doctrine by the escort designated as defender.

Preceding Page Blank

Since an escort is more likely to kill an incoming vice a crossing target, we assume an Aegis ship PSSK of .96 and .84 respectively. A mean of .91 is used for PSSK. This is consistent with other unclassified studies which estimate an Aegis ship's single shot effectiveness between the ranges of 0.7 and 0.8 [Ref. 8, p. 34]. Inputs for non-Aegis ships are assumed to be lower, due to their lower detection and targeting capabilities. Ship PSSKs are given in Table 5. Components in Table 5 that have not been discussed yet will be detailed in what follows.

Ship Class	Min Qty	Max Qty	PSSK	# SAMs	PD PK	Days	Disp (tons)	RCS
CG-47	0	2	0.91	70	0.503	90	9,600	49,896
DDG-51	0	2	0.91	50	0.503	90	8,300	24,998
DDG-993	0	1	0.75	56	0.394	90	9,800	51,515
FFG-7	0	2	0.64	21	0.394	90	3,700	33,069
Arsenal Ship				284			20,000	36,750

Table 5: Model input values for the escorts

3. Number of Escort SAMs

SAM load out for each escort was calculated using the notional VLS load out data from Ref. 10. On the average, 55.5% of the missiles per ship were allocated to SAMs and 44.5% were allocated to strike. Table 5 shows the number of SAMs (# SAMs) used throughout this thesis. This leaves 56, 40, 45, 37, 228 strike weapons for the CG-47, DDG-51, DDG-993, FFG-7, and arsenal ship respectively.

4. Cost

The marginal cost of protecting an arsenal ship comes from defense features contained within the arsenal ship and from escorts which provide protection services. Both must be charged to the overall cost of the arsenal ship program. The model considers costs on an annual lifecycle basis. The annual lifecycle cost is obtained by dividing procurement cost by lifecycle length (in years) and adding the annual expected O&S cost.

Ship procurement data was obtained from Data Search Associates' annual "U. S. Weapon Systems Costs" publications [Ref. 12, Ref. 13]. Operational and Support costs are published by the Naval Center for Cost Analysis' "Visibility and Management of Operating and Support Costs (VAMOSC)" Program Office [Ref. 16]. This thesis assumes each escort has a lifecycle of 30 years. The procurement cost is determined by taking the total program cost (sum of RDT&E, unit cost, and other costs) and dividing by the number of ships procured. O&S cost is the total O&S costs (includes operational, manpower, maintenance and indirect O&S costs) for all ships in the class divided by the number of ships in the class. Table 6 gives a breakdown of costs for each escort in 1996 dollars using a 3% inflation rate when required.

Ship	Procurement Cost ($)	O&S Cost ($)	Life Cycle Years	Annual Lifecycle Cost ($)	Annual Escorting Cost ($)
Tico (CG47)	$1,237,038,179	$28,814,948	30	$70,049,554	$14,681,619
Burke (DDG51)	$1,063,900,596	$21,500,976	30	$56,964,329	$11,939,099
Kidd (DDG993)	$469,262,553	$25,156,840	30	$40,798,925	$8,551,008
Perry(FFG7)	$298,513,074	$17,077,516	30	$27,027,952	$5,664,763

Table 6: Escort annual lifecycle cost input in dollars per year.

5. Displacement

Full-load displacements in tons, provided from Jane's Fighting Ships [Ref. 14] are input (see Table 5). These values contribute to determining staying power as defined earlier. A baseline estimate of the arsenal ship's displacement is required. The baseline is a stripped-down, typically-designed, naval ship such as an amphibious ship. Accepted arsenal ship displacement predictions are 20-40 thousand tons. We choose 20 thousand tons to be conservative in our staying power calculations. The weight of hardening and other features are not added, since their contribution to staying power is contained in their PKs. This avoids a 'double dipping' effect.

6. Escort Time

How long (in days per year) will escorts be required to provide the survivability level desired for the arsenal ship? The method of deployment and operational tempo will significantly affect this value. For example, if the arsenal ship deploys continuously with a carrier battle group (CVBG), the cost of escort services should contain the costs of all escorts in the CVBG. Escort time would be shared between the arsenal ship and carrier since both are receiving the same services. We assume there will be times when the CVBG will split into smaller task groups — this is frequently the case today. The arsenal ship, however, could deploy overseas like the maritime prepositioned force (MPF) ships. In this case, a round number of 90 days per year could be used as the length of time the arsenal ship will require protection; otherwise it would be pier side or training in non-hostile waters. Our methodology charges the escorts services solely to the arsenal ship. It is evident that the fraction of the escort's life time (hence cost) that should be charged to the arsenal ship task force is both a sensitive parameter and difficult to estimate. We chose 90 days per year. Therefore, one-quarter of the escort's life cycle cost is used for tradeoff analysis.

7. Mission Time

This is the percentage of escort time a ship actually conducts its AAW mission. Beside providing escort services, an escort requires time to conduct maintenance, training and replenishment which conflict with the AAW mission. The arsenal ship is only charged for the time AAW services are provided, and not the full time the escort is attached to the arsenal ship. We use 85% — the same as the CNO (N86) *Surface Combatant Force Level Study, Requirements for Joint Warfare* which assumed a 15% "logistics factor" [Ref 10, p 4-9].

8. Self Defense

As discussed in Chapter III, a defender is designated from the task force to provide first layer AAW defense with a SAM firing doctrine of shoot - shoot. Any leakers at that point are assumed to be too close to engage with systems other than self defense soft kill and hard kill point defense.

Most studies consider a defense in-depth methodology where all layers of defense have a chance of countering the incoming missile attack. This equates to ASCM kill probabilities of 99% or higher [Ref. 8, Ref. 15]. These studies, however, assume that enough time and space exist for an Aegis system, for example, to detect and engage with all its systems. Naturally, the hits on ships depend on more variables than test range point defense effectiveness.

In this thesis, a layered point defense is calculated as a parallel system. A CG-47 class ship has ESM/SRBOC, CIWS and ESS (each of these is explained later). The total PK is given in the equation below.

$$PK_{Total} = 1-(1-P_{ESS})(1-P_{ESM})(1-P_{CIWS})$$

$$PK_{Total} = 1-(1-.7)(1-.325)(1-.575) = .914$$

Assuming that incoming missiles will be detected in time for only one SAM salvo per missile by the defender, each ship will defend against leakers with its point defense systems. In the case of the escorts, the defense in depth methodology is applied (computing defense systems in parallel), but an effectiveness weight value is applied. The weight value applied comes from the Schulte thesis [Ref 5], since his data showed that for all the missiles shot at defended ships after 1982, there was a 45% chance of a hit. The study discovered that all but one missile which did not hit the ships, were defeated by passive defense systems. The 45% chance of being hit equates to a 55% effectiveness of point defense systems (soft and hard kill). Many other variables such as scenario, surprise, time of detection, environmental factors, and human factors affect the outcome of a missile attack. These "fog of war" variables are inherent in the combat data reported by Schulte. Consequently, our 55% effectiveness weight factor for the combined point defense systems is considered realistic.

Applying the effectiveness weight value results in the ship's final adjusted point defense (PD) PK, which does not overestimate self defense capabilities. The results for all escorts are given in Table 7.

$$PK_{Adjusted} = (PK_{Total})(weight) = (.914)(.55) = .503$$

Ship	Esm/SRBOC	ESS	CIWS	Total PK	Eff Weight	Adjusted PK
CG-47	0.325	0.7	0.575	0.914	0.55	0.503
DDG-51	0.325	0.7	0.575	0.914	0.55	0.503
DDG-993	0.325	n/a	0.575	0.713	0.55	0.394
DD-963	0.325	n/a	0.575	0.713	0.55	0.394
FFG-7	0.325	n/a	0.575	0.713	0.55	0.394

Table 7: Point defense effectiveness values (PD Pks) for the escort ships.

9. Radar Cross Section (RCS)

This input represents the escorts' signature emissions (detailed in the definition of stealth in Appendix B) that incoming missiles can target on. RCS is used as a representation of stealth. Assuming that the ship design is balanced — that is to say no detectable emission is grossly more discernible than another — the RCS approximation method is sufficient for our analysis. As discussed in Chapter III, this input is the basis on which leakers are allocated amongst ships. The actual numbers are not crucial for the model, but rather the ratios between platforms and arsenal ship are important.

Since a ship's RCS is classified, this thesis will use length, freeboard and mast height to compute representative RCS values. Figure 8 shows how RCS is actually calculated . Using this methodology, RCS is the sum of the superstructure area

Figure 8: Escort RCS

(1/2 x Length x Mast Height) and hull area (Length x Freeboard). RCS figures are given in Table 8. Assumed arsenal ship dimensions are 700ft length, 15ft freeboard and 75ft mast height. This suggests that our notional arsenal ship will ride low in the water. The Arleigh Burke Class destroyers were designed with stealth features, so the RCS used is assumed 1/2 that of a typical destroyer built without stealth.

Ship	Length	Freeboard	Mast Height	RCS
Tico (CG47)	567	20	136	49,896
Burke (DDG51)	505	25	148	24,998
Kidd (DDG993)	563	20	143	51,515
Perry(FFG7)	453	15	116	33,069
Arsenal Ship	700	15	75	36,750

Table 8: RCS inputs into model based on Length, Freeboard and Bridge Height.

Additional RCS is incurred when topside systems are added to the base arsenal ship. For each feature considered, there is an 'additional RCS' input. Our RCS values are based on the equation below. Height values were obtained from the Naval Warfare Publications (NWP-65 series).

$$\text{Additional RCS} = (\text{Length}_{\text{Arsenal}} \times \text{Height}_{\text{Feature}}) \times 0.5$$

B. SUSCEPTIBILITY FEATURES

Soft kill, hard kill, and defensive measures can be employed to reduce the susceptibility of the arsenal ship. Soft kill features can confuse or 'draw off' an incoming missile before or after it has detected and targeted the ship. To defeat an incoming missile which has detected and targeted the arsenal ship, active defense features can be used. Every system placed on the arsenal ship has associated pros and cons. The pro is its effectiveness to soft kill or hard kill the incoming missile. The cons are the cost of the system and the additional RCS (implying loss of stealth) it adds to the ship.

1. Soft Kill Systems

Passive defense systems are those that reduce detectability or prevent an incoming missile from hitting the ship by means other than physical destruction. Reducing detectability is the same as increasing the effectiveness of stealth. Soft kill is the destruction or confusion of an incoming missiles' sensors or guidance system. This causes the missile to be drawn off target or renders it ineffective to function as designed, thus missing the ship. Electronic counter measures (ECM) and Electronic support measures (ESM) with CHAFF function in this manner.

a. Stealth

Reducing RCS through 'bending' metal from the traditional 90 degree right angles or use of radar absorbing materials; reducing IR through innovative ways of cooling engine exhaust or ship surfaces; reducing visual through camouflage or decreasing the amount of ship above the waterline; reducing acoustic through innovative designs in mechanical systems and sound and vibration control; and reducing EME through employing EMCON tactics are areas the user needs to consider employing on the arsenal ship. Again, for stealth there are pros and cons to be input; effectiveness and cost. Stealth is a survival multiplier. It is well known that reducing a ship's RCS will greatly increase the effectiveness of chaff on an incoming missile. The same applies to IR reduction and TORCH deployment. But the extent of the effectiveness in combat is not known.

The cost of stealthing a ship is relatively unknown. Starting without a base ship design further complicates the problem. If stealth is to be considered as a whole, then the user must aggregate the affects of each signature reduction and their cost. If each system is considered separately, then an estimate of effectiveness and cost is required. In the absence of detailed estimates and to simplify the complication, this thesis will consider that stealthing a 20 thousand ton arsenal ship will have an affect of reducing our notional stealth signature, RCS, to 1/8 the original RCS for a 20 percent cost increase. Table 9 shov s the values as they are input into the model. AAW PK is a multiplier that effectively "destroys" (renders ineffective) seven out of eight ASCMs that would otherwise hit the arsenal ship. The 'additional RCS' input is not a factor, since stealthing a ship does not add more RCS signature to the ship.

Nomenclature	Min Qty	Max Qty	AAW Pk	Procurement	Annual O&S	LifeCycle	Annual Cost	Additional RCS
Stealth	0	1	0.875	$90,000,000	$0	35	$2,571,429	0

Table 9: Stealth values input into the model

b. ESM/CHAFF

ESM provides the ship a 'heads up' of an incoming missile. The system then alerts the ship to the incoming and is able to automatically fire CHAFF/TORCH if it is configured to do so. Evasive maneuvers and seducing missiles with CHAFF or TORCH rounds are methods to draw off an incoming missile.

As discussed in the escort self defense section, this thesis uses 32.5% for the ESM/CHAFF soft kil obability. This figure is considered an underestimation, since Schulte's analysis of ASCM effectiveness in the littoral finds that "Softkill measures employed against anti-ship missiles were extremely successful, seducing or decoying every missile it was used against. In every engagement where a defender was alerted and deployed softkill measures, every missile salvo was entirely defeated" [Ref 5, p. 35]. O&S cost per year is $500K per year (FY95 dollars adjusted to FY96 using 3% inflation rate). This cost is the average of all ESM systems (SLQ-32 V1 - 4). Procurement data for a SLQ-32 V2 and 6 MK36-6 rocket launchers is $4.34M. For additional RCS, we estimated a height of 3 feet for this feature. The values input into the model are given in Table 10.

Nomenclature	Min Qty	Max Qty	AAW Pk	Procurement	Annual O&S	LifeCycle	Annual Cost	Additional RCS
Esm/Srboc	0	1	0.325	$7,025,000	$500,000	35	$700,714	1050

Table 10: ESM/SRBOC model input values

2. Hard Kill Systems

Active defense systems are those that apply physical force to destroy or damage an incoming missile to the extent that it does not hit the ship, or reduces the damage the missile causes (i.e., causes the missile to explode before hitting the ship). These systems can acquire and attack the incoming missile at the 10-20nm range like the rolling airframe missile (RAM) or evolved sea sparrow (ESS), or at shorter ranges like the CIWS (0-1nm). The 5"54 gun loaded with an IR round also has a small ASCM capability. We did not consider using the 5"54 gun option since it is assumed there will not be enough time to acquire, target, load and fire the gun in a littoral engagement. New advanced systems like Ram accelerator CIWS, improved SMs, etc. can be tested for their overall effects in arsenal ship survivability and battleforce sustainability.

a. RAM

The RAM (RIM-116) series surface-to-air missile is designed to provide defense against cruise missiles. It is a box-type launcher on the outer deck of a ship and is propelled by a modified Sidewinder missile rocket motor, guided by a Stinger missile seeker and carries a Sidewinder warhead. It is a fire and forget missile. In this thesis 0.7 is used as the effectiveness against an incoming target [Ref. 15]. In this case only one system will be considered. Since cost data is not available, it is estimated that the lifecycle cost for RAM would be $10M amortized over 35 years. Annual O&S cost is assumed to be similar to that of CIWS, $378,000. The total annual lifecycle cost is then $661,714. For additional RCS we assume this feature has a total height of 15 feet. Table 11 shows the RAM values used in the model.

Nomenclature	Min Qty	Max Qty	AAW Pk	Procurement	Annual O&S	LifeCycle	Annual Cost	Additional RCS
RAM	0	0	0.7	$10,000,000	$376,000	35	$661,714	5250

Table 11: RAM model input values.

b. ESS

The Evolved Sea Sparrow (ESS) is based upon the current AIM-7M Sea Sparrow missile. It includes improvements in the rocket motor, aerodynamic control system and auto-pilot while remaining a vertically launched missile. This feature allows for the missiles to be stored in the arsenal ships VLS cells, thus reducing top side 'clutter'. The MK 91 FCS has a dual "headlight" configuration of antennas, one for transmitting and the other for receiving. ESS requires active guidance from the ship. This thesis will use 0.7 as the effectiveness against an incoming target. [Ref. 15]. Since cost data is not available, an $8M is estimated for procurement. O&S costs are assumed similar to that of

ESM/CHAFF, $376,000 per year. The total annual lifecycle cost is then $604,571. For additional RCS, we assume this feature has a height of 10 feet. Table 12 shows the ESS values used in the model.

Nomenclature	Min Qty	Max Qty	AAW Pk	Procurement	Annual O&S	LifeCycle	Annual Cost	Additional RCS
ESS	0	1	0.7	$8,000,000	$376,000	35	$604,571	3500

Table 12: Evolved Sea Sparrow model input values

c. ***Vulcan Phalanx CIWS***

The MK 15 Close-in Weapon system is a short range defensive system effective against air targets between (0-1nm). If so equipped, the arsenal ship will contain the required number of CIWS systems to give the ship 360 degree coverage. Each CIWS is a self contained system and is procured and supported as such. Whether there are one, two, three or four mounts on the ship, will be reflected in the cost of the total system. In the model, the quantity for CIWS should be '0' for no CIWS, or '1' for a set of two mounts. The AAW PK for CIWS is 0.575 which is the average effectiveness against four different types of missiles [Ref. 15]. The O&S cost data for two mounts is $752,000 from the Center for Naval Cost Analysis VAMOSC program office [Ref. 16]. Procurement cost is $12,391,400 dollars for two mounts obtained from Data Search Associates [Ref. 12]. The total annual lifecycle cost for two mounts is then $1,106,040. For additional RCS, we assume this feature has a height of 10 feet for each mount. Table 13 shows the CIWS values used in the model.

Nomenclature	Min Qty	Max Qty	AAW Pk	Procurement	Annual O&S	LifeCycle	Annual Cost	Additional RCS
CIWS	0	1	0.575	$12,391,400	$752,000	35	$1,106,040	7000

Table 13: CIWS model input values

C. VULNERABILITY

These are features that reduce vulnerability after a ship has been hit. All features represented in this section will be taken in parallel to combat the incoming missiles. Another way of saying this is that they "will decrease post-hit vulnerability." [Ref. 11]. The features listed in this area are considered to work in parallel to reduce the effects of a hit. Features considered are component redundancy, component location, passive damage suppression, active damage control, component shielding and component elimination. In the case of the arsenal ship this can equate to the following areas:

- Hull girder strengthening
- Blast hardened bulkheads
- Magazine protection/mass detonation protection
- Fully redundant combat systems
- Double ended propulsion
- Ship size

The effect of varying vulnerability features in cruisers and destroyers was explored in a NSWC Carderock study. This study explores increasing survivability by decreasing the ship's vulnerability. The arsenal ship design could benefit from a study like this. The Ship Vulnerability Model (SVM) "which provides probability for loss of ship functions after hits by selected threat classes" could benefit the arsenal ship designers prior to the ship being built. [Ref. 7]

For each feature considered, an effectiveness value (AAW Pk), cost and additional RCS input are required. Since the data in Ref. 7 is classified, the authors have chosen to use an aggregate hardening input. Our hardening package is assumed to provide the arsenal ship with a threefold increase in staying power. The cost is assumed to be 35% of the base ship cost. It is assumed that hardening does not increase the ship's RCS signature. See Table 14 for our model input for hardening.

Nomenclature	Min Qty	Max Qty	AAW Pk	Procurement	Annual O&S	LifeCycle	Annual Cost	Additional RCS
Hardening	0	1	0.6667	$157,500,000	$0	35	$4,500,000	0

Table 14: Hardening model input values

D. SURPRISE ATTACK/NO SURPRISE ATTACK OPTION

If the surprise option is not toggled, then the task force is assumed to be aware of every incoming hostile missile. Toggling surprise attack assumes that the first missile attack will be a surprise and will penetrate the task forces defenses with a probability of 1. The second missile attack is also assumed to be a surprise, but only has a guaranteed penetration of .5. All missile attacks which follow will be subject to a task force which is now alert and defending against incoming missiles at full effectiveness. This option originates from a littoral scenario where the battle space is small and the enemy launches without any indications or warnings. This is a real problem. Our task forces can rarely operate in a status of weapons free. The first indication of hostilities may be a surprise missile launch.

E. AGGREGATE INCREASED DETECTABILITY

If the increase in RCS attributed to each feature placed on the arsenal ship is unknown or the crude approximations presented in this thesis are not considered reasonable, then this option is recommended. Employing this model option assumes that any hard kill or soft kill features placed on the arsenal ship will cut the stealth effectiveness in half. Topside systems increase detectability by creating increased radar return, light, heat and smoke. By how much is difficult to determine, and it is complicated when stealthing is involved, but the effect is probably greater than the increase in RCS. Our simple solution is to activate the "reduced RCS" feature when the arsenal ships defenses are employed.

F. LIMITING THRESHOLDS

These inputs define desired survivability and cost. If both limiting criteria can be met, they system produces feasible solutions. This thesis uses .8 as the survivability criterion and 50 million dollars for the maximum annual system cost.

G. SUSTAINABILITY INPUTS

The inputs for the Sustainability model are identical to the Survivability model's inputs, as shown below.

1. Arsenal Ship Design

The user chooses the arsenal ship's design by hardwiring the desired soft kill, hard kill, and vulnerability features in the Excel spread sheet. The design can be a configuration produced using the Survivability model, or another that the user designates. From our runs of the Survivability program, we have decided to run varying designs of the arsenal ship. The inputs for each run will be discussed in Chapter V.

2. Task Force Composition

The Task Force can contain any number of escorts. The user decides based on the scenario being examined. This can be done by hardwiring the escort minimum and maximum values.

3. Limiting Criteria

In the Sustainability model, the only limiting criterion is the arsenal ship's survivability level. This value is the percentage of staying power above OOA which the user desires the arsenal ship to retain. Once the staying power falls below this threshold, the arsenal ship is considered OOA. Our requirement will be that the arsenal ship remain above 80% of its staying power.

V. SURVIVABILITY SUB-MODEL RESULTS

We do not intend to design the arsenal ship. Chapters V and VI show examples of how the model can be used. We cannot explore all possible configurations as a typical run may yield more than 62,000 possibilities. That is just for our data. Varying parameters and introducing additional systems can increase the candidates manyfold. We present examples that we believe are realistic or interesting and perform sensitivity analysis on significant factors affecting results.

A. DESCRIPTION

Unless otherwise stated, the inputs shown in Table 15 are used in this chapter for running the Survivability Sub-model.

	Escort Ships							
Ship Class	Min Qty	Max Qty	SAM Pk	Procurement	Annual O&S	LifeCycle	Annual Cost	# ASM SAMs
CG-47	0	2	0.7	$1,237,038,179	$28,814,948	30	$70,049,554	70
DDG-51	0	2	0.7	$1,063,900,596	$21,500,976	30	$56,964,329	50
DDG-993	0	1	0.5	$489,262,553	$25,156,840	30	$40,798,925	56
DD-963	0	0	0	$468,282,553	$44,250,516	30	$59,892,801	0
FFG-7	0	2	0.4	$298,513,074	$17,077,516	30	$27,027,952	21
Arsenal Ship	0	0	n/a	$450,000,000	0	35	$12,857,143	284
	Self Def	Underway	Displace	RCS	Mission Time			
Ship Class	Pk	Days	Tons	Value	Value	Notes	Ann. Esc. Cost	
CG-47	0.503	90	9600	49838	0.85		$14,681,619	
DDG-51	0.503	90	8300	24998	0.85		$11,939,099	
DDG-993	0.394	90	9800	51515	0.85		$8,551,008	
DD-963	0.394	90	8000	51515	0.85		$12,552,833	
FFG-7	0.394	90	3700	33069	0.85		$5,684,763	
Arsenal Ship	n/a	n/a	20000	38750				
		Systems						
Nomenclature	Min Qty	Max Qty	AAW Pk	Procurement	Annual O&S	LifeCycle	Annual Cost	Additional RCS
Stealth	1	1	0.875	$90,000,000	$0	35	$2,571,429	0
Esm/Srboc	1	1	0.325	$7,025,000	$500,000	35	$700,714	1050
CIWS	1	1	0.575	$12,391,400	$752,000	35	$1,106,040	7000
ESS	1	1	0.7	$8,000,000	$378,000	35	$804,571	3500
Active Defense Effectiveness	0.55							
Hardening	1	1	0.66666667	$157,500,000	$0	35	$4,500,000	0
○ Surprise Attack Option	☑ Degrade Stealth by 1/2, thus RCS degradation column rendered in-effective					Nomenclature	Notation	Value
◉ No Surprise Attack						Incoming Missile Shots	M	100
						Missile Explosive Weight	He	250
						Minimum Survival Probability	MINPROB	0.8
						Annual Life Cycle Cost	Cost	$35,000,000

Table 15. Input Parameters for Survivability Example

B. MEASURES OF EFFECTIVENESS APPLIED

The survivability model results reflect the inputs above for a raid of 100 incoming missiles and show three different possible measures of effectiveness (MOEs) one may employ to determine the best configuration candidates. Direction from higher authorities and the user's preferences will dictate which is best for the situation.

- MOE #1: (Best Cost/Effectiveness ratio) This MOE sorts the output by the Cost/Effectiveness, subject to meeting both cost and survival thresholds — lowest is best.
- MOE #2: (Minimum Cost Subject to Survivability Constraint) This MOE sorts the output in ascending cost order.
- MOE #3: (Maximum Survivability Subject to Cost Constraint) This MOE sorts the output in descending order of survivability.

The output for each MOE is presented graphically in composite charts which show the configuration components in different colors on a bar graph. A line indicating survivability level is superimposed and read from the scale in the right margin. The graphs display the top six configurations for each MOE, first under the input conditions stated above and then for surprise attacks. The contents of Figures 9 through 14 are indexed in Table 16.

FIGURE NUMBER	MOE NUMBER	SURPRISE ATTACK
9	1 - Cost Effectiveness	Yes
10	1 - Cost Effectiveness	No
11	2 - Minimum Cost	Yes
12	2- Minimum Cost	No
13	3 - Maximum Survivability	Yes
14	3 - Maximum Survivability	No

Table 16. Index for Survivability

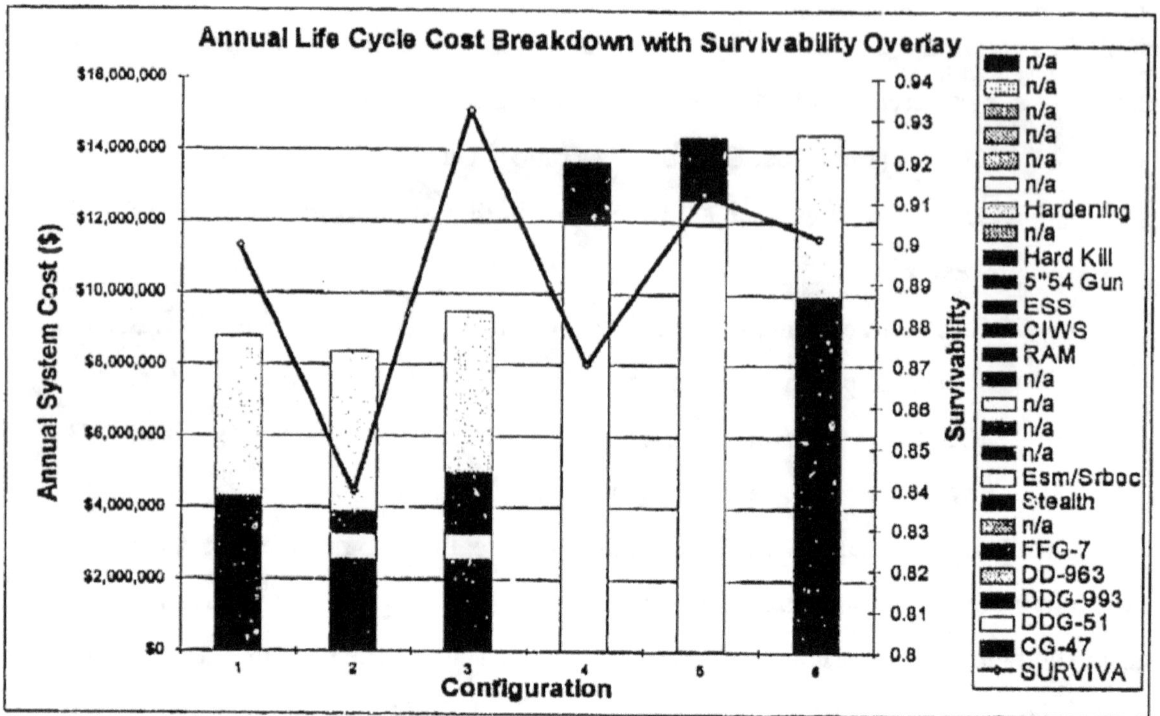

Figure 9. MOE #1: Best Cost/Effectiveness Configurations (No Surprise Attack)

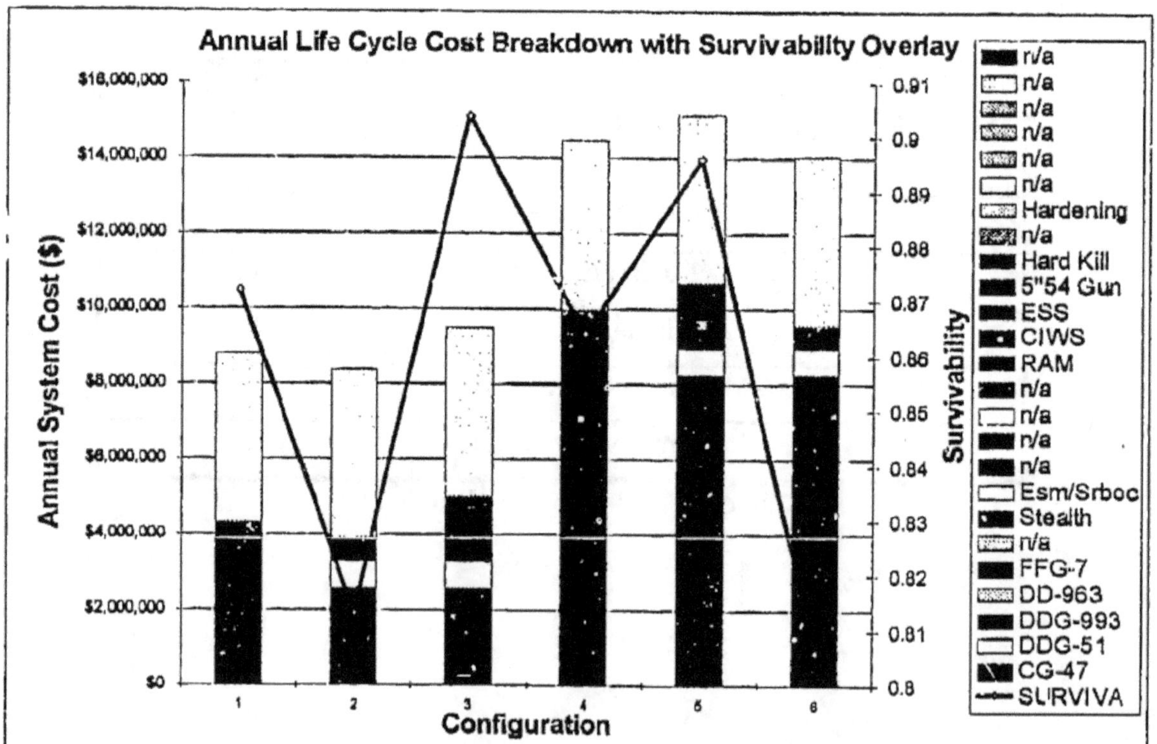

Figure 10. MOE #1: Best Cost/Effectiveness Configurations (Surprise Attack)

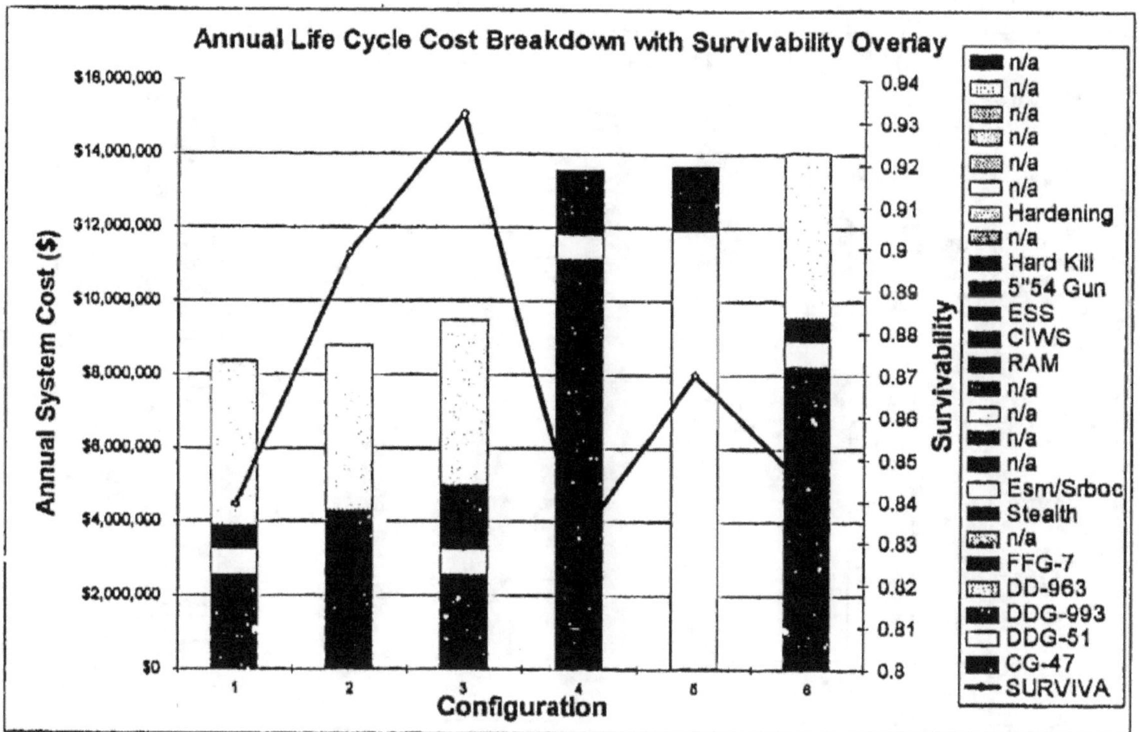

Figure 11. MOE #2: Minimum Cost Configurations (No Surprise)

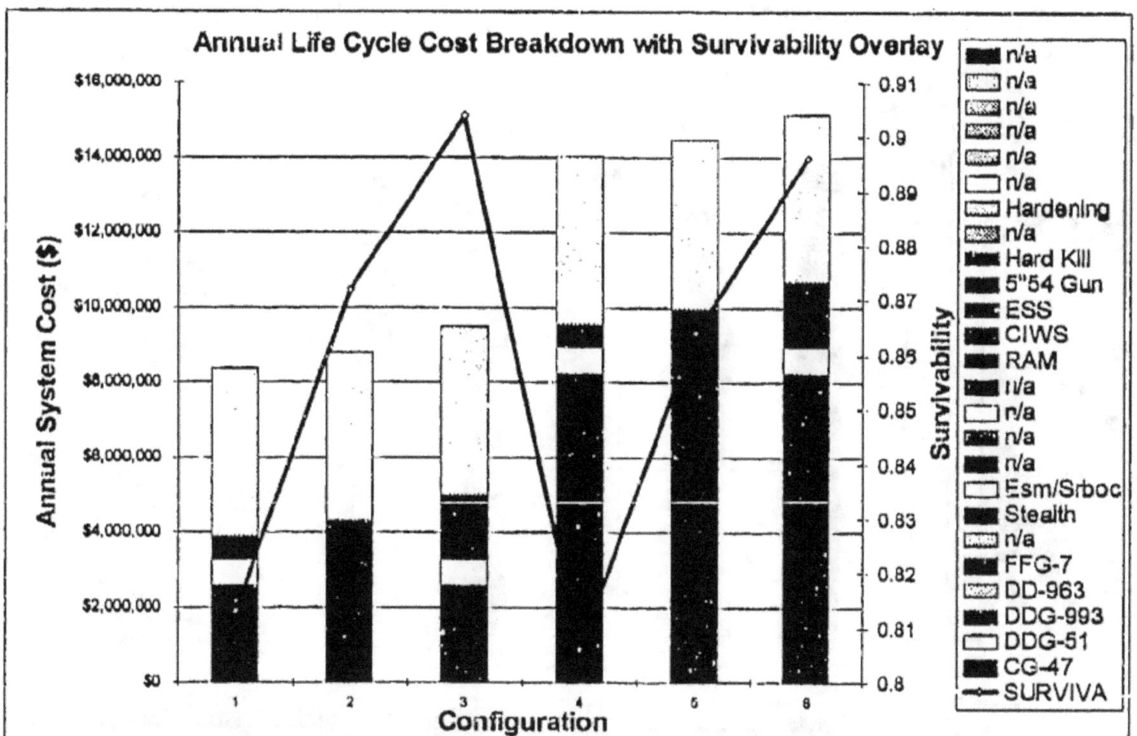

Figure 12. MOE #2: Minimum Cost Configurations (Surprise Attack)

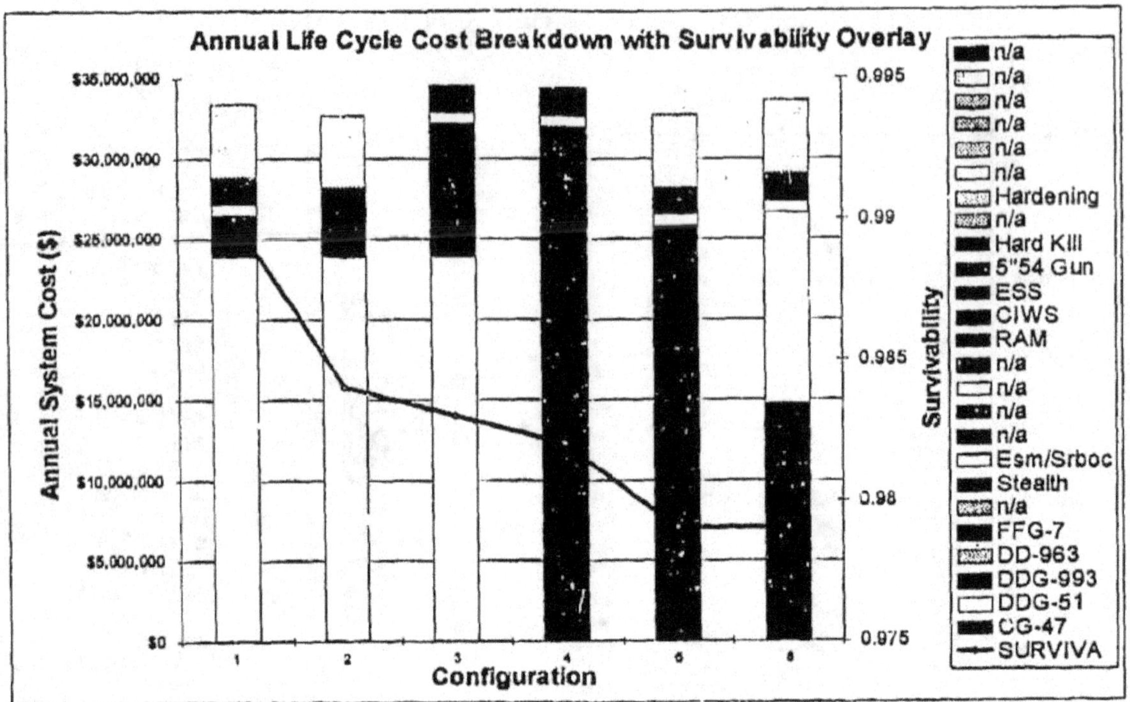

Figure 13. MOE #3: Maximum Survivability Configurations (No Surprise Attack)

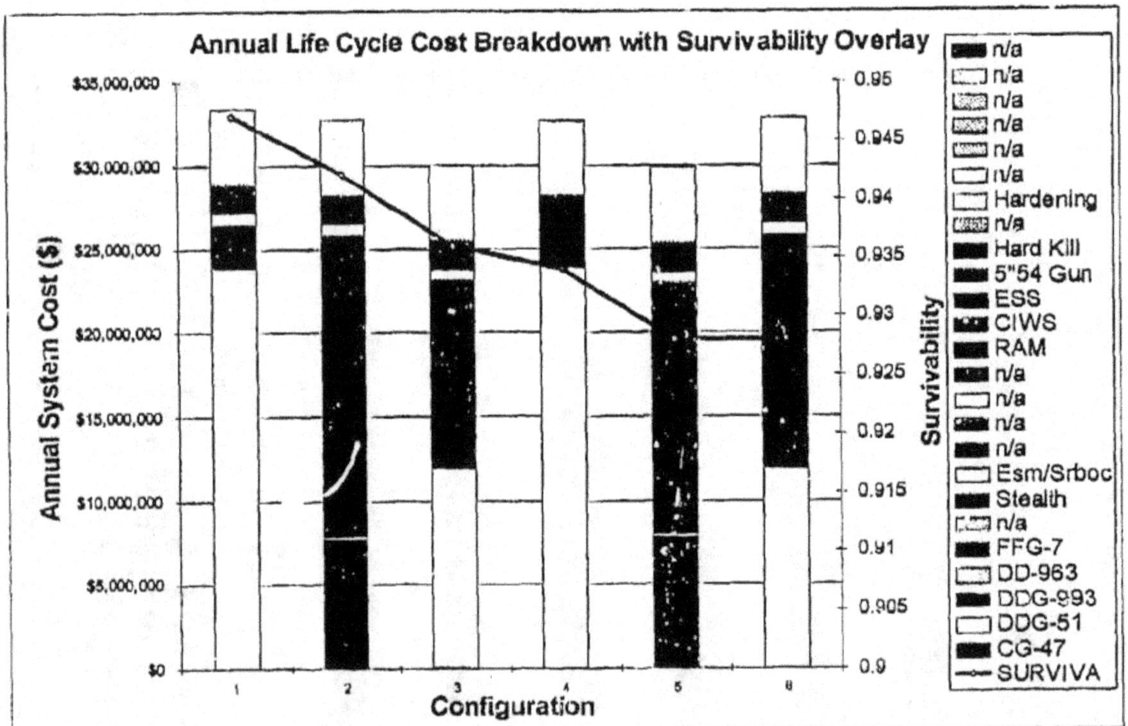

Figure 14. MOE #3: Maximum Survivability Configurations (Surprise Attack)

C. INTERPRETATION OF BEST CONFIGURATION RESULTS

1. No Surprise Attacks

Of the top six configurations for MOE #1, all six included one or more hard kill systems. Four used stealth, and four incorporated hardening. Three included soft kill, and the bottom three employed escorts. For MOE #2, the overall breakdown of desirability of features was nearly the same. In fact, four of the top six configurations are the same for MOE #1 and MOE #2. The order of preference differs, however. Stealth was added in five vice four configurations.

MOE #3 produced drastically different results. All of the top six configurations incorporated escorts and hard kill systems. Soft kill and stealth each appeared in five configurations, and hardening was present in four. Each of the top configurations from MOE #3 cost two to four times the configurations produced in the other MOEs.

2. Surprise Attacks

Recall the assumption that in a surprise attack the first missile will strike a ship and the second will have a fifty percent chance of leaking through defenses and hitting a ship. The impact is a reduction in overall arsenal ship survivability of 2.5 to 5 percentage points. The decline is greatest in configurations with escorting warships.

The astute reader, trying to ascertain patterns may notice an apparent paradox in the interrelationships of adding escorts and changing surprise options. When surprise is not present, adding escorts to a fixed configuration will increase survivability. For a surprise scenario, however, the results seem inconsistent. In some instances adding escorts improves survivability and others not. The illustration below (Figure 15) uses data extracted from Figures 9,10,13 and 14 to demonstrate the issue.

The result is counterintuitive at first. The reason is complex. In Case 1 of the illustration (Figure 15), when a surprise raid is launched on the arsenal ship escorted with an FFG-7, the arsenal ship fares worse than if it operated alone. This is because even though it incorporates stealth, the arsenal ship is more visible when operating with another ship. Making matters worse, there is only one other ship to share the damage inflicted by leakers, and a frigate is relatively small. If the frigate were larger, it would have more have more staying power and attract a larger share of the leakers. The defense afforded by the frigate is not sufficient to overcome these detractors to arsenal ship survivability. In contrast, the CG-47 and DDG-993 (in case 2 of Figure 15) have sufficient staying power and defensive capability to boost the arsenal ship's survivability even though the effectiveness of the arsenal ship's stealth is affected by operating with escorts.

Case 1: Survivability

	Base 1	Base 1 +
No Surprise	0.899	0.901
Surprise	0.872	0.864

Case 2: Survivability

	Base 2	Base 2 +
No Surprise	0.932	0.979
Surprise	0.904	0.942

Base 1: stealth, CIWS, ESS, hardening
Base 1 +: Base 1 and 1 FFG-7

Base 2: stealth, CIWS, ESS, hardening, ESM
Base 2 +: Base 2, 1 CG-47, and 1 DDG-993

Figure 15. Arsenal Ship Survivability Paradox Illustration

3. Generalizations

While each MOE is quite different in emphasis, some configuration recommendations apply universally. Hard kill systems appear in the top six configurations of each MOE regardless of surprise attack. When surprise attack applies, all top configurations incorporate stealth and hardening also.

D. SENSITIVITY

1. Stealth

Since stealth appears in most favorable output configurations, a closer look at the effects of stealth on survivability is in order. Figure 16 shows the effects of stealth for incoming missile raids of size 50, 100, and 150. The configuration includes stealth, ESM, ESS, CIWS, and hardening. The stealth factor ranges from 1 to 12 corresponding with no reduction up to a $^1/_{12}$ reduction in signature. Interestingly, the effects nearly level-off after a reduction factor of about four for the cases studied. This indicates that while stealth is important, investing past this point provides diminishing marginal returns and may not be the wisest investment.

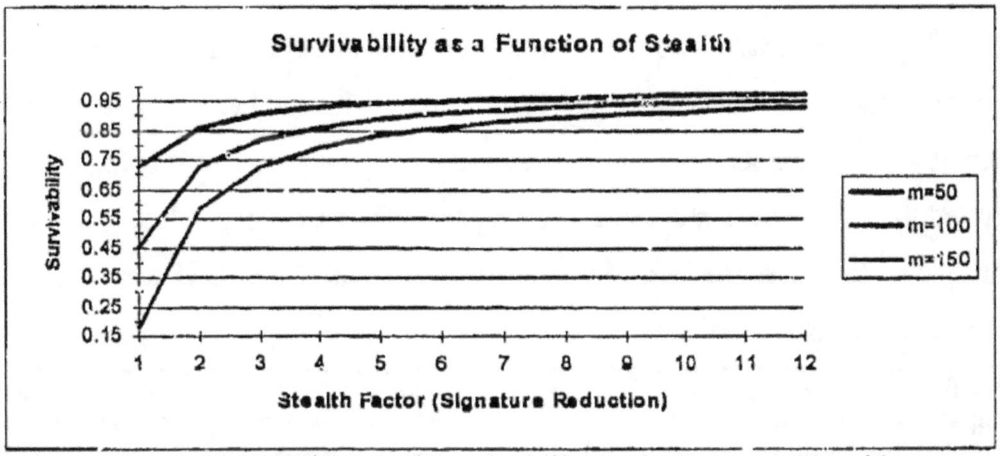

Figure 16. Sensitivity of Survivability to the Degree of Stealth

2. Hardening

Hardening also appears in the best configurations. To test the sensitivity of survivability to hardening, the degree of hardening is varied for a ship with stealth, ESM, ESS, CIWS, and hardening. The hardness factor was varied from one to six and one-half; this corresponds with no improvement of staying power over the base ship up to a six and one-half-fold improvement. The results appear in Figure 17. Again one sees a "knee" in the curve (in the neighborhood of a hardening factor of 3), thus indicating that investments in hardening are good to a point. After that, investments in other systems most-likely will affect survivability more.

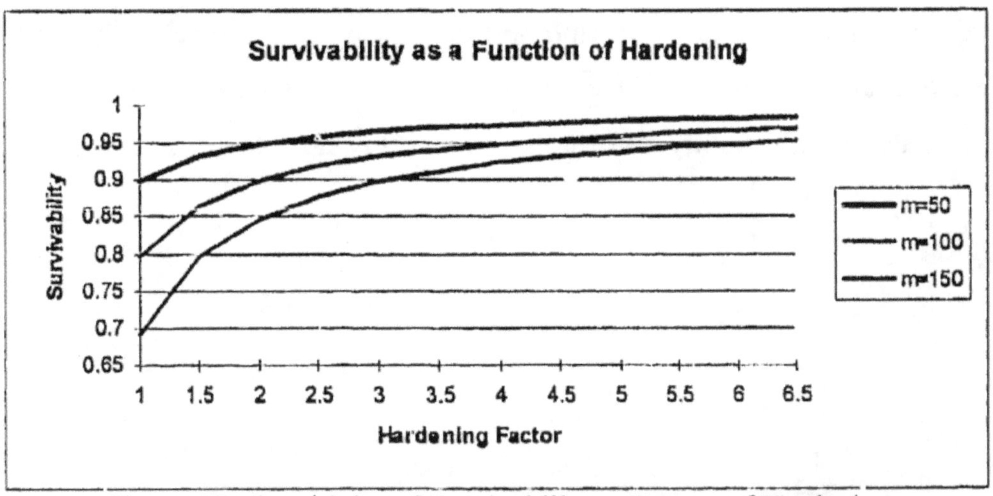

Figure 17. Sensitivity of Survivability to Degree of Hardening

3. Escort Ships

Table 17 summarizes the effects on the arsenal ship's survivability of adding one or two escorts of a particular class to the task force. Three configurations of arsenal ships are examined. The base case incorporates no defensive systems or design features; the next case includes only the ship design features of stealth and hardening; and the final case combines stealth, hardening, CIWS, ESS, and ESM. The survivability (S_A) values are generated from the model, and the values in parenthesis are the absolute increase in survivability over the unescorted case for each configuration. Figures 18 and 19 plot the data for the last two configurations.

Arsenal Ship Configuration	S_A without Escorts	S_A with one FFG-7	S_A with two FFG-7s	S_A with one DDG-51	S_A with two DDG-51s	S_A with one CG-47	S_A with two CG-47s
Base: No Defensive Systems	0 (0)	0 (0)	0 (0)	0 (0)	.191 (.191)	0 (0)	.435 (.435)
Base + Stealth + Hardening	.206	.229 (.023)	.237 (.031)	.389 (.183)	.583 (.377)	.474 (.268)	.751 (.545)
Strong Defensive Systems	.932	.933 (.001)	.934 (.002)	.946 (.014)	.962 (.030)	.953 (.021)	.977 (.045)

Table 17. Effects of Adding Escorts on Arsenal Ship Survivability (S_A)

An arsenal ship without any defensive systems is not survivable with less than two Aegis ships. A ship configured with only modest self-defense features benefits greatly from additional escorts. Both the systems on the escort and the fact that the escorts have larger signatures to draw missiles work to bolster arsenal ship survivability. An arsenal ship configured with formidable self-defense systems will naturally achieve the highest level of survivability when escorted over any other configuration. The marginal increase to survivability from adding an escort is much lower though.

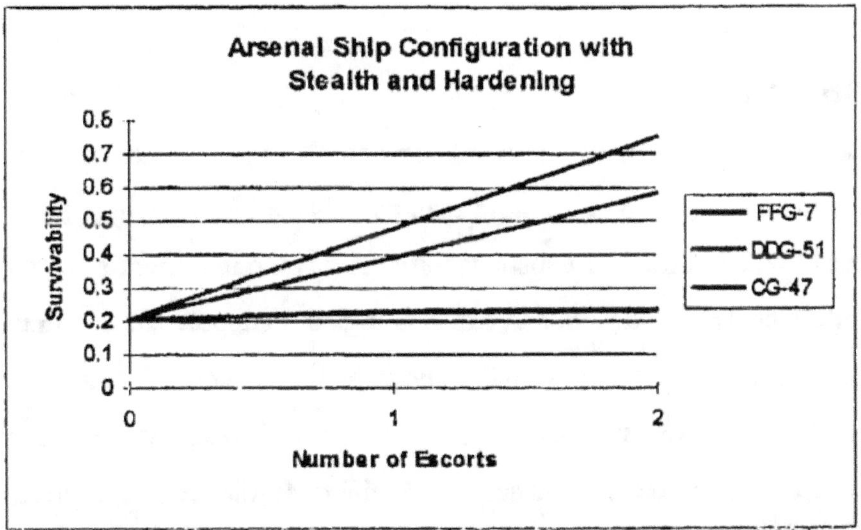

Figure 18. Sensitivity of Survivability to an Adding Escorts to Arsenal Ship with Stealth and Hardening only.

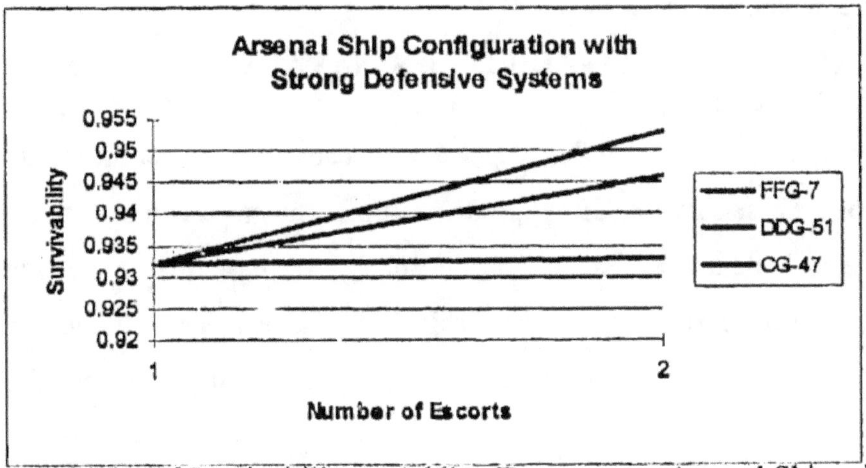

Figure 19. Sensitivity of Survivability to Adding Escorts to an Arsenal Ship with Strong Defensive Systems.

VI. SUSTAINABILITY ANALYSIS

A. DESCRIPTION

Sustainability is measured by the naval component's time on station. It needs to be emphasized that sustainability involves more than a BG's ability to conduct strikes and take hits. These are the indicators of sustainability. A naval component's sustainability results from both onboard ship design and external logistics support (fuel, ordnance, material, food, etc.). In this analysis we will review the impact of an arsenal ship's design and ordnance load on the BG's sustainability. Three areas of sustainability are examined.

- The effect of varying arsenal ship survivability on the naval component's sustainability
- BG sustainability with an arsenal ship vs. with five Arleigh Burkes
- VLS replenishment issues

B. ARSENAL SHIP SURVIVABILITY VS. NAVAL COMPONENT SUSTAINABILITY

The term "naval component" is emphasized because we are not measuring just the sustainability of the arsenal ship. Our approach assumes the naval component's sustainability is equivalent to the escorts' sustainability. We base this on the premise that a BG can remain on station without an arsenal ship, but an arsenal ship cannot remain on station without an escort present, since there are also surface and sub-surface threats which the arsenal ship requires escorts to protect against.

Our approach is to vary the arsenal ship's survivability and examine the impact on the BG's sustainability. Our baseline BG has 1 Ticonderoga Class Cruiser, 2 Arleigh Burke Class destroyers, 1 Kidd Class destroyer and 2 Oliver Hazard Perry Class frigates. This is considered the base case. Table 18 lists each modification to the base that we examine. System 1 is the base and systems 2 through 15 add an arsenal ship with varying degrees of defensive features. Running each system through the Sustainability model with an arsenal ship's limiting survivability of .8, and no surprise attack gives the results in Table 19.

System	BG Composition
1	BG
2	BG, ST
3	BG, SK
4	BG, HK
5	BG, Hard
6	BG, ST, SK
7	BG, ST, HK
8	BG, ST, Hard
9	BG, SK, Hard
10	BG, HK, Hard
11	BG, SK, HK
12	BG, SK, HK, Hard
13	BG, ST, SK, Hard
14	BG, ST, HK, Hard
15	BG, ST, SK, HK, Hard

Key
BG = 1 CG-47, 2 DDG-51 1 DDG-993, 2 FFG-7
ST = Stealth
SK = Soft Kill = ESM/SRBOC
HK = Hard Kill = CIWS, ESS
Hard = Hardening

Table 18: Sustainability Model Run Battle Group Composition

System	Missiles Escorts OOA	Missiles AS OOA	Missiles left on Arsenal	Annual Cost of Arsenal	Annual Cost with Escorts	(lower better) Cost Effect
1	108	111	0	$0	$58,440,350	0
2	176	182	150	$2,571,428	$61,011,778	14610
3	120	109	188	$700,714	$59,141,064	5839
4	210	216	94	$1,710,611	$60,150,961	8146
5	192	187	94	$4,500,000	$62,940,350	23438
6	176	186	150	$3,272,143	$61,712,493	18592
7	180	239	142	$4,282,040	$62,722,390	23789
8	176	199	150	$7,071,428	$65,511,778	40179
9	210	211	94	$5,200,714	$63,641,064	24765
10	210	232	94	$6,210,611	$64,650,961	29574
11	210	220	94	$2,411,325	$60,851,675	11483
12	210	244	94	$6,911,325	$65,351,675	32911
13	180	271	142	$4,982,754	$63,423,104	27682
14	180	371	142	$8,782,040	$67,222,390	48789
15	180	466	142	$9,482,754	$67,923,104	52682

Table 19: Model Results for the Systems identified in Table 1, an arsenal ship limiting survivability of .8, no surprise option and aggregate RCS feature

Table 19 contains the following data. 'Missiles Escorts OOA' is the number of incoming missiles it took to put the escorts OOA. 'Missiles AS OOA' is the number of incoming missiles it took to put the arsenal ship OOA with .8 probability. 'Missiles left on Arsenal' is the number of SAMs remaining on the arsenal ship when it was rendered OOA. 'Annual cost of Arsenal' is the annual system cost above the base arsenal ship cost (for which we assumed $450 million). 'Annual cost with Escorts' is the annual system cost plus the annual cost of the escorts. The 'Cost Effect' column is the annual cost of the arsenal ship divided by the number of missiles it took to put the arsenal ship OOA.

1. Measures of Effectiveness

The number of incoming missiles required to put the escorts OOA is the measure of effectiveness. The number of incoming missiles equates to 'time on station' if the incoming rate (missiles per unit time) is known. The naval component's data in Table 19 indicates systems 4, 9, 10, 11 and 12 have the greatest sustainability levels. Artificially discrete sustainability levels are introduced as a result of using the "aggregate RCS" degradation factor (refer to Chapter 4.E). The hard kill feature occurred most frequently, being in four of the five systems. Hardening and soft kill features occurred in three of the five systems each.

Stealth is not a feature in any of the five systems with the greatest sustainability. This is an artifact of allocating leakers based on a fair share of the BG's total RCS. The stealthier the arsenal ship, the lower its 'fair share' of leakers, which causes the escorts to be attrited faster, thereby reducing the naval component's overall sustainability. At this point we need to define the term balanced force. A balanced force is a BG whose ships' signature levels are roughly equivalent. It is not practical to expect all ships to have the same signature levels; however as the variance approaches zero, the missiles are allocated more uniformly (allocation per ship = 1/number of ships). With this in mind, as the Navy builds the SC-21 and future ships with lower signatures, the BG's overall signature will decline. Since a balanced force will have the greatest overall sustainability, stealthing the arsenal ship will be essential.

To discern which of these five candidates (4, 9, 10, 11, or 12) is the best, requires further evaluation of the BG's performance under additional criteria.

a. Choice by Arsenal Ship Sustainability

If we choose which of the five systems is best based on greatest arsenal ship sustainability, then systems 10 and 12 are best. In the case of system 12, another way of stating it is that the arsenal ship survives until missile 244, and so system 12 had more staying power remaining when the escorts in the other configurations were put OOA.

Though stealth is not incorporated in any of the systems of choice, the effect of stealth requires explanation in order to interpret the other system results in Table 19. After the escorts are put OOA, two methods are used to attrite an arsenal ship without escorts. When stealth is not a feature, every missile targets the arsenal ship and is subject to its point defense systems. If the arsenal ship is stealthed, the incoming missiles are reduced by the inverse of the stealth factor (1/Stealth Factor) initially. For example, a stealth factor of eight prevents seven out of eight shots from acquiring the arsenal ship. This is why systems 13, 14, and 15 require such a large number missiles to put the arsenal ship OOA.

b. Choice by Ordnance Available

Another way to determine which of systems 4, 9, 10, 11 and 12 is the best can be based on the number of arsenal ship SAMs used. Vice Admiral (Ret.) Joseph Metcalf, one of the Surface Warfare Community's former leaders, espoused that the primary mission of a surface combatant is to deliver "maximum ordnance on target." [Ref. 17, p. 36] He referred to strike ordnance, not defensive ordnance like SAMs. The more strike weapons the BG has, the greater its offensive potential, provided it can survive. In Table 19, all five systems have 94 SAMs remaining after the escorts are put OOA. As a reminder, our arsenal ship has a load out of 228 strike weapons and 284 SAMs (refer to Chapter 4.B.3 for details). Since no escorts remain to control the 94 remaining SAMs, their potential is never realized. Assume, for a moment, that all 228 strike weapons were fired from the arsenal ship and 94 SAMs remained in their cells when it had to leave station. For each of the five systems with 94 remaining SAMs, the load out could be modified to be filled with strike ordnance. This would give the arsenal ship the striking potential of 228 + 94 = 322 strike weapons vice 228. Since "ordnance on target" is the goal, all five systems chosen perform equally well.

c. Choice by Cost

Assuming for a moment that the cost of defending the arsenal ship with a BG is a sunk cost, then screening our five chosen systems by the Cost data in Table 19, indicates that systems 4 and 11 with annual lifecycle costs of $1.7 and $2.4 million, are best. System 4 incorporates hard kill features and system 11 incorporates hard and soft kill features. Systems 9, 10 and 12 contain hardening which increases their annual cost two to three times higher.

d. *Choice by Cost Effectiveness*

Another measure is cost effectiveness, which represent the cost for each unit of sustainability. The cost effectiveness column (far right) in Table 19 is calculated by taking the annual life cycle cost without escorts divided by the number of missiles to put the escorts OOA. Since a lower cost effectiveness is desired, systems 4 and 11 are the best with $8,146 and $11,483 per incoming missile respectively. Again, system 4 contains hard kill systems only, and system 11 contains hard and soft kill systems. These systems yield the best "bang for the buck."

2. Interpretation of Best Configuration Results

At this point in the analysis, caution must be taken. The means by which the best of the five is chosen is subjective. If concerned with budget, opt for the lowest cost. Another approach is to choose best "bang for the buck." Looking at individual measures separately can lead to a misinterpretation of the overall analysis. All MOEs need to be examined simultaneously, but each MOE is not necessarily equal in weight, and any weighting scheme we set can easily be countered by another. So, our stand is to rationalize the needs of the warfighter, the CINC. In order to fight and win, a BG must remain on station and overcome the enemy. This implies fighting hurt and delivering ordnance. Of the best systems (4, 9, 10, 11 and 12), systems 10 and 12 provide the BG, as a whole, with the most sustainability, and yield a potential 228 + 94 = 322 cells for missions other than BG defense. From the warfighter's perspective, first and foremost, sustainability is staying power, and systems 4, 9, 10, 11 and 12 are equal. When considering ordnance, all five systems are equal. However, when looking at the BG's sustainability as a whole (not just at the escorts) we see that system 10 affords the most overall sustainability. It is ranked 10 out of 15 for its cost and cost effectiveness, however

C. SUSTAINABILITY WITH ARSENAL SHIP VS. FIVE ARLEIGH BURKES

In order to deploy the same capacity as the arsenal ship's 512 VLS cells in theater, the Navy would need to deploy 5.7 (or 5 by rounding down to keep costs conservative) Arleigh Burkes. Running the sustainability sub-model for a BG with an arsenal ship and for a BG with five additional Arleigh Burkes, produces the results in Table 20. We chose the arsenal ship configuration which produced the most sustainable BG in Part B of this chapter (system 12). The arsenal ship has soft kill, hard kill, and hardening features, but no stealth. The cost data for one arsenal ship and five Burkes is shown in Table 21.

Category	Arsenal Ship	5 Arleigh Burkes
BG Sustainabiltiy (incoming missiles)	210	233
Potential Strike VLS Cells	228 + 94 = 322	5(40) = 200

Table 20: Sustainability Sub-Model Results for an Arsenal Ship BG and an Arleigh Burke BG. The arsenal ship's minimum survivability criteria is .8; the base BG has 1 CG-47, 2 DDG-51, 1 DDG-993 and 2 FFG-7; and a no surprise scenario was chosen.

The BG survivability levels in Table 20 indicate that the Burke BG can stay on station 10 percent longer than the arsenal ship BG. The runs indicate that the escorts in the arsenal ship BG realize their full potential, by defending the BG until placed OOA. In the Burke BG the escorts become combat ineffective by depleting their SAMs. This means their full staying power is never realized.

The runs indicate that the arsenal ship has 94 SAMs remaining when placed OOA. By replacing these with strike weapons in the load-out, the arsenal ship has a potential strike capacity of 322, while still maintaining the BG's sustainability at 210 missiles. Refer to Table 20. Conversely, the Burkes deplete all their SAMs indicating they could increase the BG's sustainability if more SAMs were available. Reallocating any of the SAM VLS cells on the Burkes would decrease the sustainability of the BG.

The cost values in the "5 Arleigh Burkes" column of Table 4 are derived using the Burke's procurement cost and O&S costs over 35 years. The arsenal ship's costs are unknown, so the following approach was taken for each area category of Table 21. Procurement cost is assumed $450 million (refer to Chapter 2.C.). Added systems cost is the sum of procurement and O&S costs times 35 years for all arsenal ship features. Personnel cost was derived by prorating the personnel cost of an Arleigh Burke by the crew size of an arsenal ship (assumed 50) multiplied by 35 years. Material cost for an arsenal ship is assumed to be equal to that of a single Arleigh Burke for 35 years. The depot cost is assumed to be equivalent to a single Ticonderoga cruiser for 35 years. [Ref. 13, Ref. 16]

Category	Arsenal Ship	5 Arleigh Burkes
Procurement	$450,000,000	$5,319,502,980
Added Systems Cost	$241,896,375	$0
Personnel Costs	$48,383,314	$1,616,002,675
Material (POL, Repair)	$253,296,120	$1,266,480,600
Total Cost	$993,575,809	$8,201,986,255
Number of Missiles	512	450
Cost per Missile	$1,940,578	$18,226,636

Table 21: Cost Comparison of an Arsenal Ship vs. Five Arleigh Burke ships in a BG

A quick comparison on the cost data in Table 21 shows that five Burkes cost 11.8 times more than an arsenal ship. The Burkes cost 9.4 times more per missile than the arsenal ship.

D. VLS REPLENISHMENT ISSUES

An arsenal ship with 512 cells is a mixed blessing. It will provide a task force with 512 more missiles in theater, thus increasing sustainability of forces on station. However, VLS replenishment at sea is still a problem, and pier side replenishment time is substantial. To put this in perspective, 512 cells with a reload rate of 5 minutes per cell (highly optimistic) would take approximately 43 continuous hours to reload. As it stands now, an Aegis ship takes 2-3 days to reload about 100 VLS cells, so it would take 10-15 days to reload an arsenal ship with 512 VLS cells. Replenishment is the Achilles' heel in VLS operations. An article published in the Fall 1988 UNREP Journal stated:

> "In wartime the enemy decides when and where we expend defensive ammo, so an ammo UNREP may be needed any time, even when the seas are rough or the decks are icy. While we may be able to rearm our aircraft carriers under these conditions, our ability to handle missiles in dollies or in VLS canisters on cruisers, destroyers, and frigates is currently extremely poor." [Ref. 18]

Our ability to UNREP VLS missiles at sea has not improved much since the article was written. The following example illustrates the potential dilemma. Our base BG without an arsenal ship encounters 25 incoming missiles, at which time the first Burke requires replenishment. If operating in the Sea of Japan it would take approximately 29 hours at 28 knots to reach Sasebo, 6 hours to reload 65 cells (assuming reload rate of 5 minutes per missile and a demand of 50 SAMs and 15 strike weapons), and another 29 hours to get back on station. This equates to 2.8 days off-station. For a multi-mission platform like the DDG-51 or CG-47, this is magnified.

It is frequently predicted that future warfare will be intense in the early days of an MRC, with the tempo leveling off after around the 10th day of conflict [Ref. 10]. If this is the case, even three days off station for our multi-mission platforms will severely reduce the sustainability of the naval component in theater. The arsenal ship provides a strong but temporary fix to the VLS replenishment problem that results from a limited number of missiles on the DDG-51 and CG-47. The arsenal ships will surely fill the gap needed for the surge in requirements at the start of a conflict. Once its missiles are expended, however, the arsenal ship will be away for a long time. It is optimistic to assume the ship can transit at 20 knots a great distance (escorted, of course), reload and return in 15-20 days, that forward logistics bases will be capable of reloading VLS quickly, and that the ordnance will be available. At today's production rate (100-120 Tomahawks per year) it could easily take two or more years worth of production to refill one arsenal ship. With so many VLS cells on the cruisers, destroyers and six arsenal ships we're looking at most missiles being at sea, and very few in reserve for replenishment.

VII. CONCLUSION

A. SUMMARY

The development of an arsenal ship represents fundamental change in naval and joint warfare. Such a vessel will operate in the littoral and be able to provide strikes ashore and theater ballistic missile defense. Additionally, weapons not previously placed on naval ships can occupy VLS cells and provide new forms of direct support for ground troops. The added capabilities, however, come at the cost of exposing the ship to sudden and potentially numerous, technologically advanced, sea and land-based missile attacks.

Our Arsenal Ship Tradeoff Analysis Model consists of a survivability and a sustainability sub-model. The survivability sub-model is a tool that may aid in the design of the ship by predicting how the ship will survive against a specified threat depending on the hard kill, soft kill, stealth, and hardening features built-in and the sevices provided by escorting surface combatants (SCs). The sustainability model considers weapon logistics and may prove useful in determining what portion of the arsenal ship's VLS cells may be devoted to offensive weapons.

B. CONCLUSIONS

Based on exercising the Tradeoff Analysis Model with reasonable weapon system effectiveness values, cost estimates, and three different measures of effectiveness, we advocate an arsenal ship design that balances offense and survivability, the likes of which have not been seen in U.S. warship design since World War II.

Justification of this conclusion is involved. The analytical basis is explained in detail on pages 42 to 48 of Chapter 5. The rationale can be summarized as follows:

- We premise that combatants operating in littoral waters will be subject to increasingly dense and sophisticated missile attacks. Among them, surprise attacks will occur.
- To carry out the multiple missions envisioned for it, the arsenal ship will be exposed to such attacks. It can be escorted intermittently and only when required, thereby saving the cost of built-in survivability. What fraction of the

cost of the escorting surface combatant to charge against the arsenal ship is crucial and depends on the proportionate time taken from other surface combatant tasks. The dollar value is nearly impossible to determine analytically. The multi-mission combatant escort cannot be regarded a free good (on the basis that no added SCs are purchased as arsenal ship escorts) because the SC may be lost in protecting the arsenal ship, in which case it is gone from every other mission, too.

- Fortunately, for a wide range of circumstances and measures of effectiveness, the analysis shows that building survivability into the arsenal ship is almost always preferred to assigning escorts, even when only a small fraction (18.6%) of the SC's life cycle cost is charged against the escort role.
- The preference for and selection of a survivability feature is less easily described. The incorporation of stealth characteristics is, by the analysis, the feature appearing in preferred designs most often. This may be a consequence of the numbers chosen. The reader is cautioned to examine the numbers in detail but is also warned that alternative numbers will be difficult to find. In particular, the reader's attention is invited to Figure 16, page 48, which strongly supports a moderate investment in stealth.
- We have gone to some pains to show that stealth, ship hardening, and defensive short range, hard and soft kill systems are complementary. For example, stealth adds nothing when the arsenal ship is firing a large missile volley, radiating, subject to air attack with bombs, surface gunfire attack, or a submarine launched torpedo attack. But, hardening retains its effectiveness in all these circumstances.
- Our overall conclusion is that stealth, ship hardening, and some set of modern point defense (hard and soft) are, in view of their modest cost in construction and operating personnel, well worth the modest cost on the margin because the arsenal ship's concept of operation requires that it be exposed to major attack.

- Further, until the arsenal ship has expended its massive offensive punch, there will be circumstances in which it should be escorted. It will probably draw the attacks by the enemy to itself in the way CVs do.
- The arsenal ship alleviates the requirement to frequently cycle SCs with fewer VLS cells from the scene of action off station to a replenishment site and back again into the action. By its presence, the arsenal ship will sustain other SCs on station. It can remain on station longer than any other ship for a given missile delivery rate, but when its weapons are expended the reload problem will take it off station for a considerable amount of time.

C. RECOMMENDATIONS

Our recommendation for arsenal ship survivability features is to incorporate stealth, point self-defense systems, and hardening into the design.

Even though hard kill systems were more commonly seen in the analytically preferred results than soft kill systems, we believe soft kill systems are essential for littoral operations. First, soft kill is synergistic with stealth and softkill effectiveness is amplified. Second, as noted, the Schulte thesis concluded that soft kill measures have been highly effective in actual combat, but hard kill systems remain largely unproven [Ref. 5]. Operating in littoral waters with current rules of engagement (weapons not free) makes the arsenal ship vulnerable to initial and sudden attacks without ample time to respond with hard kill.

Our model shows that when surprise attacks occur, hardening is a very attractive feature of ship design. The additional staying power to remain mission capable after at least one hit will help ensure that the arsenal ship's 500-plus missiles are not rendered useless by a lucky or cheap shot.

A balanced design will seek to maximize the arsenal ship's net delivered firepower over the combat life of the ship. Incorporating *all features* listed is, according to the analysis, tantamount to over-designing arsenal ship survival features, when its survival *with those features* is compared to the survival of the accompanying present-day surface combatants. Since the DDG-51 and CG-47 cost roughly the same to procure as an arsenal ship-plus-missile load-out and they will be less survivable, this may appear out of balance. However, in the future as more Arleigh Burke destroyers and SC-21 type warships enter the fleet, matching stealth and superior hardness will provide operational balance.

D. AREAS FOR FURTHER STUDY

Our thesis concentrated on the ASCM threat to the arsenal ship. While this is most likely the predominant danger, other nemeses such as submarine and mine threats should be explored. Currently ships are not designed to withstand under hull blasts from mines or torpedoes. Hardening ships displacing 10,000 tons or less against these threats is not feasible, but if the arsenal ship is in the range of 20,000 tons, then a tradeoff analysis is possible between the choices of ASW screening, active countermeasures, and hull hardening. [Ref. 7]

We studied the sustainability of the naval component only. Since the arsenal ship is a joint war fighting platform that through CEC can be utilized by ground troops as well, a theater level study of the marginal value of the arsenal ship's contribution to the ground war is in order, with emphasis on the critical events at the war's onset. With over 500 VLS cells, the ship will surely contribute to checking the enemy onslaught by threatening a massed surge or pulse of power delivered in a short time. The service performed will encompass not only destroying hardware but promoting caution and slowing down the enemy's operations as he exercises deception, concealment, and prudent behavior in general.

Currently the Navy's capability to conduct underway replenishment of VLS cells is inconsequential. VLS cells are here to stay, with a growing variety of weapons. Faster replenishment methods, be they at sea or ashore, beg for technological development.

APPENDIX A. ABBREVIATIONS AND ACRONYMS

A/C - aircraft

ARG - amphibious readiness group

ASCM - anti-ship cruise missile

ASW - anti-submarine warfare

ATACMS - army tactical missile system

BAT - battery

BDE - brigade

BN - battalion

C4I - command, control, communications, computers, and information

CEC - cooperative engagement capability

CG - cruiser

CIWS - close-in weapon system

CINC - Commander-in-Chief

CMD - command

CNO - Chief of Naval Operations

CVBG - carrier battle group

DARPA - Defense Advanced Research Projects Agency

DD - destroyer

DDG - guided missile destroyer

ECM - electronic countermeasures

EME - electromagnetic emissions

EMCOM - emissions control

ESM - electronic surveillance measures

ESS - evolved sea sparrow

EW - electronic warfare

FCS - fire control system

FFG - guided missile frigate

FY - fiscal year

IR - infrared

MRC - major regional contingency

NSWC - Naval Surface Warfare Center

O&S - operations and support

OOA - out of action

OMG - operational maneuver group

OP - operation

OTH - over the horizon

PD - point defense

PK - probability of kill

PR - program review

PSSK - single salvo kill probability

RAM - rolling airframe missile

RCS - radar cross-section

RDT&E - research, development, testing, and evaluation

SAM - surface to air missile

SC - surface combatant

SLAM - stand-off land attack missile

SRBOC - super rocket blooming chaff

TBM - theater ballistic missile

TBMD- theater ballistic missile defense

TLAM - Tomahawk land attack missile

TSSE - Total Ship System Engineering

UAV - unmanned aerial vehicle

VAMOSC - visibility and management of operational and support costs

VGAS - vertical gun advanced system

VLS - vertical launch system

APPENDIX B. DEFINITIONS

configuration - the model output which includes the collection of systems added to the base arsenal ship and the warship escorts required to protect the arsenal ship

leaker - an anti-ship cruise missile that is engaged but not defeated by defensive measures [Ref. 8]

stealth - the reduction and control of observable signatures that are exploitable by the enemy including electromagnetic emissions, radar cross section, visual (wake), infrared, acoustic, and magnetic signatures [Ref. 9]

staying power - the number of hits that a ship can absorb before being rendered not mission capable [Ref. 8]

susceptibility - the inability of a ship to avoid the sensors, weapons, and weapons effects of a man-made hostile environment [Ref. 11]

survivability - the capability of a surface ship to avoid and/or withstand a man-made hostile environment while performing its mission [Ref. 11]

vulnerability - the conditional probability of being killed given that a ship is hit [Ref. 11]

APPENDIX C. USER'S GUIDE

Arsenal Ship Tradeoff Analysis Program
User's Guide

LT Ronald Bush LT Arthur Cimiluca

Note: This program requires Microsoft's Excel spreadsheet program for Windows.

1. **Installation**

   A. Create a directory on hard drive: **C:/ arsenal**.

   B. Copy all files from 3.5 inch floppy disk to **C:/ arsenal**. Your directory should contain the following files:

alcom22.exe	batsust.exe
alcom22.pas	batsust.pas
arsenal.exe	icon_h.dll
arsenal.pas	icon_y.dll
arsenal.xls	

Note: additional text files will be created in the **C:/ arsenal** directory when the program runs.

2. **Running to solve for configurations**

   A. Open the file **arsenal.xls**.

   B. Examine the first sheet (labeled **Arsenal**). Sample data, which represents the best unclassified information the authors could locate at the time of writing, is provided in all cells.

   C. Update cost, kill probability, radar cross section, or ordnance load data if required.

   D. Set the minimum and maximum ranges for each type of escort ship and candidate system. If a system or escort is not listed, add it in the appropriate section by typing the name over an **n/a** cell and entering the required characteristic data.

E. Select **No Surprise Attack** or **Surprise attack Option.**

F. Select **Degrade Stealth by 1/2...** unless detailed radar cross section data is available for all ships and systems examined.

G. Enter threat data for **Days**, **Attacks/Day**, and **Missile/Attack**.

H. Change **missile explosive weight** if considering other than Exocet equivalents.

I. Enter the **Minimum Survival Probability threshold** of interest.

J. Enter **Annual Life Cycle Cost** limit.

K. Click on **Step # 1** button for the Survivability Sub-Model (Update).

- Answer yes to the "Replace existing 'FIGURES.TXT'?" prompt.

- Answer yes to the "Replace existing 'ARSENAL.XLS'?" prompt.

L. Highlight and double click on **Step # 2 button** (Enumerate).

M. Highlight and double click on **Step # 3** (Run Pgm) button.

- Note the number of configurations that meet your criteria and close the text box.

N. Click on **Step # 4** (Display) button.

- The **All Data** sheet will be displayed. It lists all the configurations that meet your input requirements. Normal spreadsheet functions, such as sorting, can be performed to organize the output data as desired.

- Also shown is a graph of survivability verses cost. This graph is useful in identifying interesting configurations that warrant closer examination. The graph's properties may require modification if the entire range of the budget is not displayed or if all points are not displayed over the entire x-axis range. To adjust the graph, first double click on it.

(1) Then to modify the number of data points, double click on the plotted curve. Select **Names and Values**, and observe the **Y-Values** line. The last number after the "$" represents the number of data points, overwrite it with the number of configurations that met your criteria (from step 2.M). Then select the **X- Values** tab, overwrite the last number after the "$" on the **X-Values** line as above, and click on the **OK** button to exit Format Data Series.

(2) To change the cost scale, double click on the cost values and change the maximum, minimum, and increment values as desired. After changes are made click on the **OK** button.

3. **Displaying Configurations**

Up to six configurations can be displayed in a graphical representation of cost, survivability, and individual systems and ships in the **All Data** sheet. There are two options for displaying results.

- Option 1: The user may sort the data using the spreadsheet function in Excel and display the top six results by pressing the 'Display first 6 results' button.

- Option 2: The user may determine which configurations to display from either the graph or the spreadsheet data. Select a configuration by highlighting its entire row and copying it into the **Details** sheet. Copy the first configuration into line 3 of the **Details** sheet and subsequent configurations into lines 10,17,24,31, and 38. Scroll down the **Details** sheet to see the composite graph.

This graph may be tailored to the user's needs. Adjustments to the cost scale (left) or survivability scale (right) can be made by double clicking on the desired axis or axes and changing the scale as in step 2.N.(2) above.

4. **Printing Graphs**

The options are numerous; three ways are described below.

A. Select *Print*; then *Print Selected Sheet*.

B. Activate a graph by double clicking on it. Select *Print*; then *Print Selected Chart*. This method prints the graph on an entire sheet of paper and affords the user some formatting options.

C. Highlight a graph by clicking on it. Then *copy* it to the windows clipboard and *paste* it into a compatible program which can be printed.

Note: Perhaps the best way to import a graph to another program is by using the *Paste Special* under the *Edit* menu and selecting *Paste as Picture*.

5. **Running to solve for Sustainability**

The same procedures apply as for "Running to solve for configuration." The step number and difference are given below.

2.D. Hardwire the desired escorts and candidate systems by setting **Min Qty** and **Max Qty** the same.

2.G. Not required. Sustainability program will solve for number of missiles.

2.I. Note: Min Survivability threshold is for the Arsenal Ship, not for the escorts. Each escort will fight until OOA, while the Arsenal Ship will fight until it falls below the threshold.

2.N. There is no Step #4 button

3. To display data use Notepad or any other text compatible viewer and open the file c:\arsenal\history.txt.

APPENDIX D. ARSENAL SHIP TRADEOFF ANALYSIS PROGRAM

LIST OF REFERENCES

1. Office of the Chief of Naval Operations, *Promulgation of the Arsenal Ship Concept of Operations*, Memorandum, Serial N863D/6U654802, dtd. 11 April 1996.

2. DefenseLINK, *Arsenal Ship Program Selects Phase I Contractors*, NEWS RELEASE, Reference Number: No. 422-96, 11 July 1996.

3. Chief of Naval Operations Executive Panel, *Future Navy Game II*, ltr. dtd. 17 July 1 1996.

4. Department of the Navy, *Forward...from the Sea*, Washington, D.C., November 1994.

5. Schulte, J. C., *An Analysis of the Historical Effectiveness of Antiship Cruise Missiles in Littoral Warfare*, NPS Masters Thesis, Monterey, CA, September 1994.

6. Humphrey, R. L., "Damage and Losses of Warships in Modern Combat," Presentation ORSA/TIMS Joint Meeting, Las Vegas, NV, May 1990.

7. Hansen, I. and T. D. Burton, *Destroyer/Cruiser Architecture for Reduced Vulnerability (U)*, Naval Surface Warfare Center, Carderock Division Report NSWCCD-C-TR-67--96/16, Bethesda, MD, June 1996. draft (CONFIDENTIAL)

8. Hughes, W. P., Jr., *The Military Worth of Staying Power (U)*, Naval Postgraduate School Report NPS-OR-94-008, Monterey, CA, May 1994. (CONFIDENTIAL)

9. Goddard, C. H. et al., *How Much Stealth?*, Naval Engineers Journal, Vol. 108, No. 3, pp. 105-116, May 1996.

10. Office of the Chief of Naval Operations (N86), *Surface Combatant Force Level Study (SCFLS), Requirements for Joint Warfare*, August 1995.

11. Ball, R. E. and Calvano, C. N., *Establishing the Fundamentals of a Surface Ship Survivability Design Discipline*, Naval Engineers Journal, pp. 71-74, January 1994.

12. Data Search Associates, U. S. Weapon Systems Costs, 1990, Tenth Edition, Fountain Valley, CA 92728-8361.

13. Data Search Associates, U. S. Weapon Systems Costs, 1995, Fifteenth Edition, Fountain Valley, CA 92728-8361.

Preceding Page Blank

14. Janes Fighting Ships 1996-97, Janes Information Group Limited, pp. 15-17.

15. Calvano, C. N. , et al., *Large Capacity Missile Carrier (CMX)*, USNPS Report NPS-ME-94-004, December 1993.

16. Naval Center for Cost Analysis, *Navy Visibility and Management of Operating and Support Costs (VAMOSC), Shipboard Systems Report*, Arlington, VA., May 1996.

17. Metcalf, J., *Revolution at Sea*, U.S. Naval Institute Proceedings, January 1988.

18. *The Most Challenging UNREP Problem*, Editorial, UNREP Journal, Fall 1988.

INITIAL DISTRIBUTION LIST

1. Defense Technical Information Center 2
   8725 John J. Kingman Rd., STE 0944
   Ft. Belvior, VA 22060-6218

2. Dudley Knox Library 2
   Naval Postgraduate School
   411 Dyer Rd.
   Monterey, CA 94943

3. Professor Wayne P. Hughes, Code OR/HI 2
   Department of Operations Research
   Naval Postgraduate School
   Monterey, CA 94943

4. Professor Charles N. Calvano, Code ME/CA 1
   Department of Mechanical Engineering
   Naval Postgraduate School
   Monterey, CA 94943

5. Distinguished Professor David Schrady, Code OR/SO 1
   Department of Operations Research
   Naval Postgraduate School
   Monterey, CA 94943

6. Research Office 1
   Attn: Associate Provost David W. Netzet
   Naval Postgraduate School
   Monterey, CA 94943

7. Carderock Division 1
   Naval Surface Warfare Center
   Attn: Mr. Ib Hansen, Code 174
   Bethesda, MD 20084-5000

8. Mr. Robert Keane Jr. 1
   Carderrock Division, NSWC
   Bethesda, MD 20084-5000

9. Defense Logistics Studies Information Exchange 1
U.S. Army Logistics Management Center
Fort Lee, VA 23801

10. Deputy Chief of Naval Operations (Logistics) 1
Attn: CDR Robert Drash, Code N422C
2000 Navy Pentagon
Washington, DC 20350-2000

11. OCNO (Surface Warfare) 1
Attn: Code N86B
2000 Navy Pentagon
Washington, DC 20350-2000

12. OCNO (Expeditionary Warfare) Code N85T 1
Attn: Dr. Frank Shoup
2000 Navy Pentagon
Washington, DC 20350-2000

13. Mr. Leigh Ebbert 1
John Hopkins University
Applied Physics Lab
John Hopkins Rd.
Laurel, MD 20723-6099

14. Commander Naval Doctrine Command 1
Attention: Mr. Samuel Leeds
1540 Gilbert St.
Norfolk, VA 23511-2785

15. Center for Naval Analyses 1
Attention: Vice ADM W. H. Rowden
4401 Ford Ave.
Alexandria, VA 22302-0268

16. Center for Naval Analysis 1
Attention: John V. Hall
4401 Ford Ave.
Alexandria, VA 22302-0268

17. Mr. Christopher Ryan Jr. 1
Arsenal Ship Joint Program Office
4301 N. Fairfax Dr. Suite 630
Arlington, VA 22203

18. LT Ronald S. Bush 2
9322 E. 6000 N. Ave.
Grant Park, IL. 60616

19. LT Arthur E. Cimiluca, Jr. 2
15 Old Wood Road
Bernardsville, NJ. 07924

NAVAL POSTGRADUATE SCHOOL
Monterey, California

THESIS

THE ARSENAL SHIP CONCEPT: VULNERABILITES TO SPECIAL OPERATIONS

by

Christian A. Dunbar
and
Dino Pietrantoni

December, 1997

Thesis Advisor: Wayne P. Hughes, Jr.

REPORT DOCUMENTATION PAGE

Form Approved OMB No. 0704-0188

Public reporting burden for this collection of information is estimated to average 1 hour per response, including the time for reviewing instruction, searching existing data sources, gathering and maintaining the data needed, and completing and reviewing the collection of information. Send comments regarding this burden estimate or any other aspect of this collection of information, including suggestions for reducing this burden, to Washington headquarters Services, Directorate for Information Operations and Reports, 1215 Jefferson Davis Highway, Suite 1204, Arlington, VA 22202-4302, and to the Office of Management and Budget, Paperwork Reduction Project (0704-0188) Washington DC 20503.

1. AGENCY USE ONLY *(Leave blank)*	2. REPORT DATE December 1997	3. REPORT TYPE AND DATES COVERED Master's Thesis

4. TITLE AND SUBTITLE THE ARSENAL SHIP CONCEPT: VULNERABILITIES TO SPECIAL OPERATIONS	5. FUNDING NUMBERS
6. AUTHOR(S) Dunbar, Christian A. and Pietrantoni, Dino	
7. PERFORMING ORGANIZATION NAME(S) AND ADDRESS(ES) Naval Postgraduate School Monterey, CA 93943-5000	8. PERFORMING ORGANIZATION REPORT NUMBER
9. SPONSORING / MONITORING AGENCY NAME(S) AND ADDRESS(ES)	10. SPONSORING / MONITORING AGENCY REPORT NUMBER

11. SUPPLEMENTARY NOTES

The views expressed in this thesis are those of the author and do not reflect the official policy or position of the Department of Defense or the U.S. Government.

12a. DISTRIBUTION / AVAILABILITY STATEMENT Approved for public release; distribution unlimited.	12b. DISTRIBUTION CODE

13. ABSTRACT *(maximum 200 words)*

The United States Navy has solicited proposals for a revolutionary class of ship, the Arsenal Ship. Despite reduced funding for the project, the concept is still viable for future development. We show how the development of a new unparalleled weapon system or platform will evoke a response by potential adversaries, based on capabilities and asset investment, by unconventional means. The Arsenal Ship is a formidable threat, yet is vulnerable relative to other high value units. These reasons make it an inviting target across the spectrum of conflict. This thesis will describe threats that are usually overlooked and examine the Arsenal Ship's vulnerability to them. In addition, we will show how these vulnerabilities arise as the Arsenal Ship operates through the range of geographic areas. Further, this thesis describes possible strategic and tactical defensive actions to enable the Arsenal Ship to counter these unconventional threats. Each recommended action has a direct implication upon the engineered design and the proposed Concept of Operations (CONOP). In addition, the recommendations will influence the strategy for employing any future platform based on the Arsenal Ship concept, anywhere in the world.

14. SUBJECT TERMS Arsenal Ship, ARSHIP, Maritime Special Operations, Special Operations, Combat Swimmer, VBSS, Visit Board Search and Seize, Unconventional Warfare.	15. NUMBER OF PAGES 122
	16. PRICE CODE

17. SECURITY CLASSIFICATION OF REPORT	18. SECURITY CLASSIFICATION OF THIS PAGE	19. SECURITY CLASSIFI- CATION OF ABSTRACT	20. LIMITATION OF ABSTRACT
Unclassified	Unclassified	Unclassified	UL

NSN 7540-01-280-5500

Standard Form 298 (Rev. 289)
Prescribed by ANSI Std. 239-18
298-102

**THE ARSENAL SHIP CONCEPT:
VULNERABILITES TO SPECIAL OPERATIONS**

Christian A. Dunbar
Lieutenant, United States Navy
B.S., University of Notre Dame, 1993

Dino Pietrantoni
Lieutenant, United States Navy
B.S., United States Naval Academy, 1991

Submitted in partial fulfillment of the
requirements for the degree of

MASTER OF SCIENCE IN DEFENSE ANALYSIS

from the

**NAVAL POSTGRADUATE SCHOOL
December 1997**

Authors: ______________________
Christian A. Dunbar

____
Dino Pietrantoni

Approved by: ______________________
Wayne P. Hughes, Jr., Thesis Advisor

____
Gordon H. McCormick, Second Reader

____
Maurice D. Weir, Chairman
Special Operations Curriculum Committee

iv

ABSTRACT

The United States Navy has solicited proposals for a revolutionary class of ship, the Arsenal Ship. Despite reduced funding for the project, the concept is still viable for future development. We show how the development of a new unparalleled weapon system or platform will evoke a response by potential adversaries, based on capabilities and asset investment, by unconventional means. The Arsenal Ship is a formidable threat, yet is vulnerable relative to other high value units. These reasons make it an inviting target across the spectrum of conflict. This thesis will describe threats that are usually overlooked and examine the Arsenal Ship's vulnerability to them. In addition, we will show how these vulnerabilities arise as the Arsenal Ship operates through the range of geographic areas. Further, this thesis describes possible strategic and tactical defensive actions to enable the Arsenal Ship to counter these unconventional threats. Each recommended action has a direct implication upon the engineered design and the proposed Concept of Operations (CONOP). In addition, the recommendations will influence the strategy for employing any future platform based on the Arsenal Ship concept, anywhere in the world.

TABLE OF CONTENTS

LIST OF FIGURES

X

LIST OF TABLES

I. INTRODUCTION

For any new and innovative weapon system or platform, the question of capabilities versus vulnerabilities are always at the forefront of discussion. This thesis will address critical vulnerabilities to special operations targeting that the Arsenal Ship concept and development has left unattended. Chapter II will define the proposed "concept of operations" (CONOPS) and physical design considerations of the Arsenal Ship (ARSHIP). This will then lead into the proposed role that the ARSHIP will hold in American foreign policy. Chapter II also defines the terminology of "special operations", as it will be used academically in our thesis. That chapter will also introduce the reader to the military and political utility of special operations. This will evolve into a discussion of why states would conduct special operations in lieu of, or to augment general-purpose forces. This will illustrate the attractiveness of the ARSHIP as a target for special operations.

Chapter III will categorize the outcomes that potential threats might attempt to achieve. These categories, coupled with the dissimilar geographic areas where the ARSHIP faces different defense challenges, will be used to show the potential threats that the ARSHIP will face. From this, a comprehensive list of possible courses of action that can threaten the ARSHIP will be compiled. After a brief description of each attack mode, factors that affect the defense of the ARSHIP will be discussed. This will lay the groundwork to demonstrate how special operations can effectively circumvent traditional defensive measures and achieve the previously described mission objectives.

Chapter IV will be an in-depth analysis of five special operations courses-of-action that are, in the case of the ARSHIP, more predominant and highly potent. The in-depth analysis will describe the characteristics of ingress/egress and actions at the objective. These will be described as modes of action and objectives of action. Chapter IV will further discuss the mitigating factors that affect the effectiveness of the aggressor and the defense of the ARSHIP. By understanding the interplay between the two, defense options will be formulated.

Chapter V will address possible strategic and tactical defense options that should be implemented for the ARSHIP. The defense options are formulated with unconventional threats in mind, but their value against conventional attacks should not be discounted. Most of the strategic measures are to dissuade an adversary from conducting an unconventional attack through deterrence and deception. The proposed defense options will have an effect on the engineering design, the proposed CONOPS, and the strategy utilized in the future employment of the ARSHIP concept.

This thesis was initiated prior to the formal decrease in funding of the ARSHIP concept. Its initial intent was to become an integral part of the iterative process of refining the ARSHIP's CONOPS and physical design. We have continued its pursuit based on the importance of educating future planners and designers who might propose similar concepts of a large, highly automated and low manned warship.

II. THE ARSHIP AND SPECIAL OPERATIONS

A. THE ARSENAL SHIP

The ARSHIP is a proposed U.S. naval warship that can be loaded with 500 missiles in a vertical launch system (VLS). The payload of missiles can be a combination of cruise and/or anti-air missiles. They will be able to receive targeting data and launched either from onboard the ARSHIP or remotely from another platform. The missiles will be used for theater missile defense (TMD), anti-air warfare (AAW), and deep strike operations. Ideally, the mission of the ARSHIP concept is for the vessel to be positioned within its weapons' range to an adversary and act as a tool for U.S. coercive diplomacy.

The ARSHIP will be a technologically advanced ship, designed to operate with a reduced crew of less than 50. The vessel's size will be approximately the length of two football fields and have a sustained maximum speed of 22 knots. Full automation and redundant networking will eliminate system failures with every system onboard. The command and control system will be fully integrated and utilize the most advanced technology available. Even damage control (DC) sensors and actions will be automated.

Proposed CONOPS scenarios have the ARSHIP out to sea for extended periods of time and it will either deploy as a stand-alone platform or be accompanied by a conventional surface combatant. To operate alone, the ARSHIP will rely on its stealth design for concealment. It will not be equipped with self-defense weapons systems. Along with its low visual profile, the technological features of its stealth design will reduce its electromagnetic and acoustical transmissions. Defense of the ARSHIP depends on the success of these technological advances to conceal its location and allow it to operate freely under this cloak of invisibility. Another CONOPS proposal is to have an unmanned version of the ARSHIP controlled by electronic tether from another ship operating within line of sight (LOS).

B. UNDERSTANDING THE ARSHIP'S ROLE

The ARSHIP will, herein, be described on two levels. First as a weapons platform, that is deployable and moves the system of weapons from geographic region to region. Secondly as a weapon system that fires missiles, commanded either remotely or locally, projecting power ashore. The ARSHIP can be likened to a ballistic missile submarine, a weapon system by design as well as a deployable platform. This distinction will lead to the understanding that this new ship is an important *new target*, different in critical ways from other surface combatants. This will increase the range and importance of threats that the ARSHIP will encounter.

The ARSHIP will be capable of delivering an unparalleled level of distant precise destruction. This potential destruction is a step change in current capabilities which is unrivaled by its conventional peers. Some refer to it as a capital ship while others view it merely as a floating ammunition barge. Regardless of the proponents'/opponents' views, the bottom line is that it will have unparalleled influences on U.S. foreign policy because it is a means for exacting unsurpassed tolls on an enemy's infrastructure and military capability. This places many of the U.S.' conventional competitors at a disadvantage and position of asymmetrical weakness. Consequently, this new ship warrants an opponent's heightened attention across the spectrum of conflict.

C. DEFINITION OF SPECIAL OPERATIONS

Scholars have debated the definition of special operations, developing definitions that primarily assist their own theories. The common denominators of many of these definitions boil down to three basic characteristics. Special operations are militaristic operations that are:

1) small in scale; 2) planned to have the effect of high leverage; and 3) unorthodox.[1] The usefulness of this definition is in its universal applicability beyond the narrow paradigm of Western thought.

Each characteristic must be present to distinguish a militaristic operation as a special operation in the context of war and politics. The first condition, small in scale, separates a special operation from an unconventional operation carried out in the conduct of maneuver warfare, such as an airborne assault, or amphibious landing. The second condition is that the operation is designed and has the potential to impose an effect much larger than its investment. Inherent in the condition of high leverage is that this effect is a product of planning and not of happenstance. Finally, no special operation would be "special" if it wasn't different from the main stream of tactics in the context of the conflict. Ergo, a special operation must be unorthodox.

Who conducts special operations? In an attempt to enlighten the reader to special operations, a comprehensive list of potential actors that are available to any group, organization, or state to conduct a special operation has

[1] Arquilla, John, Editor, From *Troy to Entebbe: Special Operations in Ancient and Modern Times*, University Press of America, Inc., 1996.

been established. First, there are elite state-sponsored military units that are highly trained in conducting small unit tactics that fill a void that general-purpose forces cannot. Secondly, trained militia and mercenary groups that are for hire throughout the world are recognized. These mercenaries are, in their own way, elite units based on special training and recruitment. Thirdly, organized crime and terrorist organizations that may be solicited or coerced into working on behalf of a group or state are prime candidates for conducting special operations. Fourth, is the inclusion of any civilian who is a resource and can be commissioned or politicized to organize or act in proxy for a state in the conduct of special operations. Lastly, is the admission that general-purpose forces, when utilized in a definitive manner can also execute special operations. This chapter will conclude with the answer to *why* a power would conduct a special operation.

D. COMPARATIVE ADVANTAGES OF SPECIAL OPERATIONS

In the decision to conduct a special operation, a question arises: To what extent do external and internal constraints and opportunities shape the relative advantages of conducting special operations in lieu of conventional operations. To attempt to understand the decision-making process between the two, we propose the use of two concepts of utility, simple cost-benefit analysis and our concept of singularity. Singularity occurs when there is only one option available or capable of achieving a specific goal.

Utility theory is an intricate weighing of costs and benefits in the context of the decision-maker's philosophy as well as relative resource levels. Context plays a key role in all levels of this method of analysis. Each level is dependent on current capabilities and resource levels. Utility theory is applicable in the concept of military and political decision-making and operational selection. These two concepts are deliberately and subconsciously utilized when making decisions. Cost-benefit analysis can be carried out on many levels: politically, socially, militarily, and economically.

The first level of operation selection is the military level of cost-benefit analysis. The key resources of any military are funding, personnel, high value assets, and equipment. On the cost-benefit scale of personnel investment, a special operation tilts the balance in its favor over conventional applications of force. This pertains to the size of the operation and the advantage of investing small numbers to achieve large effects. Considering the investment of high value assets such as aircraft, ships, or submarines in support of a special operation is also an important issue. This is where much of the potential deliberation is carried out. The threat of loosing a high value asset in order to carry out a small operation is measured against the expected outcome. For example, due to the number of helicopters that the U.S. had in inventory, the investment of helicopters in Desert One was a minimal factor based on the mission's possible success or failure. On the other hand, placing a SSN/SSBN into harms way in order to insert a special operation force could be seen as a risk not worth the best outcome.

The second level of operational selection via cost-benefit analysis is the political level. There are two identifiable constraints faced by a state or organization at this level, internal and external. Internal constraints apply to a state's constituency's responses to executing a special operation. The methods of execution, the expected and resultant outcomes, and the appropriateness of the operation are key perceptions that are analyzed by these audiences. External measures apply to the international response to special operations. These two measures are political and social in nature. This is to say that although a state may utilize a special operation to achieve a political goal, the world population has a voice beyond simple representative politics. This voice can be heard through product boycotts, demonstrations, and direct terrorist actions. Political ramifications, intended and unintended, take many forms ranging from political isolation in the international community to military action by coalition governments. These measures illustrate the highest gains and losses that a state can achieve or suffer in the arena of international relations.

The third level of analysis is the economic level. This takes into account the cost and availability of assets, rather than the operational and strategic value of assets. Economics can decide how much of an investment a group or state is willing to put into equipping a standing army with armored vehicles, tanks, aircraft, and large ships. Special operations forces require only a small percentage of the funding needed for a standing army. However, the bottom line with regard to economics is still the military effectiveness of the armed forces that a state is willing to build and sustain.

Singularity is distinguished by the uniqueness of an operation relative to other available means. Levels of uniqueness are based on the congruence between internal and external constraints and the ability of the forces to conduct the mission within those boundaries. The employer's specifications or the situation's demands are seemingly less tangible. These characteristics of uniqueness qualify an operation as a singular option when it is the only available method of high precision, surprise, speed, relative size, survivability, non-attribution, or success. The ability of a force to "fit" such a tightly constrained operation is characteristic of special operations.

Constraints can be categorized as internal or external. Internally, there are "rules of engagement" (ROE), "commander's intent" along with "specified and implied tasks" that dictate the uniqueness of forces and operational method selected. For example, a power may want to abduct a high-ranking official without causing any large-scale damage, without occupying terrain, all within a specific time period. There is little cost-benefit analysis applied in this decision process, especially in those states or organizations that have standing trained special operations forces. The specified uniqueness called for in this case results in only one option for strategic personnel extraction, a small covert force.

Externally, the defense posture of the opponent, the weather, the terrain, and the timeliness or urgency of an operation can dictate the necessary uniqueness of the operations selected. For example, an opponent that is heavily fortified may let down his guard in the perception of friendly or neutral activity in the area. This would apply well in commercial fishing areas where navies and coastal patrols are active. A special operation may utilize commercial or leisure craft to navigate undetected in plain sight to execute their mission. Further examples can be extended to those trained forces that are comfortable operating in challenging weather elements. Darkness, fog, wind, and high sea states may be deemed too dangerous for operations to be conducted by general-purpose forces. There are external constraints that allow a force or course of action to be the only possible choice. Some of these include the probability of surprise, success, and survivability.

These characteristics of singularity are more pronounced at the higher levels of operational and strategic force employment. This is apparent due to the high leverage that can be brought to bear by a special operation. Additionally, singularity establishes a niche for special operations in the most politicized form of warfare, the use of special force in executing foreign policy. This is considered the highest form of strategic utility.

Applying this chapter's methodology for classification and understanding the utility of operations, will assist in understanding the objectives, and therefore the ways and means of conducting special operations against the U.S.' new attempt at high technology, the ARSHIP.

III. POTENTIAL THREATS FACING THE ARSHIP

Having established the ARSHIP's position as an instrument of United States foreign policy and using defined terminology and methodology as points of departure, this chapter describes potential courses of action that the enemy can use to counter the ARSHIP.

The first section in this chapter proposes that there are five distinct categories of aggressive objectives that would mitigate the effectiveness of the ARSHIP. These objectives will define which courses of action the adversary may take against the ARSHIP. Resultant is a comprehensive list of these courses of action. This chapter then briefly reviews their objectives and potential for employment and success. This comprehensive list will be reduced in Chapter V to only include courses of action that are special operations and the shortened list will be examined in detail. Concluding the chapter is a discussion of mitigating factors that assist or hamper the success of the special operation being conducted. Chapter V will also utilize these factors to critically review the shortened list of courses of action and the defensive ability of the ARSHIP.

A. ENDS OF THE THREATS

Objectives of combating the effectiveness of the ARSHIP are categorized into five possible categories. These five objective assist in defining the possible courses of action (COA) that a state or group may take against the ARSHIP. They are:

1. Kill the ARSHIP
2. Control or displace its geo-location beyond its weapons' range
3. Reduce the rate of fire
4. Reduce the probability of hit
5. Reduce the effect of hits

The first category, kill the ARSHIP, is self-explanatory. It means the *permanent* effect of sinking the ship or destroying critical components of the ARSHIP's structure that cannot be shored up or replaced. There are no variable intensities of this objective category. Terms such as 'soft kill' will not apply to this category. Effects such as temporarily effecting maneuver and firing will fall under other categories.

The second category, control or displace the ARSHIP's geo-location beyond its weapons' range, is a temporary effect that can be broken into gradations of success based on the attacker's conflict requirements. This category includes a wide variety of actions. Possible COA include:

1. Piracy, seizure, and hostage

2. In-port attack on critical nodes
    a. Propulsion
    b. Weapons loading
    c. Manning
    d. Structural damage

3. At sea:
    a. Threat based maneuver
    b. Propulsion damage
    c. Structural damage

The third category, reduce the rate of fire of the ARSHIP's missiles, is again one with degrees of effectiveness. Reducing the rate of fire can be described as reducing to a zero rate of fire, simply reducing the rate below a certain threshold, or reducing the rate from the ominous possibility of near simultaneous launch of 500 missiles at different targets. Potential methods of achieving these goals include but are not limited to:

1. Distraction: To harass the crew or defense escorts forcing
    a. maneuver rather than fire control
    b. damage control rather than fire control
    c. defense rather than fire control
2. Disruption: To act to disrupt the fire control system that is dependent on automation

a. disrupt its external communications

b. disrupt its internal fire control

c. disrupt its use of mechanical VLS

The last set of disruption points identifies critical nodes rather than courses of action.

To achieve a reduced probability of hit, an adversary can effect the ARSHIP or the potential target set. Doing this, the adversary can take preventative, preemptive or reactive measures. Actions the adversary can take regarding the potential target set includes but is not limited to:

1. making critical nodes mobile
2. utilizing camouflage to conceal their position from overhead sensors
3. utilizing false targets to draw fire from real targets

Actions an adversary can take proactively toward the ARSHIP to attempt to effect the probability of a hit include but are not limited to:

1. disrupting targeting communications data stream
2. replace the targeting communications data stream
3. taking control of onboard guidance systems

Reducing the effectiveness of the hit of a missile is purely a preventive measure. This would include target hardening with concrete, tunnels, and bunkers.

B. MEANS OF POSING A THREAT

1. Medium Considerations

An adversary can attack a target through one or a combination of five mediums, air, land, sea, space, and the electromagnetic spectrum. Air, land and sea are the main mediums in which these actors can attempt to operate directly against the ARSHIP. The categories of space and the electromagnetic spectrum indirectly affect the ARSHIP, mostly in the realm of sensors and communications.

2. Methods of Approach

Each of the three primary mediums (air, sea, and land) can be used for direct physical attack. Attacks in a number of different ways including sub-surface, maritime and terrestrial surfaces, and air attacks. Sub-surface attacks are delivered from below the surface of the water. This makes visual and electronic detection and defense more challenging. Electronic detection depends on environmental conditions and capability of the equipment and individuals manning it. Without a means to visually detect the underwater threat; such as air bubbles, water turbulence, or any other factor associated with the attackers' movement; their approach will proceed without warning and catch the vessel by surprise and unprepared to defend the attack.

Maritime surface threats may be easier than sub-surface threats to detect, but they are serious enough to raise concern. The surface contact must be detected and assessed before it is perceived as a hostile threat. A small fishing vessel may be perceived as innocently proceeding about its job when its true intention is to close on the ARSHIP and attack it. The ability to discern between innocent and aggressive movement requires personal and professional judgment before reporting it. This delay in reporting gives the attackers additional time to accomplish their desired intention.

Electronically, discrimination between land and threatening contacts on land is nearly impossible for a maritime vessel. Visual detection of a threat is limited to the range and bearing of vision from an individual's vantage point. Land areas located near waterway and around harbors are usually highly populated. Clutter associated with urbanization, such as buildings and vehicles, can be advantageous to an attacker. Projectiles can be launched from land units as well as an assault force, when the ship is tied to a pier.

Air attacks are not limited to certain lanes of approach. An air attack can come from any direction. This requires 360-degree, 3-dimensional surveillance for detection. Another advantage is that higher speeds can be achieved during an approach than in the other types of attacks. Lower speed approach can also be an advantage to the attacker. Physically, they can fly below the detection equipment's operating azimuth caused by the earth's curvature. Additionally, the equipment may not classify a contact as hostile when aircraft approach below the automated parameters of the search system. Much like the threat of falsely perceived innocence of maritime surface threats, aircraft can proceed along established commercial air corridors then diverting quickly to threaten the ARSHIP.

3. Geographic Considerations

By choosing which outcome best fits their strategy, the adversary will choose the best time and place for their attack. The tactics which potential adversaries will utilize to pursue the ARSHIP will be guided by its location. Any ship has vulnerabilities whether *inport* or at sea. However, the ARSHIP's reduced crew size, CONOPS and streamlined topside are factors that make it more vulnerable than most other surface combatants. Therefore, our analysis of the ARSHIP will be shaped according to the geographic areas of vulnerability, categorized as *inport*, *open ocean*, and *confined water*.

Inport is the area located along a waterway or harbor near a land facility where the vessel is tied to a pier or anchored in the harbor. Factors associated with a vessel *inport* are its positioning and defense posture. At anchor or pierside, the vessel is relatively stationary and cannot get underway without proper engineering preparation. Without prior threat warning, the vessel's defense posture is usually at a lower state than if at sea or operating in the vicinity of known threats. For example, a ship *inport* does not have the necessary equipment operating to detect attack and must rely on sailors working and posted about the deck to observe threatening intentions from an adversary. *Inport* protection and readiness are usually less than when underway.

Open ocean is the operation area that is devoid of land, shallow water, miscellaneous high volume shipping traffic, and other types of hazards to navigation. In *open ocean*, a vessel can maneuver freely and unimpeded. In addition, the crewmembers are active and the defense posture is higher. All vital equipment is properly manned and operating.

Confined water is an area where a vessel is restricted in its ability to maneuver while transiting or operating in the vicinity of littoral waters in which there is "clutter." Clutter is the presence of other vessels, land, shallow water, oil rigs, air traffic, electromagnetic traffic, and other obstacles that become too numerous to track. In *confined water*, the ability to discern friendly from hostile contacts is challenged. The defense posture is higher compared to *inport*, but some radiating equipment may not be allowed to operate due to the close proximity to land. Additionally, individuals that normally man detection sensors may be conducting other duties such as "sea and anchor detail."

C. INITIAL LIST OF POSSIBLE THREATS

Having already addressed where and how a threat will be used, our thesis will now develop a comprehensive list of possible threats. A warship is the weapon platform that carries the weapon system, such as a missile system. The missile system includes all associated tracking and firing equipment as well as the weapon, the missile itself. The missile is used to directly destroy the target.

The threats to the ARSHIP are limited only to the adversaries' imagination and ability to carry them out. Table 1 illustrates our list of the entire spectrum of possible threats. These potential threats are grouped according to their means of attack. The geographic areas of vulnerability are used to discern which threats are capable of operating against the ARSHIP in that arena.

	INPORT	*OPEN OCEAN*	*CONFINED WATER*
Sub-surface			
Submarine		X	X
Mini-Sub	X		
Manned Torpedo	X	A	A
Mines	X		X
Sub-surface/Surface (M)			
Combat swimmer	X	A	A
Surface (Maritime)			
Small Boats			
Recreational/Leisure	X		X
Combat Raiding Craft	X	A	X
Other Small Craft	X	A	X
Surface (Both Maritime and Terrestrial)			
Assault Force	X	A	A
Gunfire	X	A	X
Air			
Helicopter	X	A	X
Planes			
Ultralite	X	A	X
Glider	X	A	X
RPV	X	A	X
Private/Commercial	X	X	X
Parachute	X		
Missile Attack	X	A	X

X = no assistance from additional weapon platform required.
A = assistance from additional weapon platform (aircraft or larger vessel) required.

Table 1. Comprehensive Listing of Threats to the ARSHIP

1. Submarine

Submarines, due to their size and maneuvering constraints, are assets that are restricted to movements in more open and deeper waters. Their undetected approach is impossible in all but deep-water ports and is, therefore, suited to *open ocean* or *confined water* attacks. As a weapon system, the submarine can directly attack the ARSHIP by using torpedoes, the goal being to sink the vessel or render it unable to maneuver and/or perform its mission. In *confined water*, the submarine's ability to discriminate between the ARSHIP and surrounding contacts is a problem it must overcome. The submarine can be used as a weapon platform for other types of attack. Combat swimmers and small inflatable boats can be "locked out" of a submarine at depth or deployed while on the surface. The submarine then is an ingress/egress platform for these courses of action.

2. Mini-Sub

A form of submarine that is suited for *inport* attack is the mini-sub. This smaller version can operate in shallow water, approach a target virtually undetected, and severely damage a target.

Similar to the submarine, the mini-sub is operated by highly trained personnel and therefore, is limited in its use by States that have the necessary resources. The U.S. Navy SEALs operate a mini sub, the Swimmer Deliver Vehicle (SDV), as do the countries such as Britain, Bahrain, and most notably North Korea. Recent press releases have demonstrated the North Korean's willingness to use the mini-sub for ingress into South Korea.

3. Manned Torpedo

Like the mini-sub, a manned torpedo can operate *inport,* and *confined water.* As a weapon, the manned torpedo is similar to a conventional torpedo with the exception that guidance requires a human pilot to align it with the target. The pilot can either eject or ride it until contact with the target and detonation. Many terrorist groups are more than willing to undertake a suicide mission in order to achieve a desired goal. The pilots of the manned torpedo must be properly trained in the skills required to breathe compressed air and to navigate underwater. This will eliminate many smaller, unsophisticated countries and organizations.

4. Mines

Any country or organization can procure and deploy a mine. Mines can float on or below the surface or dwell on the bottom. Sub-surface mines are virtually visually undetectable. Most mines do not have guidance systems to direct them to a desired target. However, smart mines can detonate on certain specific gravities of metal hulls, on certain acoustical signatures, on a specific number of repetitions of either occurrence, and upon command. A general-purpose mine must rely on physical contact to trigger detonation. Neither general-purpose mines nor smart mines are precision strike weapons unless they are in the vicinity of a desired target and command detonated.

5. Combat Swimmer

A sub-surface or surface threat of a combat swimmer can range from a single swimmer to a group working together as a team. The ARSHIP must be immobile to conduct such an operation due to the fact that the maximum speed of swimmer is 2 knots. To function as a weapon, the combat swimmer can attempt to strap explosive to his body and conduct a suicide mission against the ARSHIP. As a weapon system, the swimmer or team's goal can be to attach an explosive charge to the ship's submerged structures or climb aboard. A team has a better chance of neutralizing the crew in order to control the vessel and its payload.

The process of swimming or diving may also be used as ingress/egress methods to reach land or another vessel before conducting an attack. In addition, the adversary approaching on the surface is easier to detect than if submerged. The Viet Cong demonstrated the ability to conduct combat swimmer operations in the Mekong Delta during the Vietnam War. Examples of current elite units are the U.S. Navy SEAL units and the German Kampfswimmers.

6. Small Boats

Boat attacks can be from commercial boats, recreational/leisure craft, combat raiding craft, and other small craft. These types of boats are relatively inexpensive, need little training to operate, and are common throughout the world. Any group or organization whether state sponsored or acting independently can use small boats. A boat can act as a weapon by being loaded with explosives and ram the ARSHIP. A suicide mission like this can be effective and most likely used *inport* or in *confined water*.

A small boat can appear innocent and approach very close to a vessel without raising alarm. This ability can be used not only to conceal its deployment as a weapon, but assist it as a weapon system and platform. In the role of a weapon system, a small boat will likely carry armament that will be used to directly attack the ARSHIP. As a weapon platform, a small boat can deliver an assault force to the ARSHIP or insert combat swimmers in the vicinity of their target.

7. Assault Force

An assault is when the adversary attacks the ARSHIP with intentions of boarding. Their access can be gained via land, sea, or air. The force can be stationed on land or they can reach it from the sea or air by utilizing a delivery platform. To exploit the option of a land assault, the ARSHIP must be *inport* and tied to a pier. The terrestrial assaulting force can get onboard by way of the brow, mooring lines, ladders or leap onboard. In *open ocean* or *confined water*, an ingress method must be used to deliver an assault team to the vessel.

Once they have successfully gained access, they can perform their desired mission. If used as a weapon, the assault may take the form of a suicide bomber that has explosives attached to their body. The assault force, acting as a weapon system, can also place explosive charges in specific locations to damage or destroy vital equipment. The force's objective may be the crew. By controlling or eliminating the crew, the assault force gains command over the vessel and its payload.

The assaulting force's training required depends on the mission and the desired result. Any untrained individual can conduct a suicide bombing. An intricate plan to access the ARSHIP and place explosives on vital equipment or neutralize the crew requires a highly trained force. The training must focus on coordination and knowledge of the target. Many possible methods of ingress may require additional specific training against the ARSHIP's characteristics. The assault force tactic by its complex nature may require elite forces to ensure some degree of success. Recalling *who conducts special operations*, these forces are; elite military units, trained mercenaries, or organized pirate groups (criminal).

8. Gunfire

Gunfire can come from artillery and other larger caliber weapons positioned on land or from a weapon platform at sea. Gunfire requires targeting information for accuracy. It can have a range from hundreds of yards to several miles depending on the caliber and design. The gunfire can be used to destroy equipment and render the vessel unable to conduct intended missions. Many states and organizations possess the ability to procure and fire large caliber weapons. Therefore, the use of gunfire is a threat option that can be used by any adversary.

9. Aircraft

There are a number of different types of aircraft that can pose threats to the ARSHIP including helicopters, airplanes, ultralites, RPVs, and gliders. A helicopter, by flying low and slow, can go undetected by automated electronic sensors. Once close to target, a helicopter loaded with explosives can be a weapon used to impact into the target. As a weapon system, the helicopter can attack the ARSHIP's topside equipment with guns or missiles. It can also be used as an ingress and egress platform to deliver combat swimmers, small boats and personnel into the water within range of the ARSHIP, or an assault force to the deck of the target in order to accomplish the adversary's goal.

The operating range and speed of an airplane makes it a threat to the ARSHIP across all geographic areas. However, smaller types of planes need to operate closer to land. Loaded with explosives, the aircraft can become a guided warhead. Private or commercial planes may seem unthreatening if operating along normal air traffic patterns. Therefore, their ability to not draw attention as a threat can be used to approach the target.

The airplane may also perform as a weapon system. A plane can be used to drop bombs or launch missiles. For ingress, the airplane, as a weapon platform can deliver an assault force. The cost and skill required to procure and train pilots depend on the type of aircraft being used. An ultralite, glider, or RPV require little, while a helicopters and private or commercial airplanes need more training and funding to maintain and operate. These factors make it likely that any organization or state can utilize an air threat.

10. Parachute

Utilizing the helicopter or airplane as a platform, parachute operations can be used only as an ingress method. The parachute can deliver an assault force to the land, water, or directly on the vessel. Due to the difficulty associated with attempting to land a parachutist on a moving target, parachute operations require a nearly stationary platform, limiting the parachute assaults to only when the ARSHIP is *inport*. In addition, the parachute can be used to deliver boats into the vicinity of the target to attack a vessel operating at long ranges from land. Once onboard the team operates as if delivered from any one of the other mentioned delivery methods. To conduct such an operation, minimal training is required. Parachute training is available from any source and does not need to take place in a formal military environment.

11. Missile Attack

Launched from a weapon platform or launched from a land site, a missile attack utilizes the assistance of a rocket or turbofan motor as a means to deliver a destructive warhead to the target. Missile range determines the need for a weapon platform to approach within its range in order to be effective. As a weapon, the missile can have internal targeting and navigational systems or it may require external inputs for terminal guidance. Either method relies on the ability of the adversary to properly target the ARSHIP in order to assure a direct hit. Like guided bombs, missiles would rely on a guidance system for terminal guidance. A missile's radar would be defeated by the ARSHIP's stealth qualities. A sophisticated and advanced missile is expensive to obtain and like all high priority intelligence keys, will be tracked at higher government levels. Another possible missile threat is a hand held missile system and intelligence sources are important in providing possible warnings of their procurement and use by non-state groups and organizations. Other types of projectiles, which are harder to track, are rocket propelled grenades (RPG) and recoilless rifles, which could be considered as a missile threat when, launched from a proximate platform.

D. MITIGATING FACTORS

1. Aggressor's Personnel and Equipment

There are mitigating factors that affect the defense of the ARSHIP to these threats. They are the limitations of an aggressor's personnel and equipment, the environment, and the constraints of a defensive posture for the target vessel.

Physical limitations of the aggressive personnel and the equipment employed characterize the one mitigating factor category. Personnel considerations are considered mortal in nature. They depend on physical conditioning, physical exposure to the elements and hazards to life. To execute an attack, the individuals assigned the task must have the ability to successfully complete it. The physical demands of swimming, diving, boating, packing or running extended distances require physically fit individuals. As seen with a combat swimmer or air threat, the proper equipment is required along with associated training to operate it. The equipment may be expensive, possess advanced technology. The ability of an adversary to possess, maintain, and operate some types of equipment may not feasibly be within their means. The equipment alone will not make an attack succeed, but its utilization by a properly trained individual make it an asset.

2. Environment

The next factor associated with a threat's effectiveness is the environment. Environmental considerations include illumination, weather, and medium characteristics. Elite units train in harsh conditions so as not to be limited by their constraints. However, there are still environmental properties that will always adversely effect humans and equipment.

With respect to illumination, the time of day or night has an influence on the threat's ability to succeed undetected. Additionally, during the day, crewmembers on the intended target are more likely to be alert and responsive to multiple threats. Darkness can completely conceal the approach of an adversary or severely impair the target's ability to detect several threats. The adversary's ability to operate without the aid of illumination also requires coordination, training, and additional equipment such as night vision devices. The ability to attack at night falls into the realm of special military elite units but does not exclude others from operating in it.

Weather includes visibility issues, sea state, and high winds. The factors affiliated with weather affect both the attacker and target. For every positive effect associated with it, weather has a counter effect that may provide addition obstacles to overcome. Visibility is diminished with the addition of precipitation such as rain, fog, or snow. Precipitation can degrade the ARSHIP's use of radar, visual sightings, and other forms of detection for defense. While it conceals their approach, the attacker must be able to discriminate between the target and other unimportant contacts in order to succeed.

In addition to precipitation, sea state is adversely affected by bad weather. A higher sea state makes tasks such as small boat employment and surface combat swimmer more laborious and difficult. The sea state can also assist the stealth technology of the ARSHIP, by providing more 'noise' to passive and active electronic sensors. High winds increase sea state, but are not necessary to produce higher sea states. Additionally, higher wind speed can oppose the adversary's use of small aircraft and the parachute. An ultralite, glider, or RPV also, may not be able to operate under these circumstances.

The characteristics of the medium in which the attacker must transit to reach the target are ambient temperature, and ecological activity. Cold temperature can adversely affect an individual when exposed for prolonged periods. Hypothermia in cold water and frostbite in cold air are results of extended exposure. In addition, cold temperature can dull the senses and delay reaction time. Protection from extreme weather is required for anyone wanting to operate under these conditions. Bioluminescence can expose the approach of a submerged combat swimmer, mini-sub, or small boat and alert the target. Wildlife may use the approaching medium as a habitat area and their disturbance can result in a commotion that the target can detect. An example would be an assault force approaching over land that startles a flock of birds on the ground, announcing their presence.

3. Constraints of a Defensive Posture

The defensive posture of the ARSHIP will greatly affect the likelihood of a successful attack. Defensive posture is characterized by the readiness of the crew and vessel, the current position, speed of advance and assigned tasks.

The speed of advance for the ARSHIP eliminates those assaults that require a nearly stationary target or one that travels at relatively low speed, 5 knots or slower. The higher the target's speed, the harder it is to gain access to the vessel while underway. In this respect, the vessel's speed can be used as a form of self-defense.

The ARSHIP's position also dictates the type of attack that can and cannot be made. Furthermore, its position in *open ocean* will allow it to be more defensible as well as reducing the likelihood of less robust threats. *Inport* and in *confined waters* inherently reduces the maneuverability of the ARSHIP, making it a sitting duck.

The state of readiness of the vessel and crew can be key factors for success. Readiness is a factor that is based on the perceived likelihood of direct confrontation against an adversary. The level of readiness increases as the political, strategic and tactical situation with an adversary moves across the spectrum of conflict from peace to war (graphically left to right, Figure 1.). On the farthest left side of the spectrum is peace followed by Military Operations Other Than War (MOOTW). MOOTW consists of Humanitarian Aid (HA), Non-combatant Evacuation Operations (NEO), and any other similar type of operations conducted by the U.S. military throughout the world. The next stages of the spectrum refer to Deterrence, Defense, and Compellance (War) in that order. War consists of Lesser Regional Conflicts (LRC) and Major Regional Conflicts (MRC).

Along with the escalation of the level of conflict, the proximate range from the perceived threat further influences the readiness of the crew and vessel. The level of readiness will increase as they approach weapons' ranges of an adversary. In Figure 1, the top graph shows how the level of readiness fluctuates along the spectrum of conflict. The bottom two graphs show the difference in readiness according to the range from the threat. The farther away from the perceived threat a vessel is, the lower its readiness level will be relative when it is closer to the threat. This difference will continue until a point is reached along the conflict spectrum when attack is possible anywhere and at any time.

For the two readiness curves to meet, an adversary must have the ability to launch attacks that can reach U.S. assets globally. This state of higher readiness cannot be sustained for prolonged periods and will wane over time. Therefore, as it ebbs, the heightened state of readiness is no longer a factor in the vessel's favor and becomes a vulnerability. The crew can' become desensitized to constant warnings and authentic cues of an actual attack may not be heeded.

Any advantage associated with readiness is in the attacker's favor, because he decides when, where, and how he will attack. The target's state of alert can be heightened if warning is given in advance. The readiness level is also dependent on the ARSHIP's geographic position. As shown in Figure 2, the readiness curves differ for *inport*, *open ocean*, and *confined water* operation.

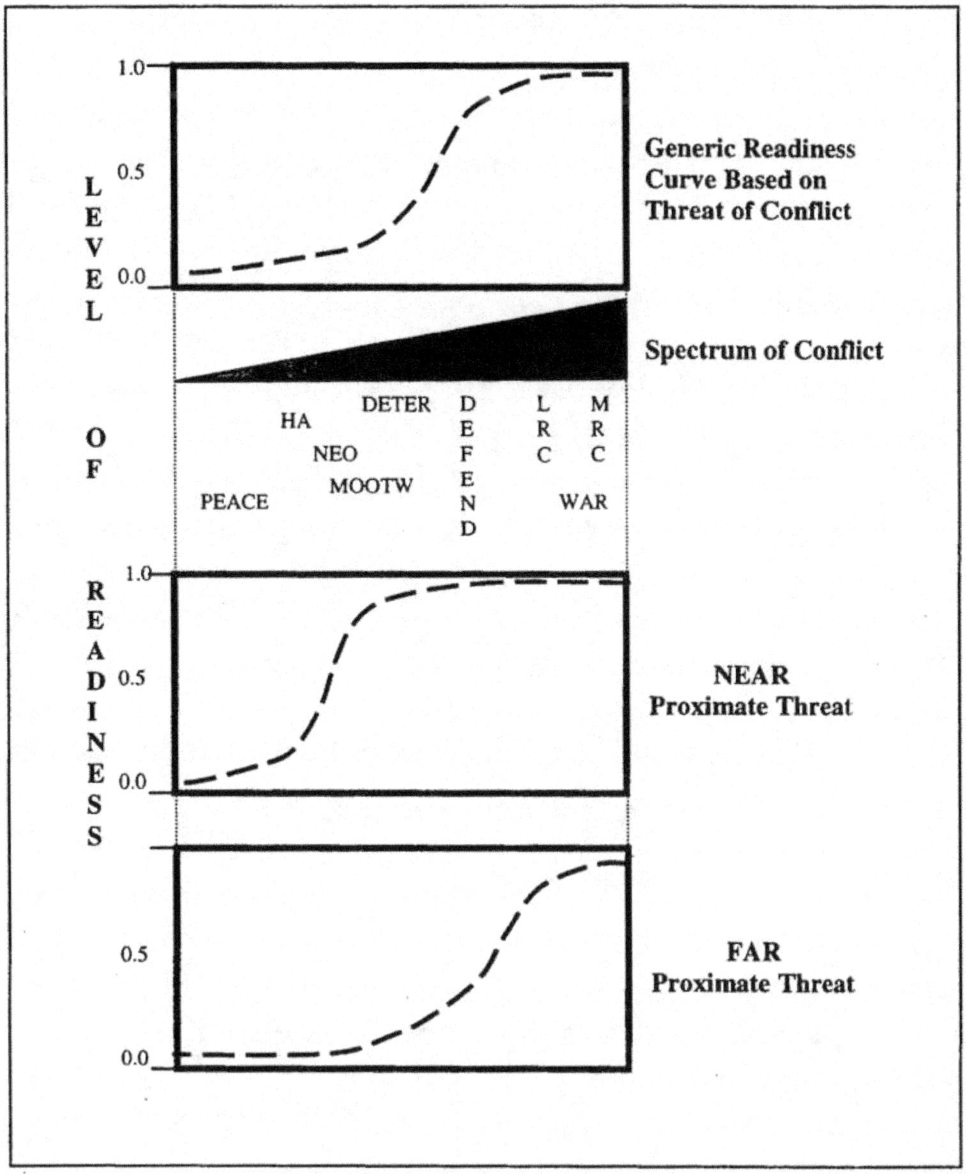

Figure 1. Readiness Curve Across the Spectrum of Conflict and Proximate Distance to Threat

Inport the preoccupation of the crew to perform routine maintenance or repair equipment along with the crew's liberty ashore moves the curve to the right.

This decreased readiness compared to the other geographic areas affords a window of opportunity for an adversary. Operating in *open ocean*, the crew is aware of their surroundings as far as the equipment can detect. Equipment is operating and manned, and the crew is focused on the tasks at hand. Levels of readiness are further shifted to the left when operating in *confined water*. Our definition of *confined water* explains this shift. The need to monitor the surroundings for maneuvering places the crew at a higher readiness level than other geographic areas. An adversary who understands the subtle shift of readiness among the geographic areas can use it as an advantage to plan and execute an attack.

Like the crew, the ARSHIP itself must be at a high state of readiness. The limits associated with the equipment affect the ability of the ARSHIP to respond to possible threats. Without special search radar, a threat built of composite material may not display a radar signature and the low speed of the threat may allow the equipment to disregard it. Therefore, the system may not recognize nor display the threat to the operator.

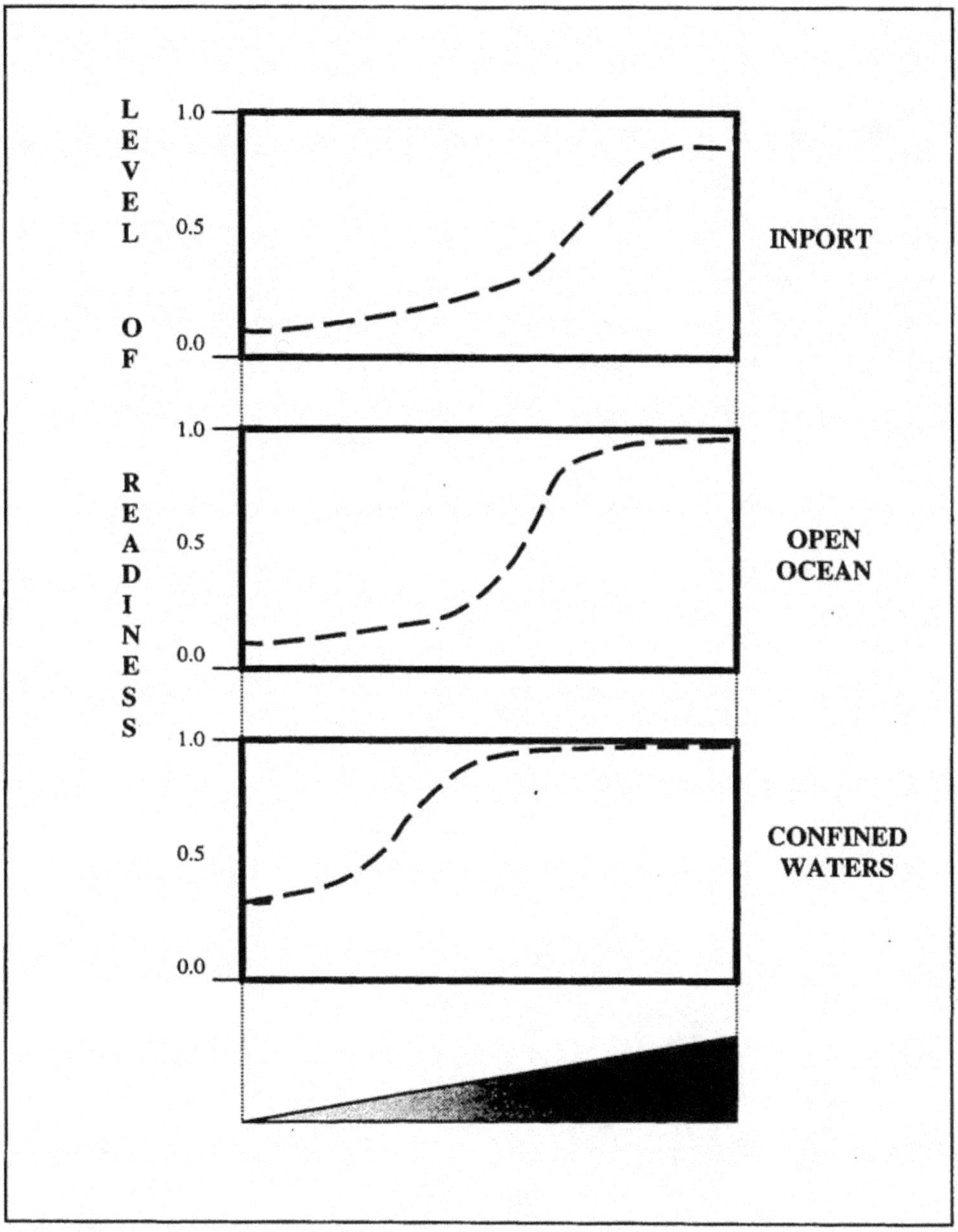

Figure 2. Readiness Curves Across the Spectrum of Conflict in the Three Geographic Areas

A key part of equipment readiness is the person operating it. The reliance on human factors requires an individual to be vigilant and most importantly properly trained. The training involves knowing how to properly operate the equipment. Every crewmember must be properly trained and familiar with all weapons and equipment available on the ARSHIP for self-defense. The weapons and equipment are only as good as the sailors using them. More than the ability to operate and repair the equipment is needed. The skill and experience of understanding what to look for is a critical part of the equation. In addition, the threat must also be reported in a timely manner.

If integral equipment is not operating, a threat may approach undetected. Often while *inport*, radar and other detection equipment are not manned or operated. The same can be said for weapon systems. During *inport* periods, a quarterdeck watch is posted along with a roving watch. The remainder of the crew is conducting maintenance, sleeping, or on liberty. If conducting "Sea and Anchor Detail" operations in *confined water*, crewmembers may be too occupied to properly monitor detection equipment.

Unlike transiting to port through *confined water*, the daily routine associated with *open ocean* operations can become monotonous and dull the senses. The designed CONOPS of the ARSHIP does not allow much slack to break the continuous schedule day after day for extended periods.

As the vessel and crew's readiness can fluctuate, the leadership that dictate the heightening of strategic and tactical readiness are also influenced by events taking place around them. The command elements, which usually think in terms of conventional threats, evaluate the situation and order the perceived level of sufficient readiness. As the conflict escalates, conventional warfare thinkers prepare for possible attack. As illustrated in Figure 3, the area above the readiness curves represents the latitude that potential attackers have relative to their target. The reason for the curves never reaching full readiness lies in the inability to properly prepare for all possible attacks, conventional or unconventional.

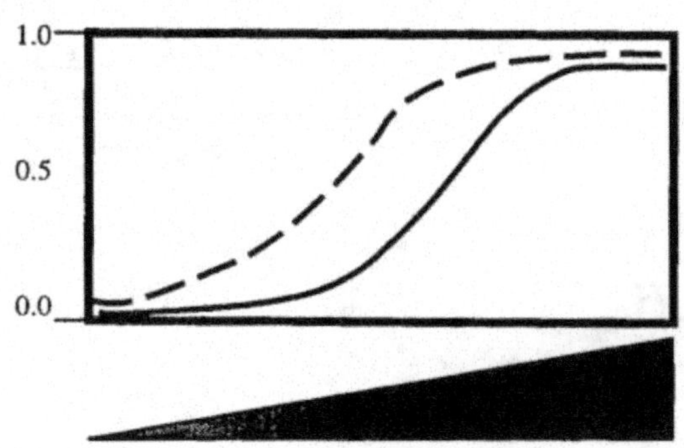

Figure 3. Actual Readiness Curve Versus Perceived Readiness Curve Across the Spectrum of Conflict

In Figure 3, the lower readiness curve depicts how conventional thinking will increase readiness. This means that the threats that might have been effective earlier in the conflict spectrum subsequently become challenged. This does not eliminate them as options, but they will face additional obstacles. The conflict moves from left to right across the spectrum. The upper curve represents how readiness should increase with the addition of thinking of unconventional threats as a feasible threat option. This separation between conventional and unconventional thinking forms an area between the curves.

Using the Figure 1 discussion to frame the situation, a conventional mindset only acknowledges the range of conventional weapons as the threat. For unconventional threats having unlimited range, the "distance" from the threat is not as great as perceived by the conventional leadership. This "distance" becomes an advantage the adversary can exploit through the utilization of long range special operations.

E. PREEMPTION AS A THREAT

Understanding why a state would conduct a special operation against the ARSHIP shows how such an operation fits into the context of the conflict. In times of relative peace, the political utility of a special operations mission is higher than during an actual armed conflict where the special operations would have tactical utility. Context poses issues that are regularly forgotten in traditional U.S. military thought. Referring back to the section on political utility and singularity, and coupling those theories with the issues of readiness across the spectrum of conflict, we suggest that a preemptive or preventative strike is one of the greatest threats to the ARSHIP.

Prevention is a form of non-attributable aggression prior to a conflict. This is a viable option that is executed under the premise of deterrence and defense. In the case of the ARSHIP, it is suspected of participating and playing a key role in a forthcoming conflict and targeted on that premise. This targeting can be accomplished long beforehand, dependent on the time to repair or replace it, thus preventing the ARSHIP from being utilized effectively in the conflict.

Preemption is aggression utilized at the outset of conflict. This kind of strike would be utilized to initiate hostilities based on a key component of the U.S.' defense or retaliation forces being neutralized. The attacker would want to employ a one-time surprise to take out key nodes in the U.S. military potential and exact the highest toll. The purpose is to raise the probability of success of the future engagement.

IV. POTENTIAL SPECIAL OPERATIONS THREATS TO THE ARSHIP

By using the definition of special operations the previous list of possible threats has been refined. The elimination of threats, such as mines, artillery and missiles, and all large pieces of military potential becomes apparent as one distinguishes conventional from unconventional. These threats can still be used as the delivery platform to position an unconventional threat within attacking distance. However, we will only focus on threats that can be used directly against the ARSHIP. The threats this chapter focuses on are: combat swimmer, mini-sub/manned torpedo, assault force, boats, and aircraft.

Chapter IV is organized in a way to illustrate precise operational options against the ARSHIP's defense, and illustrate their effectiveness and substitutability for other courses of action. This organization begins by outlining the characteristics of the operation's attack modes, its objectives (the critical nodes of the ARSHIP) and the means of meeting those objectives. The center section of the outline elaborates on the mitigating factors limiting the aggressor, the environmental constraints, and the defense constraints the specific course of action is faced with or can exploit. These will be put into perspective in the third section of each operational outline. There the operation's likelihood of success and employment, its probable effectiveness and its utility relative to other operations are examined.

A. COMBAT SWIMMER

1. Modes

SUB-SURFACE: This mode utilizes divers to approach the ARSHIP under the water. A team of divers can approach as a convoy or the team may be broken into pairs to reach the target ship on their own, so as not to become entangled with others in the dark waters. A classic example of a combat swimmer mission is the U.S. Navy SEAL mission during Operation Just Cause in Panama that destroyed Manuel Noriega's yacht. Although that mission was able to destroy the target, the size of charges that divers can carry limits the amount of damage that can be exacted on the ARSHIP.

Divers need to be launched from some sort of platform. The platform needs to bring the divers within swimming range of the target. This can be done from land, boats, or air but must be unobserved. A parachute jump or helicopter cast can effectively bring the divers or their subsequent small boats to within close proximity to their target area.

Combat swimmer operations would normally be carried out at night. However, murky waters of harbors may allow divers to approach the ARSHIP undetected in daylight. The divers breath compressed gases that allow for an oxygen rich mixture to circulate through the lungs and a carbon dioxide scrubbing system. Most re-breather closed circuit scuba units use granules of a substance that absorbs CO^2 and allows the volume of unconsumed O^2 to loop back into the system. This allows extended duration dives up to four hours. Elite forces utilize the closed circuit scuba equipment for insertion and combat swimmer missions for this reason because there is no bubble trail on the surface following the divers.

A combat swimmer mission, dűe to the speed limitations of a human, can only be accomplished while the ARSHIP is *inport*. The mission must also have sufficient intelligence to be able to locate the ARSHIP, its pier, slip or anchorage. This is important because the combat swimmer utilizes charts and tide tables to plan the route to reach the target from their submersion point. The divers go from point A to point B underwater utilizing tidal current vectors and swimmer vectors to calculate an underwater compass course. This compass course is only effective when the launch points are exact, the target is where it is supposed to be, and the two competing vectors of swimmer advance velocity and current velocity are predicted accurately. The stealth of this depends on remaining submerged throughout the mission. Coming to the surface to "peak" and see if you are on the right track is extremely dangerous in hostile *confined waters* with active patrol boats.

SURFACE: Surface swimmers have less of a tactical surprise advantage due to the fact that there are heads or other appendages breaking the surface of the water allowing for visible detection. Surface swimmers, traveling alone or in an organized team need to be launched from somewhere near the target. However, surface swimmers may have similar objectives as the submerged divers.

2. Objectives

There are two possible objectives of the combat swimmer mission. The first is to reach the ship from the water, place explosive charges on critical components, and swim away. The second is to utilize the stealth of the water to approach the ARSHIP in order board the ARSHIP from the water's edge.

DAMAGE: The objective of causing damage is straightforward. It involves surreptitiously attaching explosives to the outside of the ship on critical nodes.

Due to the relatively small size of explosives that can be hauled in by divers, the critical nodes of a steel-hulled warship are small in number. It is physically infeasible to carry enough explosives to the target via divers to 'kill the ARSHIP.' However, the critical nodes below the water line do allow for enough destruction to effect its mobility. Nodes such as the struts, which hold the shaft, can be destroyed so the weight of the shaft will bend itself, rendering the ARSHIP immobile until it can be dry-docked and repaired. Likewise, nodes such as variable pitch propellers can be warped so as to produce cavitation or high speed warbling of the ARSHIP. Both negate its stealth advantage. The packing glands are weak points that make the hull more easily breached. Shaped charges can pierce the outer hull but must be placed outside of critical electrical or propulsion spaces that are located near the external hull to have the greatest effect. All of these targets produce short-term effects that only delay the availability of the ARSHIP or its speed of advance toward an area of employment.

BOARD: The concept of the assaulting force will be discussed later in this chapter. There are similar concepts to the combat swimmer and the waterborne assault.

These will be emphasized here whereas their objectives will be discussed in Section C.

3. Mitigating Factors

Aggressor Considerations: There are many physiological constraints when divers breathe high percentages of oxygen under pressure. "O^2 toxicity" is the diagnosis for prolonged exposure to compressed Oxygen that results in nervous disorders. Unclassified but indicative numbers warning of O^2 toxicity range from 4 hours exposure when breathing O^2 at 20 feet depth to 10 minutes of exposure at fifty feet. Death can result, especially if irresponsible diving has gotten the divers within proximity of their target but the medical emergency has forced them to surface. By doing so, they have exposed the operation, costing their own lives and threatening those of the other divers. This is not a common ailment, but is a factor in the training of responsible diving tactics.

Physical limitations of the divers include the amount of explosive they can carry, the speed at which they can swim, and their exposure limits to extremely cold water. The speed of the swimmers makes it impossible for them to achieve contact with the ARSHIP while it is underway. Additionally, if the water currents are such that the swimmers cannot overcome opposing currents they will never reach their target. The extreme cold of the water will dictate the type of thermal outerwear worn to protect the divers from hypothermia. However, water exchanges heat 80% faster than air, and prolonged exposure regardless of thermal protection will eventually endanger the divers.

Characteristics of the equipment utilized for this mission profile can both add and detract from the effectiveness of the mission. One such characteristic is the compressed O^2. Normal scuba tanks hold enough air for a one to one and half-hour dive. Compressed O^2 tanks circulating through a scrubber can extend shallow dive times to four hours. Inert gases mixed with O^2 extend those physiological limits beyond four hours. This translates to a difference of underwater movement from 3000 yards to a possible distance of 7500 or 8500 yards. Even more influential is the distance the divers can swim on the surface from their launch platform to their submersion point.

Another limiting characteristic is the explosives and the detonation devices carried. Clock detonators placed on 5-10 lb. shaped charges are the community standard. Each diver can carry two of these. A sizable force could send 5 pairs to a target. This translates to around 75 lbs. of explosives placed at critical points on the ARSHIP.

Sounding impressive and actually being effective are two separate things. One substitution would be to carry traditional haversacks of explosive, which would double the amount of explosives placed under the ARSHIP, but haversacks do not have attaching mechanisms nor are they shaped charges. This quantity is not enough to break a keel and permanently sink the ARSHIP.

Environmental Considerations: The primary medium concern of these types of missions is exposure time in the water and the amount of thermal protection needed to keep the body warm for prolonged mission duration. Beyond the normal constraints of exposure, swimming in confined areas poses issues of water clarity and silhouette visibility from the surface. This is only an issue for surface surveillance during the day or a full moon where the water is clear and the bottom is contrasting the divers. One phenomenon of many bodies of water is bioluminescence; organisms in the water that glow when excited by motion. A swimmer, whether on the surface or submerged, may have the misfortune of exciting them by disturbing the water. This glow, a bright neon color, can vary in intensity. If bioluminescence qualities are high, the glow could easily be seen from long distances if near the surface, or deeper if the water is clear.

In the discussion of biology, we must consider that the swimmers share a medium in which they are no longer on the top of the food chain. Mammals and fish in the water can react adversely to the presence of a foreigner. Their reactions can simply be the unusual barking of sea lions that would be considered as early warning defense, or it could mean becoming a hors d'oeuvres for "Whitey."

In bad weather, divers can get seasick when submerged in high sea states. The constant swaying without a reference plane can cause vertigo and nausea.

Defender Considerations: The readiness of the personnel on the ship plays a small role in defense from combat swimmer operations. The primary concern in this category would be under conditions of watch standing. The watchstander must diligently be scanning the waters near the ARSHIP and monitoring any defensive measures they have set out to repel a swimmer attack. This is a tedious and boring proposition. Likewise, the principle of a swimmer attack in a "friendly" port would also seem, to a watchstander, unlikely, causing him to let down his guard.

The ship's equipment does little the defense against a combat swimmer attack. If the ARSHIP design has one, its sonar could be radiated regularly at high power to disturb the inbound divers. The amount of power that a sonar dome can emit into the water is enough to destroy eardrums and be especially uncomfortable for all semi permeable cavities of the human body. Suctions and discharges also pose interesting quandaries for divers who must travel up or down the length of the ship during the course of the mission.

A ship that anchored or moored to a buoy poses a more difficult navigational problem to the divers than a ship that is pier-side. There are few geographic references that divers can use to verify their path. Pylons, navigation buoys, quay walls, and pier lighting are excellent navigational aids to divers under the water who are attempting to find a target without breaking the surface to look. Additionally, the positioning of the ship in a well-protected and trafficked waterway poses a threat to the divers of being run over or spotted if they experience difficulties and must come to the surface. Remembering the issues of human limitations due to the speed of currents, the ARSHIP could be strategically located near river inlets or power plant cooling discharges to create a current that is impassable.

4. Utility and Substitutability

Cost and Asset Availability: The personnel required to accomplish this mission profile need to have the training to dive using closed circuit scuba, navigate underwater, handle underwater explosives. The closed circuit scuba system can be found commercially and there are many amateur and professional divers who navigate well underwater. If these personnel were to be commissioned and given the mission profiles of what to carry and where to put it, they could successfully attempt this mission. Elite units throughout the world's navies train for this mission regularly. The extent of their training may be costly, but most definitely increases their probability of effectiveness to near perfection.

The cost of a closed circuit scuba rig is around US$ 4000. A team equipped with enough explosives to do damage to a ship would then run about $60,000 to stand up. Additionally, the boarding party would be higher due to the number of personnel needed to secure the objective. This is not an inordinate amount of money relative to the likelihood of success due to the clandestine nature of the mission.

Benefit: Due to their clandestine nature and the effectiveness of their attack the combat swimmer mission has the characteristic of singularity. The effectiveness of only altering the ARSHIP's geographic location for specified time will alter the weighting that the decision-maker will ultimately lay upon its relative utility. However, given the means of infiltrating a hostile port, or distanced port with lower defenses, this mission is allows for total anonymity, total clandestine success, and it is the most dangerous threat to the ARSHIP in any port.

B. MINI-SUBS AND MANNED TORPEDOES

1. Modes

Mini-subs and manned torpedoes allow underwater transportation at speeds and distances unparalleled by kick-stroke-and-gliding (swimming). This allows attacks to be launched covertly from platforms at greater distances and with greater destructive payload than the combat swimmer. The mini-sub and the manned torpedoes can venture into waters that are shallower and less maneuverable than those of conventional submarines. Many of the mini-subs and manned torpedoes are open to the water so the drivers or transported personnel must breathe compressed air and wear thermal protection.

2. Objectives

Transport vehicles can bring explosives to the ARSHIP from greater distances. Their advantage relative to the combat swimmer, besides distance, is the amount of explosives that can be delivered. These vehicles either can trail an explosives package behind them or be laden with the explosives as in the case of the torpedo. The objective would be to place the explosive packages in a position where they would 'kill the ARSHIP' or do the most damage. The keel of a ship can easily be broken when a mini-sub trails 50 to 100 times as much explosives to the target as the combat swimmer. The manned torpedo can do the same with a terminal warhead that has carefully navigated to its target. The discussion of equipment limitations outlines the relative disadvantages the submersible transports have to combat swimmer operations partially offsetting this asymmetrical advantage.

3. Mitigating Factors

Aggressor Considerations: The same constraints present to combat swimmer are present here, except for the relative distances and speeds of advance. There are training constraints regarding navigation systems that are more sophisticated than the simple compass and chart work of the combat swimmer. All in all, the human factors involved with direct control and precision targeting are aspects that favor this mode of explosives delivery.

The fact that there is a rapidly moving transport at depth causes certain safety issues to arise concerning dive physics and Pulmonary Over-inflation Syndrome (POS). These considerations can be trained for to near perfection.

However, the utilization of poorly trained personnel can result in mission failure and death.

Breathing equipment used on these transports by the crew is similar to that used by the combat swimmer. The speed of advance and mission duration may be dictated by the physiological limits of the crew or by the amount of fuel carried. Speed of advance, the transport's size, and noise-producing electronics on board make these transports detectable to passive and active sonar. Additionally, maneuverability is decreased compared to the combat swimmer. This is important in high traffic areas and tight geographical features. The size of these transports also produces a larger silhouette in clear water.

Environmental Considerations: Daytime operations are restricted by clarity of water and possible detection from above the surface. Moonlight and water clarity affect nighttime operations to some extent. Sea state plays a larger role with these transports due to their susceptibility to swells and their use of ballasting. High sea states and large swells can cause depth changes at a rate up to 6 feet per second. This rate is dangerous to the diver breathing compressed gasses. (The suggested safe rate of ascent to avoid POS is one foot per second.)

Exposure characteristics are the same as for the combat swimmer. Water clarity is more crucial do to the size and speed of these transports. The mission is not immune to biological constraints, and bioluminescence becomes more pronounced with speed. The threat of being a shark's lunch is however, greatly decreased.

Defender Considerations: The defender considerations in the case of manned torpedoes and mini-subs are much the same as for the combat swimmer. However, the active or passive sonar if turned on and monitored in port would have a better than zero chance to detect incoming mini-subs or manned torpedoes.

If the mission was designed for terminal guidance of a warhead, then the two transports could effectively target the ship while it is underway. Optimally, this would only be done when in *confined waters* and when the target has been positively identified. Higher speeds would cause 'louder' sonar signatures of the incoming warhead and possibly reduce the effectiveness of that type of mission.

The mini-sub and manned torpedoes are less affected by swift currents. Their navigation systems allow them to find their targets more easily and precisely without needing physical references. High traffic lanes, as discussed earlier, impose navigational hazards to all submerged threats.

One large constraint is the movement of the submersible transports to the region of the target. It is easier to deliver divers and their scuba equipment to a region than a large torpedo or mini-sub. The ability of the attacker to do that will play greatly in mission selection.

4. Utility and Substitutability

Cost and Asset Availability: The personnel constraints are similar to those for a combat swimmer. However the likelihood of procuring commercial mini-subs is low. Drivers could come from the pool of experienced divers and would only need to be trained on this submersible vehicle. Relative to the combat swimmer, this mission would require less personnel to have the same or higher destructive effects; one to six man units.

Many countries have hollowed out torpedoes and added controllable steering and speed to enter this mission specialty. The crudeness of the transport will make more demands on the crew. The transport is difficult to acquire more so due to required navigational equipment rather than the submersible vehicles. However, commercial and military sonar can be easily adapted to create an onboard navigation system.

Benefit: Mini-subs and manned torpedoes are another means of clandestinely infiltrating a waterway not accessible by conventional submarines to exact a toll on the ARSHIP. The quality of singularity of these methods is due to the fact that this mission can actually deliver enough explosives to 'kill the ARSHIP.' The main differences between the combat swimmer and the submersible transports are the embedded internal and external constraints. These are based on the objective of the mission, the availability of the assets, and the underwater geography of the mission area. If the objective is to kill the ARSHIP, the submersible vehicle is the only method to bring explosives to the ARSHIP underwater.

C. ATTACK FORCE

An attacking force can come from all three traditional mediums, sea, air and land. This section will address all three modes of attack as well as their relative advantages and disadvantages. One mode of launching an assault will not be discussed in this section, the combat swimmer assault. The reason for its lack of discussion is due to its dual nature. First, the ingress of the combat swimmer assault is the same as an explosives attack. Secondly, once on board the force acts like any other, with the same objectives. The only difference is that the ARSHIP must be stationary and the swimmers climb from the surface of the water vice from a boat.

1. Modes

SEA: An attack force directly launched from a vessel inspires thoughts of swashbuckling and piracy. There are distinguishing characteristics of these courses of action. The primary distinguishing characteristic of is the size of the vessel coming alongside to launch boarders assaulting the ARSHIP. The three primary categories are small boats, large boats, and ships. Small boats are craft 30 feet in length or less. Large boats are from 30 to 90 feet, while a ship is larger than 90 feet. Sub-categories of these are distinguished by the type of the craft; commercial, leisure, or military.

The sea threat is one where boarding can only take place when direct contact of the two vessels occurs. The boarding team can climb aboard the ARSHIP via a flexible hook ladder to carry out their mission. This boarding can take place in all of the geographic areas. The sub-categories have large ramifications as to when the approaching attacker will be distinguished as a threat. On first glance, smaller vessels would warrant less alarm as they approach due to the cognitive bias of the observers on board the ARSHIP. Likewise, leisure and commercial vessels would warrant less of an immediate response. Therefore, it behooves the aggressor to select those vessels that would proceed closest to the ARSHIP without alarming the crew.

AIR: The air threat is very similar to the sea threat in that a delivery vehicle carries the assault team to the ARSHIP. Logically, this must be an aircraft with vertical lift capability, a helicopter. The assault team can be brought to the deck of the ARSHIP by two means, the helicopter landing on the deck or using ropes to execute a controlled descent from the aircraft to the deck. Modes of descending from a helicopter to the deck are fast-roping and rappelling. Fast-roping involves sliding a team down a 3 inch thick rope without hard connectors or safety harnesses. A team member wears insulated welder's gloves to hold onto the rope as it passes through their hands on descent. The control comes when the team member squeezes the rope prior to his collision with the pitching deck. This slows his downward momentum allowing for a controlled landing. A well-trained assault force can land 20 members from 60 feet using only one rope in about ten seconds.

With rappelling there are constraints of harnesses and hard attachment points to the rope that must be secured before descending and discarded upon reaching the deck. Depending on the number of ropes, a well-trained team can land 20 members from 60 feet in about 90 seconds. The advantage of rappelling is the longer length of possible descent. Fast-roping is only feasible for distances of less than 75 feet, whereas rappelling allows landing from heights up to 150 feet.

The use of gliders or parachutes to land a team onboard the ARSHIP is likely only when the ship is in port. Execution of a mass landing of parachutes from any drop height onto the ARSHIP requires great skill and training. This would only be by maneuverable parachutes, such as those commonly flown when sky diving and not static line deployment. Gliders like those used in World War II would be effective during nighttime. However, being able to land one or more gliders specifically onto a target the size of the ARSHIP would be nearly impossible.

LAND: Land assault on the ARSHIP is possible when the ship is tied to a pier. The land assault would entail the assault team either forcibly entering, clandestinely entering, or overtly entering a naval base, and then proceeding to board the vessel. The assault team would need to strike swiftly with precision and not expect to stay long in the target due to follow-on security forces being alerted. The team must utilize the element of surprise before hostilities begin. This method of insertion onto the naval base could utilize small boats, parachuting, swimming, or over land navigation to reach their objective. Overt ingress to the target can be accomplished under the ruse of pretending to be 'indigenous' personnel or activities that do not look out of the ordinary. A force could hide in a delivery truck, or wear local military fatigues executing a routine entry onto a base. Red Cell activities of SEAL Team SIX in the 1980's exposed the vulnerabilities of the United States' Naval Bases. They utilized all of the tactics described above with great success.

2. Objectives

Whether an assault force ingresses from sea, air or land, its objective is the same. The mission is dictated by the strategy, recalling the list in Chapter III of the five effects that an attacker would want to achieve on the ARSHIP. These mission drivers dictate the size of the force, the amount and type of arms carried by that force, and the specific actions at the objective.

The first objective of any assaulting force would be to neutralize the crew. First and foremost, the crew must be killed or detained. Deadly force would require the assault team to use hand-held weapons or quick-acting deadly chemical/biological agents from which the team was protected. Non-lethal force would require a more permissive environment than is normally associated with an U.S. warship.

To sink the ARSHIP, the assault team would most likely need to place huge amounts of explosives on board the vessel in strategic places. This would take time and extreme amounts of exertion. The amount of explosives necessary to sink a warship would be difficult for an assault team to bring onto the vessel without follow-on delivery. Each member of the assault team would be required to carry at least 35 lbs. of high explosives. This would severely inhibit on the mobility and efficiency of the team.

To affect the ARSHIP's geo-position, propulsion or steering machinery would need to be temporarily or permanently destroyed. The extent to which the United States Navy could repair or replace the damaged equipment would determine the effectiveness of that mission.

A major concern is that the ARSHIP could be hi-jacked and piloted by the assault team to their position of choice. Remotely piloted or not this would require the destruction of communications equipment. For those instances where a team was not able to seize the crew, they may enter into the labyrinth of passageways and proceed internally to objective points. If the team were to be able to breach into internal passageways they may be able to head directly to critical nodes, such as propulsion, steering, and electrical generators.

To affect the ARSHIP's rate of fire or probability of hit, the assaulting team can distract the crew to the point that it is more occupied with fighting than firing missiles. Moreover, if the ship's fire control is by remote targeting, the assault team need only destroy or impair the communications nodes on board. Falling short of seizing the crew or the vessel, the assault team could proceed to a secondary objective of destroying critical topside nodes. The assault force could also destroy the hydraulic mechanisms on the VLS tube hatches, which would make it impossible to launch those missiles. Topside antennas would be easy targets as well. In the instance of a portable antenna being used to replace a destroyed one, the decreased link margin coupled with a raised bit error rate may cause mis-targeting data or delay the rate of fire. The decreased link margin and raised bit error rate is due the fact the replacement antenna has considerably less gain. Any node that would permanently or temporarily slow the rate of fire of the ARSHIP can be accessed with a well-trained team once onboard, even without the seizure of the crew.

3. Mitigating Factors

Aggressor Considerations: An assault team must have enough personnel to accomplish entry and seizure of the ARSHIP. Its size will also depend on the advantage the assaulting team expects to achieve by surprise. An assault that has strategic and tactical surprise will require fewer personnel and expect less attrition than one that has lost surprise.

The amount of equipment carried by a team member should not exceed a physical and tactically effective limit. A human male may be able to pack 125 lbs. However, that weight would be cumbersome and tactically inefficient. A reasonable expectation for a load of equipment or explosives would be 35-40 lbs. Considering that most assault teams would be tethered by communications, this would exclude some members from carrying a load of explosives. Additional breaching equipment and charges would be distributed among the assault force at the expense of loading explosives. Without means of mass destruction there is only so much critical damage that the assault team can do in a short amount of time beyond seizing the crew and vessel.

Other limitations of the personnel depend on training and their competency with fast-roping, climbing aboard from craft alongside the ARSHIP, and weapons and demolitions experience. In addition, the expertise required to land a parachute on the vessel and then begin conducting an assault takes extensive training. Seemingly, the boarding of an U.S. warship requires the skill of professionally trained elite units. These skills are found in trained military, mercenary, and organized crime (pirate) units.

Necessary equipment to accomplish these methods of attack depends on the objective and mode of attack. An *inport* assault requires little additional equipment beyond whatever the team would utilize to reach their objective and breach the defenses of a naval base.

Waterborne assaults require ladders and assault craft. The limiting factor on the assault craft is its maximum speed of advance and handling characteristics. Small and large boats offer the greatest maneuverability. However, their speed of advance across high seas, their limited range, and the length of the freeboard that the assault team must climb, become limiting factors of their employment. Additionally, the size of the boat utilized is governed by the amount of personnel it is must carry, the amount of fuel it can carry and the speed of advance that is necessary to transit long distances under strict time constraints e.g., (the cover of darkness). Of note, smaller craft can be delivered by aircraft (parachute) or helicopter (cast) to within a reasonable range of the ARSHIP.

Helicopters used for an air attack need to be large enough to carry troops. Fast-ropes or rappelling ropes are easily acquired. However the helicopters are constrained by the distance they can travel when laden with troops and ammunition. This projected range from probable launch points should be considered as a factor in their defensive posturing.

Environmental Considerations: Low visibility, once the aggressor has located and identified the ARSHIP, is to the advantage of the attacker. The movement of a relatively small force, small watercraft, slow moving air vehicles, or indigenous craft and vehicles go relatively undetected in low visibility. Constant alertness inherently decreases a passive defender's perception of the threats' ability or probability to assault the ARSHIP.

During the light of day an assault is less likely to occur unless the attacker does not have the ability to locate and identify the ARSHIP at night. In this instance the attacker would approach under the ruse of being an indigenous activity. During the night, the illumination of the moon plays a great role in exposing a force. Sheltered from sight by darkness, trained units can almost always approach the ARSHIP undetected and achieve tactical surprise.

Weather considerations such as sea state will have great effects on smaller waterborne assault craft. This will determine their range and maneuverability, as well as how seasick the assault team gets before they reach the ARSHIP. Wind affects the likelihood of an approach and ability of landing a team onboard a moving ARSHIP. The helicopter may be constrained from physically landing onboard the ARSHIP, and the wind swept ropes may make the assault team fall short of their intended landing points. Additionally, sea state adds complications to roping onto a pitching deck. The rope is to maintain contact with the deck. As the deck moves upward and downward relative to the ropes hanging from a stationary helicopter, there is a danger of the ropes no longer reaching the deck when the ship is in a trough. Another problem is if the deck accelerates upward at a descending team member when the ship hits a peak. Additionally, a wet and pitching deck can make tactical movement difficult for the assault team. Despite the difficulties of conducting an assault in bad weather, that is the time when an assault is least expected and the ARSHIP is the most vulnerable.

Medium characteristics play less of a role in the assault force unless exposure to wind, rain, sea state, and ambient temperatures would render members of the team physically ill to the point that it degrades their mission or excludes them from the mission leaving the team short handed.

Defender Considerations: Readiness is an issue that has been discussed at length. The principal factor to observe is that attacks are most likely under conditions of darkness or foul weather.

The speed of advance alluded to earlier will affect missions that attempt to board a moving ARSHIP. By altering speeds and sharply changing directions the ARSHIP can affect the ability of an assault force to board as well as buy critical time to form a "security alert force" to repel boarders.

In the *open ocean,* small boats are less likely to independently transit distances farther than 75 miles under cover of darkness. Alongside a pier, the ARSHIP faces the threat of land assault that has repeatedly been executed successfully by Naval Special Warfare units. The ARSHIP should be cordoned off in more secure basing, much like submarines and special weapons.

4. Utility and Substitutability

Cost and Asset Availability: Personnel for a coordinated assault need to be highly trained unless the assault is merely a suicide mission. The suicide mission merely requires the personnel to gain access to the ARSHIP and then detonate their explosives. Elite units can most likely only carry out a smooth and efficient assault. These units are not difficult to acquire in the case of mercenary units and pirates, both of which work for money. Trained military units, however, are more difficult to acquire and take time and preparation before they are seasoned enough to attack and expect success.

The assault platform is one of the major limiting constraints. Helicopters and large commercial vessels are high-ticket items. Therefore, powers that have limited resources are far more reluctant to employ this type of mission. Small boats and indigenous boats however are more readily available for acquisition and the cost of their loss is more tolerable. The least costly of all is the land assault, requiring only a minimal investment at times.

Benefit: The prospect of seizing the ARSHIP is one that would make any aggressor salivate. Additionally, a successful attack on the ARSHIP would be politically and strategically impressive, attracting great attention. If the ARSHIP were operating in a region that was thought to be safe but nonetheless was attacked, either fleet tactics would be changed or the ARSHIP would attempt to move further from those areas of vulnerability. The latter would move the missile ranges of the ARSHIP out to sea, affecting the probability of hit as discussed in Chapter III.

Because of the ARSHIP's small crew and minimum defenses, the ability of a trained force to board and seize it seems high. The attack would be almost as simple as boarding a hostile commercial cargo ship, which pirates in Asian waters accomplish almost routinely.

The larger question is the likelihood of the approach being discovered and thwarted. As discussed earlier, small or indigenous vehicles have higher probability of success gaining proximity to the vessel undetected or unnoticed. Additionally, the capabilities granted to nodal destruction or seizure by an assault force greatly outweigh the precision of inbound missiles, torpedoes or projectiles.

This makes the threat of an assault highly productive for its investment.

D. BOAT ATTACK

The discussion of boat attacks will center on their use as a weapon platform against the ARSHIP, and not their use as a personnel delivery platform. To see how a boat can be used as a launch platform, refer to Section C. That section will also be useful to understand the limitations and capabilities of different styles and sizes of boats. 'Boats' in this section include earlier defined vessels as well as <u>commercial</u> and <u>leisure</u> vessels that may be larger than 90 feet. Boats have the ability to converge on the target with a minimal amount of suspicion. The covertness of this approach lies in its ability to appear innocent and non-threatening.

1. Modes

Along almost every waterway in which the U.S. Navy operates, there are local users that rely on that same water for their existence or recreation. These users are fisherman and merchants that move their goods by boat. Additionally, recreational boaters may be on the water enjoying themselves and their surroundings. This activity (clutter) becomes commonplace while operating in littoral waters near landmasses or on the open seas along sea-lanes. The ARSHIP may detect boats visually or electronically, but many times not perceive them as a threat. The attacker's ability to take advantage of this will allow him to approach within weapon release range.

2. Objectives

Once within range, the boat can fire at the ARSHIP with small arms, machine guns, large caliber weapons, rockets, or torpedoes. The attacker's goal dictates the targeted node on the ARSHIP. To kill the ARSHIP, the attacker would need to destroy the watertight integrity of the hull. Loaded with explosives, a boat could come alongside the vessel and detonate deadly cargo. The explosion can destroy machinery and vital equipment in the direct vicinity of the blast or breach the hull and flood the compartments. However, the explosion would have to be large for the blast and subsequent flooding to sink the ARSHIP. A torpedo attack would have the same desired effect as a boat loaded with explosives. Damage to navigation and propulsion equipment can render the ARSHIP unable to maneuver, making it either incapable of leaving port or requiring a return to port for repairs.

The goals of influencing the geo-location and the rate of fire of the ARSHIP can be accomplished by boat attacks. By using weapons such as small arms, machine guns, large caliber weapons, and rockets, the small boat can attempt to destroy or damage the vital equipment required to navigate, communicate, or launch missiles. The size of the attacking boat is dependent upon the size and weight of the weapon being launched from it.

Small arms can be used effectively to eliminate the crew and to a lesser extent damage vital equipment. Machine guns can be used against people and equipment with greater effect. Machine gunners can rake the ARSHIP's exposed topside appendages. Large caliber weapons and rockets will accomplish the same results, but these weapons possess more destructive power. A .50 caliber machine gun can cause severe damage to equipment that is vital for the ARSHIP to perform its mission. Targets might include antennas and exposed communications equipment used for the reception of information for the missiles, navigation or communications. This will result in the ARSHIP's inability to target and launch missiles in an efficient manner and hinder its ability to radio other vessels.

Random firing at the target may or may not strike vital equipment, but the psychological effects can accomplish a strategic result. Boat attacks can drive the ARSHIP from its present location as it maneuvers to avoid the actions of an aggressor.

3. Mitigating Factors

Aggressor Considerations: The skill required to operate a boat is possessed by any individual willing to do so. No formal or special training is required for smaller vessels but some commercial boats may require licenses. Unlike the driver of the boat, the individuals operating the weapons need to be properly trained. Some of the more complex weapons and systems need more training compared to firing small arms.

The only additional skills required, other than those previously mentioned, are navigation and target recognition. Electronic or "dead reckoning" navigation requires skills that are not necessarily difficult but are also not intuitive. Additionally, the individuals targeting their weapons need to be able to identify vital equipment on the ARSHIP. To be effective, they need to be able to recognize vital equipment from non-vital appendages onboard the vessel.

Equipment choice, either the boat or the weapons, depends on the aggressor's intention and resources. The size of the boats available may govern the type of weapon or weapons the attackers can select to accomplish their mission. The geographic area in which the ARSHIP is targeted also influences the choice of boat size. If an aggressor must travel out to sea for any great distance, a larger boat may be needed. The boat will need to have fuel capacity and sea-worthiness to transit the respective range and operate in the conditions of an *open ocean*. Therefore, the type of boats available to an aggressor will govern the range at which the ARSHIP can be struck. The addition of navigation and detection equipment will be required when transiting large distances to locate and attack the ARSHIP.

The choice of boat and weapon may be dependent upon the aggressor's ability to procure them. Small arms or machine guns may not provide sufficient firepower to carry out the task assigned. If the aggressors want to destroy or damage a specific piece of equipment, they may require a rocket or large caliber weapon. If a longer range or more destructive weapon is necessary to achieve a desired effect, an adversary may not have the necessary resources to possess it. If this cannot be accomplished, an alternate weapon or platform must be selected to accomplish the same desired mission.

Environmental Considerations: The ability for a boat to approach undetected is to the advantage of the aggressor. On the other hand, low visibility associated with rain, fog, or snow, affects the adversary's weapon performance due to their visual sighting characteristics.

Boat attacks may be attempted during the day or night. The advantage of a daylight attack is the ability of the aggressor to see specific targets onboard the ARSHIP. The disadvantages normally associated with overtly approaching a target in daylight are countered by the aggressor boat's ability to appear to be an innocent and non-threatening vessel. Nighttime attacks add the ability to approach visually undetected. Unfortunately, the ability to visually identify critical nodes is also degraded. An attacker may need to provide illumination with spotlights to locate vital nodes.

The use of or lack of navigational lights plays in favor of the attacking boats. By appearing to be what they are not by using false navigational lights or not using them at all, potential threats can appear to be innocent. The reliance of U.S. Naval Ships on nighttime navigational lights to classify ships can be used to the advantage of the aggressor.

Sea state has the greatest effect on boat operations. Smaller boats have difficulty operating when the sea state rises. This can cause a boat to be unable to transit long distances out to sea or even result in it capsizing. A higher sea state may require a larger boat to survive the transit to the target. The pitching and rolling associated with an increased sea state will affect the accuracy of ballistic and fire-and-forget weapons. In addition, the boat crew may lose their balance, fall and injure themselves, or be swept overboard. Another factor associated with a pitching and rolling boat is the possibility of the crew getting seasick. This may render them incapacitated and unable to conduct the operation.

Temperature also plays a role in a boat attack, although to a lesser extent than other factors. Sea spray from the boat combined with cold air temperatures can effect exposed weapons. These weapons may then be rendered inoperable or less effective than normal.

Defender Considerations: Readiness has been discussed throughout this thesis. Associated issues to be recalled include the alertness of the crew, the suspicions placed on closing vessels, and the training of watchstanders. However, in the case of boat attacks, the readiness of topside weapons plays an integral part in ship defense. Readying a .50 cal machine gun for action is a tedious endeavor entailing "head spacing and timing," not just loading, pointing and pulling the trigger. Additional items, such as surface search radar and passive sonar, are detection devices that must be monitored closely to identify potential threats from unmarked or falsely marked vessels.

The design of the ARSHIP calls for it to have a maximum sustained speed of 22 knots. Small boats possess speeds greater than this, usually in the 30+ range. This speed advantage gives the aggressor a tactical advantage over the ARSHIP. The aggressor boats can maintain their positioning relative to the target and maneuver around the vessel.

The distance from land the ARSHIP can loiter and operate dictates the size and sea-worthiness of the potential aggressor boats. This, therefore, can be used as a deterrent measure by denying the aggressor boats access to a target that is outside their maximum range.

4. Utility and Substitutability

Cost and Asset Availability: The training for a small boat attack is minimal. The boat driver will need little to no training. While the individuals using the weapons will require training on the operation and targeting of each weapon. The cost of training personnel associated for a small boat attack is of little consequence to the organization or group directing its employment.

Small boats are easily procured at a relatively low cost. A greater expense is associated with the weapon chosen for the mission. The availability of the weapons may be high, as with small arms, or more challenging, as with large caliber weapons, rockets, or torpedoes. Major costs involved in boat attacks are accrued in procuring more sophisticated boats. Boats such as commercial fishing, shipping, or leisure boats are more expensive than small inflatables or speedboats. Overall, the investment of personnel and vessel is less when compared to that of a war vessel.

Benefit: The amount of critical topside nodes on the ARSHIP is limited. Antennas and crewmembers make up the extent of critical nodes that can be destroyed with hand held weapons. In the cases of collision or direct contact detonation of high explosives, the boat attack has higher effects against internal and structural critical nodes. The operation may not damage or destroy the target, but the attempt may alter the utilization of the ARSHIP in the future. Tactically, the mission may or may not succeed, but strategically, the desired effect may be obtained, which will cause the ARSHIP to reposition through fear of other boat attacks.

The cost of the required equipment, boat and weapons, and the possible effectiveness make the small boat option a viable threat with high utility. The gains achieved through the employment of a small boat attack outweigh the losses or possible failure to accomplish a desired objective.

E. AIRCRAFT

The term "aircraft" in this section encompasses ultralites, gliders, and any private and commercial airplanes and helicopters. This aircraft discussion will concentrate on their use as a weapon against the ARSHIP and not a delivery platform for other forces as discussed in Section C.

1. Modes

Like a boat attack that can approach without raising suspicion, an aircraft must rely on this strategy to approach the ARSHIP. An aircraft may seem non-threatening if it is operating along established flight corridors. Aircraft are primarily detected through electronic means by U.S. naval warships, but an ultralite or glider is almost undetectable by radar or electronic counter-measures (ECM). These small, lightweight planes have minimal to no radar signature due to the use of composite material such as aluminum, canvas, and fiberglass. The ability to fly low and slow allows smaller aircraft to operate below automated detection thresholds of many air search radar. The ability to approach the ARSHIP in a non-threatening manner is an integral element necessary to close within range to utilize on-board weapons. In addition, many air defense missiles will not home in on an aircraft with little or no radar or infrared signature. An additional platform would need to be utilized to launch an aircraft for *open ocean* attack. Aircraft, such as ultralites, gliders, RPVs, and private and commercial planes, can also operate as a weapon, again by performing kamikaze missions on the target. Loaded with explosives, the plane can become a guided warhead. A small ultralite can be built from a relatively inexpensive kit and requires minimal skills to operate. A glider constructed of wood or composite material can operate undetected by electronic means. Like ultralites and gliders, a RPV can be constructed at a low cost and is hard to detect with electronic and visual means. A RPV as a weapon does not require a pilot onboard.

2. Objective

Once within range, the aircraft can be used as a flying warhead or to launch ordnance to accomplish its mission. The mission's goals and critical targets on the ARSHIP are similar to those of a boat attack. The differences lie in how they are conducted. As a weapon, the plane can be loaded with explosives that detonate upon impact on the ARSHIP. Historically, this *kamikaze* style mission has proven to be an effective weapon as seen in WWII. It eliminates equipment and personnel without discrimination, and is a crude weapon compared to smart munitions. The fuel carried by the plane can also cause secondary explosions and fires on the ARSHIP.

In its role of launching ordnance, the size of the plane will dictate the amount and type of munitions it can carry. The munitions can be bullets, torpedoes, bombs, missiles, or a combination of some or all of them. Bullets can be used to repeatedly strafe the ARSHIP damaging or destroying vital equipment and the crew. This method may not have as much precision as other munitions, but it can be effective.

Bombs can have proximity fuses, which detonate the bomb above deck inflicting damage to anything topside. They may explode on contact with the deck with similar results. Additionally, shaped charge bombs can penetrate the deck to a lower part of the vessel and explode in the interior of the ARSHIP. The results of such an explosion could be catastrophic to the vessel's survival. The large circular area of probability of precision associated with a conventional bomb means the chance of a direct hit is not as precise as a bomb that is guided to its target. Guided munitions or laser guided bombs (LGB) can receive terminal guidance information through an electromagnetic link. This ability was displayed during the Gulf War when U.S. warplanes made precision missile attacks against Iraqi targets. The stealth characteristics of the ARSHIP do not combat such guidance systems. LGBs can be dropped from larger platforms such as commercial aircraft.

3. Mitigating Factors

Aggressor Considerations: The ability to fly an aircraft depends on the technology associated with it. The skill required to fly an ultralite is minimal and training can be at a low level. The more advanced an aircraft is, the more training required to fly it skillfully. The skills needed for advanced aircraft make these pilots and aircraft unfavorable choices for suicide type missions. A trained pilot is an asset not easily wasted by an aggressor.

The choice of airframe depends on the aggressor's economical means and accessibility to aircraft. The aggressor may not have the financial ability to purchase a large private or commercial aircraft. Smaller planes may be all they can afford. The ultralite is not suited to carry large munitions, but it can be fitted with explosives or the pilot can carry hand grenades. Grenades can be effective as demolition of topside equipment, attrition of topside security forces, and "Thermite" grenades can be dropped on the VLS tubes and burn into the missiles from above. In the role as a guided warhead, the ultralite can fly low and slow toward the target to avoid detection.

Accessibility is not a significant issue if the adversary can afford it. The problem arises when the aggressor's objectives require an aircraft that cannot be obtained. His plans must include aircraft that can be procured and maintained.

Environmental Considerations: Visibility affects the ability of the pilots flying small aircraft to locate the ARSHIP. Smaller aircraft rely on the pilot to visually locate the target and position for attack. Visibility is an important factor to conduct aircraft attacks with planes that are not technologically advanced.

Aircraft attacks can be attempted day or night. The target must be well lit to distinguish it from nearby vessels. Effective night attacks at sea are limited to when the aggressor has the ability to identify the ARSHIP from other vessels and obstructions. Daylight greatly aids in locating and attacking the ARSHIP from the air.

Precipitation and winds associated with bad weather can conceal the approach of an aircraft. The plane must be equipped with navigation equipment and the ARSHIP must be in a known location and diverging from its last known position slowly. The same weather that conceals the aircraft's approach makes it harder to locate and attack the ARSHIP. High winds can make it impossible for the aircraft to fly and interfere with its navigation. Head winds can reduce the aircraft's range and may render it incapable of reaching the ARSHIP.

Defender Considerations: To understand the implications of readiness and positioning, refer to the discussion of these issues of the ARSHIP in the previous chapter.

The speed of advance for the ARSHIP decreases the hit probability of conventional bombs, but a plane intent on crashing into the ARSHIP with explosives will be unaffected. In addition, a plane must have the range to attack the ARSHIP if it is in the *open ocean*. A small aircraft does not have the capability to attack across a large expanse of water.

4. Utility and Substitutability

Cost and Asset Availability: The availability of trained pilots depends on the type of aircraft that is to be used for the attack. Training a pilot to use a more advanced plane will cost more than for a smaller plane that can utilize a pilot with lesser skills.

Aircraft are relatively expensive. Therefore, the cost is dependent on the aircraft used. The option to use a small aircraft such as an ultralite would reduce cost. These types of planes can be ordered and built from a kit at a relatively low cost to the aggressor, whereas, commercial jets can range into the millions. Commercial aircraft are more likely to be rented, commissioned, or hijacked for such a mission.

Benefit: The benefits are similar to those of a boat attack, but the difference lies in the speed an aircraft can operate and its method of approaching the ARSHIP in a non-threatening manner. The aircraft's speed gives the ARSHIP less time to react to an attack. In addition, the ship may not pay attention to air contacts if it has to track and maneuver constantly to avoid possible collisions with surface contacts in *confined water*. Overall, if an air contact can approach the ARSHIP via commercial or leisure flight corridors, it has a higher potential of sneaking up and swiftly striking the ARSHIP than boats. This is a likely scenario and is most effective in *confined water* near well-established flight paths.

V. CONCLUSIONS

A. DEFENSIVE MEASURES

After the discussion in the previous section, the question that remains to be answered is whether or not there is a way for an Arsenal Ship, with reduced manning and meager self-defenses, to counter an unconventional threat. There are both, strategic and tactical, defense measures that the ARSHIP can employ to deter, defend against, or lessen the effects of the above threats. A strategic defense measure is one that requires changes beyond the organic assets of the vessel. Strategic defense options affect the ARSHIP's CONOPS and political and naval strategies for its employment. A tactical defense measure is one used by the ARSHIP itself for self-defense. Tactical defense measures will influence the design of the ARSHIP. Some of these defense measures will enhance the defenses against conventional warfare threats as well.

B. STRATEGIC DEFENSE OPTIONS

Strategic defense options for this technological target are based in psychological operations (PSYOP) and deception, two actions of warfare that doctrinally fall under special operations in the United States. An option in this category is the use of phantom ships, created electronically or physically, to deter, confuse, distract, or ambush the forces operating against the ARSHIP. The physical phantom ships can be constructed with an inexpensive and buoyant material such as wood, fiberglass, plastic, foam, or rubber. This can and should be augmented with radar and infrared reflectors and an additional infrared (IR) source. Deterrence, the first possible method of conducting conflict resolution in this case, takes two forms. Phantom ships could introduce a false number of ARSHIPs to the operating environment. This action can keep the opponent from attacking because of the perceived decreased probability of kill due to the numbers of ARSHIPs present. By presenting a larger than truthful number of targets the ARSHIP's probability of survival automatically increases. Quite simply, the opponent does not foresee that enough damage will be inflicted to the ARSHIP armada to affect the overall outcome of conflict, and the ARSHIP merely becomes a target of opportunity when it presents itself.

The second form of possible deterrence by this course of action would be to generate false escort ships, giving the impression that the ARSHIP is a well-defended target. The perception of escort combatants will lessen the chance of the ARSHIP appearing as an easy target. Unfortunately, the use of false escorts will counter actions and technology utilized to hide the ARSHIP.

When considering the effect of these false ships, there are psychological effects that come to light. The first effect is produced by using these ships as a means to dissuade the enemy from contemplating or executing an attack. Essentially, these false ships will act as weapons sinks that draw off an enemy's weapon system or confuse his ability to discriminate between the false targets and the ARSHIP. The psychological effects of distraction and confusion can cause an adverse psychological effect on opposing forces. The forces conducting the attack will interpret their efforts as wasteful and more risky. This not only affects those forces operating directly against the ARSHIP, but also on the operational and strategic planners. Across this spectrum, the feeling of inadequacy caused by second guessing and internal dissension can exact high tolls on the upper echelon of leadership.

The last application of these phantom ships is for ambush. Reminiscent of past actions on the high seas, the Q-ships of WWI looked like regular merchant vessels and were employed to entice German U-boats. Once the U-boats surfaced to engage the 'ill-defended merchant', the Q-ship would drop its facade with deadly results. The Q-ships were designed and equipped with enough firepower to destroy a submarine attacking on the surface. This type of PSYOP had a wide range of effects. In addition to eliminating the attacking adversary, it demoralized the enemy's forces. Operationally, one successful ambush by a false ARSHIP would cause a reshaping of tactics employed against the ARSHIP. In the case of the Q-ships, it led to German U-boats only conducting submerged attacks, thus lowering their success rate. Strategically, a successful ambush could lead to questioning the utility of attacks against the ARSHIP or which assets to employ.

C. TACTICAL DEFENSE OPTIONS

1. CONOPS

By addressing the ARSHIP's CONOPS and making changes to its proposed geographic operating areas the ARSHIP can effectively lessen attacks from smaller air and boat threats. Placing the ARSHIP 75-100 miles from land, this can exclude small gunboat and inflatable boat attacks due to their range limitations. The 75-100 mile range distance from land only moves the ARSHIP to sea by 10% of the total effective range of cruise missiles. This does not, however, exclude aircraft dropped small boats or larger boats but does lessen the threat from a certain target set.

This distance from land also places the ARSHIP away from normal air traffic lanes. Its reduces the likelihood of an air threat diverting from a 'normal' track and attacking without being noticed in a timely manner and countered.

Even, when not expecting attack or without indications of a potential boarding the ARSHIP should travel at speeds above 20 knots. These speeds effect the wind over the topside of the decks, making helicopter boardings slightly more hazardous and make small boat boardings more difficult because of the hull wake.

The consideration of escort ships and/or fast attack boats could be considered as deterrents and defense platforms augmenting the ARSHIP's meager defenses. However, these counter the stealth capabilities of the ARSHIP. The escorts, by their necessity to be near the ARSHIP to extend their defensive umbrella, will give away the area of operations of the ARSHIP.

2. Gun Mounts and Missile Platforms

Along with the strategic options, several tactical defense options must be considered for self-defense. The addition of gun mounts and platforms for the firing of shoulder fired missiles can be used to defend against many of the unconventional threats. The gun mounts should accommodate M-60, .50 caliber, or Vulcan chain machine guns. These machine guns, with their high caliber, long range, and high rates of fire, can be used to repel a surface combat swimmer, an assault force, a small boat, and slow aircraft attack. The missile platforms can be used to fire RPGs, shoulder-fired air defense missiles, or recoilless rifles. The tradeoff considered in external mounts is the expansion of the ARSHIP's radar cross section. The mounts and platforms can be retractable to reduce radar returns and maintain the stealth characteristics of the ARSHIP concept.

3. Crew

With a reduced crew size, everyone aboard the ARSHIP must become an integral part of the ship's self defense force. The armory should be stocked for all options. Its inventory should include pistols, automatic rifles, shotguns, body armor, concussion, illumination and chemical grenades, 40mm grenade launchers, RPGs, AT-4s, and Stingers. Proper training is required of all crewmembers to become efficient on all weapons. The range of possible threats warrants additional training, not only in weapon proficiency, but also against unconventional tactics. The U.S. Navy SEALs, who are trained in forcible boarding, search, and seizure of water vessels can teach counter tactics. A properly trained and manned crew is a must to defend the ARSHIP from the surface and air threats.

4. Detection Equipment

To detect approaching threats, supplementary equipment may be needed. The addition of large sensor equipment systems such as active sonar may not be feasible. However, remotely operated cameras would assist in visual detection and isolation of threats. A single individual can monitor the camera system to observe the topside spaces, ship's sides, and the water adjacent to the ARSHIP. Forward looking infrared (FLIR) cameras and other forms of night vision technology are essential to help reduce the threat of night attack. Additionally, there must be a real time link to overhead sensors and companion vessels that may detect threats over the horizon from the ARSHIP to allow for proper warning.

5. Communications

Communications plays a key role in the effective employment of the ARSHIP. Some defensive options along these lines will be discussed under the section titled "deceptive options." One additional consideration beyond the automated communications system is the use of a phased array antenna built into the structure of the ship. More conventional considerations have already taken into account link margin and protection of EHF SATCOM.

6. Deceptive Options

Other possible tactical defense options rely on deception. Similar in thought to the strategic use, the tactical use of deception will be to nullify the attack on the ARSHIP. The application of a visual form of camouflage, reminiscent of the paint schemes used on World War II warships, can be applied to the ARSHIP without altering the stealth properties gained through technology. Due to the dependence of special operations forces on visual identification, the paint schemes can aid in the ability of the ARSHIP to elude or deceive visual detection. The paint can camouflage the vessel with the surrounding sea or make it appear as another ship, such as a commercial or leisure vessel. To bolster the effects of deceptive paint, the use of false targets onboard the ARSHIP can be utilized.

Deceptive superstructure or equipment in conjunction with camouflage could effectively distract enemy fire from vital equipment and spaces. Dummy targets could include high profile antennas and communications equipment, superstructure for the operations of ship's maneuvering or crew berthing, and false weapon systems. Likewise, false weapon systems on the deck to deceive an adversary could also deter a possible attack from special operations forces due to perceived self-defense capabilities. Again, the trade off between the technology of stealth and the possible radar signature of adding false structures to the ARSHIP must be evaluated.

7. *Inport* Defense

The ARSHIP is extremely vulnerable when *inport*. Defensive measures that can thwart the combat swimmer are fishing nets with audible or visible warnings, active sonar, active suctions and discharges, high speed currents, mammals, and active patrol boats. Each pose their own strengths and weaknesses. Nets in the water need to provide a barrier that the divers cannot sneak under or cut through without activating the warning devices. These nets will also need to be deployed and recovered every time the ARSHIP leaves and returns to port. The nets also must be maintained and checked often. The other options have been touched on in Section A of Chapter IV.

One additional method of defending against the combat swimmer operation is to utilize concussion grenades over the side of the ship to over-pressure the water and the semi-permeable cavities of the divers. This is effective, but only when the divers are near the ship. This use of grenades excludes the use of trained mammals for swimmer defense.

D. SUMMARY

The ARSHIP will be engineered for stealth against electronic detection and classification. The proposed low cost tactical defensive measures are vital to the enhancement of its defense capabilities achieved through technology.

This thesis has proposed that the ARSHIP is most vulnerable during times of relative peace and at long distances from perceived threats. Chapter III proposed that the ARSHIP, due to its strategic importance in a conflict, is a prime target for preventative and preemptive attacks. Due to the ARSHIP's dependence on technological 'hiding' and not on defensive firepower, potential attackers must not be allowed to approach it within weapon release range. Of primary importance, is the ARSHIP's reliance on Joint Vision 2010's concept of *full dimensional protection*. Historically, the United States' military has been caught off guard, at home and abroad. This can be seen from the attack on Pearl Harbor, the Marine barracks in Beirut, and more recently, the Kobar Towers in Saudi Arabia. Special operations have the ability to strike without warning and at long distances. The United States' opponents **will** utilize this option in the future.

Additionally, special operations will be utilized under conditions of singularity. When all other conventional attempts against the ARSHIP have failed or the means to face the ARSHIP conventionally are not available to the adversary, conventional operations will be ceased and give way to innovative special operations. These operations will be more clandestine, illusive, and most likely more successful. Their success will stem from the cognitive bias of the conventional naval warrior and the innovative spirit of the aggressor.

This thesis has shown the viability of possible unconventional threats that must be addressed in making the ARSHIP concept a successful reality. Any group or organization, which feels threatened by U.S. foreign policy and its implementation of the ARSHIP, may acquire the means to attack the ARSHIP via special operations. Conventional thinking, which dictates platform design and utilization, must adapt to a world in which unconventional threats to our most valuable military capabilities is becoming more predominant.

INITIAL DISTRIBUTION LIST

1. Defense Technical Information Center 2
   8725 John J. Kingman Rd. Ste 0944
   Fort Belvoir, VA 22060-6218

2. Dudley Knox Library 2
   Naval Postgraduate School
   411 Dyer Rd.
   Monterey, CA 93943

3. Wayne P. Hughes 1
   Senior Lecturer, Operations Research Curriculum
   (Code OR/Hl)
   Naval Postgraduate School
   Monterey, CA 93943-5000

4. Gordon H. McCormick 1
   Academic Associate, Special Operations Curriculum
   (Code CC/Mc)
   Naval Postgraduate School
   Monterey, CA 93943-5000

5. Jennifer Duncan 3
   Center for Terrorism and Irregular Warfare
   (Code CC/Jd)
   Naval Postgraduate School
   Monterey, CA 93943-5000

6. Christian A. Dunbar 2
   755 Amador Ave
   Seaside, CA 93955

7. Dino Pietrantoni 2
   131 Laura Lane
   Elizabeth, PA 15037

NAVAL POSTGRADUATE SCHOOL
Monterey, California

THESIS

ARSENAL SHIP AUTOMATION AND MANNING ANALYSIS

by

Thomas E. McNerney, III

March 1997

Thesis Advisor: Wayne P. Hughes, Jr.

Approved for public release; distribution is unlimited.

REPORT DOCUMENTATION PAGE			Form Approved OMB No. 0704-0188
Public reporting burden for this collection of information is estimated to average 1 hour per response, including the time for reviewing instructions, searching existing data sources, gathering and maintaining the data needed, and completing and reviewing the collection of information. Send comments regarding this burden estimate or any other aspect of this collection of information, including suggestions for reducing this burden to Washington Headquarters Services, Directorate for Information Operations and Reports, 1215 Jefferson Davis Highway, Suite 1204, Arlington, VA 22202-4302, and to the Office of Management and Budget, Paperwork Reduction Project (0704-0188), Washington, DC 20503.			
1. AGENCY USE ONLY (Leave Blank)	2. REPORT DATE March 1997	3. REPORT TYPE AND DATES COVERED Master's Thesis	
4. TITLE AND SUBTITLE ARSENAL SHIP AUTOMATION AND MANNING ANALYSIS			5. FUNDING NUMBERS
6. AUTHOR(S) McNerney, Thomas E. III			
7. PERFORMING ORGANIZATION NAME(S) AND ADDRESS(ES) Naval Postgraduate School Monterey, CA 93943-5000			8. PERFORMING ORGANIZATION REPORT NUMBER
9. SPONSORING / MONITORING AGENCY NAME(S) AND ADDRESS(ES)			10. SPONSORING / MONITORING AGENCY REPORT NUMBER
11. SUPPLEMENTARY NOTES The views expressed in this thesis are those of the author and do not reflect the official policy or position of the Department of Defense or the U.S. Government.			
12a. DISTRIBUTION / AVAILABILITY STATEMENT Approved for public release, distribution is unlimited			12b. DISTRIBUTION CODE

13. ABSTRACT (Maximum 200 words)

The Arsenal Ship concept of operations is unique. The Arsenal Ship provides a remote magazine for other joint warfare systems to utilize, with limited ability to defend itself. Ultimately, it resembles a combat logistics ship designed to sail into harm's way ready to provide the initial "punch" as required. Therefore, it should be minimally manned by employing the most cost-effective technology. With the requirement to reduce crew size, a new approach to manning is required. This thesis provides an alternative approach to manning by identifying the most cost-effective investment in automation commensurate with reducing crew size to the lowest feasible level.

14. SUBJECT TERMS Arsenal Ship, Manning, Automation, Crew size, Deployment cycle			15. NUMBER OF PAGES 89
			16. PRICE CODE
17. SECURITY CLASSIFICATION OF REPORT Unclassified	18. SECURITY CLASSIFICATION OF THIS PAGE Unclassified	19. SECURITY CLASSIFICATION OF ABSTRACT Unclassified	20. LIMITATION OF ABSTRACT UL

NSN 7540-01-280-5500

Standard Form 298 (Rev. 2-89)
Prescribed by ANSI Std. 239-18
298-102

Approved for public release; distribution is unlimited.

ARSENAL SHIP AUTOMATION AND MANNING ANALYSIS

Thomas E. McNerney, III
Lieutenant, United States Navy
B.S., Jacksonville University, 1989

Submitted in partial fulfillment of the
requirements for the degree of

MASTER OF SCIENCE IN OPERATIONS RESEARCH

from the

NAVAL POSTGRADUATE SCHOOL
March 1997

Author: Thomas E. McNerney, III

Approved by: Wayne P. Hughes, Jr., Thesis Advisor

Ronald L. Brown, Second Reader

Frank C. Petho, Chairman
Department of Operations Research

ABSTRACT

The Arsenal Ship concept of operations is unique. The Arsenal Ship provides a remote magazine for other joint warfare systems to utilize, with limited ability to defend itself. Ultimately it resembles a combat logistics ship designed to sail into harm's way ready to provide the initial "punch" as required. Therefore, it should be minimally manned by employing the most cost-effective technology. With the requirement to reduce crew size, a new approach to manning is required. This thesis provides an alternative approach to manning by identifying the most cost-effective investment in automation commensurate with reducing crew size to the lowest feasible level.

TABLE OF CONTENTS

x

LIST OF ABBREVIATIONS AND ACRONYMS

ASRD	Arsenal Ship Requirements Document
BAQ	Basic Allowance for Quarters
BAS	Basic Allowance for Subsistence
CG	cruiser
CIC	combat information center
CONREP	connected replenishment
CONUS	continental United States
CTS	central training site
DARPA	Defense Advanced Research Projects Agency
DDG	guided missile destroyer
FOB	forward operating base
FY	fiscal year
IMAV	intermediate maintenance availability
LCC	life cycle cost
LOS	length of service
LSM	Landing Ship Medium
LSMR	Landing Ship Medium Rocket
MED	Mediterranean Sea
NBCF	Navy Billet Cost Factor (Model)

NEA	North East Asia
OPTEMPO	operational tempo
SBCM	Simplified Billet Cost Model
SC	surface combatant
SWA	South West Asia
TSSE	Total Ship System Engineering
VAMOSC	Visibility and Management of Operations and Support Costs
VERTREP	vertical replenishment
VHA	Variable Housing Allowance
WQSB	Watch, Quarter and Station Bill

EXECUTIVE SUMMARY

As intensifying national concern over the federal deficit continues to impact the budget for national defense, an Arsenal Ship design that is affordable to build, maintain, and operate must be implemented. Historically, personnel costs have had great impact on the life cycle cost of ships. Reduced manning reduces ship life cycle cost, reduces the number of personnel placed in harms way, and limits the impact on the Navy Manpower Plan. Therefore, an analysis of manning is required to determine the minimum optimal manning level that supports mission requirements. This thesis provides an alternative approach to manning by establishing the most cost-effective investment in automation commensurate with reducing crew size to the lowest feasible level. Before manning can be analyzed, factors such as locations of Forward Operating Bases, location of homeport, ship deployment cycle, crew deployment cycle and special evolution manning requirements must be determined.

This thesis selects two Forward Operating Base locations, Diego Garcia and Guam, as rational locations considering transit time and security. San Diego was chosen as the homeport because of its distance from both Forward Operating Bases. A "steady state" deployment cycle was designed in which four Arsenal Ships can be continuously forward deployed with another undergoing an overhaul in homeport for a total of five ships. This deployment cycle also insures 100 percent coverage of two theaters. The crew deployment cycle also resembles a "steady state" process in which crews train as a team and then forward deploy for 16 weeks. Special evolutions including flight operations, underway replenishment, and General Quarters were also considered. Damage control was determined as the limiting factor for reduced manning.

Four manning structures are identified using the Naval Postgraduate School's Total Ship Systems Engineering Team design as a basis. These structures were used to develop a 30 year life cycle cost curve based on crew size. A simple spreadsheet model was designed as an alternative to the Navy Billet Cost Factor Active Component Cost Estimation Model to assist in determining a 30 year manning life cycle cost curve. This curve is similar to an economic supply curve. Automation cost curves were then formulated. The 30 year manning life cycle cost curve and the automation cost curve were then utilized to determine an optimal reduced crew target size as well as an optimal investment in automation.

I. INTRODUCTION

A. PROBLEM STATEMENT

The Arsenal Ship concept of operations is "a revolutionary approach to warfare" [Ref. 1]. Likewise, the design concept is a revolutionary approach to Naval shipbuilding. It is unlike any previous, or current, United States Navy ship design concept. Naval warships are normally designed by the Navy and built by private contractors. Shipbuilders submit proposals for construction based on the Navy's design. The shipbuilder with the "best" proposal is awarded the contract.

The new design concept encourages shipbuilders to design an independent version of the Arsenal Ship in accordance with the Arsenal Ship Requirements Document (ASRD). One of the requirements the ASRD establishes is a maximum crew size of 50. In the past, ships were built with the latest technology and the manning level was determined based on manpower requirements to maintain and operate the equipment. With the requirement to reduce crew size on the Arsenal Ship, a different approach is required. This thesis provides an alternative approach to manning by establishing the most cost-effective investment in automation commensurate with reducing crew size to the lowest feasible level.

B. BACKGROUND

In July, 1996, the Defense Advanced Research Projects Agency (DARPA) awarded five design teams Phase One contracts to design the Arsenal Ship in accordance with the ASRD. They were given six months to produce an initial proposal. On

January 10, 1997, three of five contractors were awarded Phase Two contracts to finish designing a more detailed Arsenal Ship. Next year a single contractor will be selected to build the prototype. This program is managed by the Arsenal Ship Joint Program Office for DARPA [Refs. 2 and 3].

1. Requirements

The Arsenal Ship Concept Executive Summary identifies three theaters displaying international tension which require overseas presence: North East Asia (NEA), South West Asia (SWA), and the Mediterranean (MED). Force structure requirements for the Arsenal Ship are based on continuous coverage of SWA and near continuous coverage of the MED [Ref. 2].

A primary design requirement is to reduce crew manning through the use of innovative manning concepts. Specific topics to be addressed in this area include the following [Ref. 2]:

1. Elimination of unnecessary functions
2. Reduction of on-board and Navy-wide manpower needs while maintaining effective operations, readiness and ship safety
3. Effective damage control
4. Reduction of life cycle costs (LCC)

2. History

The Arsenal Ship concept is not entirely new. Similar ships were in service during WWII and Korea. The Landing Ship Medium Rocket (LSMR) was manned with a crew of 110, including 5 officers. Converted from the hull of the Landing Ship Medium (LSM), the LSMR was equipped with up to 105 rocket launchers which included 8,000 rockets. Nicknamed "Little Ship Many Rockets", they were designed to deliver a barrage of fire support during an amphibious assault [Ref. 4]. In the late 1950's and early 1960's almost all of the 48 LSMRs were sold [Ref. 5]. Although the Arsenal Ship's mission is not solely to provide fire support during amphibious assaults, the concept is essentially the same: One ship transporting a plethora of firepower into an area of potential conflict.

C. PREVIOUS STUDIES

Using existing technology, an Arsenal Ship could be designed to operate with the minimum manning required under existing law and practice for merchant ships. In 1983, the Committee on Effective Manning was established by the Marine Board of the National Research Council to investigate the manning of merchant ships. The committee reported a statutory minimum crew of ten [Ref. 6]. The committee's minimum manning composition, along with equivalent Navy personnel, is listed in Table 1.

Merchant	Navy Equivalent	Quantity
Licensed Master	Commanding Officer	1
Qualified Mate	Qualified Officer of the Deck	3
Qualified Deck Sailor	Qualified Boatswain Mate	3
Licensed Engineer	Qualified Engineering Watch Officer	3

Table 1. Minimum Merchant Manning Comparison to Navy Personnel

In 1989, the Coast Guard requested an assessment of the effect of smaller merchant crews on maritime safety from the Marine Board of the National Research Council. Their product, Crew Size and Marine Safety [Ref. 7], includes a thorough investigation into the 1980's innovations that allowed crews of more than 30 persons to be reduced to crews of less than 20. They also developed a model to provide an estimate of minimum manning levels. The model was validated using data from a mixed product tanker and a container ship.

The minimum manning level for an Exxon® mixed product tanker was calculated using the model. The tanker had an initial crew of 18 persons. The model calculated the manning requirement could be reduced to 15 crew members. Exxon® agreed a minimum crew of 15 would be ideal. It was also noted that a crew of 15 exceeded the minimum manning requirement of 11 persons to fight an engine room fire on this type of ship.

In September, 1993, Gregory Hildebrand [Ref. 8] completed a thesis analyzing the cost benefit of a reduced active duty crew as an alternative to a civilian manning for Combat Logistic Force Ships. He found the Navy training requirements produced Navy sailors equally qualified as their civilian counterparts. Selective manning was determined to be an integral element in reducing crew size. As long as truly competent professionals are selected to fill the billets, crew reduction is feasible. Interestingly enough Hildebrand concluded a reduced Navy crew is more economical than a civilian crew, which contradicts [Ref. 9].

The Naval Postgraduate School Total Ship System Engineering (TSSE) team completed an Arsenal Ship design in December, 1996. The team modified a T-AO 201 Class fleet oiler by upgrading the engineering plant and adding weapon systems. Through automation, remote sensors and cameras, the manning requirement is reduced to 44 crew members. The engineering plant is upgraded to satisfy the sustained speed requirement of

22 knots and to provide electrical power required by the additional weapon systems [Ref. 10].

The Smart Ship initiative is another aspect of manning that must be recognized. The Smart Ship concept reduces manning through common sense approaches to manning and innovations in technology. The USS Yorktown (CG 48) is the test platform for the Smart Ship program. Matthew Fleming is completing a thesis [Ref. 11] at the Naval Postgraduate School which concludes that reduced manning is risky. "A 0.54 percent cost savings by risking the readiness of all combatants and the United States Navy seems imprudent" [Ref. 11]. There are two major differences between the Smart Ship initiative and the Arsenal Ship. First, the Smart Ship is a combatant designed to fight. The Arsenal Ship is a support ship designed to deliver a large amount of readily available ordinance. Second, the Arsenal Ship is designed to reduce the number of personnel placed in harms way. Minimizing the number of personnel at an ammunition depot is prudent. These conceptual differences make the approach to manning very different.

D. PROBLEM DEFINITION

"The Arsenal Ship concept is simple—build a ship with massive firepower which increases the capabilities of Joint Forces already in theater" [Ref. 1]. Now comes the difficult task, designing an Arsenal Ship that is affordable to build, maintain and operate. Historically, personnel costs have had great impact on LCC. Given the basic design structure, an analysis of manning is required to determine the minimum optimal manning level required to support mission requirements. In order to reduce crew size to its lowest feasible level, the most cost-effective investment in automation must first be determined.

The Arsenal Ship's mission is to provide an arsenal of weapons to end users within a theater of operations. "The Arsenal Ship is an ominous platform because it will have one principal mission: to inflict overwhelming, punishing damage" [Ref. 12]. Because the Arsenal Ship's mission resembles a fleet logistics ship more than a combatant, the manning concept is very different from a combatant. Senior ranking, highly trained, professional personnel must be selected to man the Arsenal Ship. Cross training must be extensive across all ratings. Therefore, this analysis is specialized for the Arsenal Ship and is not applicable to other warships, for example DDG-51 or SC-21. At its point of departure, this thesis utilizes the TSSE Arsenal Ship design, with minor alterations, to analyze reduced manning for the Arsenal Ship.

II. PROBLEM DESCRIPTION

A. LOCATIONS

Before explicit manning can be analyzed, location of Forward Operating Bases (FOB) and homeport must be considered. FOB and homeport locations directly impact ship deployment cycles and crew deployment cycles, which affect overall manning levels.

1. Forward Operating Bases

The Arsenal Ship Executive Summary identifies six possible FOB locations. FOBs will be used for Intermediate Maintenance Availabilities (IMAV), crew turnovers, replenishing provisions and fueling.

Transit times from FOBs to the theaters of interest are replicated from [Ref. 1] in Table 2. Forward-deploying ships to two locations vice three will reduce FOB overhead costs and the number of Arsenal Ships required. After examining transit times between the proposed FOBs, it becomes apparent that forward-deploying the Arsenal Ship to Diego Garcia and Guam is a rational decision. Transit time to all three theaters is minimized (24.2 days) when operating from Diego Garcia, making it the best overall FOB. The ability to arrive in NEA within about four days is also beneficial. Diego Garcia is a secure location since it is under long term lease from Great Britain. The ability to arrive in NEA within three days and the fact that it is a U.S. Territory also make Guam a prime FOB site. Assigning each FOB a primary and secondary theater of operations will ensure continuous coverage of at least two theaters. At 20 knots, it requires less than five days for each Arsenal Ship to transit to its primary theater. It requires less than 15 days to transit to each other's primary theater. The Arsenal Ship can enter the Mediterranean Sea within 20 days from Guam and within 10 days from Diego Garcia. However, most areas of instability can be covered without transiting the Suez Canal. Furthermore, Arsenal Ships returning from the continental United States (CONUS) are within seven days of the Mediterranean Sea if homeported on the east coast.

Forward Operating Base	North East Asia	Mediterranean Sea	South West Asia
Augusta Bay, Sicily	18	1	8
Diego Garcia	10	10	4.2
Guam	3	19	13
Yokosuka, Japan	2	20	14
Souda Bay, Crete	18	1	7
Sasebo, Japan	0	18	12.5

Table 2. After Ref. [1] Transit Days From Forward Operating Base at 20 Knots

2. Homeport

The homeport is the location for all overhauls and the Central Training Site (CTS). It is also the homeport for the families of the deploying crews. This allows the deploying crew to be home during training. In an effort to lower LCC, one location should be selected as the homeport for all Arsenal Ships.

The primary Naval bases considered in this thesis are located in Norfolk, Virginia and San Diego, California. When transit time to and from the FOBs and the theaters is the primary concern, the homeport of choice is San Diego. The transit time from San Diego to Diego Garcia and San Diego to Guam is 2/3 of the transit time from Norfolk. This thesis did not take any other factors involved with locating Naval bases into account.

B. MAINTENANCE

1. Overhaul

Overhauls must be routinely conducted because of the limited maintenance accomplished while deployed and the short IMAV periods. The overhaul will be conducted by an overhaul crew trained specifically for Arsenal Ship overhauls. The overhaul crew, which is assigned by the Arsenal Ship Squadron Commander, will be responsible for maintaining a Quarterdeck Watch, Fire Party and Security Force. Overhauls will be managed by the Arsenal Ship Squadron Commander in homeport. Therefore, this thesis assumes overhauls will be conducted in San Diego, California.

2. Intermediate Maintenance Availability

Each ship spends one month in an IMAV to complete urgent repairs and all routine maintenance with a periodicity of 90 days or more. The Arsenal Ship design should include concepts which increase periodicity of maintenance requirements beyond the normal deployment length of the Arsenal Ship. In this thesis, maintenance required while deployed is reduced by increasing periodicity between maintenance requirements to 90 days. Maintenance with a periodicity of less than 90 days must be performed by the deployed crew.

C. SHIP DEPLOYMENT CYCLES

To develop a "steady state" for Arsenal ships to operate, the following ship deployment cycle is selected. This deployment schedule requires five Arsenal Ships. Ships will deploy for 84 day periods with overlapping intervals and will be relieved on station. After nine deployments each ship will have four weeks to return to homeport for an overhaul. The percentage of time a ship is in each phase of a deployment cycle is provided in Table 3. The transit phase is the time allotted to transit to and from overhauls conducted in homeport. Each Arsenal Ship is operational 60 percent of the time using this method. Three of five ships are deployed at any one time. An additional ship is in a maintenance availability and can be deployed within a few days depending on the type of maintenance in progress. This provides 100% coverage of two theaters of operation at all times with a ready recall ship available. Appendix A provides a five ship deployment schedule. The initial forward deployments in Appendix A are irregular lengths allowing a steady state process to develop.

Phase	Time in Phase
Transit	4.4%
IMAV	17.8%
Deployed	60.0%
Overhaul	17.8%

Table 3. Percentage of Time in Each Phase

D. CREW DEPLOYMENT CYCLES

The following crew rotation is selected to resemble a "steady state" process. Crews are homeported in San Diego, California. A typical crew tour is outlined in Table 4. Sixteen weeks of initial training is followed by four weeks of team training conducted in a simulator at the CTS. After qualifying in the simulator, the crew flies to the FOB for crew turnover. After a one week turnover, the crew has two weeks to prepare the ship for deployment. The crew then deploys for 12 weeks to its primary theater of operations. Upon returning from deployment to the FOB, the crew initiates an IMAV and has one week to turn over the ship to the next crew. Once the crew returns to homeport, each crew member receives 16 weeks of combined stand down and specialty training. The type of specialty training is determined by the needs of the Navy and the professional development of the sailor. Specialty training does not necessarily have to pertain to the Arsenal Ship. The crew is reunited after the additional training phase and attends a 4 week team trainer in the Arsenal Ship simulator. After requalification, the deployment cycle begins again. The crew returns to homeport after the 16 week deployment and is detached from Arsenal Ship duty.

Phase	Length
Initial training	16 weeks
Team training	4 weeks
Deployed	16 weeks
Additional training	16 weeks
Team training	4 weeks
Deployed	16 weeks

Table 4. Typical Arsenal Ship Crew Tour

The crew on a ship scheduled for overhaul has four weeks of additional time at sea while returning the ship to homeport. Following arrival in homeport, they are relieved by the overhaul crew. If the crew returning from deployment has completed two deployments, they may request to stay onboard and augment the overhaul crew. At the end of the overhaul, the overhaul crew turns the ship over to the scheduled deploying crew. The scheduled deploying crew transits to the primary theater, completes a scheduled deployment and returns the Arsenal Ship to a FOB.

Because crew size is small and every individual is important, every effort is made in the schedule to promote high morale, crew stability, and comprehensive individual and team training. These standards should not be compromised for efficiency or cost savings as such savings were gained by minimizing the crew size to start with.

E. SPECIAL EVOLUTIONS

1. Flight Operations

Flight operations will require full crew participation. Flight operations will be utilized for all transfers at sea. Flyaway teams will be required to fly via fixed wing aircraft to the closest aircraft carrier for further transfer via helicopter to the Arsenal Ship. The following flight deck manning is proposed:

1. On Scene Leader
2. Team Leader
3. LSE
4. Nozzleman
5. Hoseman (2)
6. Hotsuitman (2)
7. Chock and Chainman (2)

The Watch, Quarter and Station Bill (WQSB), Appendix B, outlines all manning positions during flight quarters.

2. Underway Replenishment

The Arsenal Ship will be designed for underway replenishment. However, connected replenishment (CONREP) will not normally be required. The Arsenal Ship will carry more than adequate fuel to remain deployed for 90 days. CONREP is manpower intensive and should be avoided if possible. The following UNREP station manning is proposed:

1. Rig Captain
2. Signalman
3. Linehandler (7)

Vertical replenishment (VERTREP) will be the normal replenishment method. A non-organic helicopter supplied by the replenishment ship will be utilized. VERTREP is much less manpower intensive than CONREP. VERTREP manning is the same as flight operations manning except the chock and chainmen are cargo handlers.

The WQSB, Appendix B, outlines all manning positions during UNREP.

3. General Quarters

General Quarters will be required for two specific instances. Major damage control and weapons launching will require entire crew involvement at their specialty locations.

a. Launch

Since launching weapons is the Arsenal Ship's mission, it is the most important evolution the ship will perform. General Quarters will be set to maximize the probability of success.

b. Damage control

Major damage control is the most manpower intensive evolution the crew will be required to perform. To insure maximum survivability, total crew involvement and cross-training will be required. The following damage control team manning is proposed:

1. On Scene Leader
2. Team Leader
3. Nozzleman (2)
4. Hoseman (4)
5. Plugman (2)

The WQSB, Appendix B, outlines all manning positions during General Quarters.

4. Concurrent Special Operations

Concurrent special operations are not possible with the proposed manning. General Quarters for launching does not support damage control. Once damage is anticipated or realized, the Arsenal Ship will be required to transition to the damage control posture. A damage control posture includes the use of self defense weapons, but it does not include offensive weapon firing.

III. ANALYSIS

A. INTRODUCTION

In order to estimate the most cost-effective investment in automation, manning costs incurred over the Arsenal Ship's expected 30 year service life must be determined. To develop a manning LCC curve suitable for evaluating tradeoffs between manning levels and automation, a minimum of two alternative manning structures are required. This thesis develops four such manning structures. A simplified model, designed by the author and calibrated to the Navy Billet Cost Factor (NBCF) Model results, is used to estimate LCCs for each of the manning structures. The estimated costs of the four manning structures are used to calculate a life cycle cost curve using regression analysis. This cost curve is then used to examine the tradeoffs between manning levels and automation and to estimate the most cost-effective investment in automation.

B. MANNING STRUCTURES

The Arsenal Ship manning cost curve is determined using four different manning structures. One manning structure was determined by the TSSE design team, while the other three structures were determined by the author. Table 5 is a comparison of the four manning structures.

1. Structure One: Minimum

The first manning structure is based on the minimum manning level required by existing law and practice for merchant ships as discussed in Chapter I.C. The Arsenal Ship is fully automated and all functions of the ship can be monitored from one location.

Manning Comparison				
Rank/Rate	Structure 1 Minimum	Structure 2 Proposed	Structure 3 TSSE	Structure 4 Maximum
CDR	1	1	1	1
LCDR	1	1	2	2
LT	3	3	6	7
BM1	1	1	1	1
BM2	2		2	2
DC1			1	1
DC2			1	1
EM1	1	1		
EM2		1	2	2
EMC			2	2
EN1			3	3
EN2		2	1	1
ENC	1	1	2	2
ET1		1	2	2
ET2		2	2	2
ETC		1		1
EW1			1	1
EW2			1	3
GMM1			1	1
GMM2			1	1
GMM3			4	4
HM1			1	1
MS1		1	1	1
MS2		1		
MS3			1	1
OS1		1		
OS2		2		2
QM1		1		
QMC			4	4
RM1			1	1
SM1		1		
Total	10	22	44	50

Table 5. Arsenal Ship Manning Comparison

This structure does not require special manning for flight operations, underway replenishment or General Quarters. The crew's responsibility is to monitor the Arsenal Ship's systems for safety.

2. Structure Two: Proposed

The second manning structure is a notional estimate proposed by the author. This manning structure is based on a three section watch rotation with continuous watches on the bridge, in combat information center (CIC), in sensor control, and in engineering. This structure satisfies the manning requirements for flight operations, underway replenishment, and General Quarters as discussed in Chapter II.E. The lower bound for this manning structure is determined by the manning requirements for damage control since it is the most manpower intensive evolution. Once a lower bound was established, the matching of duties to personnel became a matter of job related abilities and the author's experience. The ratings selected for the duties involved are not the focus of this thesis. Because specific manning requirements cannot be determined until the Arsenal Ship design has been accepted, the pay grades selected are the real focus. The manning requirement for boat operations is eliminated by not embarking a small boat. The manning requirement for a special physical security force is eliminated by training all personnel, arming all watchstanders and providing access to additional weapons at each watchstation. The WQSB, Appendix B, provides additional details about the proposed manning structure.

3. Structure Three: TSSE

The billet structure developed in the TSSE report is used for the third manning structure. This structure is based on a four section watch rotation. Billets are separated into two categories: Watchstanders and Special Evolutions. The WQSB for the TSSE manning is reproduced as Appendix C. Additional manning for boat operations and physical security is included in this structure.

4. Structure Four: Maximum

The fourth manning structure is a notional estimate in which personnel were added to the TSSE billet structure to achieve the maximum allowed crew of 50 personnel. One officer, one chief petty officer and four second class petty officers were added to the TSSE billet structure.

C. NAVY BILLET COST FACTOR MODEL

1. Background

The NBCF, Active Component Cost Estimation Model was utilized to calculate the initial LCC estimates for the four manning structures. The model was developed for the Bureau of Naval Personnel by SAG Corporation in 1995, and is based on cost data extracted from the Office of the Secretary of Defense pay tables, allowance tables and other Navy sources. The base year costs are in FY 94 dollars. The pay tables used were effective January 1, 1995. FY 93 end strength was used as the current inventory and budget data was extracted from the FY 93 - FY 94 President's Budget [Ref. 13].

An estimated cost for each rating and pay grade is calculated using a weighted average approach. The following excerpt from the *Navy Billet Cost Factor: Active Component Cost Estimation Model Operations Manual* [Ref. 13] explains how the average basic pay is determined.

> Average basic pay is determined by first multiplying the inventory of members of all grades, lengths of service (LOS), and ratings by the basic pay rate. This figure is then summed across LOS for each pay grade and rating and divided by the total inventory for that pay grade and rating.

This yields the average annual basic pay for each grade by rating.

The model includes the following elements of manpower cost:

1. Military Compensation (Basic pay, Basic Allowance for Quarters, Basic Allowance for Subsistence and Variable Housing Allowance)
2. Retired Pay Accrual
3. Selected Reenlistment Bonus
4. Special Pays (Hazardous Duty Pay, Flight Pay, Sea Pay etc.)
5. Training
6. Enlisted Recruiting
7. Medical Support (Military Hospitals and Civilian Health and Medical Program for the Uniformed Services)
8. Other Benefits (Death Gratuities, Deserter Rewards, Survivor Benefits, Clothing Allowances etc.)
9. Permanent Change of Station
10. Officer Acquisition
11. GI Bill
12. Separation Costs

2. Initial Life Cycle Cost Results

The 30 year LCC per ship was calculated for each manning structure using the NBCF Model. The LCCs are based on constant FY 94 dollars for the entire 30 year life cycle. The results are provided in Table 6.

Manning Structure	Crew Size	NBCF Model Results (Millions)
Minimum	10	23.41
Proposed	22	43.79
TSSE	44	84.61
Maximum	50	94.68

Table 6. NBCF Model Results

Figure 1 provides the linear regression utilizing the results from the model. The regression equation is:

$$30\text{ Year Life Cycle Cost} = (1.80 * \text{ Crew size}) + 4.98 \qquad (1)$$

$$r^2 = 0.9997;\ n = 4;\ p = 0.0002$$

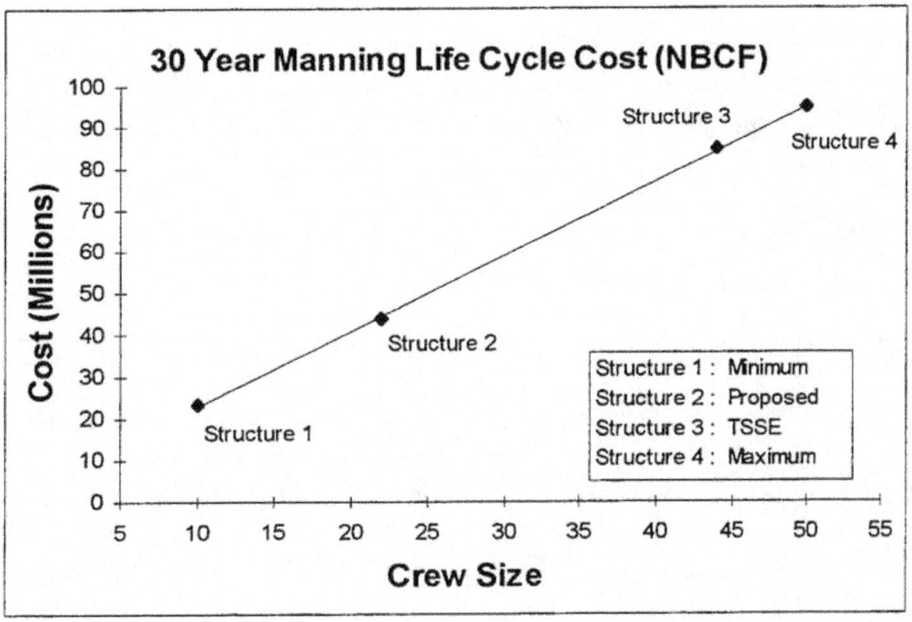

Figure 1. Initial Navy Billet Cost Factor Model Results

D. SIMPLIFIED BILLET COST MODEL

1. Background

The second model is a Simplified Billet Cost Model (SBCM) that was developed by the author as a simplified and updated model to better suit the purposes of this thesis.

The SBCM excludes data elements not required by this thesis and uses the maximum rate per pay grade instead of a weighted average calculated from a previous end strength. Since this thesis assumes that the Arsenal Ship will be homeported in a specific location, namely San Diego, the SBCM was designed to utilize specific housing allowances rather than the Navy-wide weighted average used by the NCBF Model. The SBCM can be updated to accommodate housing allowances for any location. Finally, the SBCM includes 1997 cost data obtained from the *1997 Navy Times Pay Chart* [Ref. 14]. The resulting SBCM utilizes an Excel® spreadsheet to calculate billet costs based on Basic Pay, Basic Allowance for Quarters (BAQ), Basic Allowance for Subsistence (BAS), Variable Housing Allowance (VHA) for San Diego, Retired Pay Accrual and various projected annual inflation rates. Retired Pay Accrual is assumed to be 36.4% of basic pay as in [Ref. 8].

2. Initial Life Cycle Cost Results

For comparison, the LCCs for each manning structure were calculated using the SBCM. Pay tables [Ref. 15] effective January 1, 1995, were used to correspond with the data in the NBCF model. To represent constant dollars, no annual inflation rate was used. The 1993 VHA (with dependents) rates [Ref. 16] for San Diego were used realizing that they probably do not match the weighted average VHA rate in 1993. The results are provided in Table 7.

Manning Structure	Crew Size	SBCM Results (Millions)
Minimum	10	19.20
Proposed	22	35.01
TSSE	44	69.26
Maximum	50	77.37

Table 7. SBCM Results

Figure 2 provides the linear regression utilizing the results from the model.. The regression equation is:

$$30 \text{ Year Life Cycle Cost} = (1.48 * \text{ Crew size}) + 3.71 \tag{2}$$

$$r^2 = 0.9989;\ n = 4;\ p = 0.0005$$

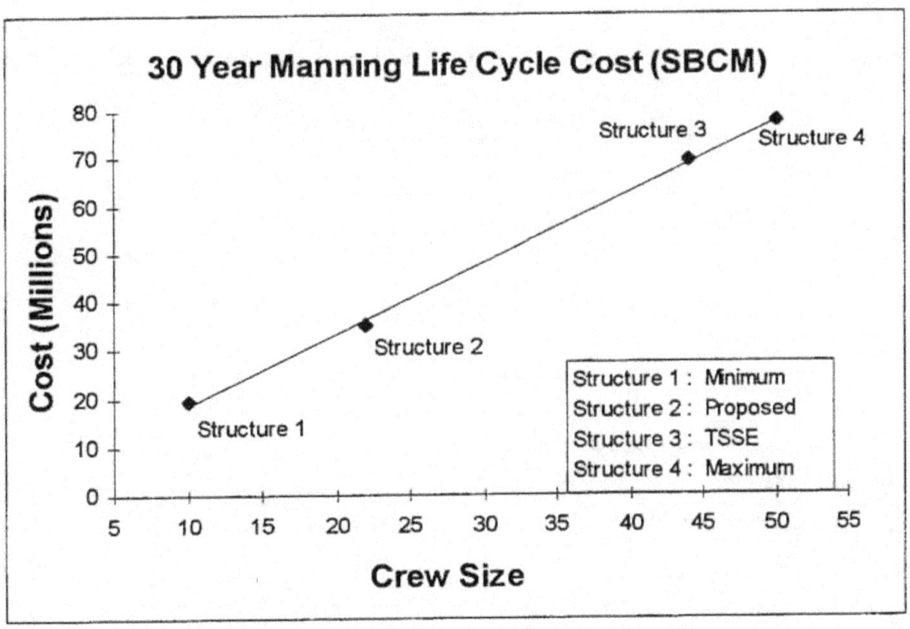

Figure 2. Initial Simplified Billet Cost Model Results

E. MODEL COMPARISON

As can be seen in Figure 3, both models produced similar slopes for the 30 year manning LCC curve with constant dollars. The difference between the slopes is 0.32. The

SBCM underestimates manning LCCs because accession costs, healthcare costs, specialty pay, etc. are not included in the model. To include an estimator for these costs and calibrate the SBCM, the absolute difference between the models was calculated for each manning structure. The proportionate difference was then calculated, to reduce the compounding difference between the models as manning increased, using the NBCF Model. The results are provided in Table 8. The mean proportionate difference is $\overline{X}=$ 0.1861, with a standard deviation of $s = 0.0097$. Since the distribution of the estimated factor is unknown, Chebyshev's inequality is used to determine a confidence interval. A correction factor of 0.15 ($\overline{X} - 3s$) is used to maintain 94 percent confidence that the true factor is not less than the estimate. Equation 3 shows the relationship between the NCBF Model Estimate (Calibrated SBCM) and the SBCM.

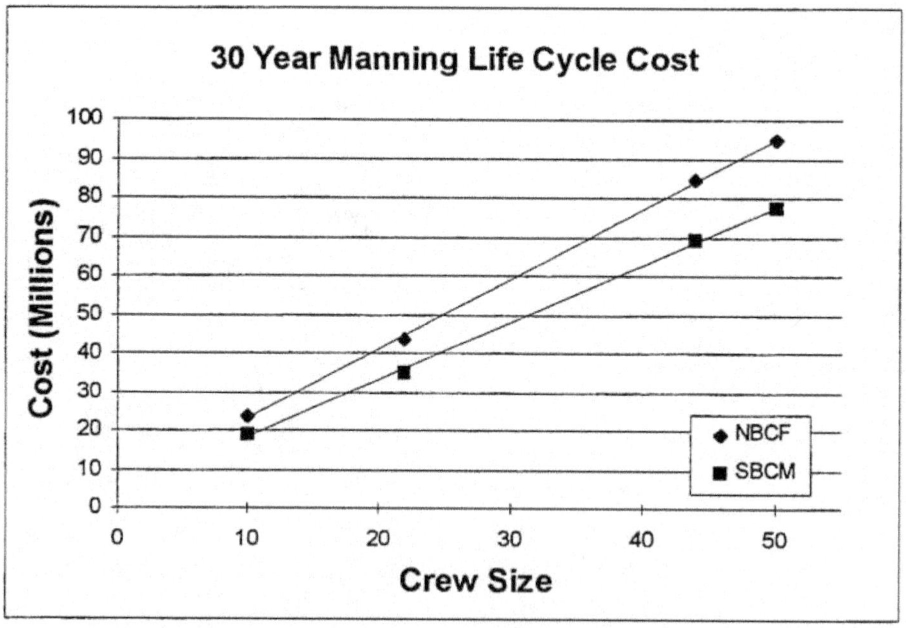

Figure 3. NBCF and SBCM Comparison

Crew Size	Model Results: NBCF	Model Results: SBCM	Difference Between NBCF and SBCM	Proportionate Difference[1]	NBCF Model Estimate	Difference Between NBCF and NBCF Model Estimate
10	23.41	19.20	4.21	0.1797	22.78	0.63
22	43.79	35.01	8.78	0.2005	41.53	2.26
44	84.61	69.26	15.35	0.1815	82.16	2.45
50	94.68	77.37	17.31	0.1828	91.79	2.89

[1] Proportionate Difference = Difference Between NBCF and SBCM / NBCF Results

[2] NBCF Model Estimate = (0.15 * MBCF Model Result) + SBCM Result

Table 8. SBCM Calibration Utilizing the NBCF Model

NBCF Model Estimate = (0.15 * NBCF Model Actual) + SBCM Results (3)

Since the NBCF Model Estimate is approximately equal (within 6%) to the NBCF Model Actual Result, substituting the NBCF Model Estimate for the NBCF Model Actual in Equation 3 and solving for the NBCF Model Estimate yields:

NBCF Model Estimate = 1.18 * SBCM = Calibrated SBCM (4)

The results from the NBCF Model , SBCM, and NCBF Model Estimate are compared in Figure 4. The calibrated 30 year LCC curve is :

Calibrated 30 Year Life Cycle Cost = (1.75 * Crew size) + 3.74 (5)

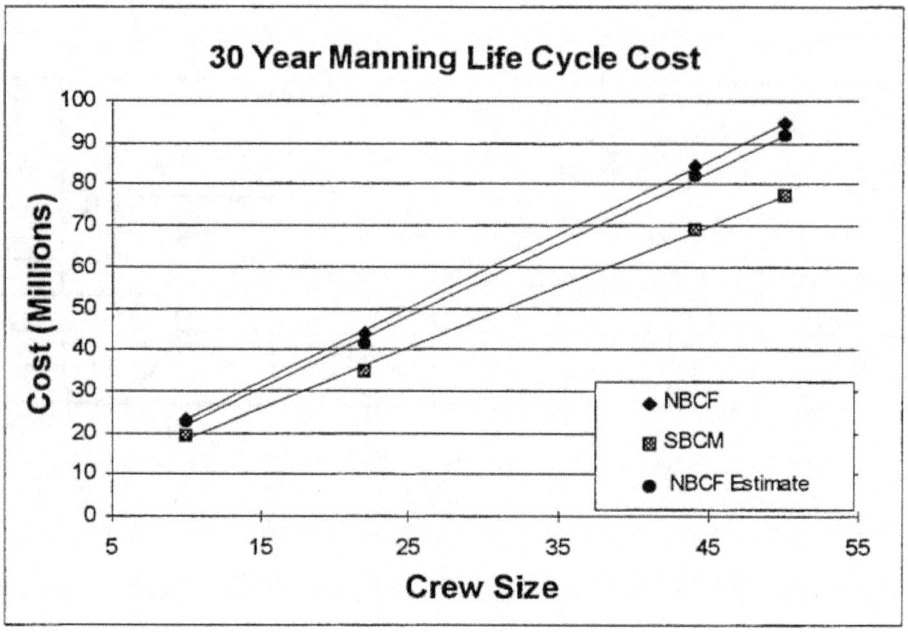

Figure 4. NBCF Model, SBCM, and NBCF Model Estimate Comparison

Since the SBCM underestimates the LCC, the maximum amount to spend in automation will also be underestimated. This provides a built-in buffer ensuring the maximum amount to spend on automation is not overestimated. The SBCM has the advantage of being a spreadsheet model, which makes it easier to use. Furthermore, it can be updated to include San Diego VHA rates and 1997 cost data which makes it more accurate for the purposes of this thesis. Therefore, the calibrated SBCM was used to produce the 30 year manning LCC curve using current year dollars and pay charts, which is the basis for Arsenal Ship manning analysis in this thesis.

F. INDIVIDUAL ARSENAL SHIP MANNING ANALYSIS

Once the 30 year manning LCC curve is determined, it can be used to calculate the cost of adding an additional billet to the billet structure of a single Arsenal Ship. The slope of the curve is the cost per billet over the 30 year service life of the Arsenal Ship.

The maximum amount available to invest in automation, while remaining cost-effective, is determined by calculating the difference between the LCC of the current manning level and the LCC of the proposed manning level and multiplying by the cost per billet obtained from the slope of the LCC curve. For example, using the calibrated SBCM result (Equation 5), the estimated maximum investment in automation to offset a reduction in manning from 44 (Structure 3) to 22 (Structure 2), is $38.5 million.

G. ARSENAL SHIP TOTAL PROGRAM MANNING ANALYSIS

Since maintenance and training are supported ashore, the Squadron Staff and Overhaul Detachment in homeport as well as the Maintenance Detachments at the FOBs must also be considered. As the number of deploying personnel is reduced, the number of support personnel ashore must increase if maintenance requirements remain the same. In some cases, maintenance requirements may increase as the maintenance responsibility is shifted ashore. In order to calculate manning costs, a proposed manning structure had to be designed. Appendix B details a proposed Squadron Staff and Overhaul Detachment as well as a proposed Maintenance Detachment for the purpose of this thesis.

Now that a model to determine the cost of additional manning per ship has been developed, the total program cost of adding an additional billet to the Arsenal Ship can be calculated. Since this thesis assumes a crew deploys from a FOB every four weeks and it takes 36 weeks to cycle a crew, as discussed in Chapter II, nine crews are required to man five Arsenal Ships. Therefore, each additional billet costs the total Arsenal Ship program nine times the cost of adding a single billet to one ship.

H. AUTOMATION COST

In addition to calculating the LCC curve for manning, the cost curve for manning vs. automation is useful. Minimizing the combined LCC for manning and the cost curve for automation identifies an initial crew target size and the optimal investment in automation. The automation cost curve should be calculated by each contractor and supplied as part of their bid proposal since it is dependent on how their design implements innovations in automation. The Navy currently has no data base for estimating a manning vs. automation cost curve. The Naval Center for Cost Analysis maintains the Navy's historical data using the Navy Visibility and Management of Operations and Support Costs (VAMOSC) [Ref. 13]. VAMOSC does not contain the data required to calculate an automation cost curve since there are no systems specifically designed for the purpose of reduced manning.

The civilian sector does maintain automation cost data. McDermott Shipbuilding Incorporated provided automation cost data for a 524 ft. container ship. Their designs include unmanned engine rooms, with crew sizes ranging from 13 to 19 personnel. The manning structure for a crew of 19 is provided in Table 9. The cost of automation, to reduce a crew by approximately 20 personnel, is $275,000 to $325,000 in material and 5,000 man hours of labor [Ref. 14]. Since a container ship does not man weapon systems, the cost of automation is in updating engineering and bridge control stations.

The TSSE design team calculated a unit cost of $487 million for the Arsenal ship. This includes the acquisition cost of the T-AO, combat systems upgrades, and automation costs using Smart Ship Implementation costs [Ref. 9]. To estimate the cost of automating

Job Description	Quantity
Captain	1
Chief Engineer	1
1st Officer	1
1st Engineer	1
2nd Officer	2
2nd Engineer	2
Radio Officer	1
Boswainmate	1
Cook	1
Fitter (cook help)	1
Seaman	4
Motormen	3

Table 9. Manning Design by McDermott Shipbuilding Inc.

the Arsenal Ship, the acquisition cost ($167 million) and the Vertical Launch System cost ($215 million) are subtracted from the unit cost since neither cost contributes to automation. The remainder, approximately $105 million, is the estimated cost of automating the Arsenal Ship under the TSSE design. Not all of the $105 million is directly related to automation cost. Therefore, this is an overestimate of automation cost.

IV. RESULTS

A. INDIVIDUAL ARSENAL SHIP MANNING RESULTS

1. Constant Dollars

A 30 year manning LCC for a single Arsenal Ship was calculated using the calibrated SBCM along with 1997 pay charts and 1997 VHA rates for San Diego. Table 10 provides the results from the model.

Manning Structure	Crew Size	Calibrated SBCM Results (Millions)
Minimum	10	23.75
Proposed	22	43.21
TSSE	44	85.43
Maximum	50	96.20

Table 10. Calabrated SBCM Results

Figure 5 provides the linear regression utilizing the results from the model. The regression equation is:

$$\text{30 Year Life Cycle Cost} = (1.83 * \text{Crew size}) + 4.40 \qquad (6)$$

$r^2 = 0.9990$; $n = 4$; $p = 0.0005$

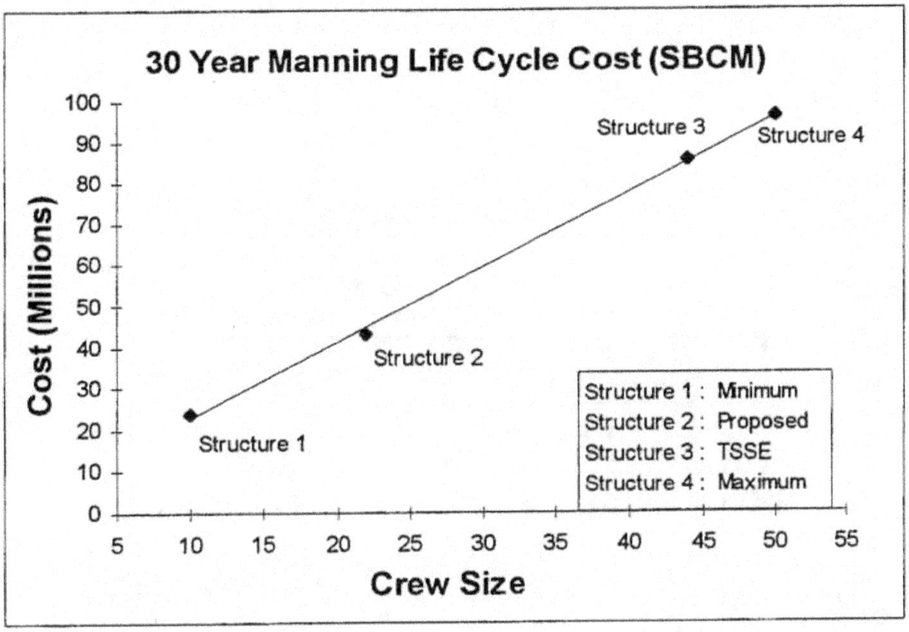

Figure 5. Simplified Billet Cost Model Results, Constant FY 97 Dollars

The 30 year manning LCC curve slope of 1.83 remains consistent with the cost curves obtained from the NBCF and the SBCM in Chapter III. Using Equation 6 above, the maximum investment in automation (constant FY 97 dollars) to offset a reduction in manning from 44 (Structure 3) to 22 (Structure 2) is $40.3 million. When considering the construction of a single Arsenal Ship, using the TSSE design, it would be cost-effective to spend an additional $40.3 million in automation to reduce the crew size by 22 crew members. The total TSSE design cost would increase from $487 million to $527 million, which is below the maximum allowed acquisition cost of $550 million [Ref. 9].

2. Inflation

Since automation cost is considered an initial investment cost in this thesis, inflation does not affect automation cost. However, inflation does affect manning cost throughout the 30 year service life. Therefore, a 30 year manning LCC for a single

Arsenal Ship was calculated using the calibrated SBCM with an annual three percent inflation rate, 1997 pay charts and 1997 VHA rates for San Diego. Table 11 provides the results from the model.

Manning Structure	Crew Size	Calabrated SBCM Results (Millions)
Minimum	10	37.66
Proposed	22	68.52
TSSE	44	135.49
Maximum	50	152.56

Table 11. Calabrated SBCM Results with 3% Annual Inflation

Figure 6 provides the linear regression utilizing the results from the model. The regression equation is:

$$30 \text{ Year Life Cycle Cost} = (2.91 * \text{Crew size}) + 6.98 \quad (7)$$

$$r^2 = 0.9990; n = 4; p = 0.0005$$

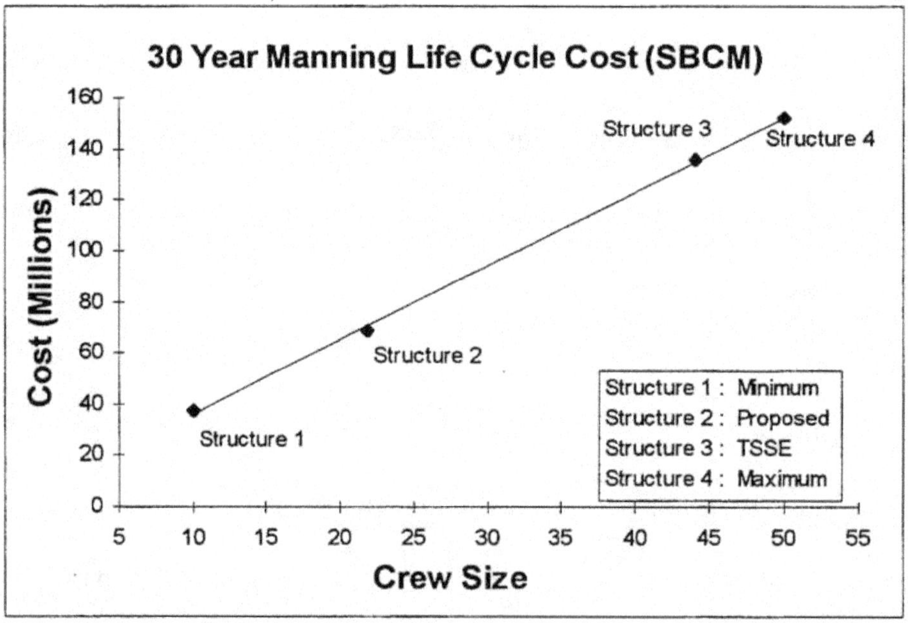

Figure 6. Simplified Billet Cost Model Results, 3% Annual Inflation

Using the TSSE design and Equation 7, it would be cost-effective to spend $64 million on automation to reduce the crew size from 44 to 22 crew members. The total TSSE design cost would increase from $487 million to $551 million, which is one million dollars more than the maximum allowed acquisition cost.

B. ARSENAL SHIP TOTAL PROGRAM MANNING RESULTS

1. Additional Billet Costs

Now that the cost of additional manning per ship has been determined, the impact of adding an Arsenal Ship billet to the entire Arsenal Ship program can also be calculated. As discussed in Chapter III.G, each additional Arsenal Ship billet costs nine times the cost of a single billet. Since inflation affects manning cost, Equation 7 is utilized. The slope from Equation 7 is multiplied by nine to calculate the cost when adding an Arsenal Ship billet to the Arsenal Ship program. Therefore, the cost of adding one Arsenal Ship billet is $26 million ($2.85 million * 9). The maximum cost-effective investment in automation to reduce the crew size by 22 crew members is $572 million. Consequently, $114.4 million would be available for automation in each of the five ships. This method to estimate the investment in automation is valid as long as the number of ships and the number of crews required to man the ships are known.

2. Annual Manning Costs

Using the calibrated SBCM, Equation 6 (constant FY 97 dollars), and the manning structures detailed in Appendix B, the estimated manning cost of a Squadron Staff and Overhaul crew is $7.7 million per year while the estimated manning cost of each FOB maintenance detachment is $5.6 million per year. Therefore, the total annual program manning cost is approximately $32 million for five Arsenal Ships with manning Structure 2 and two FOBs. Figure 7 displays a proportional representation of the annual total program manning cost.

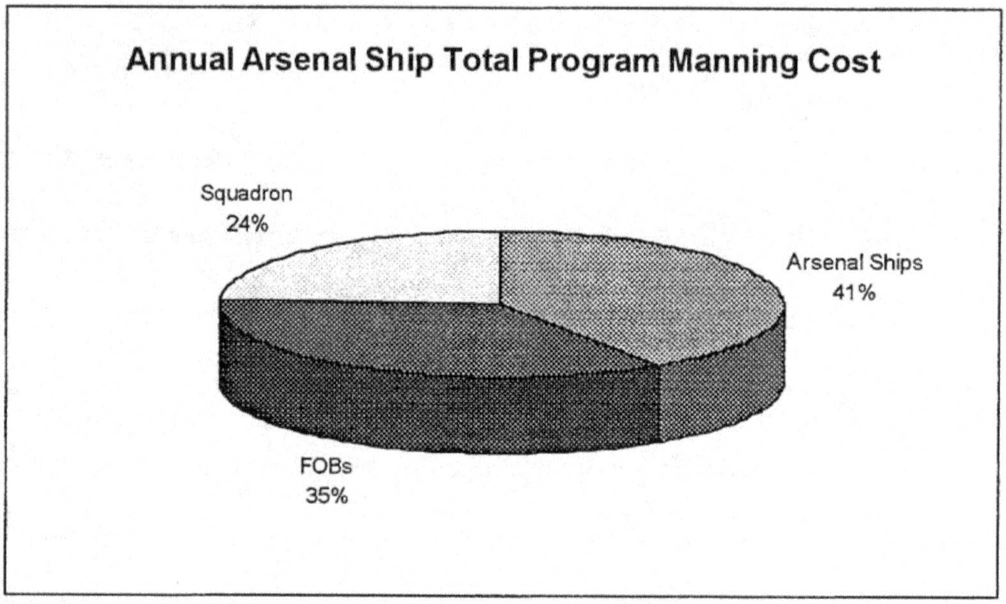

Figure 7. Proportional Annual Total Program Manning Cost Using Structure 2

C. AUTOMATION RESULTS

Before the automation cost curves are discussed, a modification to the 30 year LCC must be identified. Since the automation cost curve will reflect the cost of automation for one ship, the 30 year manning LCC must be scaled to reflect the mean cost of manning per ship. This thesis identified nine crews for five ships. Therefore, the LCC curve must be factored by 9/5. The modified 30 year LCC curve is:

$$\text{Mean 30 Year Life Cycle Cost} = (5.24 * \text{Crew size}) + 12.56 \quad (8)$$

1. Linear Automation Cost Curve

After identifying a 30 year LCC curve and an automation cost curve, the combined cost curves can be used to identify an initial target crew size and optimal investment in automation. Since generalized automation cost data is unavailable and the cost curve depends on the particular technology utilized, the author developed an automation cost curve for illustration purposes only. Assuming the TSSE design can be manned by a crew of 100 with minimal automation cost ($1 million) and the estimated automation cost for a crew of 44 calculated in Chapter III.G ($105 million) is reasonable, the automation cost to reduce manning by 54 crew members (100 - 44) is $104 million. This yields an automation cost slope of 1.9. Using this slope and the TSSE design cost of $105 million for a crew of 44 , the following linear automation cost curve was formulated:

$$\text{Automation Cost (millions)} = (-1.9 * \text{Crew size}) + 189 \qquad (9)$$

Treating the mean LCC curve as a supply curve and the linear automation curve as a demand curve, the optimal investment in automation can be determined by adding the two curves. The minimum point on the combined curve (called the average total cost curve in economics) is the optimal investment in automation. The mean LCC curve (Equation 8), linear automation cost curve (Equation 9) and the combined cost curve are depicted in Figure 8. In this example, the target crew size is 0, or the minimum point of the combined cost curve. This corresponds to a 30 year manning LCC of $13 million. The optimal investment in automation is $189 million. The total combined investment in manning and automation is $202 million per ship. The results indicate that a linear automation cost curve is inaccurate at the tails.

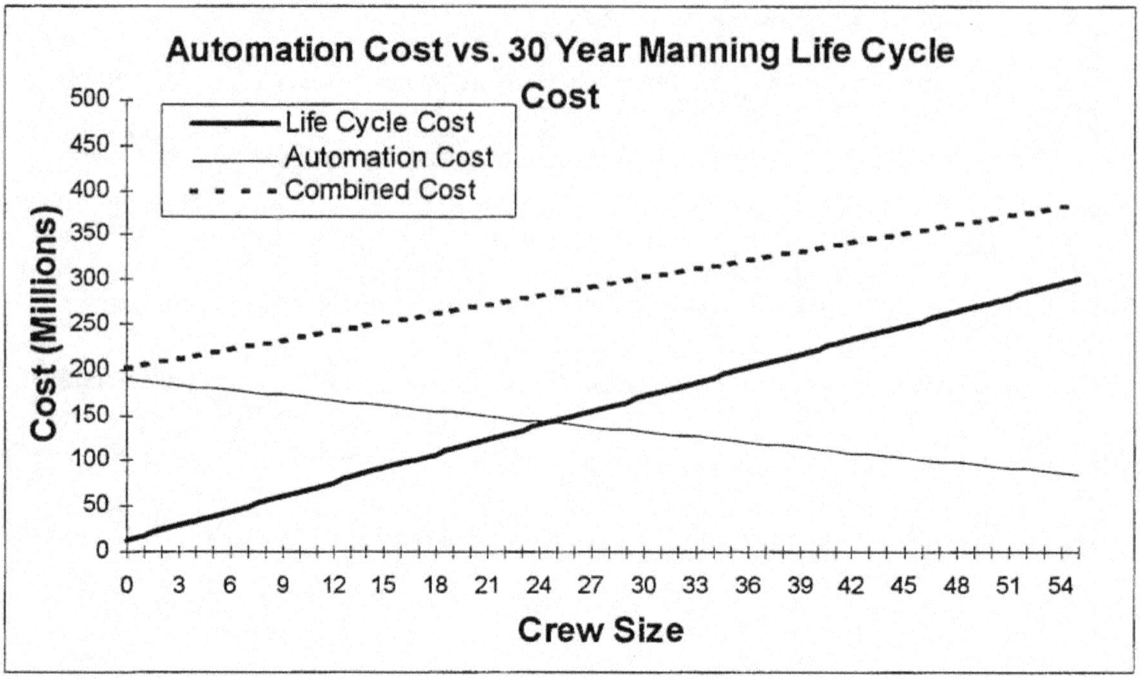

Figure 8. Linear Automation cost vs. Mean 30 Year Manning Life Cycle Cost

2. Exponential Automation Cost Curve

As the crew size approaches zero, the expected cost of technology increases dramatically. Therefore, the automation cost per man increases rapidly as well. On the other hand, as crew size approaches a level in which a four section watch rotation can be implemented, the cost of automation approaches zero because the manning can be reduced by reducing the watch rotation. Therefore, an exponential function was also formulated because it produces more realistic results at the tails. A substantially large value, the maximum allowed procurement cost for the Arsenal Ship ($550 million), was selected as the cost for total automation (y-intercept). Lambda was determined by forcing the function through the estimated automation cost of $105 million for 44 crew members and solving for lambda. The following exponential automation cost curve was formulated:

$$\text{Automation Cost (millions)} = 550\ e^{-(\lambda * \text{Crew size})} \quad \text{where } \lambda = 0.04 \qquad (10)$$

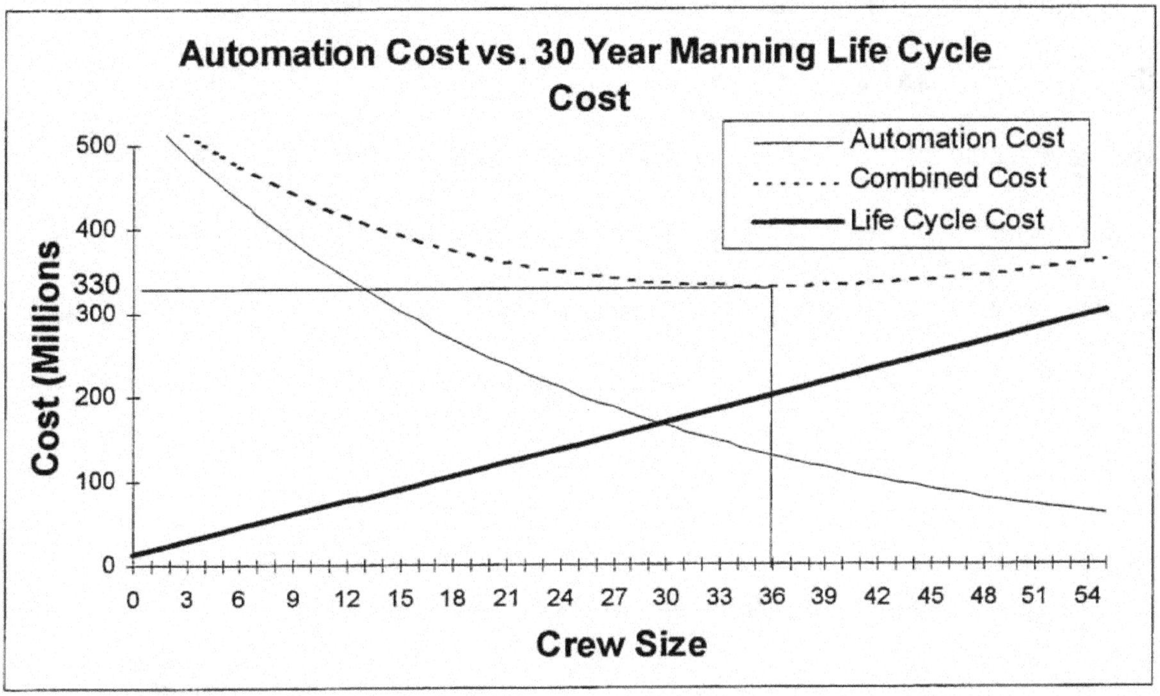

Figure 9. Exponential Automation Cost vs. Mean 30 Year Manning Life Cycle Cost

The mean 30 year manning LCC curve (Equation 8), exponential automation cost curve (Equation 10) and combined cost curve are depicted in Figure 9. The target crew size from the exponential automation cost curve is 36 for a 30 year manning LCC of $200 million. The optimal investment in automation is $130 million and the total investment in manning is $330 million per ship. However, the combined cost curve is so flat around the minimum that a manning level anywhere between 24 and 49 appears to be justified. This is only true, of course, if the exponential automation cost curve is approximately correct.

D. SENSITIVITY ANALYSIS

1. Inflation

Inflation rates tend to fluctuate, therefore, a sensitivity analysis was conducted on inflation. As discussed in Chapter IV.A.2, an increase in inflation will increase the initial investment in automation. The 30 year manning LCC curve from Equation 8, with 2%, 3% and 4% inflation rates is displayed in Figure 10. An increase in annual inflation from the baseline assumption of 3% to 4% will increase the manning LCC for the TSSE design by $47 million. This would imply that $47 million more should be invested in automation.

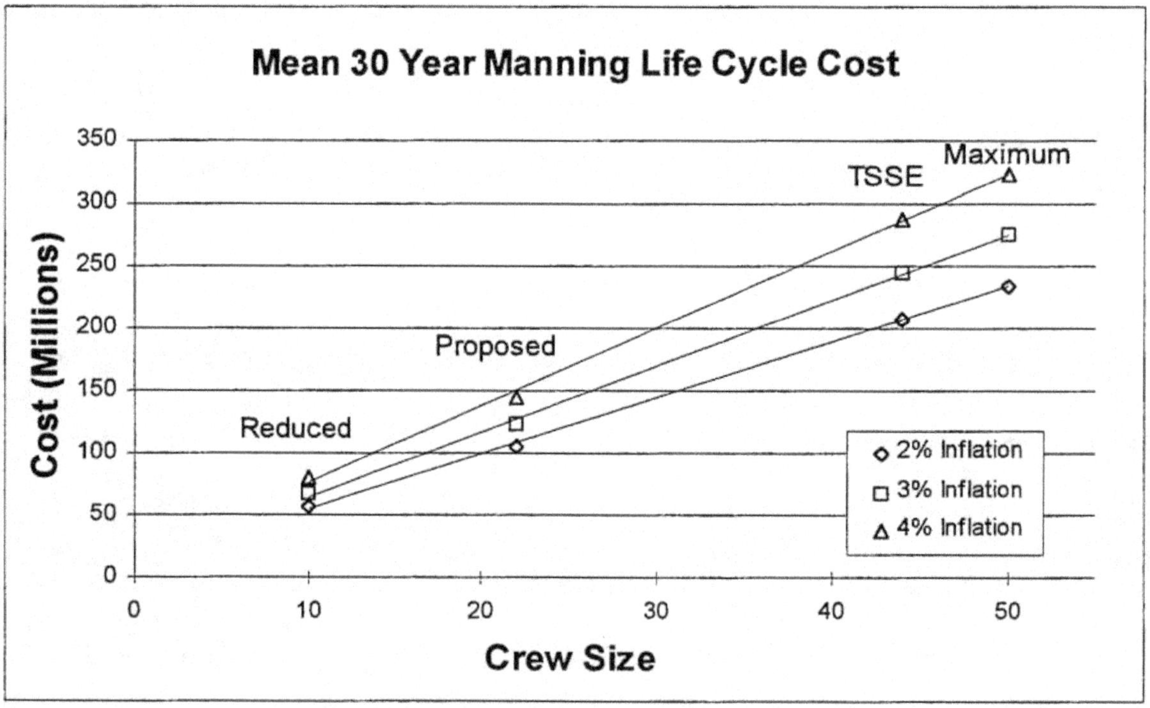

Figure 10. Inflation Sensitivity Analysis

2. Cost Curves

a. Linear Cost Curve

Sensitivity analysis was conducted on the linear automation cost curve by adjusting the slope and resolving for the intercept. The linear cost curve indicates that a combination of maximum automation and zero manning is optimal, until the magnitude of the slope is equal to the magnitude of the mean 30 year manning LCC curve as displayed in Figure 8. When these magnitudes are equal, the slope of the combined cost curve is zero as in Figure 11. In this case, the minimum combined cost is $233 million. However an unlimited number of cost-effective automation/manning combinations exists. If the magnitude of the automation cost curve slope exceeds the magnitude of the mean 30 year manning LCC curve, the linear cost curve indicates no automation, as in Figure 12. Therefore, a linear automation cost curve provides an all or nothing result and should be avoided if possible.

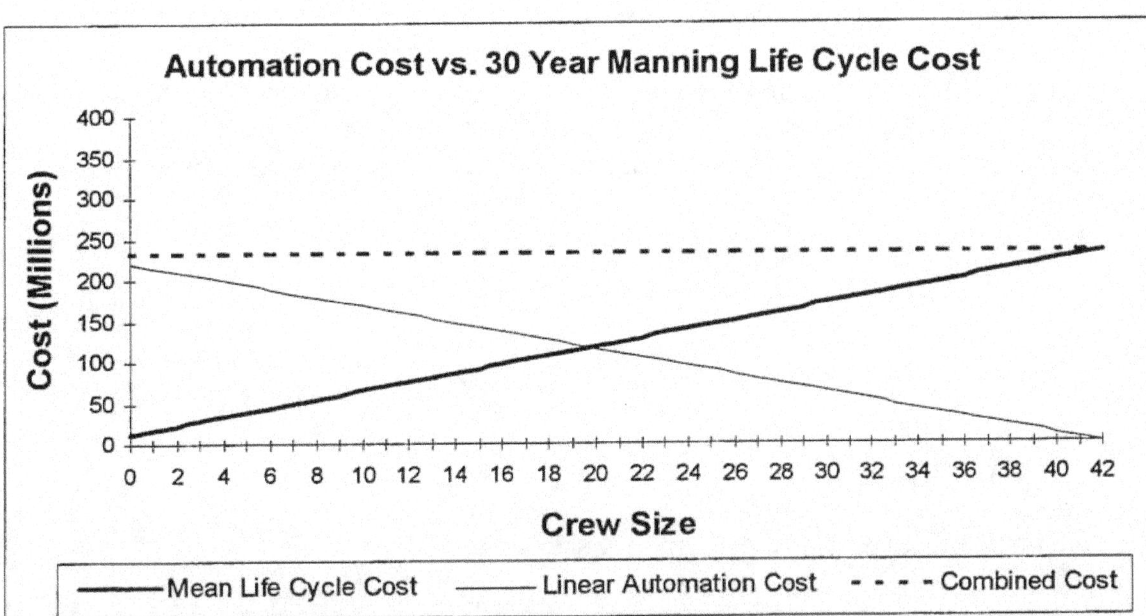

Figure 11. Sensitivity Analysis of Linear Automation Cost Curve (slope = -5.17)

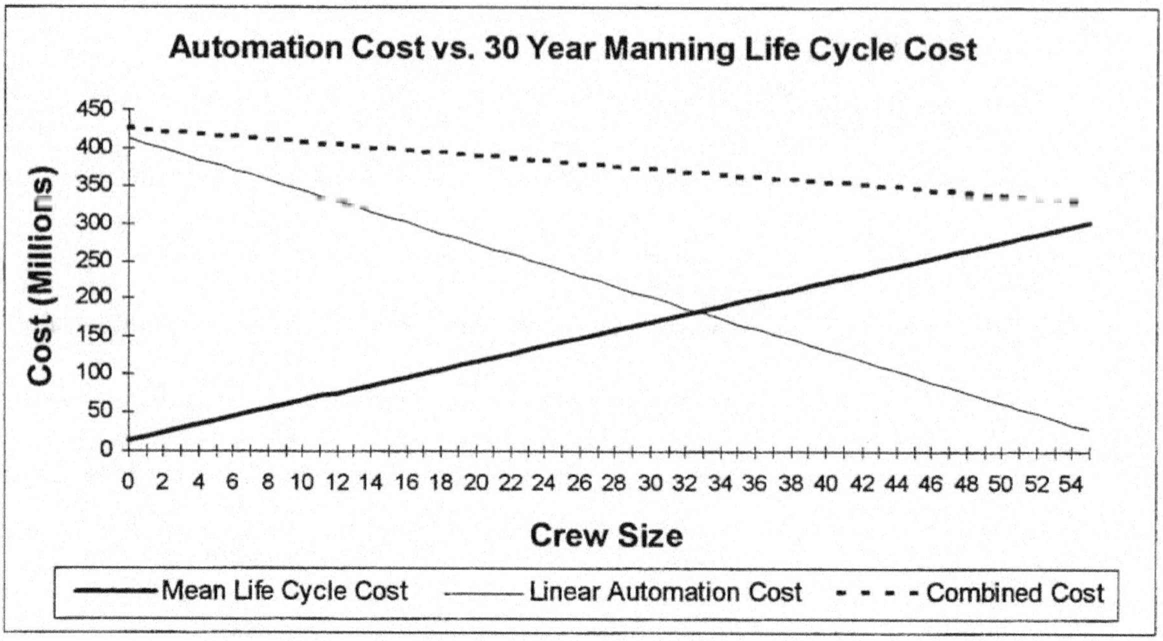

Figure 12. Sensitivity Analysis of Linear Automation Cost Curve (slope = -7)

b. Exponential Cost Curve

Sensitivity analysis was also conducted on the exponential automation cost curve by adjusting lambda $\pm$ 0.01. Since the LCC curve has an increasing slope and the exponential cost curve has a decreasing slope, the minimum of the combined cost curve is where the slopes have the same magnitude with different signs. Equation 11 is the combined cost curve using the LCC curve (Equation 8) and the automation cost curve (Equation 10) with lambda = 0.03.

$$\text{Combined Cost} = (5.24 * \text{Crew size}) + 12.56 + 550\, e^{-(\lambda * \text{Crew size})} \quad (11)$$

$$\lambda = 0.03$$

Taking the first derivative with respect to crew size yields:

$$5.24 - 16.5\, e^{-(0.03 * \text{Crew size})} \quad (12)$$

Setting Equation 12 equal to zero and solving for crew size produces a minimum crew size of 38. Following the same procedure for lambda = 0.05, a crew size of 34 is obtained.

The minimum cost and crew size can also be determined by graphing the magnitude of the slopes as displayed in Figure 13. Figure 14 displays the change in combined cost with respect to a change in lambda.. The crew size ranges from 34 to 39, while the optimal investments in automation range from \$100 million to \$171 million. Therefore, the optimal investment in automation is sensitive to the slope of the exponential cost curve; however, the target crew size is not very sensitive to changes in the slope.

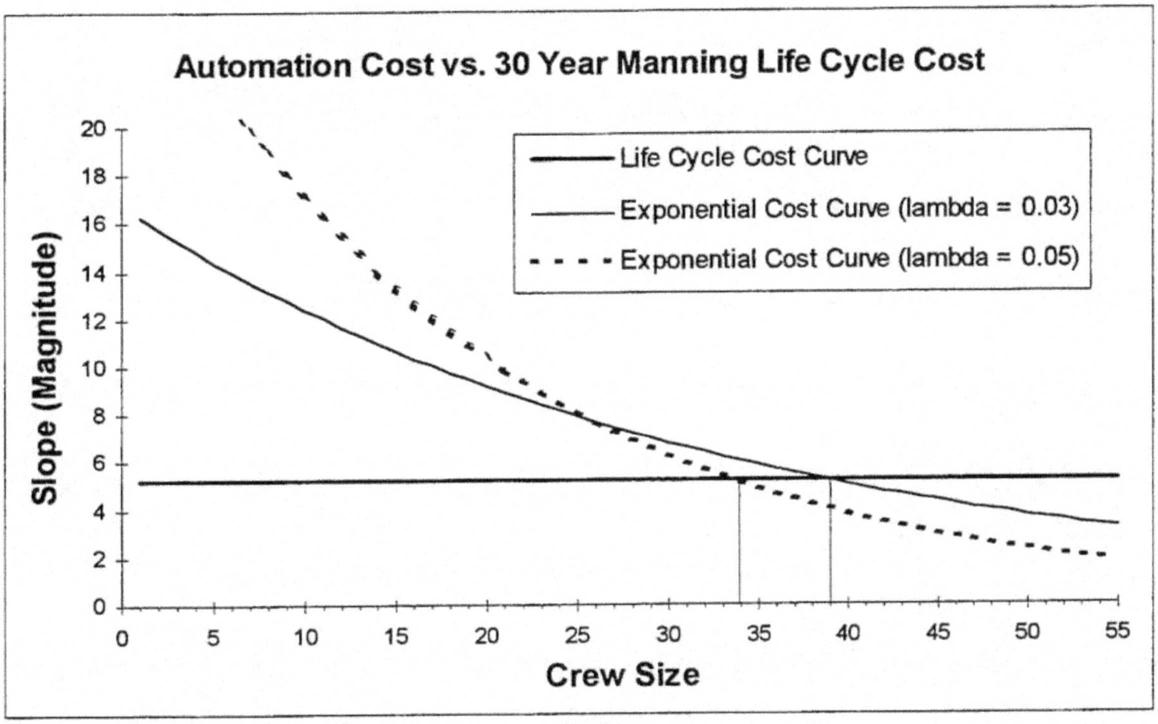

Figure 13. Sensitivity Analysis of Exponential Automation Cost Curve

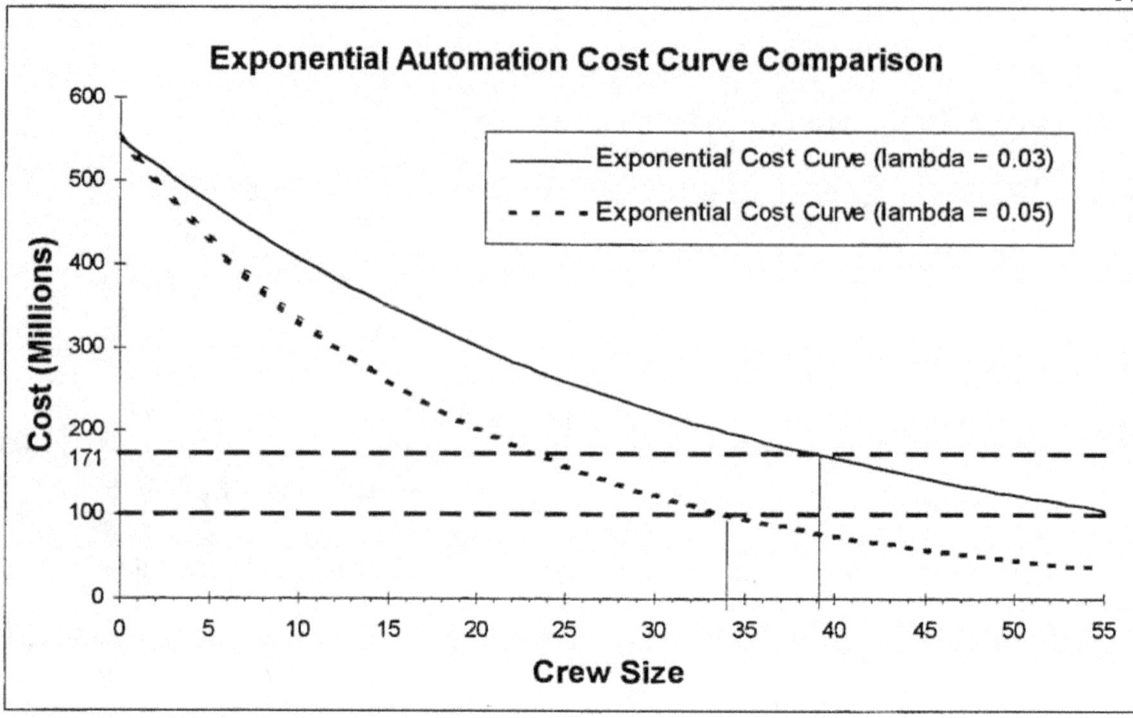

Figure 14. Exponential Automation Cost Curve Comparison

42

V. CONCLUSION

A. SUMMARY

As the intensifying national concern over the federal deficit continues to impact the budget for national defense, innovations in automation must be utilized to reduce LCCs. However, safety must not be sacrificed to achieve this goal. A tradeoff analysis between automation and manning should be conducted early in the design phase. Since the process of designing ships is changing, commencing with the Arsenal Ship, an excellent opportunity exists to incorporate manning analysis into the design phase.

This thesis identified the manning requirements and the operational requirements affecting manning as outlined in the ASRD. The TSSE design concept was adopted as a baseline for manning. To determine the required number of ships and crews, separate deployment cycles were developed for the Arsenal Ship and the Arsenal Ship crew. The manning structures for special evolutions were developed to assist in identifying minimum manning levels. Damage control manning was determined to be the overall limiting factor. The SBCM is a simplified, updated costing model designed to calculate a 30 year manning LCCs for different manning structures. The TSSE design manning structure was used as the starting point to develop three additional manning structures. Using these four structures and data from the mid 1990s, the results from the SBCM, were compared to the NBCF Model. After the SBCM was calibrated, both models produced similar results. Therefore, the SBCM was used to compute 30 year manning LCCs for the four structures using 1997 pay tables. The LCCs were then used to determine 30 year LCC curves, which were used to conduct cost tradeoff analysis between manning and shipboard automation. Finally, the 30 year LCC curve, in conjunction with an automation cost curve, can also be utilized to calculate an initial crew target size and an optimal investment in automation. This thesis is intended as an analysis tool; it does not provide the answer to minimum manning requirements for the Arsenal Ship because the cost of automation per billet saved does not exist in the Navy cost data.

B. CONCLUSIONS

The Arsenal Ship resembles a logistics platform which is intended to go in harm's way. Therefore, it should be designed for minimum manning. Manning vs. automation tradeoff analysis should be implemented in the design phase of all ships. However, before outfitting ships with the latest technological advances in automation, manning and projected operational tempo (OPTEMPO) should be analyzed. Manning structures should be compared after identifying the limiting factor or factors, which is damage control in this thesis.

Deployment cycles for crews and ships should be designed as a "steady state" process. Slack should be included to allow room for adjustment without causing a ripple effect as minor delays arise. The deployment schedule developed for this thesis requires five, not six, Arsenal Ships. Four are forward deployed at two FOBs, while one is in overhaul at homeport. This allows 100 percent coverage of two theaters of operation. The training and deployment cycle requires nine deployment crews. Four of the nine are embarked and five of the nine are in training or in transit to or from the FOBs.

The data required to support an automation cost curve is not readily available in the Navy. However, as systems are designed or modified for the purpose of reduced manning, the data should be collected and analyzed. With increasing manning cost, retrofitting ships with the technology which does not support reduced manning is ill-advised. Every billet in the Arsenal Ship should be completely justified.

C. RECOMMENDATIONS

I recommend challenging the design teams to build an Arsenal Ship that balances marginal automation cost with marginal manning cost. Efficiencies in manning must be realized as budgets are reduced. The design concept should force reductions in manning towards the lower bound, whether the lower bound is cost related or technology related.

The NBCF Model should be updated to provide more flexibility. The current model's options are limited and updating the data is a very difficult if not an impossible task. A model that cannot be updated by the user becomes inaccurate and thus obsolete quickly.

Program managers should be encouraged to conduct automation vs. manning tradeoff analysis prior to solicitation of contracts. Likewise, contractors should provide their automation vs. manning tradeoff analysis early in the contract process to ensure such analysis is a part of the design phase.

D. FURTHER STUDY

Since this thesis does not have the information to develop a valid automation cost curve, the most obvious area for further research is to compile actual or reasonable data and compute an automation cost curve as a function of watchstanding and other crew functions. Likewise, this thesis does not identify the cost of maintaining and operating an Arsenal Ship training simulator, which is an important part of the training process developed in Chapter II.

The NBCF Model could also be updated and revised to include more options. For instance, cost comparisons for different geographical regions could be conducted if the model allowed separate VHA rates to be identified rather than using a weighted average.

APPENDIX A. DEPLOYMENT CYCLES

This appendix contains the proposed Arsenal Ship deployment schedule designed by the author.

Proposed Arsenal Ship Deployment Cycle

			M1	M2	M3	M4	M5	M6	M7	M8	M9	M10	M11	M12	M13	M14
FOB																
Diego Garcia	Ship A		Transit		Deployed		IMAV		Deployed		IMAV		Deployed		IMAV	
	Crew				1				5				9			
	Ship B				Transit		Deployed		IMAV		Deployed		IMAV		Deployed	
	Crew						3				7				2	
	Ship C															
	Crew															
	Ship D															
	Crew															
	Ship E															
	Crew															
Guam	Ship A															
	Crew															
	Ship B															
	Crew															
	Ship C			Transit		Deployed		IMAV		Deployed		IMAV		Deployed		IMAV
	Crew					2				6				1		
	Ship D					Transit		Deployed		IMAV		Deployed		IMAV		Deploye
	Crew							4				8				3
	Ship E															
	Crew															

Proposed Arsenal Ship Deployment Cycle

FOB			M15	M16	M17	M18	M19	M20	M21	M22	M23	M24	M25	M26	M27
Diego Garcia	Ship A		eployed		Transit	Overhaul									
	Crew														
	Ship B		IMAV	Deployed			IMAV	Deployed			IMAV	Deployed			IMAV
	Crew		6				10				14				
	Ship C														
	Crew														
	Ship D														
	Crew														
	Ship E				Transit	Deployed			IMAV	Deployed			IMAV	Deploye	
	Crew				8				12				16		
Guam	Ship A													IMAV	
	Crew														
	Ship B														
	Crew														
	Ship C		Deployed			IMAV	Deployed			IMAV	Deployed			Transit	
	Crew		5			9				13					
	Ship D			IMAV	Deployed			IMAV	Deployed			IMAV	Deployed		
	Crew			7				11				15			
	Ship E														
	Crew														

Proposed Arsenal Ship Deployment Cycle

FOB			M28	M29	M30	M31	M32	M33	M34	M35	M36	M37	M38	M39	M40
Diego Garcia	Ship A														
	Crew														
	Ship B			Deployed		IMAV		Deployed		Transit				Overhaul	
	Crew			18				13							
	Ship C									Transit		Deployed		IMAV	
	Crew											17			
	Ship D														
	Crew														
	Ship E			IMAV		Deployed		IMAV		Deployed		IMAV		Deployed	
	Crew					11				15				19	
Guam	Ship A		eployed		IMAV		Deployed		IMAV		Deployed		IMAV		Deploye
	Crew		7				12				16				20
	Ship B														
	Crew														
	Ship C					Overhaul									
	Crew														
	Ship D		IMAV		Deployed		IMAV		Deployed		IMAV		Deployed		IMAV
	Crew				10				14				18		
	Ship E														
	Crew														

Proposed Arsenal Ship Deployment Cycle

FOB			M41	M42	M43	M44	M45	M46	M47	M48	M49	M50	M51	M52	M53
Diego Garcia	Ship A														
	Crew														
	Ship B														
	Crew														
	Ship C		eployed		IMAV		Deployed		IMAV		Deployed		IMAV		Deploye
	Crew		1				25				21				25
	Ship D														Transit
	Crew														
	Ship E		IMAV		Deployed		IMAV		Deployed		IMAV		Deployed		Transit
	Crew				23				19				23		
Guam	Ship A			IMAV		Deployed		IMAV		Deployed		IMAV		Deployed	
	Crew					24				20				24	
	Ship B					Transit		Deployed		IMAV		Deployed		IMAV	
	Crew							26				22			
	Ship C														
	Crew														
	Ship D			Deployed		Transit					Overhaul				
	Crew			22											
	Ship E														
	Crew														

Proposed Arsenal Ship Deployment Cycle

FOB		M54	M55	M56	M57	M58	M59	M60	M61	M62
Diego Garcia	Ship A									
	Crew									
	Ship B									
	Crew									
	Ship C		IMAV	Deployed			IMAV	Deployed		
	Crew		29				33			
	Ship D	Deployed			IMAV	Deployed			IMAV	
	Crew	27			31				35	
	Ship E	Overhaul								
	Crew									
Guam	Ship A	IMAV	Deployed			IMAV	Deployed			
	Crew	28				32				
	Ship B	eployed		IMAV	Deployed			IMAV		
	Crew	6		30				34		
	Ship C									
	Crew									
	Ship D									
	Crew									
	Ship E									
	Crew									

Arsenal Ship Availability

Phase	Time
IMAV	17.8%
Transit	4.4%
Deployed	60.0%
Overhaul	17.8%
	100.0%

Three of five ships deployed at any time.
100% coverage of two contingencies.

APPENDIX B. PROPOSED MANNING

This appendix contains the proposed Arsenal Ship, Maintenance Detachment, and Squadron Staff and Overhaul Detachment manning structures designed by the author.

Proposed Arsenal Ship

Watch, Quarter and Station Bill						
					Genral Quarters	
Rate/Rank	Billet	Watch	UNREP	Flight Operations	Damage	Launch
CDR	CO		Bridge	Bridge	Bridge	CIC
LCDR	XO/NAV	OOD	Bridge	Bridge	Bridge	Bridge
LT	OPS	OOD	SO (UNREP)	SO (FLT DECK)	Bridge	Bridge
LT	CSO	OOD	Bridge	Bridge	CIC	CIC
LT/LDO	ENG	EOOW	CCS	CCS	CCS	CCS
ENC		EOOW	MER 1	MER 1	MER 1	MER 1
EM1	ELO	EOOW	MER 2	MER 2	MER 2	MER 2
EN2		Operator	MER 1	OSL	OSL	MER 1
EN2		Operator	MER 2	Team Leader	Team Leader	MER 2
EM2		Operator	CCS	CCS	Electrician	CCS
BM1		JOOD	Rig Captain	LSE	Bridge	Bridge
SM1		JOOD	Signalman	Nozzleman	Nozzleman	Bridge
QM1		JOOD	Bridge / Helm	Bridge / Helm	Bridge / Helm	Bridge / Helm
OS1		COMMS	COMMS	COMMS	COMMS	COMMS
OS2		COMMS	Linehandler	Hoseman	Hoseman	CIC
OS2		COMMS	Linehandler	Hoseman	Hoseman	CIC
ETC	EMO	Sensor Control	Sensor Control	Sensor Control	Sensor Control	Sensor Control
ET1		Sensor Control	Linehandler		Nozzleman	Sensor Control
ET2		Sensor Control	Linehandler	Hotsuitman	Hoseman	Sensor Control
ET2		Sensor Control	Linehandler	Hotsuitman	Hoseman	Sensor Control
MS1	Cook	Galley	Linehandler	Chock and Chain	Plugman	Galley
MS2	Baker	Galley	Linehandler	Chock and Chain	Plugman	Galley

(Definitions included on the following page.)

Acronym	Definition
CCS	Central Control Station (Engineering)
CO	Commanding Officer
COMMS	Communications (Combat Information Center)
CSO	Combat Systems Officer
ELO	Electrical Officer
EMO	Electronics Maintenance Officer
EOOW	Engineering Officer of the Watch
JOOD	Junior Officer of the Deck
LSE	Landing Signalman Enlisted
MER	Main Engine Room
OOD	Officer of the Deck
OPS	Operations Officer
OSL	On Scene Leader
SO	Safety Officer
UNREP	Underway Replenishment
XO/NAV	Executive Officer/ Navigator

Maintenance Detachment (FOB)

Enlisted	
Rate/Rank	Quantity
BMC	1
BM1	2
BM2	4
BM3	6
BMSN	6
DCC	1
DC1	1
DC2	4
DC3	3
DCSN	3
EMC	1
EM1	1
EM2	2
EM3	4
ENC	1
EN1	2
EN2	4
EN3	6
ENSN	2
ETC	1
ET1	2

Enlisted	
Rate/Rank	Quantity
ET2	4
ET3	6
EW1	1
EW2	2
GMGC	1
GMG1	2
GMG2	4
GMG3	4
GMMC	1
GMM1	2
GMM2	4
GMM3	4
PNC	1
PN2	2
PN3	1
SKC	1
SK1	2
SK2	3
SK3	5
YN1	1
YN2	1

Officers		
Rank	Billet	Quantity
CDR	CO	1
LCDR	XO	1
LCDR	SUPPO	1
LT	1st LT	1
LT	EMO	1
LT	ENG	1

Enlisted	109
Officers	6
Total	115

Squadron Staff and Overhaul Detachment (Homeport)

Enlisted	
Rate/Rank	Quantity
BMCM	1
BMCS	1
BMC	1
BM1	3
BM2	4
BM3	8
BMSN	8
DCCS	1
DCC	1
DC1	2
DC2	4
DC3	3
DCSN	3
DKC	1
DK1	1
DK2	1
DK3	1
EMC	1
EM1	1
EM2	2
EM3	4
ENCS	1
ENC	1
EN1	2
EN2	4
EN3	6
ENSN	2
ETCS	1
ETC	1
ET1	2
ET2	4

Enlisted	
Rate/Rank	Quantity
ET3	6
EW1	1
EW2	2
GMGC	1
GMG1	2
GMG2	4
GMG3	4
GMMC	1
GMM1	2
GMM2	4
GMM3	4
HTC	1
HT1	1
HT2	2
HT3	1
MAC	1
MA1	1
MM1	1
MM2	1
MM3	1
PNCS	1
PN1	1
PN2	2
PN3	1
SKCS	1
SKC	1
SK1	2
SK2	3
SK3	5
YN1	1
YN2	1

Officers		
Rank	Billet	Quantity
CAPT	CO	1
CDR	XO	1
LCDR	CSO	1
LCDR	MO	1
LCDR	OPS	1
LCDR	SUPPO	1
LT	1st LT	1
LT	ADMIN	1
LT	AMO	1
LT	DISBO	1
LT	EMO	1

Enlisted	136
Officers	11
Total	147

APPENDIX C. TOTAL SHIP SYSTEM ENGINEERING DESIGN

This appendix contains a reproduction of the Naval Postgraduate School's Total System Engineering design team's Watch, Quarter and Station Bill [Ref. 10].

		Billet				Normal	Combat	Bridge		Helo	Boat	Damage
	Crew	Number	Title	Rank	Quals	Steaming	Steaming	Evolutions	UNREP	Operations	Operations	Control
												Physical
												Security
Admin	25	Adm - 501	Corpman	E-6	HM							
	26	Adm -502	Cook	E-6	MS							
	27	Adm -502	Cook	E-4	MS							
Labor	28	LF - 501	Deck Force	E-6	BM	Normal Work	Normal Work	Normal Work	Safety	LSE	Safety	SAT
Force	29	LF - 502	Deck Force	E-5	BM	Normal Work	Normal Work	Normal Work	Winch	Tiedown	Coxswain	SAT
	30	LF - 503	Deck Force	E-5	BM	Normal Work	Normal Work	Normal Work	Linehandler	Tiedown	Boat	SAT
	31	LF - 504	HME Force	E-6	DC	Normal Work	Normal Work	Normal Work	Fuel Sampler	OSL/AFFF		OSL/AFFF
	32	LF - 505	HME Force	E-6	EN	Normal Work	Normal Work	Normal Work	Linehandler	Team Leader	Engineman	Hoseman
	33	LF - 505	HME Force	E-5	EM	Normal Work	Normal Work	Normal Work	Linehandler	Hoseman	Linehandler	Hoseman
	34	LF - 506	HME Force	E-5	DC	Normal Work	Normal Work	Normal Work	Linehandler	Hoseman	Linehandler	Team Leader
	35	LF - 507	Combat Force	E-6	ET	Normal Work	Normal Work	Normal Work	Signals	Hotsuitman		Hoseman
	36	LF - 508	Combat Force	E-6	ET	Normal Work	Normal Work	Normal Work	Phonetalker	Hotsuitman		Plugman
	37	LF - 509	Combat Force	E-5	ET	Normal Work	Normal Work	Normal Work		Plugman		SAT
	38	LF - 510	Combat Force	E-5	ET	Normal Work	Normal Work	Normal Work				SAT
	39	LF - 510	Combat Force	E-6	GMM	Normal Work	Normal Work	Normal Work				SAT
	40	LF - 510	Combat Force	E-5	GMM	Normal Work	Normal Work	Normal Work				SAT
	41	LF - 510	Combat Force	E-4	GMM	Normal Work	Normal Work	Normal Work	Linehandler	Phonetalker		SAT
	42	LF - 510	Combat Force	E-4	GMM	Normal Work	Normal Work	Normal Work	Linehandler			SAT
	43	LF - 510	Combat Force	E-4	GMM	Normal Work	Normal Work	Normal Work				SAT
	44	LF - 510	Combat Force	E-4	GMM	Normal Work	Normal Work	Normal Work				SAT

Table III-9. Watch, Quarter and Station Bill for Special Evolutions

	Crew	Billet Number	Title	Rank	Quals	Normal Steaming	Combat Steaming	Bridge Evolutions	UNREP	Helo Operations	Boat Operations	Damage Control Physical Security
Command	1	Ops - 100	Captain	O-5/6	SWO							
	2	Eng - 100	Maintenance Officer	O-4	LDO							
Watch	3	Ops - 201	OOD	O-3	SWO	OOD #1	OOD #1	OOD #1	OOD #1	OOD #1	OOD #1	OOD #1
	4	Ops - 202	OOD	O-3	SWO	OOD #2	OOD #2	OOD #2	OOD #2	OOD #2	OOD #2	OOD #2
	5	Ops - 203	OOD	O-3	SWO	OOD #3	OOD #3	OOD #3	OOD #3	OOD #3	OOD #3	OOD #3
	6	Ops - 204	OOD	O-3	SWO	OOD #4	OOD #4	OOD #4	OOD #4	OOD #4	OOD #4	OOD #4
	7	Ops - 301	JOOD/lookout	E-7/8	QM	JOOD #1	JOOD #1	JOOD #1	JOOD #1	JOOD #1	JOOD #1	JOOD #1
	8	Ops - 302	JOOD/lookout	E-7/8	QM	JOOD #2	JOOD #2	JOOD #2	JOOD #2	JOOD #2	JOOD #2	JOOD #2
	9	Ops - 303	JOOD/lookout	E-7/8	QM	JOOD #3	JOOD #3	JOOD #3	JOOD #3	JOOD #3	JOOD #3	JOOD #3
	10	Ops - 304	JOOD/lookout	E-7/8	QM	JOOD #4	JOOD #4	JOOD #4	JOOD #4	JOOD #4	JOOD #4	JOOD #4
	11	Com - 201	CSO	O-4	SWO	CSO #1	CSO #1	CSO #1	CSO #1	CSO #1	CSO #1	CSO #1
	12	Com - 202	CSO	O-3	SWO	CSO #2	CSO #2	CSO #2	CSO #2	CSO #2	CSO #2	CSO #2
	13	Com - 301	COMMS	O-3	RM	COMMS #1	COMMS #1	COMMS #1	COMMS #1	COMMS #1	COMMS #1	COMMS #1
	14	Com - 302	COMMS	E-6	RM	COMMS #2	COMMS #2	COMMS #2	COMMS #2	COMMS #2	COMMS #2	COMMS #2
	15	Com - 401	SD	E-6	EW	SD #1	SD #1	SD #1	SD #1	SD #1	SD #1	SD #1
	16	Com - 402	SD	E-5	EW	SD #2	SD #2	SD #2	SD #2	SD #2	SD #2	SD #2
	17	Eng - 201	EOOW	E-7	EN	EOOW #1	EOOW #1	EOOW #1	EOOW #1	EOOW #1	EOOW #1	EOOW #1
	18	Eng - 202	EOOW	E-7	EN	EOOW #2	EOOW #2	EOOW #2	EOOW #2	EOOW #2	EOOW #2	EOOW #2
	19	Eng - 203	EOOW	E-7	EM	EOOW #3	EOOW #3	EOOW #3	EOOW #3	EOOW #3	EOOW #3	EOOW #3
	20	Eng - 204	EOOW	E-7	EM	EOOW #4	EOOW #4	EOOW #4	EOOW #4	EOOW #4	EOOW #4	EOOW #4
	21	Eng - 205	EO	E-6	EN	EO #1	EO #1	EO #1	EO #1	EO #1	EO #1	EO #1
	22	Eng - 206	EO	E-6	EN	EO #2	EO #2	EO #2	EO #2	EO #2	EO #2	EO #2
	23	Eng - 207	EO	E-5	EN	EO #3	EO #3	EO #3	EO #3	EO #3	EO #3	EO #3
	24	Eng - 208	EO	E-5	EM	EO #4	EO #4	EO #4	EO #4	EO #4	EO #4	EO #4

Table III-8. Watch, Quarter and Station Bill for Watchstanders

APPENDIX D. SIMPLIFIED BILLET COST MODEL RESULTS

This appendix contains the Excel® spreadsheet format utilized for the SBCM including the data and the results.

Reduced

36.40%

Paygrade	Quantity	Basic Pay	Sea Pay	BAQ	BAS	VHA	Retirement	Paygrade Cost	Total Cost
E1	0	900.90	0.00	361.50	0.00	218.50	327.93	1808.83	0.00
E2	0	1010.10	0.00	361.50	0.00	218.50	367.68	1957.78	0.00
E3	0	1196.70	0.00	379.80	0.00	200.20	435.60	2212.30	0.00
E4	0	1394.70	160.00	408.00	0.00	197.27	507.67	2667.64	0.00
E5	2	1731.30	350.00	469.20	0.00	231.73	630.19	3412.42	6824.85
E6	2	2040.00	450.00	521.70	0.00	275.37	742.56	4029.63	8059.26
E7	1	2794.80	500.00	564.60	0.00	310.28	1017.31	5186.99	5186.99
E8	0	3106.50	520.00	608.10	0.00	305.70	1130.77	5671.07	0.00
E9	0	3478.50	520.00	659.70	0.00	317.87	1266.17	6242.24	0.00
O1	0	2170.80	280.00	490.50	154.16	315.55	790.17	4201.18	0.00
O2	0	2751.60	280.00	548.70	154.16	332.69	1001.58	5068.73	0.00
O3	3	3708.60	290.00	642.60	154.16	356.76	1349.93	6502.05	19506.15
O4	1	4287.90	300.00	776.70	154.16	420.45	1560.80	7500.01	7500.01
O5	1	5128.80	340.00	881.10	154.16	451.56	1866.88	8822.50	8822.50
O6	0	6285.60	380.00	914.10	154.16	435.47	2287.96	10457.29	0.00
O7	0	7154.40		1015.20	154.16	429.00	2604.20	11356.96	0.00
	10								55899.7536
									x12
									670797.0432

Annual Inflation Rate	30 Year Manning LCC
2%	27212947.58
3%	31913448.18
4%	37621610.41

Proposed

Paygrade	Quantity	Basic Pay	Sea Pay	BAQ	BAS	VHA	Retirement (36.40%)	Paygrade Cost	Total Cost
E1	0	900.90	0.00	361.50	0.00	218.50	327.93	1808.83	0.00
E2	0	1010.10	0.00	361.50	0.00	218.50	367.68	1957.78	0.00
E3	0	1196.70	0.00	379.80	0.00	200.20	435.60	2212.30	0.00
E4	0	1394.70	160.00	408.00	0.00	197.27	507.67	2667.64	0.00
E5	8	1731.30	350.00	469.20	0.00	231.73	630.19	3412.42	27299.39
E6	7	2040.00	450.00	521.70	0.00	275.37	742.56	4029.63	28207.41
E7	2	2794.80	500.00	564.60	0.00	310.28	1017.31	5186.99	10373.97
E8	0	3106.50	520.00	608.10	0.00	305.70	1130.77	5671.07	0.00
E9	0	3478.50	520.00	659.70	0.00	317.87	1266.17	6242.24	0.00
O1	0	2170.80	280.00	490.50	154.16	315.55	790.17	4201.18	0.00
O2	0	2751.60	280.00	548.70	154.16	332.69	1001.58	5068.73	0.00
O3	3	3708.60	290.00	642.60	154.16	356.76	1349.93	6502.05	19506.15
O4	1	4287.90	300.00	776.70	154.16	420.45	1560.80	7500.01	7500.01
O5	1	5128.80	340.00	881.10	154.16	451.56	1866.88	8822.50	8822.50
O6	0	6285.60	380.00	914.10	154.16	435.47	2287.96	10457.29	0.00
O7	0	7154.40		1015.20	154.16	429.00	2604.20	11356.96	0.00
	22								101709.43
									x12
									1220513.16

Annual Inflation Rate	30 Year Manning LCC
2%	49513874.55
3%	58066420.96
4%	68452404.60

TSSE

Paygrade	Quantity	Basic Pay	Sea Pay	BAQ	BAS	VHA	Retirement (36.40%)	Paygrade Cost	Total Cost
E1	0	900.90	0.00	361.50	0.00	218.50	327.93	1808.83	0.00
E2	0	1010.10	0.00	361.50	0.00	218.50	367.68	1957.78	0.00
E3	0	1196.70	0.00	379.80	0.00	200.20	435.60	2212.30	0.00
E4	5	1394.70	160.00	408.00	0.00	197.27	507.67	2667.64	13338.20
E5	10	1731.30	350.00	469.20	0.00	231.73	630.19	3412.42	34124.23
E6	12	2040.00	450.00	521.70	0.00	275.37	742.56	4029.63	48355.56
E7	6	2794.80	500.00	564.60	0.00	310.28	1017.31	5186.99	31121.92
E8	2	3106.50	520.00	608.10	0.00	305.70	1130.77	5671.07	11342.13
E9	0	3478.50	520.00	659.70	0.00	317.87	1266.17	6242.24	0.00
O1	0	2170.80	280.00	490.50	154.16	315.55	790.17	4201.18	0.00
O2	0	2751.60	280.00	548.70	154.16	332.69	1001.58	5068.73	0.00
O3	6	3708.60	290.00	642.60	154.16	356.76	1349.93	6502.05	39012.30
O4	2	4287.90	300.00	776.70	154.16	420.45	1560.80	7500.01	15000.01
O5	1	5128.80	340.00	881.10	154.16	451.56	1866.88	8822.50	8822.50
O6	0	6285.60	380.00	914.10	154.16	435.47	2287.96	10457.29	0.00
O7	0	7154.40		1015.20	154.16	429.00	2604.20	11356.96	0.00
	44								201116.868
									x12
									2413402.416

Annual Inflation Rate	30 Year Manning LCC
2%	97907100.37
3%	114818623.21
4%	135355524.27

Maximum

Paygrade	Quantity	Basic Pay	Sea Pay	BAQ	BAS	VHA	Retirement (36.40%)	Paygrade Cost	Total Cost
E1	0	900.90	0.00	361.50	0.00	218.50	327.93	1808.83	0.00
E2	0	1010.10	0.00	361.50	0.00	218.50	367.68	1957.78	0.00
E3	0	1196.70	0.00	379.80	0.00	200.20	435.60	2212.30	0.00
E4	5	1394.70	160.00	408.00	0.00	197.27	507.67	2667.64	13338.20
E5	14	1731.30	350.00	469.20	0.00	231.73	630.19	3412.42	47773.92
E6	12	2040.00	450.00	521.70	0.00	275.37	742.56	4029.63	48355.56
E7	7	2794.80	500.00	564.60	0.00	310.28	1017.31	5186.99	36308.91
E8	2	3106.50	520.00	608.10	0.00	305.70	1130.77	5671.07	11342.13
E9	0	3478.50	520.00	659.70	0.00	317.87	1266.17	6242.24	0.00
O1	0	2170.80	280.00	490.50	154.16	315.55	790.17	4201.18	0.00
O2	0	2751.60	280.00	548.70	154.16	332.69	1001.58	5068.73	0.00
O3	7	3708.60	290.00	642.60	154.16	356.76	1349.93	6502.05	45514.35
O4	2	4287.90	300.00	776.70	154.16	420.45	1560.80	7500.01	15000.01
O5	1	5128.80	340.00	881.10	154.16	451.56	1866.88	8822.50	8822.50
O6	0	6285.60	380.00	914.10	154.16	435.47	2287.96	10457.29	0.00
O7	0	7154.40		1015.20	154.16	429.00	2604.20	11356.96	0.00
	50								226455.5984
									x12
									2717467.181

Annual Inflation Rate	30 Year Manning LCC
2%	110242423.83
3%	129284630.79
4%	152408977.67

FOB

Paygrade	Quantity	Basic Pay	Sea Pay	BAQ	BAS	VHA	Retirement (36.40%)	Paygrade Cost	Total Cost
E1	0	900.90	0.00	361.50	0.00	218.50	327.93	1808.83	0.00
E2	0	1010.10	0.00	361.50	0.00	218.50	367.68	1957.78	0.00
E3	11	1196.70	0.00	379.80	0.00	200.20	435.60	2212.30	24335.29
E4	39	1394.70	160.00	408.00	0.00	197.27	507.67	2667.64	104037.99
E5	34	1731.30	350.00	469.20	0.00	231.73	630.19	3412.42	116022.39
E6	16	2040.00	450.00	521.70	0.00	275.37	742.56	4029.63	64474.08
E7	9	2794.80	500.00	564.60	0.00	310.28	1017.31	5186.99	46682.88
E8	0	3106.50	520.00	608.10	0.00	305.70	1130.77	5671.07	0.00
E9	0	3478.50	520.00	659.70	0.00	317.87	1266.17	6242.24	0.00
O1	0	2170.80	280.00	490.50	154.16	315.55	790.17	4201.18	0.00
O2	0	2751.60	280.00	548.70	154.16	332.69	1001.58	5068.73	0.00
O3	3	3708.60	290.00	642.60	154.16	356.76	1349.93	6502.05	19506.15
O4	2	4287.90	300.00	776.70	154.16	420.45	1560.80	7500.01	15000.01
O5	1	5128.80	340.00	881.10	154.16	451.56	1866.88	8822.50	8822.50
O6	0	6285.60	380.00	914.10	154.16	435.47	2287.96	10457.29	0.00
O7	0	7154.40		1015.20	154.16	429.00	2604.20	11356.96	0.00
	115								398881.2972
									x12
									4786575.566

Annual Inflation Rate	30 Year Manning LCC
2%	194182176.70
3%	227723322.38
4%	268454792.68

Homeport

Paygrade	Quantity	Basic Pay	Sea Pay	BAQ	BAS	VHA	Retirement (36.40%)	Paygrade Cost	Total Cost
E1	0	900.90	0.00	361.50	0.00	218.50	327.93	1808.83	0.00
E2	0	1010.10	0.00	361.50	0.00	218.50	367.68	1957.78	0.00
E3	13	1196.70	0.00	379.80	0.00	200.20	435.60	2212.30	28759.88
E4	44	1394.70	160.00	408.00	0.00	197.27	507.67	2667.64	117376.20
E5	38	1731.30	350.00	469.20	0.00	231.73	630.19	3412.42	129672.08
E6	23	2040.00	450.00	521.70	0.00	275.37	742.56	4029.63	92681.49
E7	11	2794.80	500.00	564.60	0.00	310.28	1017.31	5186.99	57056.86
E8	6	3106.50	520.00	608.10	0.00	305.70	1130.77	5671.07	34026.40
E9	1	3478.50	520.00	659.70	0.00	317.87	1266.17	6242.24	6242.24
O1	0	2170.80	280.00	490.50	154.16	315.55	790.17	4201.18	0.00
O2	0	2751.60	280.00	548.70	154.16	332.69	1001.58	5068.73	0.00
O3	5	3708.60	290.00	642.60	154.16	356.76	1349.93	6502.05	32510.25
O4	4	4287.90	300.00	776.70	154.16	420.45	1560.80	7500.01	30000.02
O5	1	5128.80	340.00	881.10	154.16	451.56	1866.88	8822.50	8822.50
O6	1	6285.60	380.00	914.10	154.16	435.47	2287.96	10457.29	10457.29
O7	0	7154.40		1015.20	154.16	429.00	2604.20	11356.96	0.00
	147								547605.2164
									x12
									6571262.597

Annual Inflation Rate	30 Year Manning LCC
2%	266583501.50
3%	312630549.76
4%	368548853.68

LIST OF REFERENCES

1. Office of the Chief of Naval Operations, *Promulgation of the Arsenal Ship Concept of Operations*, Memorandum, Serial N863D/6U654802, April 11, 1996.

2. Chief of Naval Operations N86, *Arsenal Ship Requirements Document*, Washington, D.C., January 17, 1996.

3. Baltimore Sun, *Pentagon Taps 3 Firms to Vie to Build Warship*, section 11C, p. 1, January 11, 1997.

4. *Little Ships Many Rockets*, All Hands, vol. 468, p. 45, 1956.

5. Silverstone, Paul H., U.S. Warships Since 1945, pp. 109-110, Naval Institute Press, 1987.

6. Committee on Effective Manning et al, Effective Manning of the U.S. Merchant Fleet, National Academy Press, 1984.

7. National Research Council, Crew Size and Marine Safety, National Academy Press, 1990.

8. Hildebrand, Gregory A., *Modified Navy Verses Military Sealift Command Operation of Combat Logistics Force Ships*, Naval Postgraduate School Masters Thesis, Monterey, CA, September 1993.

9. Rost, Ronald F. et al., *Civilian Manning of Combat Logistics Force Ships: The Potential for Cost Savings*, Center for Naval Analysis, Alexandria, VA, July 1990.

10. Baumann, G. et al., *An Arsenal Ship Design*, Naval Postgraduate School Student Report, Monterey, CA, December 1996.

11. Fleming, Matthew, *The Impact of Technology on Reduced Manning forUS Navy Combatants*, Naval Postgraduate School Masters Thesis, Monterey, CA, March 1997.

12. Wright, Richard L., *Potent and Punishing*, Surface Warfare, pp. 21-23, January/February 1997.

13. SAG Corporation, *Navy Billet Cost Factor: Active Component Cost Estimation Model*, Falls Church, VA, April 12, 1995.

14. Navy Times, *The 1997 Navy Times Pay Chart*, December 16, 1996.

15. Navy Times, *The 1995 Navy Times Pay Chart*, September, 26, 1994.

16. Navy Times, *The 1993 VHA Rates*, December, 21, 1992.

17. Naval Center for Cost Analysis, Navy Visibility and Management of Operating and Support Costs (VAMOSC), Shipboard Systems Report, Arlington VA, May 1986.

18. Electronic Mail from Ray Francis, McDermott Shipbuilding, Morgan City, Louisiana, http://www.mcdermott.com/sig/msi/index.html, February 3, 1997.

INITIAL DISTRIBUTION LIST

1. Defense Technical Information Center 2
   John J. Kingman Rd., STE 0944
   Ft. Belvoir, VA 22060-6218

2. Dudley Knox Library 2
   Naval Postgraduate School
   411 Dyer Rd.
   Monterey, CA 94943-5101

3. Professor Wayne P. Hughes, Code OR/HI 2
   Department of Operations Research
   Naval Postgraduate School
   Monterey, CA 94943

4. CDR Ronald L. Brown, Code OR/BR 1
   Department of Operations Research
   Naval Postgraduate School
   Monterey, CA 94943

5. Professor Michael G. Sovereign, Code OR/SM 1
   Department of Operations Research
   Naval Postgraduate School
   Monterey, CA 94943

6. Professor Charles N. Calvano, Code ME/CA 1
   Department of Mechanical Engineering
   Naval Postgraduate School
   Monterey, CA 94943

7. Distinguished Professor David Schrady, Code OR/SO 1
   Department of Operations Research
   Naval Postgraduate School
   Monterey, CA 94943

8. OCNO (Surface Warfare) 1
   Attn: Code N86B
   2000 Navy Pentagon
   Washington, DC 20350-2000

9. OCNO (Expeditionary Warfare) Code N85T.. 1
2000 Navy Pentagon
Washington, DC 20350-2000

10. Naval Center for Cost Analysis 1
Suite 400, West Tower
1111 Jefferson Davis Hwy
Arlington, VA 22202-4306

11. Center for Naval Analysis.. 1
4401 Ford Ave.
Alexandria, VA 22302-0268

12. LT Thomas E. McNerney, III.. 1
351 SE 6th Ave.
Pompano Beach, FL 33060

THE ARSENAL SHIP

ACQUISITION PROCESS EXPERIENCE

Contrasting and Common Impressions from the Contractor Teams and Joint Program Office

ROBERT S. LEONARD JEFFREY A. DREZNER GEOFFREY SOMMER

NATIONAL DEFENSE RESEARCH INSTITUTE

RAND

The research described in this report was sponsored by the Defense Advanced Research Projects Agency. The research was conducted in RAND's National Defense Research Institute, a federally funded research and development center supported by the Office of the Secretary of Defense, the Joint Staff, the unified commands, and the defense agencies under Contract DASW01-95-C-0059.

Library of Congress Cataloging-in-Publication Data

Leonard, Robert S.
The arsenal ship acquisition process experience : contrasting and common impressions from the contractor teams and joint program office / Robert S. Leonard, Jeffrey A. Drezner, Geoffrey Sommer.
p. cm.
"Prepared for the Defense Advanced Research Projects Agency (DARPA) by RAND's National Defense Research Institute."
"MR-1030-DARPA."
Includes bibliographical references (p.).
ISBN 0-8330-2690-9
1. United States. Navy—Procurement. 2. United States. Navy—Weapons systems—Costs. 3. Warships—United States—Design and construction. 4. Shipbuilding—United States. I. Drezner, Jeffrey A. II. Sommer, Geoffrey, 1957– . III. United States. Defense Advanced Research Projects Agency. IV. National Defense Research Institute (U.S.). V. Title.
VC263.L42 1999
359.6 ' 212 ' 0973—dc21 98-52862
CIP

Published 1999 by RAND
1700 Main Street, P.O. Box 2138, Santa Monica, CA 90407-2138
1333 H St., N.W., Washington, D.C. 20005-4707
RAND URL: http://www.rand.org/
To order RAND documents or to obtain additional information, contact Distribution Services: Telephone: (310) 451-7002; Fax: (310) 451-6915; Internet: order@rand.org

An Executive Committee consisting of:

Assistant Secretary of the Navy (RD&A)
Director of Surface Warfare (N86)
Director, DARPA
Commander, NAVSEA
Chief of Naval Research

will review the program at major decision milestones to evaluate the validity of program cost thresholds and provide re-direction as necessary.

Term of Agreement:

It is expected that this MOA shall remain in effect for the duration of the demonstration program. Early termination of the program due to funding unavailability, lack of legal authority or other reason beyond the control of the parties shall be a basis for termination of this MOA. Any termination shall be preceded by consultation among the parties.

DARPA -- Arsenal Ship Lessons Learned

31 December 1997

Distribution Statement: Unlimited Distribution

Charles S. Hamilton
Capt USN
Arsenal Ship Program Manager

1.0 -- Executive Summary

1.1 -- Introduction

This report captures many of the lessons learned in executing the first two phases of the Arsenal Ship program, a period of about 22 months. It focuses on acquisition process-oriented lessons, rather than individual consortia technology initiatives which are proprietary in nature. This report was written by the Arsenal Ship Joint Program Office (biographies at Tab M), and the views expressed do not necessarily reflect DARPA or Navy concurrence.

The first two phases of the Arsenal Ship Program provided an excellent return on the Navy's and DARPA's investment. These two phases successfully demonstrated that industry, involved early in the ship design process, could develop an optimum mix of performance capabilities that could be accommodated within affordability constraints; successfully demonstrated teaming between combat system integrators and shipyards; and introduced innovative concepts in reduced manning, automated damage control, topside integration, and modular design.

1.2 -- Background

The basic requirement for the Arsenal Ship, established in a joint Navy DARPA Memorandum on March 18, 1996 (Tab A), was to satisfy joint naval expeditionary force warfighting requirements in regional conflicts by providing the theater commander with massive firepower, long range strike, and flexible targeting and possible theater defense through the availability of hundreds of vertical launch system (VLS) cells. To meet this warfare requirement affordably, the Arsenal Ship concept and design was to be straightforward and simple. Detailed requirements and concept of operations were defined in separate documentation (nine pages total), however, key elements for the Arsenal Ship included:

- Provide approximately 500 VLS calls, with the capability to launch Navy and joint weapons to support the land campaign;

- Integrate the combat system with Cooperative Engagement Capability (CEC) links to serve in, or as, the off-board control;

- Appropriate ship design features for survivability and ship self defense which could be incorporated at a later date if needed;

- Low ownership Costs through the use of innovative maintenance and operational methods, procedures, and technologies;

- Crew size not to exceed 50 personnel. The design objective will be to minimize crew size to the maximum extent below 50 which is technically feasible.

In the face of limited budget levels, the use of acquisition reform initiatives and streamlined contracting methods were paramount to meet the basic requirements of the Arsenal Ship in an affordable manner. To accomplish this, a non-acquisition category demonstrator ship was to be developed, in the water and ready for testing by October 2000, and which would have been convertible to a fleet asset at a future date.

In addition, cost was viewed as an independent variable, and early industry involvement with the development of a cooperative industry-government team was viewed as key to achieve Arsenal Ship goals. To minimize cost, off-the-shelf systems were to be used exclusively. Any development of new systems required the approval of ASN (RD&A). The cost of acquiring the first ship was not to exceed $541 million including the cost of concept development and competition. These funds were to be provided jointly by the Navy and DARPA with contributions of $371 million and $170 million respectively.

The non-ACAT Arsenal Ship demonstration program was created to evaluate sea-based massed precision firepower, while minimizing the risks in acquisition of approximately six ships. To ensure that the program remained affordable, a firm acquisition cost threshold for the production ships was established (Unit Sailaway Price $450M goal, $550M cap in FY96 $). A corresponding Life Cycle Cost threshold was also established (Operating and Support annual cost of $13.7M/ship [roughly a third of DDG-51]. This program was conducted using DARPA's Section 845 Agreements Authority so as to allow industry wide latitude in satisfying the Navy's requirements within this threshold. Agreements were structured to allow tradeoffs between cost and performance. Program success was to have been judged by the extent to which the Arsenal Ship met operational requirements.

A second purpose for this demonstration program was to accelerate the Navy's ongoing acquisition reform activities focused on buying improved ships at a lower cost. To this end, the joint program was to focus on exploiting DARPA's culture and experience in prototyping system programs. The Navy and DARPA anticipated the production Arsenal Ship contracts would serve as a model for future streamlining.

This joint Navy/DARPA demonstration program was conducted under DARPA lead, as articulated in a joint 28 May 1996 memorandum (Tab B), with an envisioned transition of leadership to the Navy in the testing and production stages of the program. The program was managed by a joint Navy/DARPA program office with the Program Manager reporting to DARPA. A small program office was mandated. DARPA, Naval Sea Systems Command

(NAVSEA), and the Office of Naval Research (ONR) each provided two billets. It was expected that the program office would grow to a maximum of three billets each as the program grew to maturity. The Navy developed a concept of operations (CONOPS) (Tab C), and a Ship Capabilities Document (SCD) (Tab D), which the program office used to guide the Phase I and II trade studies conducted by industry. The Program Manager developed a program plan including major decision milestones, and a DARPA to Navy program transition plan.

Organizationally, the Arsenal Ship Program Manager reported to two committees for guidance and direction: the Steering Committee and the Executive Committee.

The Steering Committee included:

Director, TTO -- DARPA (Chair)
Deputy Assistant Secretary of the Navy (DASN, Ships)
Assistant Director, TTO for Maritime Programs -- DARPA
Director, Surface Warfare Plans/Programs/Requirements Branch
- OPNAV (N863)
PEO for Surface Combatants
Office of Naval Research (ONR33)

The Executive Committee included:

Director, DARPA (Chair)
Assistant Secretary of the Navy (RD&A)
Director of Surface Warfare (N86)
Commander, NAVSEA
Chief of Naval Research

The program was divided into six phases. Phase I was a six month Concept Definition effort for multiple industry consortia. Phase II was a twelve month Functional Design effort, originally for two consortia (expanded to three based on the robust designs of 3 teams. Phase III was to be a 33 month Detailed Design and Construction of the Demonstrator Ship, by one consortia. Phase IV was to be a twelve month Demonstration/testing phase, followed, if successful, by exercising the Phase V(a) Production option, Phase V(b) Conversion of the Demonstrator to production option, and the Phase VI Service Life Support of the class for the lifetime of the ships. The initial industry funding and schedule is arrayed below:

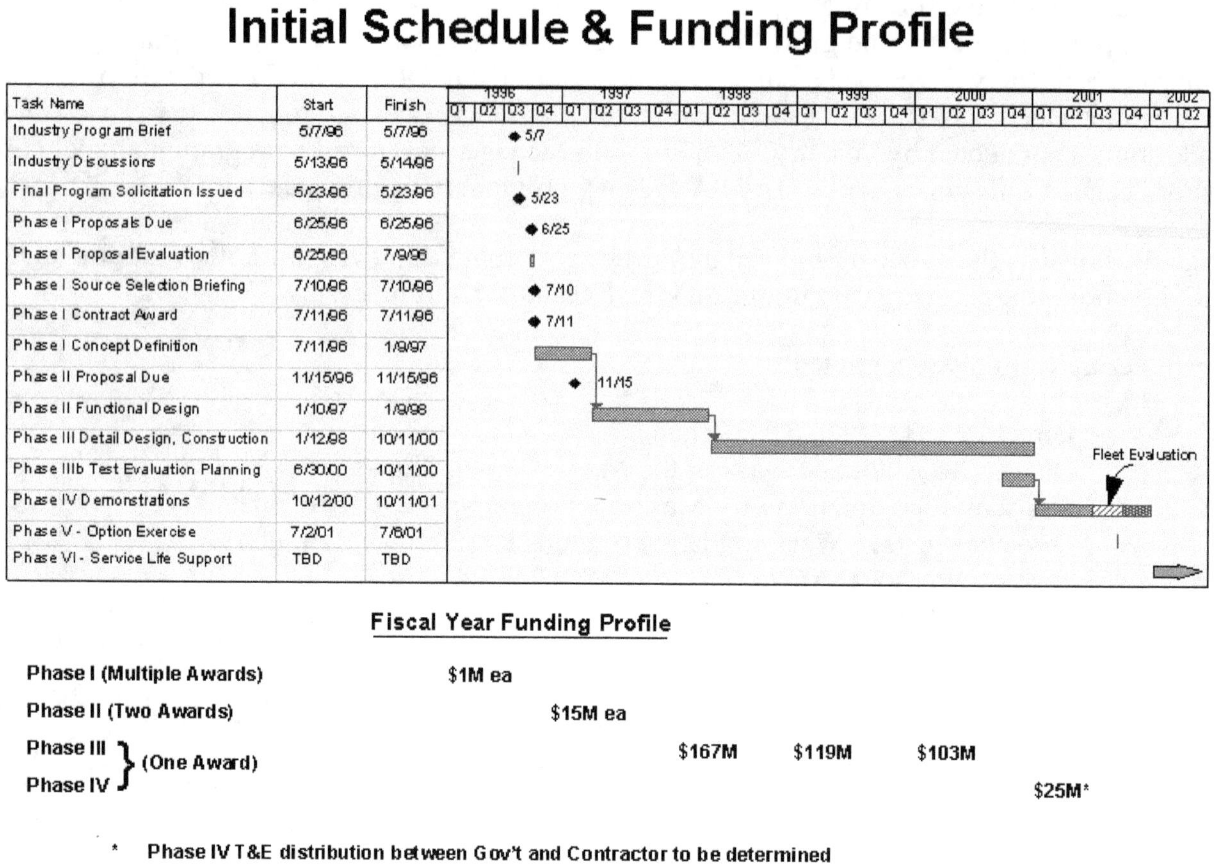

Figure 1 -- Initial Schedule & Funding Profile

Five Teams were awarded Phase I agreements on July 11, 1996, under the 10 USC 845 Other Agreements Authority, with the potential to execute through Phase VI of the program. This Other Agreements Authority excepted the program from the Federal Acquisition Regulations (FAR) and any military specifications and standards. The scope of Section 845 Prototyping Authority is explored fully in a DARPA General Counsel October 24, 1996, memorandum (Tab E). USD(A&T) implementation of 845 Other Agreements Authority within the Services (Section 804 of the FY97 Authorization Act) is included at Tab F.

Three Teams were awarded Phase II agreements for Functional Design effort on January 9, 1997. In April 1997, the original Arsenal Ship Demonstrator concept was expanded by the Navy and DARPA to include risk reduction efforts for the Surface Combatant of the 21st Century (SC-21) (Tab G) and the Demonstrator's name was changed to the Maritime Fire Support Demonstrator (MFSD) (Tab H) to reflect this expansion.

The FY98 Appropriation Act appropriated $35 million of the requested $150.2 million for Maritime Fire Support Demonstrator. On 24 October 1997, the Secretary of the Navy reluctantly determined that continuation of MFSD into Phase III and further phases was unachievable (Tab I). Phase II activities were completed and the Arsenal Ship Joint Program Office closed operation on 31 December 1997 (Tab J).

1.3 -- Lessons Learned

1.3.1 ***Acquisition Streamlining Works.*** The process being followed by Arsenal Ship demonstrated a 50% reduction in acquisition time for the design portion of the ship compared to the traditional design approach. The Industry Teams were prepared to complete the detail design and build the ship in 33 months, again about half the normal time. This was primarily enabled by using an industry led acquisition operating under Section 845 authority, with industry having full trade space and responsibility for the design.

1.3.2 ***Price As Established (PAE) Spurs Innovation and Drives Down Acquisition Cost But Increases Risk.*** Price as established (PAE) was the approach taken throughout the Arsenal Ship Program. Price here means industry cost to manufacture and a reasonable profit (or return of investment to the company). A price goal was established early in the program. All designs were monitored by the contractor teams to ensure an affordable item at a price including all non-recurring and recurring costs to manufacture and a reasonable return on corporate investments or in other words, profit. All other aspects of the government desired capabilities were tradable against the price goal. This is different from Cost as an Independent Variable (CAIV). Cost as independent variable (CAIV) assumes two things; first, the government is monitoring and controlling the trade decisions to ensure affordability, and second, cost is only one of several factors to be traded. Requirements creep is most certain. CAIV in present government literature means a government program manager yardstick to consider trades against requirements.

After several months, all competitive teams were to fix their own price against their design or PAE. During later phases more trades could result. The contractor team should have responsibility for all trade decisions and be encouraged to use the trade space within the government's desired capabilities to ensure an acceptable and capable product that meets PAE. PAE focused the design trades on mission essential performance, without requiring the use of legacy systems as GFE (Government Furnished Equipment). This resulted in designs that were engineered at a total system level, highly integrated, commercially-based, and low cost, while at the same time introducing new systems with some attendant risk. Industry Teams were prepared to meet the goal of $450M for the average production ship cost, about 1/3 less than early Navy estimates for the production ship. Risk reduction programs were key to achieving the PAE goals and worked effectively. Under the Arsenal Ship acquisition strategy, substantial elements of the risk were transferred from Government responsibility to Industry. In short, competing teams concluded "zero risk is unaffordable."

1.3.3 ***Low Manning is Readily Achievable.*** Manning is the largest single factor in life cycle cost, and accounts for roughly 40% of the annual operating and support costs of Navy ships. Reduced manning is the key to lower life cycle costs. The enablers to achieve lower manning among the Industry Teams included: commercial ship operating and maintenance practices, insertion of COTS technologies such as integrated bridge system; nested, remote sensors; and high levels of controls and C4I integration. The Arsenal Ship goal of less than 50 crew was easily achieved by the Industry Teams (who averaged 22 crew). Preliminary Navy estimates for Arsenal Ship were 269 crew.

1.3.4 ***An Industry Design Competition Could be More Meaningful Than a Government AOA (Assessment of Alternatives).*** Industry Teams produced an array of design solutions that were achievable and affordable, with better cost information than is possible within the Government's data base. Since the designs were optimized for each Team's production capabilities and facilities, industry design alternatives were available at lower cost than Government designs, and had been measured against producibility metrics, so designs were at a much higher state of maturity.

1.3.5 ***Industry is Fully Capable of Designing and Developing Complex Navy Ships.*** The majority of design and production skills needed to produce Navy ships was assembled as a natural part of the teaming of the full service consortia. Industry Teams were able to find needed technical expertise for almost all design areas in the commercial marketplace. In selected cases (principally survivability, and weapons effects), some Government R&D Center expertise was obtained by industry under individual contract. During the downselect process, losing Team members became available to join the winners to develop even stronger Teams. This was encouraged in the Agreement between the Government and the Industry Team.

1.3.6 ***Significant Cost Savings in Development and Acquisition Programs can Only be Achieved Through Program Stability.*** Industry's willingness to invest its time, talent and resources to compete for 845 type agreements is strongly influenced by the clarity and stability of the program, as well as the ability to realize return on investment. The Arsenal Ship Joint Program Office left the CONOPS and key elements of the Arsenal Ship unchanged through the first two phases; this stimulated about a $5 million per team investment (Consortia IR&D) in Phase I, and approximately $15 million per team investment in Phase II. The potential for achieving a reasonable return on investment for production Arsenal Ships, as well as future production SC-21 vessels rounded out Industry's motivation for their investment strategy.

1.3.7 ***Technology is Already Available for Breakthrough Performance.*** Technologies still "under study" by the Government were readily incorporated in the Arsenal Ship designs based on COTS products. Areas of particular strength include: reduced manning; automation; information systems; communications/connectivity; propulsion machinery; fire fighting; maintenance and logistics. Effective passive survivability and signature reduction technologies derived from previous Government programs are also available in the Marine and Aerospace industry. Combined, these resulted in improved performance with major savings for both acquisition and service life costs at little technical risk.

1.3.8 ***Minimal Government Direction is a Key Factor to Success.*** In a typical Government acquisition program, extensive specifications, cost and schedule requirements, and oversight (DoDI 5000.2, as interpreted by program offices and oversight staffs) can result in an overly constrained environment and unachievable objectives. Further, large Government program offices provide multiple opportunities for redirection, levying of additional requirements, confusion and delay in administering acquisitions. Arsenal Ship, with a minimal set of technical objectives (9 pages of goals with no thresholds), coupled with an office size of six Government employees, kept industry's trade space open, communications direct, and delays to a minimum. Significant trust and mutual respect was a direct consequence of the more open and coherent dialog with Industry.

1.3.9 ***Adequate Time is Needed for Industry Team Formation and Growth.*** Much energy was spent by industry in the first two months of Phase I sorting out Team membership and Team relationships. At the start of Phase II, the average Team size increased from about 50 to over 200 people in a short time frame, which caused different design maturation rates among the Teams. The phases of the program were structured to allow seamless transition from one phase to the next. Industry did not take full advantage of the downselect decision period to position themselves for the next phase. Had they done so, the Phase transition would have been much smoother. Without this, an additional 2 months for Phase II would have permitted all three consortia to fully mature their functional design.

1.3.10 ***Competitive Design Solutions are a Package Deal.*** All aspects of the Industry Team's competing designs were not equal in performance, but each Team met the PAE goal without containing any fatal design flaws. Achieving a notional "best possible ship" by having the Government attempt to integrate various aspects of each Industry design would have resulted in

(a) holdback of the most innovative ideas from competing Teams

(b) a Government directed solution, undermining the industry-based design responsibility and

(c) a program that would not meet its PAE goal.

The Government must understand (and participate in the underlying trades which lead to the winning design), and then accept the solution as the best compromise that can be achieved within the PAE constraint.

1.3.11 ***Section 845 permits "fly before buy" for Naval Ships with no lost time to full production.*** The Defense Science Board's 1996 Task Force on Defense Acquisition Reform (Phase III) highlighted the time to field the average major weapon system at 16 to 18 years. Surface Navy ships have been no exception. Traditional surface combatant acquisition programs require five years to complete concept, functional and contract design, after a five to seven year research and development cycle. Following award of lead ship, production of the first vessel averages five years. Two years after lead ship award, the follow ship contract is let, when the lead ship has barely started construction, committing the Navy to procurement of many ships before the first one is even delivered.

Using Section 845 authority, and involving industry at the inception of the program, cut the development cycle in half. The Arsenal Ship demonstrator would have been available for testing after just four years of design and production. The production decision could then be based on actual test data, allowing a ship to be acquired in a "fly before buy" fashion, reducing cycle time and accelerating technology insertion into the fleet.

1.3.12 ***R&D Funding must be properly balanced with production costs.*** Industry's limit of $389M for development of the demonstrator ship was extremely challenging as compared to the average production cost goal of $450M. Industry pushed some software development to the production ship to reduce demonstrator ship non-recurring costs, with attendant reduction in demonstrator capabilities and increased production risks.

1.3.13 ***Industry lead for acquisition is a permanent "fork in the road".*** Industry's freedom to balance the PAE tradespace equation necessarily includes their ability not to choose Government developed systems. This has implications for the Government's R&D investment strategy, the infrastructure currently in place to support subsystem development (Participating Managers [PARMS]), and equipment configuration control (outsourced and privatized to industry or centralized with the Government). The net effect could be to lower total Life Cycle Costs by attacking infrastructure as well as taking advantage of the commercial marketplace.

2.0 -- Program

2.1 -- Acquisition Strategy

2.1.1 *Background*

The DARPA High Altitude Unmanned Aerial Vehicle Program (Tier II+) provided the model for the Arsenal Ship acquisition strategy. Tier II+ was the first to use DARPA's Section 845 authority, granted to DARPA in the FY 94 Authorization Act.

Like Tier II+, the Arsenal Ship acquisition strategy was a multi-phase competitive procurement, with each phase reducing the number of competitors and increasing the maturity of the design for both the Demonstrator and production ships. The central element of the strategy was to use competition, with each of the Industry Teams committing to an irrevocable offer for the production ships at the end of Phase II. Phase I's Concept Design was awarded to all qualified full service Teams who could potentially perform all phases of the program. Phase II was to be awarded to two Industry Teams for the Functional Design (similar to Contract Design in maturity). The downselect decision instead awarded three Industry Team agreements based on three very robust Industry designs. Phase III, detail design and construction for the Demonstrator, as well as a priced option for five production ships, was to be awarded to one Industry Team.

Industry was encouraged to form teams to include all of the requisite analysis, design, build capability, and life cycle support required to execute the entire program. The intent was to encourage the creation of a Team where individuals applied their expertise irrespective of company affiliation. This could manifest itself as a joint venture, a limited partnership, or other corporate structure. The classic Prime/Sub relationship was not precluded, but alternative teaming arrangements were suggested as a way to break down corporate stove pipes, as well as achieve efficiencies and save costs.

The use of the Other Transactions Authority eliminated most of the procurement regulations, including those in FAR/DFARS (see Tab E). This allowed a streamlining of the process and shortening of the schedule, especially the time taken to conduct source selection by the Government. It should be noted that even though most of the "rules" are waived by the authority, a review of DoDI 5000.1 showed that the spirit and intent of DoD acquisition policy was being followed. Specific procedures prescribed in DoDI 5000.2 were not emulated.

Concept of Operations (CONOPS) and Ship Capabilities Documents (SCD) were provided as goals (Tabs C and D). The stated philosophy of "no requirements" effectively put responsibility for cost and performance on each Industry Team, and gave them the trade space necessary to achieve PAE goals.

A PAE range of $450M to $550M was established for the average Unit Sailaway Price (USP) of the five production ships. USP and the manning limit of 50 people were treated as the program's only hard requirements.

2.1.2 ***Lessons Learned***

2.1.2.1 ***Effective competition motivated the Industry Teams to meet the program goals of cost and performance.*** Clearly the three Phase II Teams felt the competitive pressure of the other consortias and worked hard to create a competitive advantage. It is doubtful that consortia full effectiveness or corporate CEO level participation could have been achieved without the benefit of competition.

2.1.2.2 ***Industry could effectively form teams containing the requisite capability to perform the contract.*** The three Teams awarded Phase II agreements created effective organizations that operated mostly as a single unit. However, the members did not lose all of their corporate identity. This was manifest in the degree to which individual company cost data was closely guarded.

2.1.2.3 ***The shortened source selection process allowed the multi-phase process to work.*** Historically, the largest issue with pursuing a multi-phase program was the schedule penalty mandated by a lengthy source selection process, often lasting over six months. For Arsenal Ship, slightly less than two months were scheduled for both the Phase II and III downselects, with the selection date cast in concrete and put on all of the principals' schedules months in advance. Government access to Team's designs throughout their development, rather than only at source selection time, was critical to being able to execute the evaluation on a short schedule. The climate of rapid prototyping within DARPA, the willingness to explore new concepts, and the focused support by Director, DARPA, greatly facilitated this process.

2.1.2.4 ***One of the advantages of using the Agreements Authority was deleting "protest avoidance" from the process.*** Many of the rules established by the acquisition community over the years to avoid litigation have become cumbersome and detrimental to an efficient process. Although the integrity of the individual Arsenal Ship Team design and business data were jealously guarded, the "fairness" rules [1], for example, were eliminated, a fact which greatly improved the ability of the Government to support and communicate with the individual Teams.

1 ***The so-called "fairness" rules require the government to provide the answers to one team's questions to the other teams. Being a version of "Other Transactions", Section 845 agreements are not protested in the General Accounting Office. Protests to the Agency and protests to Court (where the standard is illegality or arbitrariness) while available under Section 845, are deemed more predictable and less threatening than the GAO process.***

2.1.2.5 ***When cost is the only requirement, it is essential that both the development budget and PAE goal be achievable and balanced.*** The $450M PAE goal for average production USP turned out to be a good number and universally supported by industry. Industry's limit of $389M for development of the demonstrator ship was extremely challenging as compared to the average production cost goal of $450M. Industry pushed some software development to the production ship to reduce demonstrator ship non-recurring costs, with attendant reduction in demonstrator capabilities and increased production risks.

2.1.2.6 ***A streamlined acquisition process can significantly reduce the time and cost of bringing the product to market.*** The process being followed by Arsenal Ship demonstrated a 50% reduction in acquisition time for the design portion of the ship compared to the traditional design approach. The Industry Teams were prepared to complete the detail design and build the ship in 33 months, again about half the normal time. This was primarily enabled by using an industry led acquisition operating under Section 845 authority, with industry having full trade space and responsibility for the design.

2.1.2.7 ***One power of Section 845 Authority was in changing the mindset of the participants.*** The acquisition process currently in place incentivizes risk averse behavior, and stifles innovation. True acquisition reform and streamlining is possible if all the rules are taken away first and then reinserted on a case basis, such as for security and explosive safety. If each exception to the conventional rules must be justified, streamlining will not occur. Section 845 lets good business decisions govern.

2.1.2.8 ***Developing a long term vision and fully disclosing it to industry at the outset sets the stage for the entire program.*** In order for industry to plan for and invest in a program it has to know the Government's plans. Industry can seek this through the back door or Government can put the word out openly and consistently. Arsenal Ship sketched the program vision in two industry day sessions early, and then reinforced the vision themes at each industry interaction.

2.1.2.9 ***Allowing industry to set requirements increases design innovation.*** By not invoking any requirements, the Government gave Industry the total trade space for the overall design and the subsystems selection. Consequently, the use of new technologies integrated in different ways was facilitated, especially in the areas of communication, information systems and topside integration. Opening the development filters permitted the next logical step from the Goldwater Nichols Act of 1986; joint warfighting design at program inception.

2.2 -- Contracts (Agreements) and Legal

2.2.1 *Background*

2.2.1.1 *Key Events*

Phase I

Draft Solicitation and Industry Briefing	7 May 96
Industry Discussions	14 May 96
Final Phase I Solicitation	23 May 96
Phase I Proposals Received	25 June 96
Phase I Selections Announced	11 July 96
Negotiations Concluded (Agreements executed -- five Teams)	Aug 96

Phase II

Draft Solicitation	13 Sept 96
Final Phase II Solicitation	3 Oct 96}
Phase II Proposals Received	15 Nov 96
Phase II Selections Announced	10 Jan 97
Negotiations Concluded (Agreement mods executed (3 Teams)	Feb 97

Phase III

1st Draft Solicitation	16 Jun 97
Draft Phase III Model Agreement Modification	15 Jul 97
Commence Phase III Agreement Mod negotiations	Aug 97
2nd Draft Solicitation	1 Aug 97
Final Phase III Solicitation	15 Sept 97
Cancellation of Program/Solicitation	30 Oct 97
Phase III Proposals Due	14 Nov 97
Phase III Award Scheduled	16 Jan 98

2.2.1.2 ***Approach.*** The contracting approach for the Arsenal Ship program was to utilize P.L.103-160 Section 845, Other Transactions Authority. This authority was provided to DARPA by the Congress in FY94 and has since been passed on to the military services by PL 104-201 Section 804. DARPA had experience with using 845 authority on other R&D prototyping projects. The Arsenal Ship Program's acquisition strategy was modeled after DARPA's Tier II+ program for Unmanned Aerial Vehicles. Use of Section 845 Authority allows for the procurement of prototypes outside FAR and DFARS regulations. Additionally, the Arsenal Ship Program was designated as a "non-ACAT" program (Tab A), relieving it from the requirements of DoDI 5000.2. The Contracting Officer was free to negotiate terms and conditions specific to the individual needs of the program.

There were several motivating factors that support use of Section 845, including: 1) enticing non-traditional DoD companies, at both the prime and subcontract level, into the business; 2) facilitating insertion of commercial technology; 3) encouraging innovation, and; 4) decreasing the overall time and cost to design, build and deliver a ship.

The solicitations for each Phase of the program contained a "model agreement" with notional terms and conditions appropriate for that particular phase. Industry Teams were free to propose their own terms and conditions to the agreement based upon their particular circumstances. Agreements were then negotiated with each Team individually.

2.2.2 ***Lessons Learned***

2.2.2.1 ***Non-traditional DoD companies are willing to do business with the Government using Section 845 Other Transactions.*** During Phase I of the program, one non-traditional company participated at the prime contract level. They proposed the use of a unique construction technique to which it owns patents. Use of Section 845 definitely provided an avenue for this

Team to propose use of its proprietary technology because FAR/DFARS rights were not invoked during the competitive process. Additionally, non-traditional DoD companies also participated at the subcontractor level. Hopefully, this will lead to new business relations between traditional and non-traditional DoD companies and facilitate the introduction of the latest commercial technologies into Government acquisition.

2.2.2.2 ***Industry retention of most data rights under Section 845 facilitated innovation.*** Section 845 allowed for modified data rights which encouraged innovation during competitive phases. Industry Teams were allowed to retain data rights in the event they were unsuccessful during any downselect process. Teams were eager to put forth their best efforts without the fear that their innovations would be taken by the Government and distributed to other Teams. The Government clearly stated its intent with regard to data rights at the onset of the program -- i.e., should the Team be chosen to construct the MFSD/Arsenal Ships, the Government wants rights to all data necessary only to maintain, modernize and support the ships. Rights to data would be defined by what the data is used for, instead of who paid for or prepared the data, as is the case under FAR/DFARS data rights clauses. Defining rights by their usage, as opposed to whether the Government paid for the rights, was believed to be a better approach to facilitating proper use of the data.

2.2.2.3 ***The intent to eliminate Government Furnished Equipment (GFE) facilitated innovation during the competitive phases of the program and could dramatically reduce the Government's risk of claims.*** No GFE was envisioned under the Arsenal Ship Program. This allowed for maximum design flexibility and took the Government out of its GFE responsibility role. This approach allowed for extensive review of new ways to satisfy the Government's technical requirements while potentially reducing production, maintenance and installation costs. Elimination of GFE placed the Industry Team in charge of the final product and greatly reduces opportunities for claims related to poor specifications, late GFE, inoperable GFE, etc. Some Teams discussed the potential use of GFE for particular systems if it appeared that it would be more economical for the Government to purchase a system through a quantity buy. The Arsenal Ship program was concluded before the full implications of such an approach could be explored or negotiated.

2.2.2.4 ***Access to data created under traditional Government contracts for legacy systems is extremely difficult.*** Because of the Arsenal Ship's operational interdependency on Government legacy systems already in the fleet, access to data on these legacy systems was necessary in order for industry to perform design trades and engineering development. Access to data on Government legacy systems proved to be extremely difficult and time consuming despite extensive efforts by the Industry Teams and the ASJPO. The experience indicates that despite data rights provided by the FAR/DFARS, Government offices either questioned the Team's rights/ability to utilize any of Government's data or simply refused to provide it to other Government agencies for use on their programs. As a result, industry had to make assumptions regarding legacy system operations or even in some cases reverse engineer aspects of their design to ensure compatibility with legacy systems. Such action is time consuming, inefficient, potentially costly and underscores that traditional data rights clauses are counter productive.

2.2.2.5 ***The implications of an "irrevocable offer", which fixed the cost of production units, provided tremendous leverage to the Government and incentive to the Industry Teams to***

design to cost. At the conclusion of Phase II, all competing Teams were to provide an irrevocable offer to build five production Arsenal Ships at a fixed price. The detailed design of this Arsenal Ship would be defined during the Phase III construction of the MFS Demonstrator. Industry Teams were driven to propose the best capability for the money as a result of competition. However, they were also driven to develop a design that was realistic for the cost as a result of the irrevocable offer. If the design proved too expensive, the Industry Team would lose money on the production ships. However, if the design lacked in capability, the Industry Team ran the risk of not being continued in the program. These dynamics were created to facilitate true PAE design development and credibility in cost estimating. All Industry Teams viewed the Irrevocable Offer seriously from a contractual stand-point. The Irrevocable Offer drove Industry Teams to develop the most realistic, vice optimistic, cost estimates because they were contractually committed for these estimates at a fixed price. Negotiations of contractual language pertaining to the Irrevocable Offer were extremely complex and time consuming.

2.2.2.6 ***Negotiation of Section 845 Agreement terms and conditions facilitates greater understanding between the parties.*** Use of Section 845 Other Agreements Authority, provides both Government and Industry with the opportunity to deviate from traditional DoD terms and conditions when it is appropriate for the circumstances. With such opportunity comes a great responsibility to fully understand all technical and business aspects from which the Industry Team is operating. It allows both sides to analyze why a particular procurement can or cannot be performed under traditional terms and make exceptions to such terms if it will facilitate development under the program that will ultimately benefit both the Government and Industry. As a result, negotiations may become very complex and require understanding of industry's motives and concerns in order to determine what is in the best interest of the Government. It is extremely important that legal and technical personnel, as well as contracting personnel are actively involved in the negotiation process.

2.2.2.7 ***Effective implementation of Section 845 authority required close teamwork among the Government's contracting, legal and technical team members.*** The lack of established rules or regulatory guidance inherent in the use of Section 845 authority necessitated close communication between the Government contracting, legal and technical team members. Two major attributes contributed to this successful working relationship. First, the small size of the program office allowed for rapid access to the necessary team members, especially the program manager, for discussion and decision. Second, the team members in the ASJPO were specifically selected for their ability to be open-minded, flexible and willing to experiment with the authority granted under Section 845. These attributes were particularly important in the contracting and legal team members, whose personnel have been traditionally viewed as impediments. The support provided by DARPA General Counsel and DARPA Contracts Management Office, was invaluable.

2.3 -- Role of Government

2.3.1 *Background*

2.3.1.1 *ASJPO Relationship to Industry Teams*

ASJPO objectives differed by program phase. During Phase I, the strategy for engagement was guided by the objective to permit uninhibited concept exploration and innovation by Industry Teams. To this end, Government involvement was structured as "objective" in nature, that is, data or calculation results were made available to the Teams but no advice of any kind was provided. The resulting industry-led efforts produced the desired significant innovation as characterized by:

- Major system innovations
- Independent, highly competitive cost estimates
- Introduction of commercial practices
- Aggressive schedules
- Effective teaming of systems integrators and shipbuilders

These accomplishments, coupled with the compressed schedule, effectively locked in the innovative concepts at the total system level and most of the associated life cycle cost benefits expected.

The objective for Phase II was to ensure successful functional design development that would result in highly desirable competitive proposals for construction. This objective demanded a more pro-active involvement to ensure the development of a ship that the fleet would evaluate positively in all areas. This approach could be taken without jeopardizing the Phase I success because:

- Remaining trade-offs were primarily at the subsystem level with limited cost impact

- Operational (user) input is more important as design detail unfolds

- Government had specific areas of expertise not generally available from industry (e.g., survivability)

- Under Section 845 Agreements, industry was not required to take Government advice

This last competitive phase was the best opportunity to encourage industry to improve weakness in their programs, improving the chances that the "all or nothing" operational test would prove a success and move us into the Arsenal Ship production during Phase V. Thus, ASJPO conveyed its views as a continuing "report card" during the process, rather than when it was too late after the final downselect. In addition, with such a tight schedule, time was of the essence in letting industry know where to focus additional effort rather than experiment with paths that have proven unsound in the past. ASJPO's role during Phase II was thus to:

- Interact in a supportive way on a minimally invasive basis during reviews, technical interchange meetings, and internal team meetings. Coach each Team to be the best they can be within the context of their approach.

- Encourage consideration of out-of-the-box opportunities for technology insertion and business practice reform. Provide real-time feedback.

- Facilitate access for Teams to Government offices, information and software.
- Continuous involvement of ASJPO personnel in the design reviews and open discussions increased understanding of the approach, allowed for identification, disclosure and correction of major deficiencies and shortened the evaluation process following formal proposal submission. It also increased productivity on both sides.

ASJPO questions got the Industry Teams thinking, and stretched their decision space. Great care was taken not to imply direction. The ASJPO core members quickly embraced this new role and used it to great advantage. Training external Government personnel brought in to provide analysis and technical advice to adjust to this non-traditional role was an on-going process. This new role on the part of the Government was viewed as crucial to the success of the program.

The Industry Team members faced a similar cultural shift. It took some initial convincing on the part of the ASJPO before the Industry Teams believed that the program office was serious in pursuing a more flexible and commercial-like way of doing business.

During Phase III it was anticipated that a very close working relationship with the single winning Industry Team would evolve, including on-site ASJPO representatives who would answer questions and facilitate getting Government data, coordinating program functions with other Government activities (such as WSESRB, fleet operational testing, INFOSEC security, testing ranges, etc.), and perhaps even participate directly in software development via the Government R&D Centers. However, the responsibility for defining the best product for the fleet would remain with the Industry Team.

2.3.1.2 *ASJPO relationship to support Contractors and Government R&D Centers*

ASJPO recognized prior to the initiation of Phase I that specific expertise in analysis of ship motions, survivability, signature measurement, C4ISR, Life Cycle Management and ship production was needed to augment the Program Office. Contractual arrangements were negotiated with several private companies including: SYNTEK, Vail, SPC, and LMI. Additionally, task statements were put in place at NSWC Carderock, NSWC Dahlgren (to include Coastal Systems Station), NAWC China Lake (to include Point Mugu), NSWC Port Hueneme Division East Coast Operation, NCCOSC NRaD, JHU/APL, and NRL, to provide support to the ASJPO. These particular private industry Government Laboratory personnel were identified as source selection assistants, and were barred from interacting with Industry IPTs. All personnel included in internal ASJPO activities submitted Financial Disclosure Statements which were screened and approved by DARPA General Counsel for potential conflict of interest.

2.3.1.3 *Government R&D Centers relationship to Industry Teams*

ASJPO made available to the Industry Teams, government services that could not be obtained from private industry. A Phase I Memorandum of Agreement (MOA) (Tab K) was negotiated between NSWC and ASJPO that outlined the allowable interaction with industry during the first phase of the program. This allowed the Laboratories to provide objective products and services, non exclusively.

In Phase II, the relationship between the Industry Team and the Laboratories was expanded (Tab L) to include lab participation on Industry IPTs. Great flexibility was provided by the MOA; however, because of R&D Center concern with the legal restriction that they not compete with private industry, R&D Center personnel were slow to start work on industry IPTs. To alleviate this problem during Phase II, funding ($450k per Team) was made available at the R&D Centers directly from ASJPO to support each Industry Team. Industry Teams developed Statements of Work for this support since the Government itself (ASJPO) was providing the funds, the R&D Centers felt much more at ease with industry requests. This arrangement avoided any competition with industry and worked extremely well.

2.3.1.4 *ASJPO and Industry Team relationship to PARMs*

Within the trade space available to the Industry Teams to satisfy Arsenal Ship design goals, was the opportunity to select legacy Navy systems. The use of legacy systems in many cases was more attractive than developing new ones based on investment required, and because these systems have passed the WSESRB, OPTEVFOR and other safety/operational checkpoints in other developmental programs. These systems could be obtained either as Contractor Furnished Equipment (CFE), or purchased through the Navy PARMS. PARMs (PARticipating Managers) are the Government program offices responsible for individual Navy systems such as ATWCS, CEC, and SSDS. Systems of primary interest to Arsenal Ship Teams were ATWCS, AFATDS, and CEC.

The PARMs were receptive to making hardware available to the Arsenal Ship Teams, but expressed concerns about requests for the release of source code software to Industry Teams. They cited the difficulty in maintaining configuration control, plus the apparent liability of the PARM for performance anomalies in his system but was not under its control. As a consequence, PARMs were unwilling to release Government owned software for Arsenal Ship program use.

These concerns were the subject of an Arsenal Ship Offboard Systems Working Group, discussed in the C4I section of this report. Although not resolved during Phase II, it was hoped that a good working relationship could be developed during Phase III when only a single Industry Team would have interest in modifying source code. Initiatives in ATHENA working groups would likely have facilitated this process.

2.3.2 *Lessons Learned*

2.3.2.1 *A small Government program office is sufficient when the total system design responsibility is in industry's hands.*

The six Government personnel and six technical SETA proved adequate to monitor the consortias' progress, without imposing excessive oversight. By maintaining a daily consortia calendar across the teams, ASJPO could be physically present at the majority of design interactions by the teams. By locating access to their Integrated Product Design Environments (IPDE) at ASJPO, real time accessing of trade studies, IMS, IMP, 2D and 3D CAD/CAM models permitted virtual presence for those periods when ASJPO could not be physically present.

2.3.2.2 ***Industry welcomes openness of Government feedback.*** Industry was solicitous of Government feedback, to avoid revisiting errors of previous design efforts (which cost time and money), and to hear directly from the customer on likes/dislikes. They welcomed the openness of ASJPO comments on their progress. This relationship was enabled by two aspects of the Agreement; no claim provisions and no unilateral Government direction.

2.3.2.3 ***Providing expertise and "lessons learned" to Industry is a new Government role.*** Much as it was difficult for industry to accept full design responsibility at the beginning of Phase I, it was equally difficult for the Government to provide advice rather than give direction. Government retains a wealth of knowledge and operational experience that is indispensable to industry's design/build process. ASJPO called on technical specialists to provide briefings on survivability, combat systems interfaces, and Joint warfighting C4ISR. Access to intelligence reports, studies, and government designs were also facilitated by ASJPO. The guidance to Government personnel was that information exchange was acceptable, but that design mandates were not.

2.3.2.4 ***Government intervention was required in obtaining data, access, and equipment.*** Industry does not have authority to compel PARMs or System Command headquarters to provide information and access to data or equipment. ASJPO frequently had to exercise such authority.

2.3.2.5 ***Safety and security cannot be delegated to industry.*** Industry's responsibility for and integration of all aspects of the design/build product requires a special interaction with the Government Program Office in the case of safety and security. The Program Office is ultimately responsible for personnel and ordnance safety. Extant Government agencies (WSESRB and INFOSEC, for example) do not respond to private industry, but are chartered to respond to the Program Manager. In this special relationship, the Government must fully understand the design, and certify its safety.

2.3.2.6 ***Direct funding of R&D Centers by the Program Office is essential to performance of some tasks.*** Due to legal restrictions on Government R&D Centers "not to compete" for services available from private industry, direct Government funding is needed to provide flexibility for some industry tasking. For example, combat system engineering is available from a variety of industry sources; however, for a specific system, this may not be true. In this case, the Government must identify work-arounds to facilitate competition.

2.3.2.7 ***The Government team should have expertise, insight into, and understanding of Teams' business decisions as well as technical decisions.*** In some cases the Industry Teams made design decisions because they were either the developer of a system/software, because they could not get detailed information about a system/software developed by others, or because their real market for a particular system is another program. The Government team needs to understand the motivations which underlie specific design decisions in order to assess if those design trades yield the best solution.

2.4 -- Funding

2.4.1 *Background*

The funding for the Arsenal Ship program was initially laid out in the Memorandum of Agreement (MOA) for the Joint Navy/DARPA Arsenal Ship Demonstration Program between Mr. Larry Lynn (Director, DARPA), Mr. John Douglass (ASN(RDA)), and RADM D.J. Murphy, USN (N86) in May 1996 (see Tab B). The cost of the R&D program for the Arsenal Ship Demonstrator was determined to not exceed $520 million including the cost of concept development and competition. The Navy DARPA joint funding profile was:

	FY96	**FY97**	**FY98**	**FY99**	**FY00**	**FY01**	**Total**
Navy	$4.0	$25.0	$141.0	$90.0	$80.0	$10.0	$350.0
DARPA	$1.0	$15.0	$47.0	$50.0	$36.0	$21.0	$170.0

In the same MOA the Navy agreed to provide its share of the funds to DARPA at the beginning of each fiscal year.

In January 1997, during the Phase I to Phase II downselect, three team's designs were determined to be of such value as to warrant their continuation into Phase II. The cost of adding a third Industry Team to Phase II was calculated to be $21M which the Navy agreed to fund. The $21M was split funded between FY97 ($9M), and FY98 ($12M).

During the PR-99 review process, the Navy redistributed $50M from FY98 to FY99; the overall Navy total of $350M remained the same. A dialog with industry determined that this revised funding stream could be accommodated without impacting the delivery and test schedule of the Demonstrator.

2.4.2 ***Lessons Learned***

2.4.2.1 ***Funding Navy Research and Development (R&D) Centers/University Affiliated Research Centers (UARC) to Support Industry was necessary.*** Money sent to R&D Centers from ASJPO on behalf of the Industry Teams was necessary to maintain each Team's competitive edge and to allow for specific support. R&D Centers were allowed to perform any kind of support work without regard to "non-competitive" rules of accepting money directly from industry.

2.4.2.2 ***The ability of a University Affiliated Research Center (UARC) to support Industry needs to be broadened.*** Industry relationships with UARCs are much more difficult to establish than those Government Laboratories with DBOF funding. UARC's, much more than DBOF laboratories have been concerned with maintaining their appearance of objectivity with its many customers. The legal barriers of industry UARC interaction need to be more fully explored in the future to enable a much needed dialog.

2.5 -- Source Selection

2.5.1 ***Background***

The Arsenal Ship Program utilized "best value" in determining the successful offerors for each phase of the program. A new solicitation was released for each competitive phase of the program that contained the source selection criteria applicable to that particular phase. The entire solicitation, including source selection criteria, was released to industry in draft form to allow them an opportunity to provide input and ensure clarity.

Source Selections for all Phases of the Arsenal Ship Program were performed in accordance with the approved Source Selection Plans. A new plan was prepared and approved for each consecutive phase of the Arsenal Ship Program. The Source Selection Plans were the same in structure and format to those prepared under traditional Government source selections. All source selection criteria and any information regarding its interpretation were contained in the solicitation. The source selection plan did not provide any elaboration of the selection criteria beyond that which was shared with industry in the solicitation.

The Arsenal Ship source selection approach was to perform an integrated technical, financial and management review of each Team's proposal. All source selection participants were required to attend one of several briefing sessions that outlined the source selection process, criteria and schedule. Source selection participants were divided into teams based upon their area of expertise. However, there was cross-over of personnel on all teams (i.e., some technical team members were also cost or management team members, etc.). The contracting, legal and technical team members all read and commented on all parts of the proposals and all participated in the discussions and selection of the ultimate awardees. Each team had a designated team leader who coordinated the team's evaluation efforts. Team leaders met briefly each morning to address questions or issues that may have affected the other teams' efforts and ensured a free exchange of ideas and information across the technical, financial and management disciplines.

In addition to the written proposal, oral presentations were held with industry approximately three weeks after receipt of proposals. Dates for oral presentations were assigned to each Industry Team by random drawing. All Teams were provided equal time to present highlights to their proposal. Oral presentations provided the ASJPO with the opportunity to clarify aspects of each Team's proposal. The solicitation specifically stated that oral presentations were not an opportunity to revise proposals and that the Government did not intend to request revised proposals as a result of the oral presentations. All oral presentations were held at the Team's facility and were video taped, with a copy to the Government. Only ASJPO personnel attended these oral presentations.

Evaluations from all source selection participants were consolidated into team briefings that were presented to the Source Selection Evaluation Board (SSEB). The SSEB then provided these team briefings to the Source Selection Advisory Committee (SSAC), along with its recommendations for award, over a period of two days. The SSAC was provided an opportunity to comment and question the results of the SSEB. A consolidated briefing, including the comments and recommendations of the SSAC, was then delivered by the SSEB to the Source Selection Authority (SSA). The SSA made his decision based upon this briefing within a day.

After the award announcement, detailed debriefings were held with both successful and unsuccessful offerors covering cost, management and technical issues.

2.5.2 *Lessons Learned*

2.5.2.1 ***Detailed debriefs to successful offerors provided Teams full insight into perceived strengths & weaknesses which allowed them to improve their efforts in subsequent phases and reduce risk for the program.*** The ASJPO provided detailed debriefs to all offerors, including successful offerors. The source selection brief given to the SSA was used as the basis for detailing strengths and weaknesses in the Teams' proposals. The detailed debriefs to the successful offerors proved extremely beneficial at enlightening the Teams regarding areas where the Government perceived excessive technical risk, questioned costs or the management approach. All Teams made an extensive effort at redressing questionable areas noted in the debriefs. This led to improved designs and overall risk reduction for the program. Industry commented that they felt these debriefs were extremely beneficial, and the best they had received in twenty years.

2.5.2.2 ***Cross-over of source selection personnel (i.e., technical, financial and management) is imperative to evaluating proposals.*** The Arsenal Ship Program allowed industry to develop its own ship design. Because each Team's design, management and production approach differed, cost estimating had to be very individualized. The financial analysis team's results were very dependent on the findings of the technical team with regard to the design being proposed and the production methods utilized. Likewise, the management team's results were important to both the technical and financial teams' analysis. Insight into the findings of the financial and management teams provided the technical team further understanding of the proposal and potential risks. All source selection personnel were provided full access to all parts of the Industry Teams' proposals, including cost data. All source selection participants agreed that this full and open approach led to a greater understanding of what was being proposed, eliminated many questions and guessing, allowed for better identification of risks, and facilitated a true best-value decision.

2.5.2.3 ***Continuous interaction with Industry Teams throughout the design phases facilitated a short source selection schedule.*** The Phase I source selection for the Arsenal Ship Program was completed within three weeks from the time of receipt of proposals to the announcement of successful offerors. Phase I source selection criteria's main focus was that offerors assemble a team fully capable of performing all aspects of the Arsenal Ship Program (i.e., trade-off analysis, design, construction, and support). Six proposals were received and five agreements were awarded. However, the Phase II source selection criteria required detailed analysis of each Team's concept design in order to determine which represented best value to the Government. The Phase II source selection was completed within two months (Nov 15-Jan 11) from the time of receipt of proposals to the announcement of successful offerors. Given the large amount of information presented and the detailed analysis required, this schedule presented a great challenge. The ASJPO believes this schedule was made possible due to the intense amount of interaction that occurred between the Teams and the ASJPO while industry was developing their concept design. Technical, cost and management approaches were continuously discussed prior to and during the Phase II and III periods of performance. As such, the ASJPO was provided great insight into what the Team's proposal would contain. Risks were continuously assessed and discussed. Proposals presented by the Teams built upon work that had been presented throughout the year.

2.5.2.4 ***The ASJPO's ability to strictly adhere to schedules established at the onset of the program greatly enhanced program credibility and the commitment of the Industry Teams.*** In the Phase I Arsenal Ship Program Solicitation, the Government provided a top level program schedule addressing all phases of the program. This schedule indicated start/end dates for each phase and indicated release dates for solicitations, due dates for proposals and announcement dates of successful offerors for each phase. Every program milestone was met as originally scheduled for the eighteen months of Phases I and II. The ASJPO received tremendous feedback from the Industry Teams regarding its ability to establish a schedule up front and adhere to it. The Government's ability to manage its own portions of the program schedule facilitated the Industry Teams' ability to plan their own program schedule with Team members and subcontractors and execute accordingly without fear of delay on the part of the Government, which is all too typical in the majority of Government programs.

2.6 -- Testing and Certification

2.6.1 *Background*

Early in the Arsenal Ship program, the program office decided to pursue substitute methods for the traditional processes of test and evaluation and ship acceptance. As a consequence, provisions were included in the Phase II solicitation that required the Industry Teams to develop processes and procedures by which the Arsenal Ship Demonstrator and the Production Arsenal Ship would be certified as "fit and safe for the intended service." The purpose of this part of the Arsenal Ship "experiment" was to define and implement a streamlined process that would achieve the intent of Navy T&E and INSURV at less cost and in less time.

The Arsenal Ship Program Agreement required the Industry Team to provide the Government a Preliminary Certification at the end of Phase III. The Preliminary Certification signaled the end of Phase III and the readiness of the Demonstrator to proceed into Phase IV, Demonstration. A Final Certification report was also required of the Team. This report was to detail the test and evaluation results of the Phase IV Demonstration; define the capabilities of the Demonstrator, and verify the military utility of the Arsenal Ship. Receipt of the Final Certification Report signified the completion of Phase IV. The Government would accept the Demonstrator upon receipt of the final report or upon completion of deficiencies whichever was later.

ASJPO placed responsibility on Industry to identify and analyze the constituent elements of the current Navy test & evaluation and acceptance process and find acceptable alternatives that reduce time and cost. One cost reduction strategy was to reduce redundancy in tests and inspections. Self certification, as an example, is one way to do that. By the end of Phase II each Team had developed a plan and a proposed program by which they would "certify" both the Demonstrator and the Production Arsenal Ships.

2.6.2 *Lessons Learned*

2.6.2.1 ***An alternative to the traditional T&E and Acceptance process is possible as part of ship acquisition and that process can result in less cost and time without compromising safety, reliability, or performance.*** The work completed by the three Industry Teams, including preparation of a Demonstrator Test Plan, and a plan for Certification by the end of Phase II work, supports this conclusion. Inherent in the meaning of "certification" as "fit and safe for intended service" is a process by which all tests, analyses, inspections, etc. essential to verification that specific performance and safety requirements are met, are identified and completed. This then requires the Industry Team to establish traceability among all performance/safety requirements and the certification events necessary for their verification.

2.6.2.2 ***Some certifications are best performed by Government bodies.*** Certain certifications currently performed by Government activities, e.g., weapons safety by WSESRB, were deemed essential and had no acceptable alternatives. Other certifications, e.g., that for helicopters, were reviewed with an eye to change in order to improve performance. In the latter case, the rule makers were willing to make changes as long as safety principles were retained. There was mutual agreement between ASJPO and the Industry Teams that these special cases would be incorporated into each Team's certification plan.

2.6.2.3 ***Certain unique range and test capabilities are owned by the Government (e.g., Eglin and Point Mugu missile test ranges) and it makes no sense to invest in any new facilities or replicate existing capabilities.*** DoD policy is supportive of industry use of Government ranges. Industry Team interaction and discussions with managers of various Government test facilities were positive and indicated that industry could avail themselves of these facilities in cooperative arrangements with Government. At facilities that are not unique, range managers will be challenged to control costs in order to remain competitive.

2.6.2.4 ***Classification Societies represent a resource to provide some cost effective inspections, analyses and commercial certifications.*** Classification Societies such as the American Bureau of Shipping (ABS) can perform certain hull and mechanical system certifications but do not have the background, experience nor the personnel to address uniquely military systems such as combat systems. ABS has developed a proposed Certification plan, which identifies by SWBS group, systems to be certified, appropriate standards to be applied, and recommended certification agents. The ABS work lends confidence to a movement towards commercial certifications.

2.6.2.5 ***Early involvement by Navy T&E activities in Industry-led IPTs is desirable.*** Currently, operational ship test planning expertise and knowledge resides in the Government and most range testing done for ships has been coordinated by military and Government civilians; therefore, these individuals can bring a wealth of information to the design and development process. This has to be controlled by the program office to insure that inefficient methods and procedures are not perpetuated and that the Government interaction has value added to the Industry Teams. The role of COMOPTEVFOR in demonstrations designed to support production decisions should be determined early in the program. OPTEVFOR participation is not essential to execute an effective program; however, their participation in Industry Team-led IPTs will bring operational credibility and experience to the process.

2.6.2.6 ***The issue of Live Fire Test (LFT&E) requirements and how they are handled within an Industry Team-led environment needs to be addressed.*** Arsenal Ship as a non-ACAT program was a special case not bound by LFT&E legislation. Other programs subject to this legislation will have to address the best mechanism to satisfy this requirement. In particular, some of the current requirements to perform vulnerability assessments, which could dictate design choices, may run counter to unconstrained performance trade space.

2.6.2.7 ***Some Government furnished items may be appropriate in the context of operational test and evaluation and demonstrations.*** Future ship acquisition programs will likely strive to minimize Government furnished equipment (GFE); however, the Navy should supply military or other unique, Government-owned items directly to the Industry Team if that is cost effective and facilitates meeting schedule milestones. Examples include missiles and other ordnance for demonstrations that may not be easily obtained by an Industry Team.

2.6.2.8 ***Ship Designation, "public" or "private," determines the applicability of U.S. Coast Guard and other international regulations.*** An early determination needs to be made whether a ship is a public or private vessel. Private vessels must comply with U.S.C.G. regulations. 46 USCA 2109 exempts public vessels from these same regulations. U.S. Coast Guard jurisdiction is determined by ownership and ownership in turn implies indemnification and liability. For Navy Ships, DoD (Navy) initiates a waiver of compliance for all life saving, navigation, and other regulations. Regulations for carrying/packaging hazardous materials are handled by a different organization within the Department of Transportation called the Research and Special Projects Administration (RESPA).

2.6.2.9 ***Early establishment of Industry T&E IPTs and Government participation in them is important.*** Active Government participation in Arsenal Ship T&E IPTs proved to be very beneficial and an effective way to introduce Government input and thinking into the Industry test and certification process. Early involvement of T&E personnel on other IPTs and in the design development process is essential, particularly when testing is part of a risk reduction program. Risk reduction tests and demonstrations need to be tracked and monitored at the Program office level.

2.6.2.10 ***User participation and Fleet Exercise planning in Industry tests and demonstrations should be coordinated by the program office.*** The Government program office should arrange for the availability and participation of active duty military and Government civilian personnel to support T&E and demonstration activities. The program office should also coordinate ship participation in fleet exercises.

2.7 -- Office Operations

2.7.1 *Background*

2.7.1.1 *Organization*

A small program office was an essential element of the ASJPO acquisition reform approach. A large number of people would have had a stifling affect on industry's innovation just from the sheer magnitude of opinions that would be expressed by such a group and industry's intense desire to please its customers. The make up of the office dictated by the original Memorandum of Agreement of 28 May 1996 among DARPA, ASN(RDA) and N86 specified six Government employees, growing to a maximum of 9 during Phase III. DARPA, NAVSEA, and ONR were to "contribute" (long term loan) 3 persons each, thus no billets beyond that for the Program Manager at DARPA were created for the program office.

Additional private industry personnel were collocated in the Program Office and dedicated full time. The companies, Vail Research and Technology Corporation and SYNTEK Inc., were barred from any contractual relationship with the competing Arsenal Ship Industry Teams. This approach is very much in keeping with past DARPA program office structures.

The total on-site staff by early Phase II (spring 1997) was as follows:

Program Manager	Capt. Charles Hamilton	DARPA
Contracts Negotiator	Cindy Shaver	NAVSEA (02)
Technical Director	Kit Ryan	NAVSEA (03)
C4I Engineer	Jack Frink	ONR (IPA from JHU/APL)
Combat Systems Eng.	Terry Sheehan	NSWC/Dahlgren
Survivability Engineer	John Grizzard	ONR
Admin assistance	8 people	Vail Research and Technology Corp.
Technical assistance	6 people	SYNTEK
Total on-site	20 people	

Significant off-site staff support was provided by:

General Counsel	Rick Dunn	DARPA
Assistant General Counsel	Dianne Sidebottom	DARPA
Contracts Management Directorate	John Ablard	DARPA

Support personnel were fully representative of the program office to the Industry Teams and to other Government activities; close integration of the support contractor team with the Government Program Office while avoiding personal services and inherently governmental function issues proved possible through open and continuous communication and clear understanding of roles. The support team consisted of:

Hull, Mechanical Electrical Engineer	Robert Percival	SYNTEK
Ship Operations and Support Eng.	Dr. Bob Johnson	SYNTEK
Systems Engineer	Ron Kiss	SYNTEK
Test and Certification Engineer	Jack Turner	SYNTEK
Operational Effectiveness Engineer	Bruce Dyer	SYNTEK
C4I Supporting Engineer	Mike Yermakov	SYNTEK
Ship Cost Engineer	Dave Schwiering	Vail
Office Manager	Jennifer Bennett	Vail
Financial Manager	Jennifer Sowa	Vail
Security Manager	Jim Booth	Vail
Computer Manager	Jack Stevens	Vail
Graphic Artist	Jill Judge	Vail

Administrative/Contracts Assistant	Wendy Turner	Vail
Travel Assistant/Receptionist	Frances Horner	Vail

2.7.1.2 ***Information and document control***

On-site capability for handling documents through the Secret level was maintained. Front door access required a special access badge issued by DARPA. Only the above on-site personnel were issued these badges.

An acknowledged special access compartment was established to handle compartmented data at a site close by which was already functioning for this purpose. This greatly facilitated handling such data, discussions and even travel arrangements.

A classification guide was developed with the objective of making design information easily available to designers and production people for the life of the program. Thus, ship lines were made FOUO, for example. Even so, some of the Industry Team personnel were inadequately cleared in advance for sensitive data, which hindered Team communication and, sometimes, decision making. Structural design and signature analysis (and their effect on ship systems engineering) suffered from some of these restrictions on personnel.

2.7.1.3 ***Communication via IPDE***

The goal of keeping in close contact with the technical progress of the Industry Teams necessitated regular communications. In addition to calls, email, and normal deliverables (which were few), numerous informal meetings were held on specific topics with individual Teams. While few in number during Phase I (perhaps 1 per month on each topic), these mushroomed into perhaps 1 per week on each topic during Phase II. It was hoped that use of an internet-based engineering database system would enable much of this activity to take place without travel, but this did not turn out to be the case. A combination of travel and IPDE monitoring provided adequate insight.

2.7.2 ***Lessons Learned***

2.7.2.1 ***Integrated Product Design Environment (IPDE) requires a cooperative collaborative effort between Government and Industry.*** The IPDE works best if all the users (including the Program Office) can make recommendations on security, software selection, modem speeds, and architecture/equipment needs. This should not be construed as an "automatic" at the initiation of a design effort.

2.7.2.2 ***"Open orders" need to be issued to facilitate rapid response on travel.*** Meeting dates were in constant flux due to natural program dynamics of each Industry Team. ASJPO was constantly making last minute travel plans and changes to accommodate this. Near the end of Phase II, DARPA instituted new travel rules that required more lead time than was available for many of these meetings. To be responsive, open orders, with limitations, would greatly ease the administrative burden on the travel system. In addition, a portion of travel funding should be left at the parent organizations to expedite the travel process for those outside the Program Office.

2.7.2.3 ***Extensive ASJPO use of email for internal communications worked well.*** Portable computers allowed on-line access while on travel and during off-hours. A secure ASJPO Web Page and encrypted e-mail capability for source-sensitive materials would have improved the process. Secure video teleconferencing capability was underutilized but recommended for future efforts.

3.0 -- Technical

3.1 -- Systems Engineering

3.1.1 *Background*

The Arsenal Ship Program Office, recognizing the complexity of the ship being developed as well as the challenging schedule established to complete the design and construction, expected the Industry Teams to rely heavily on a systems engineering approach. A key aspect of this program was the decision by the Government to turn the systems development process over to industry from the earliest stages of the process. This decision challenged industry to develop and design the optimum mix of performance capabilities within the production and life cycle cost affordability constraints.

Related to the application of the systems engineering approach were the use of an Integrated Product Development Environment (IPDE) and Integrated Product Teams (IPT).

Critical drivers of the process were the use of Price As Established (PAE) and the extremely low manning goal established for the ship. The low cost goal established for the Demonstrator ship, compelled the Teams to focus on major system engineering objectives and tradeoffs.

3.1.2 *Lessons Learned*

3.1.2.1 ***A systems engineering approach is essential.*** A disciplined systems engineering process based on defined goals, thresholds and trade space is essential. It creates the environment for true innovation, and enables industry to solve the complex design problems associated with the MFSD. Providing industry with unprecedented freedom to develop their own solution to a ship design without the application of Systems Engineering would not likely yield a successful outcome. Systems engineering in and of itself was no guarantee that the "best" solution was derived, but its discipline helped the program management understand better whether the approach was complete, and whether the right priorities were assigned and the right questions asked.

3.1.2.2 ***Adequate time is needed for Industry Team formation and growth.*** Much energy was spent by industry in the first two months of Phase I sorting out Team membership and Team relationships. At the start of Phase II, the average Team size increased from about 50 to over 200 people in a short time frame, which caused different design maturation rates among the Teams. The phases of the program were structured to allow seamless transition from one phase to the next. Industry did not take full advantage of the downselect decision period to position themselves for the next phase. Had they done so, the Phase transition would have been much smoother. Without this, an additional 2 months for Phase II would have permitted all three consortia to fully mature their functional design.

3.1.2.3 ***An Integrated Product Development Environment (IPDE) and Integrated Process Teams (IPT's) were beneficial to design success.*** IPDE and IPT's contributed in very powerful ways to expedite the process and to ensure that the process did not neglect any critical aspects of the design. In one Teams case, the IPDE allowed geographically dispersed members of the Team and the Government to work with the most current information available at all times. This was essential to permitting the level of concurrency necessary to meet their constrained schedule. The IPDE also facilitated the ability to move quickly from stage to stage in the design evolution. The Team that excelled in using IPDE tended to make faster progress because information transfer among Team members occurred at a more rapid rate. Similarly the IPT's ensured that interdisciplinary teams were focused on the most critical parts of the overall design problems and that complete solutions were considered. To ensure that the IPT's themselves did not become "stove-piped", collocation, cross membership of key individuals and the IPDE were utilized.

3.1.2.4 ***Freedom to make trades enabled achievement of cost and schedule goals.*** Giving Teams nearly infinite flexibility to make trades enabled them to achieve low cost goals and schedule constraints. The Teams reinvented the sequence of steps in the ship design and construction process to meet the cost and schedule goals of the program. Certain subsystems were eliminated because their lead times would not meet the schedule.

3.1.2.5 ***A design competition performed by industry may be an acceptable substitute for a Government run COEA/AOA.*** A PAE-based program trades cost, performance, and schedule. The competitive Industry Teams had the freedom to seek the "best value" solution. Competing Industry designs effectively encompassed the range of real solutions to the CONOPS. Not only were they varied in size and capability, but the solutions were also realistic, in that they were actually procurable. This contrasts with typical Government conducted COEA studies that identify hypothetical solutions that are not validated, particularly with respect to cost or producibility.

3.1.2.6 ***Industry Team control of the full trade-space is essential to achieving total ship integration within the PAE goals.*** Previously unthinkable trades crossing many boundaries, such as trading propulsion capability for lines of code in the communications suite were discussed. In Arsenal Ship, the PAE process forced significant innovation: three new launchers; improved software processes; COTS insertion and refresh, and software reuse are some examples. Through industry IPTs, a total ship engineering approach was applied for a balanced hull and topside design, trading off signatures, communications systems, navigation systems, weapons and cost.

3.1.2.7 ***Industry Teams are fully capable of producing new combatant ship designs.*** The design products delivered over the course of the program were excellent, and comparable in overall quality to previous in-house Navy designs. Industry was extremely effective in assembling a mix of private and public sources, marine and aerospace, to cover the full range of technical disciplines.

3.1.2.8 ***Design commercialization requires new systems architecture not only for information systems, but also for mechanical systems, to permit use of COTS systems and equipment.*** One of the best examples of this innovative thinking involved methods of shock mitigation for selected equipment. It was obvious that militarizing all equipment would have eliminated the opportunity to use commercial-off-the-shelf items. By focusing on shock isolation schemes a more cost effective approach was achieved.

3.1.2.9 ***Early application of budgets for cost and performance was essential.*** Budget and performance allocations to established goals and thresholds are facilitated by the discipline inherent in the system engineering process. The teams who were most successful at setting cost and performance budgets and sticking to them became the most competitive.

3.2 -- Survivability

3.2.1 Background

The survivability of the Arsenal Ship was broken into three categories: susceptibility, vulnerability, and recoverability. Susceptibility was defined as the probability of the ship being detected, targeted, engaged and hit by a threat weapon. Vulnerability was defined as the probability of the ship being damaged to the point of losing mission capability, propulsion, or being sunk. Recoverability was defined as the capability of performing fire fighting, damage control, and recovering mission capabilities after incurring damage from a threat attack. Recovering mission capability (= "fighting hurt") was not a program goal. These definitions were given to all Industry Teams early in Phase I as a guide to how the ASJPO was to analyze and evaluate Industry designs.

Since the Industry Team was responsible for the ship design, the ASJPO did not define a specific set of threats through a System Threat Assessment Report (STAR). Instead, the ASJPO provided references and sources for threat information within the Defense Intelligence Agency (DIA), Office of Naval Intelligence (ONI), and NAVSEA. Each Industry Team was then allowed to request threat information directly from these sources. In many cases, the ASJPO helped facilitate the transmission of threat information to the Teams. None of the Teams asked for an identical set of information, however, the threat assessments provided similar results. The top level design goals for each Team, derived from these threat assessments, were not drastically different; however the approaches taken in the ship designs to achieve these goals were drastically different and showed unique and innovative solutions.

In addition, the Teams were provided with points of contact for Government ship survivability expertise. The Industry Teams were provided an opportunity to contract directly with Government R&D Centers to hire Government experts and to take advantage of Government computer models and simulation tools. In the survivability area, most of this activity occurred between industry and NSWC Carderock for weapon effects analysis and design. In Phase I the Teams were only allowed to contract for "objective services". The NSWC engineers were allowed to provide lessons learned from previous work and provide computer analysis of industry designs. In Phase II, all three Industry Teams contracted with NSWC Carderock for exclusive IPT support for vulnerability reduction analysis and design. Personnel supporting the Industry Teams in Phase II were "fire-walled" from each other to prevent inadvertent release of competition sensitive information. A separate set of engineers at NSWC Carderock provided direct support to the ASJPO. The management of personnel resources to allow NSWC Carderock to provide support to both industry and the ASJPO required up front planning and policy making.

The ASJPO Survivability Team included members from ONR, NAVSEA, NSWC Carderock Division, and System Planning Corporation (SETA support to DARPA/TTO). By the end of

Phase I, and for Phase II Source Selection, the team consisted of seven people: The ASJPO Survivability Manager, one magnetic signature engineer, one electromagnetic (RCS/IR) signature engineer, two ship acoustic signature engineers, and two vulnerability engineers. After Phase II Source Selection was complete, the team was increased to eleven with the addition of one electromagnetic signature engineer, two damage control/fire fighting engineers, and one composites/materials engineer. The team of eleven was maintained throughout Phase II and all would have participated in the Phase III Source Selection effort. With this diverse group, the ASJPO was able to review all aspects of industry's survivability designs, provide pertinent information on Navy ship survivability activity, and leverage off ship survivability research and development efforts throughout DoD. Because most of the team members simultaneously worked on other ship programs, lessons learned from other programs were easily transferred into the Arsenal Ship program.

One dilemma a program office faces when industry is given total responsibility is that analysis tools used can differ from one competitor to the next. Often there is no standard by which to calibrate these tools for comparison sake. In order to discount this problem, the ASJPO tasked several Government organizations to perform independent analysis of the three industry designs. The Naval Research Laboratory was tasked to perform RCS/IR signature predictions and run CRUISE MISSILE engagement simulations against expected threats. NSWC Dahlgren Division Coastal System Station, Panama City, FL was tasked to perform mine susceptibility studies using the Total Mine Simulation System (TMSS) and Naval Mine Warfare Simulation (NMWS) tools. NSWC Carderock Division was tasked to perform ship vulnerability analysis against projected warhead threats using the Ship Vulnerability Model (SVM). By using the same models to analyze all three ships, the ASJPO was able to obtain a fair comparison of ship performance that would have been used as part of the downselection process.

3.2.2 ***Process Lessons Learned***

3.2.2.1 ***An integrated Survivability team and approach allows for an optimized design that provides improved ship survivability.*** By working all ship survivability aspects together, trade-offs and optimization can be achieved to provide a more survivable, cost effective design. By working the different survivability areas separately, complementary and optimized designs will most likely not be achieved, compromising ship survivability and affordability.

3.2.2.2 ***Ship vulnerability/weapons effects hardening expertise, technology development and analysis capability resides largely in the Navy R&D Centers.*** System design and integration capabilities are the Industry Team's area of expertise. Early involvement in the Functional Design Phase of the Navy R&D Centers working with the Industry Teams is vital. During Phase I, the ASJPO detected a weakness in the structural survivability area on all five Industry Teams. Industry has never had motivation to develop ship vulnerability/weapons effects design expertise as an in-house capability. Early in Phase II, the ASJPO, in conjunction with NSWC Carderock, conducted a survivability course for the three Phase II Teams. This course allowed the Government to share Navy ship survivability methodologies, technologies, and lessons learned. All three Teams used the information, as well as NSWC Carderock on their IPT's, to improve their ship designs.

3.2.2.3 ***In the case of signature control and damage control, technology exists to achieve the signature and automation levels desired.*** However, cost-effective designs and large-scale demonstrations are needed to prove the concepts. Without large-scale demonstrations, technologies will go from the R&D Center straight to a ship without sufficient development, causing high risk and costly implementations. The Arsenal Ship Demonstrator, or Maritime Fire Support Demonstrator, would have provided the Government and Industry an opportunity to show at a large-scale that affordable signature control and automated damage control/fire fighting are technically feasible. The demonstrator ship would have allowed for advanced concepts to be implemented and tested, ultimately providing low risk, cost effective solutions to future ship programs. Without these engineering and demonstration efforts, future ship programs will continue to implement stove-piped, costly, and inferior systems.

3.2.2.4 ***In order to transition Government developed technologies to an industry lead design effort, a program office must identify pertinent technologies and communicate their availability to the Industry Teams.*** The three Industry Teams were very receptive to Government R&D efforts and used a variety of recently developed Navy technologies in their designs. Throughout Phases I & II, the ASJPO Survivability Team briefed the Industry Teams on a variety of technology programs and developments including the Integrated Magazine Protection System and surface ship acoustical tiles. The Teams used the information provided by ASJPO to enhance their designs.

3.2.2.5 ***The Industry design teams favored components and/or subsystems that were already developed or whose development was being funded by the Government.*** If a technology or system development was being funded by the Government, the Industry Teams would, in most cases, include those in their design instead of incurring the expense of internal development. This is important in answering the question of how to integrate Government developed systems into designs totally controlled by industry. The answer is, "It happens naturally because the Industry Teams do not want to add unnecessary cost to their system, especially during a competition."

3.2.2.6 ***It is important for key industry participants to have appropriate security clearances and access to information early in the mission analysis/concept design process.*** During Phase I and the early stages of Phase II, many industry personnel, especially in the shipyards, did not have Top Secret clearances. Much of the data and analysis regarding ship susceptibility is held at that security level. Because of the mismatch, many key personnel were not privy to some of the trade studies and design goals. This caused requirements flow-down without being able to communicate requirements justification. In some cases, requirements were derived without being able to execute the appropriate trade studies and analysis.

3.2.3 ***Product Lessons Learned***

3.2.3.1 ***Defining a realistic program goal at a top level (e.g., passive survivability through use of signature control) and giving industry the freedom and resources to choose the approach and achieve the goal, will provide products that meet or exceed performance goals and cost goals.*** Given a top level goal, freedom to use industry developed technologies and processes, and reasonable program resources (time and money), industry can conceive, design, and produce product solutions faster, cheaper, and better than internal Government led design teams.

3.2.3.2 *It is imperative that the Government, in executing a source selection, be able to compare the ship designs and performance characteristics using a common set of analysis tools.* With multiple Industry Teams doing ship design, a variety of design, modeling, and simulation tools are used. Some of these tools have been validated against test data, some have not. In order to provide a common analysis environment, the ASJPO tasked NRL, NSWC/DD CSS, NSWC/CD, and NAWC-WD Pt Mugu to do analysis of the three ship designs independent of the Industry Teams. This enabled the ASJPO Source Selection Group to do comparisons of the three ships with common analysis tools.

3.2.3.3 *The fidelity of inputs available for independent Government evaluation varied between designs.* Because of the varied design approaches and maturity levels, the information available to the ASJPO for on-going evaluation and source selection were not equal from all the Teams. The ASJPO had to develop analysis approaches which normalized the fidelity of the designs for its analysis effort.

3.3 -- Combat System/Command, Control, Communications, Computers and Intelligence (CS/C4I)

3.3.1 *Background*

ASJPO CS/C4I objectives during these two phases were: to understand each Industry Team's Combat System and Command, Control, Communications, Computers, and Intelligence concept; evaluate it; rank each concept relative to the Arsenal Ship CONOPS; and assess its performance relative to Arsenal Ship Missions and consistency with the DoD C4I Vision. An additional ASJPO objective was to facilitate interaction and coordination among Government organizations and between the Teams and Government organizations to maximize each Team's potential for developing a successful CS/C4I functional design

Responsibility for requirements derivation and system engineering structure and development of a Functional Design resided with the Industry Teams. ASJPO provided a top level CONOPS and Ship Characteristics Document (SCD) as the basis for Industry to derive CS/C4I performance requirements. As such, ASJPO adopted an oversight philosophy based on an interactive and supportive approach on a minimally invasive basis during reviews, technical interchange meetings, and internal meetings with each Industry Team.

3.3.2 *Organization*

Figure 3-1 presents the ASJPO CS/C4I Assessment Team (CCAT) structure which consisted of a Core Team supported by advisors in CS/C4I functional areas. The Core Team itself is part of ASJPO and served not only to integrate analysis and assessments of CS/C4I functions, depicted at the bottom of the pyramid, but also to integrate those products with other functional areas (HM&E, T&E, Risk, Cost, etc.) to support an overall assessment of each Industry Team's Arsenal Ship design.

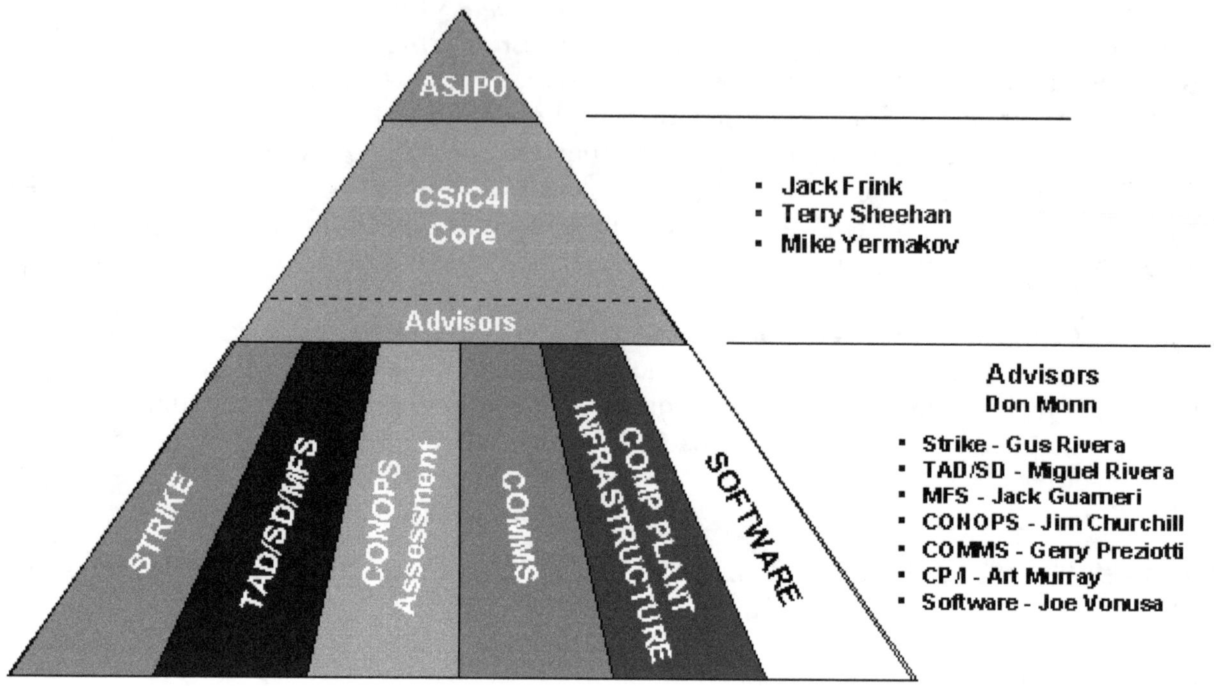

Figure 3-1. -- CCAT Structure

The CCAT Core Team attended all program reviews and encouraged consideration of out-of-the-box opportunities for technology insertion and business practice reform. Additionally, the Core Team provided the interface with other functional design areas (e.g., HM&E) within the ASJPO.

Table 3-1. -- CCAT

ASJPO
SSC-SD (formerly NRaD)
NSWC/DD
JHU/APL
SYNTEK
WEPTAC (China Lake)
NSWC/PHD/ECO
SPC

The organizations provided about 10 technical advisors to the ASJPO CCAT are shown in Table 3-1. They received documents and attended two meetings with each Industry Team related to CS/C4I to gain understanding of the requirements, concepts, and functional trade-offs studies that led to CS/C4I design baselines.

ASJPO encouraged direct Industry liaison with Government program offices developing or responsible for systems that the Teams wished to integrate or consider as part of their design. However, ASJPO was a facilitator for interaction when required.

3.3.3 *ASJPO/Government Interactions*

Because all the Industry Team concept proposals resulting from Phase I had impact on off-board systems that would require both resource and schedule coordination within the Government, an Arsenal Ship Off-board Systems (ASOS) working group (Table 3-2) was established.

Program/Activity	Resource Provider
Arsenal Ship	ASJPO
AEGIS	PEO-SC/AP
ATWCS	PMA-282
AFATDS	PM-FATDS (Army)
JMCIS/JMCOMS	SPAWAR
CEC	PEO-TAD/C
ATACMS	SSPO
NSFS	PMS-429

Table 3-2. -- Arsenal Ship Off-board Systems Working Group

The ASOS provided a forum for exchanging information between Government organizations on both technical and programmatic issues. For example, system baseline descriptions including funding and schedule profiles were presented. The ASOS also provided a mechanism for negotiating policies for interaction between a represented activity/program, ASJPO, and Arsenal Ship Teams.

Communications between CCAT members working at R&D Centers, etc. were open. However, communications of industry related business-sensitive information was strictly controlled. Additionally, advisors supporting the Core Team were limited to interaction on a strictly technical basis and as a rule attended only ASJPO-scheduled Industry meetings. Additionally, ASJPO limited access to Industry Team IPDEs to the Core Team.

3.3.4 *Government C4I Assessment Activities.*

The work effort to support CS/C4I was structured into two general work task assignments: Warfighting Assessments; and Technical Assessments.

- The warfighting assessments consisted of a CONOPS review in the context of scenarios in both simulated and seminar style wargames. Feedback from cross service operators was obtained on viability and impact to current and evolving doctrine/tactics. Industry Teams were encouraged to participate in the process to improve their designs by integrating operator input into their design process.

- The technical assessments were focused on traceability of industry-derived (based on the Government's CONOPS and SCD) requirements to the CS/C4I concept and functional design and a "best-value" analysis of the designs. The CS/C4I architectures were broken down into four main elements: mission; communications; support infrastructure; and software. Each element was assessed for feasibility, dependence on other programs, and impact on off-board systems or effect on infrastructure (as in the case with various COMM architectures). Additionally, an assessment based on broad-based DoD Joint Operations goals, such as Joint Vision 2010, Concept for Future Joint Operations, and Netted Targeting was conducted.

For each Industry Team, a system block diagram and functional requirements trace was captured and updated periodically as the Teams' designs matured. Additionally, associated technical and development schedule data was captured (when available) and maintained to keep the CCAT informed of progress toward Phase II deliverables.

3.3.5 *General Lessons Learned*

3.3.5.1 *Core Team interactions with Industry stretched their decision space.* By questioning old approaches, suggesting new ways enabled by technology, and challenging the Industry Teams to find innovative and creative solutions, ASJPO increased the range of design options considered.

3.3.5.2 *The notion of availability of Government legacy system information, expertise, and software to competitive Teams was not universally accepted.* Some program offices, Government R&D Centers and UARCs delayed or refused access to information and software. Specifically, issues of control, future development agent, and configuration management were surfaced. This is a serious issue for Industry led acquisition programs that will need resolution in the future.

3.3.6 *CONOPS Assessment*

In keeping with the goals and objectives of Joint Vision 2010, Warfighters across DoD were integrated into the assessment process through a series of seminar and simulated wargames. The wargaming activity brought together ASJPO, Industry Teams, and Joint Service representations into a structured and "competition secure setting" to understand and evolve industry's concepts based on user assessments and feedback. It addition, it provided a mechanism for ASJPO to identify those factors the warfighters considered important to rank the industry designs. The active duty participants were generally at the O4 -- O5 level and their insights and vision were beneficial to each of the Teams as well as to ASJPO.

3.3.6.1 *CONOPS Assessment Process Description*

The Weapons and Tactics (WEPTAC) Facility at China Lake, California was selected as the site for the wargames. Its remote location allowed the participants to focus with minimal interruptions.

Three basic tactical scenarios were developed to exercise the Arsenal Ship designs and programmed into the resident computer-assisted gaming facility. Each participant was assigned an operational command role and located in a closed game room where appropriate displays, information and communications channels were available to conduct operations. For the first games, each Industry Team had a day to explain their Concept of Operations, advise the participants on how to fight their Arsenal Ship concept, observe the games, and collect feedback from the participants. At the end of each day, and at the end of the week, Government-only sessions were held to discuss the results and collect feedback from the participants. Each scenario was played in a separate week over the summer.

A final Government-only week consisting of open seminar discussions of specific issues and comparisons of the Teams' approaches was held.

3.3.6.2 *CONOPS Lessons Learned*

3.3.6.2.1 *Involving Joint warfighters in exercising the Operational Concepts of the Teams was very positive for the Teams and for ASJPO.* Exercising each Industry approach in a wargame environment provided immediate feedback to the Team as to military utility, impact on current and future doctrine and employment methods, and helped them to refine their concepts. Interaction among the operators from three services was important to get a Joint perspective. ASJPO got comparative feedback from the warfighters on the approaches. Another benefit occurred when operators returned to their duty stations able to address both the pros and cons of the Arsenal Ship concept.

3.3.6.2.2 *The Government only seminar game at the end of the process provided the most feedback to ASJPO.* Candid discussions and comparisons of approaches by the warfighters could be made only when no Industry Team members were present. Early games were valuable to provide feedback to the Teams and to train the participants on the approaches.

3.3.6.2.3 *There is a need for improved capability to assess and contrast future warfighting concepts.* The CONOPS assessment process highlighted the need for future improvements, such as: real-time gaming, higher fidelity simulations and live system participation. The Joint and Maritime Battle Centers should be included in the future.

3.3.6.2.4 *All the Teams demonstrated the ability to invent new ways of conducting warfare that are operationally meaningful.* The Industry Teams conducted effective mission area analysis to support CONOPS development and requirements generation. New command structures and organizational relationships were considered, including Joint Vision 2010 and warfighting experiments. The Arsenal Ship vertical gun for advanced ships (VGAS) element was heavily used to support ground fires.

3.3.6.2.5 *Flexibility of operational employment was valued by the warfighter.* Designs that enabled the operation commander to use Arsenal Ship in creative ways were valued by the warfighters. In particular, they wanted to maximize the design flexibility of the existing (or planned) C4I architecture to be able to reconfigure the connectivity as needed. A primary utility of Arsenal Ship proved to be early massed precision firepower for both invasion stopping and anti-SEAD prior to arrival of traditional forces.

3.3.6.2.6 *The Sensor-to-Shooter concept was accepted and excursions with JSTARS were particularly effective.* Direct targeting, especially from JSTARS, was found to be effective in calling in Precision Guided Munitions (PGM) and calls for fire. Joint real-time battle management, including en-route and target area airspace deconfliction, was identified as a critical need as rapid response weapons become deployed in the same volume as air and ground forces.

3.3.6.2.7 ***New types of precision weapons, seekers, and warheads were desired by operators.*** With 500 Arsenal Ship VLS cells available on short notice the warfighters wanted more types, and lower cost ordnance for the cells.

3.3.6.2.8 *Warfighters prefer the opportunity to apportion assets to any echelon or force level necessary to conduct the mission.* Arsenal Ships could be considered floating magazines with missiles belonging directly to the force to which they are apportioned. A common weapon control and communications network is required to respond to fire calls at any echelon apportioned weapon control authority. This approach is desired for any missile ship supporting land attack missions.

3.3.6.2.9 *The Joint forces need to review the responsibilities for weapon control and launch authorization under "network centric" warfare concepts.* Remote trigger raised concern about the responsibility for launch authorization. Traditionally, responsibility resides with the ship's commanding officer; however, when weapons are allocated to off-board control nodes the responsibility for safe use of the weapons must be allocated off-board.

3.3.6.2.10 *The warfighters preferred a ship that can operate with minimum dependence on other ships in order to minimize disruption to their missions.* Collaborative self defense was accepted but routine dependence on off-board assets was viewed as invasive. However, the availability of Arsenal Ship with large numbers of Land Attack missiles afforded the opportunity to load the AEGIS ships with more cruise missile defense missiles.

3.3.6.2.11 *Warfighters were interested in the ability to rapidly fire missiles in order to minimize time exposed to enemy fire.* While exploring different employment concepts for the Arsenal Ship, the warfighters concluded that they wanted to minimize the time to launch for the entire mission since the ship was hard to hide while firing (flaming datum).

3.3.6.2.12 *Warfighters and engineers are concerned with security as the Navy moves to "Network Centric" operations.* With the delegation of weapons release authority to off-board systems, protection requirements against information warfare and inadvertent launch become more stringent.

3.3.6.2.13 ***Initial skepticism about the efficacy of the Arsenal Ship concept changed as the games progressed.*** In particular, as details of the passive self defense capabilities were included in the conduct of the games, more tactical employment options were tried with success.

3.3.7 *Mission Areas (TAD, STRIKE, MFS) and Combat Systems*

3.3.7.1 *Process Description*

The mission element assessed how each Team approached and met the requirements derived in each mission area (TAD/Strike/MFS). This assessment included feasibility, dependence on other programs, and impact on off-board systems. NSWC/DD was tasked to provide an assessment of Industry designs in TAD and Strike missions. Both JHU/APL and NSWC/DD provided an assessment of the Teams' MFS functional designs.

NSWC/DD, JHU/APL, and SPC were tasked to develop and maintain a functional requirements trace, system block diagram and message flow diagram for each Industry Team's design. Design impact to Off-Board systems were studied to understand and develop a basis for assessing the cost and feasibility for launch of SM-2, Tomahawk and ATACMS.

3.3.7.2 *Mission Area Lessons Learned*

3.3.7.2.1 ***The ASOS was an effective group for sharing programmatic information among Government offices involved with the Arsenal Ship project.*** In general, the working members of ASOS had senior technical responsibility for their respective projects and were interested in finding workable solutions to issues of integration with Arsenal Ship. The discussions and information exchanged had benefit beyond the Arsenal Ship since it fostered a collaborative working relationship for issue resolution, including for example, between individual Program Offices which had no other communication mechanism.

3.3.7.2.2 ***The ASOS dialogue suggested ideas shared development costs and identified opportunities for implementing new capability in the fleet.*** The Arsenal Ship Industry Teams were prepared to upgrade Tomahawk Weapon Control System Software to automate functions. This software could be applied in other surface combatants as well, which could effectively reduce manning on those ships. As another example, the Arsenal Ship program and Special Projects Office planned on combining efforts to accelerate introduction of ATACMS capability into the Navy. SPO planned to integrate the missile into VLS and Arsenal Ship planned to develop the weapon/launch control systems.

3.3.7.2.3 ***The Teams selected system designs based on cost and performance rather than on internal Government program pressures.*** The Teams universally endorsed the Army's AFATDS system on Arsenal Ship rather than pursuing the in-house Navy LAWS system.

3.3.7.2.4 ***Some PARMs were reluctant to work with three Industry Teams.*** Several of the PARMs deferred serious consideration of Industry Team designs and implications for their system development until Arsenal Ship downselected to one developer.

3.3.7.2.5 ***ASJPO sponsored feasibility studies at the R&D Centers assisted in identifying the off-board system impacts resulting from Arsenal Ship system concepts.*** This information assisted in evaluation of Industry Team remote trigger concepts and ensured consequences to off-board systems were described and accounted for by the Teams.

3.3.7.2.6 ***Industry can determine the most effective use of legacy systems within their design concepts.*** Integration of legacy systems was accomplished by one of three general approaches:

1. The Team used the functionality and algorithms of an existing system but redesigned the software in an object-oriented environment using modern software languages (e.g., Standard Missile launch control).

2. Legacy Systems were integrated into the computing and control infrastructure with few changes by developing a "wrapper" around the system. This "wrapper" approach makes it easier to incorporate new baselines of legacy systems with minimal changes over the life of the ship. This approach simplifies configuration management, minimizes programmatic coupling, and is still able to capture future upgrades.

3. The Team chose not to utilize the legacy system and built their own because the available system did not support enough of the Arsenal Ship unique requirements. (e.g. integrated survivability control system).

3.3.7.2.7 ***Fire support deconfliction must be addressed at the Joint force level to ensure rapid and safe response to call-for-fire.*** Deconfliction is determined by allocated no-fly and no-fire zones rather than supporting real-time deconfliction decisions, which are necessary with the increase of sea-based long range fire support weapons.

3.3.7.2.8 ***The availability of large volumes of inexpensive ordnance remains a priority of the Army and Marine Corps.*** ATACMS launches are not currently considered available for call fires by USA and USMC. The Marines consider low cost weapons more accessible for direct and general fire support missions. There is a need for analysis to determine the overall cost effectiveness of massive amounts of inexpensive dumb ordnance versus smaller amounts of extremely expensive precision ordnance.

3.3.7.2.9 ***The operator can be taken out of the loop for weapon control systems.*** The issue still remains whether the technology is advanced enough to automate target selection, identification, and battle damage assessment functions, all of which were to be performed off-board Arsenal Ship. Command authority issues should never be automated but can be delegated to rapid response systems.

3.3.8 *Communications*

As an entity enabled by the Information Age, the Arsenal Ship relies on robust connectivity to obtain targeting, mission data and weapons release authority.

3.3.8.1 *Process Description*

Resources at System Planning Corporation (SPC), SPAWAR Systems Center, San Diego, (SSC-SD, formerly NRaD), and The Johns Hopkins Applied Physics Laboratory (JHU/APL) were tasked to track and technically assess the evolving communications designs for ASJPO. They developed common format block diagrams of the functional architecture, conducted engineering analyses, and identified technical and operational risk areas. One purpose was to understand the information flow, independent of the particular communications channel employed, from source to ultimate destination.

An effort was initiated at SSC-SD to configure their C4I R&D Centers to replicate the connectivity proposed by the Teams. This effort was terminated in August 1997 when it became apparent that the original objective to make quantitative measurements was not feasible this early in the design cycle.

ASJPO interactions with the Teams encouraged them to consider diversity of communications paths and an ability to accommodate future structures of Joint Force connectivity over the life of the ship.

3.3.8.2 ***Communications Lessons Learned***

3.3.8.2.1 ***The rapidly changing connectivity designs made it difficult to replicate them in a R&D Center.*** The lead time to configure hardware sets, and computer models, made it difficult to replicate and assess the evolving connectivity designs. In addition, proposed new hardware elements and message types were not available. The process of asking detailed questions to support R&D Center testing of their concept, however, did encourage the Teams to become more specific in their designs.

3.3.8.2.2 ***Some design trades, such as new TADIL messages and formats, are outside the control of the Teams.*** Operational issues, such as availability of satellite communication bandwidth, modified or new TADIL message formats, and future JTIDS stacked net configurations were not under the control of the Teams. They made assumptions to support their connectivity concepts, which would have matured jointly in the following phases.

3.3.8.2.3 ***Many options exist for future joint connectivity to support Arsenal Ship-enabled distributed warfighting.*** No single unified approach to joint connectivity was found. The Teams produced approaches derived from their CONOPS which were consistent with today's connectivity options and which could support projected ones. Communications path innovation was limited since onboard assets had to be compatible with offboard systems. It was in the automation of onboard assets and handling of data that opportunities for different approaches were found.

3.3.8.2.4 ***Industry has much more capability to quickly develop and field a total integrated system for communications.*** The Teams were effectively designing a Joint Maritime Communications system without relying on the Navy for the components or software development. Industry has more freedom to select from a broader range of COTS, GOTS or new development systems than the services. There is less "stovepiping" of systems design due to the pressure to cut costs and improve overall performance.

3.3.8.2.5 ***Non-Conventional threats such as Information Warfare, conventional Electromagnetic Pulse (EMP), and High Power Microwave (HPM) were not highlighted.*** Distributed warfighting places increased reliance on availability and security of communications. Evaluation of the threat of unconventional means of disturbing the transfer of information should be undertaken.

3.3.9 ***Information Infrastructure***

Large systems increasingly rely on Information Technology and are typically designed with an integrated, internal digital communications capability linking computers together into a collaborative distributed processing environment. This shared infrastructure was recognized as essential to the proper functioning of the Arsenal Ship.

3.3.9.1 ***Process Description***

Perhaps here, more than the other areas, the Teams were encouraged, primarily by their internal PAE process, to adopt commercial practices and develop a design that embraced open system architectures. No formal assessment methodology was employed and normal reviews and design documents provided disclosure of their approaches.

In all cases, innovative and robust designs were proposed that fell into the mainstream of commercial computing practices and represented major increases in capability over present ships. Ubiquitous high speed integrated digital communications (voice, data, video) and general purpose redundant computer plants to support all electronic functions on the ship, including sensors for damage control and condition based maintenance, were designed.

3.3.9.2 ***Information Infrastructure Lessons Learned***

3.3.9.2.1 ***Teams demonstrated a capability to adopt commercial open-architecture practices and adapt them to a ship design.*** Government open system standards such as DII/COE were acknowledged and selectively adopted but the main emphasis was on commercial open system approaches in areas of operating systems, networks, computer hardware, and protocols.

3.3.9.2.2 ***Multi-level security for integrated networks is an important issue.*** The traditional physical segregation technique for data at different security levels is incompatible with the basic philosophy of shared networks and has a major cost impact. This is an issue bigger than Arsenal Ship and new techniques, procedures, and regulations are needed and are being addressed by appropriate agencies.

3.3.9.2.3 ***Information System designs are scaleable and have the potential for reuse.*** Typically computer networks and distributed processing designs are scaleable either upwards or downwards.

3.3.10 ***Software Development Process***

Computer software is an area of increasing concern and cost in large system development and integration. Studies by the Software Engineering Institute and others have concluded the development and support processes are critical to producing high quality computer programs.

3.3.10.1 ***Government Process Description***

NSWC/PHD/ECO was tasked with reviewing and assessing the process of software development proposed by each Team. The Teams were each invited to host a meeting where they described their approach to a small set (4 or 5) of ASJPO advisors. Team attention was focused on understanding, rethinking, and refining their approach to software development in an environment without imposed Government process specifications. Process innovations were encouraged but not pushed.

Each of the Teams described their methodology for Arsenal Ship software development. They described their successes on past projects and process improvements proposed for AS. Internal software process audits, SEI ratings, and other relevant information were presented to convince ASJPO that their approach was credible.

3.3.10.2 Software Lessons Learned

3.3.10.2.1 ***The Government approach had the desired effect of highlighting the importance of software and its development process.*** The request for disclosure of software process spurred the Teams to focus attention on this area and consider commercial innovations. Specialists from the commercial software marketplace or SEI should be invited to participate as advisors in future competitions.

3.3.10.2.2 ***Relief from Military Standards permitted a more appropriate software development process.*** Industry Teams were allowed to rely on their own, proven software development methods, rather than be forced to apply rigid Mil Standards, such as 1679A, 2167A, 498, etc. Most took a modified 498 approach that retains control while not overburdening the development process. A demonstrated development process plus having controls and metrics to ensure tracking, consistency and quality are the critical components. With commercial certification at SEI level 3 or higher, Industry Teams are very capable of governing their own software process.

3.4 -- Ship Hull, Mechanical and Electrical Design (HM&E) & Ship Production

3.4.1 ***Background***

Each of the competing design teams used a slightly different philosophy but all were similar in their motivation to satisfy both the requirements and the cost goals at the highest performance level. Survivability and mission capability of carrying and remotely launching 500 missiles dictated the HM&E design.

Each Team's HM&E design reflected the experience of the shipyard that was a member of the Team. Overall, the hull design starting point was from another program or typical of the shipyard's product line (e.g., commercial ship, aircraft carrier or destroyer). That initial hull concept decision then influenced a lot of the other design trades for the ship systems.

Initially none of the shipyards had complete early stage design capability or naval architecture talent available. Each Team engaged a naval architecture firm as part of their effort.

All of the designs were heavily driven by the magazine protection systems. The one week Survivability Course provided by the Government at the beginning of Phase II marked the start of this effort. In addition, the Teams brought the NSWC Vulnerability Group onboard, some to a greater extent than others. The trades driven by the physics of the problem were very clear and each Team chose a different approach consistent with their hull design concept.

All of the Teams integrated their production engineering and planning early in the process and used design innovations to try to improve the production process. This was motivated by the requirement to provide an irrevocable offer as part of the Phase III proposal. This effort included developing a build strategy, module break concept and process layout in some detail.

All of the Teams focused on using some form of commercial specifications and materials. All used commercial grade with no HY-series steel.

The mission profile of both a high speed transit and a 90 day loiter speed on station suggested some form of electrical propulsion or auxiliary propulsion system.

All of the Teams used commercial specifications for most of the equipment and relied on some form of rafting or isolation to protect the equipment from shock or environmental effects.

3.4.2 ***Process and Production Lessons Learned***

3.4.2.1 ***For industry to significantly alter the way they do business (as was done for the Arsenal Ship Program), they must create project teams largely separate from the parent organization and its associated culture and bureaucracy. Further, these project teams must be empowered to implement change.*** Large corporations have a deeply ingrained culture in which it is difficult to deviate from the norm. It is simply not possible for an entire company to change on short notice for a single project. All of the Arsenal Ship Industry Teams enjoyed considerable autonomy and respite from much of the bureaucracy of the parent companies. The Teams must have full ownership and responsibility for design, fabrication, assembly and test.

3.4.2.2 ***Repeatability and Commonality significantly enhance productivity.*** Navy-designed ships have relatively little "commonality" or inherent production repeatability. The Arsenal Ship Industry Teams made extensive use of repeat structural segments, mirroring of the ship (port and starboard sections identical), and common modules.

3.4.2.3 ***Significant industry cost and schedule savings can be realized through the integrated efforts of engineering, production, procurement, logistics and cost estimating personnel.*** Typical industry practice involves stove-piped efforts of engineering, production, procurement, logistics and cost estimating personnel. This practice provides for less than optimum products and processes resulting in increased costs and greater cycle times. Integrating the various groups can yield significant gains in cost and schedule. For example, as engineering conducts the design, drawings are produced that optimize production, equipment can be chosen to optimize procurement options, supportability can more readily be designed-in up front, and the impact of various design options on acquisition and life cycle costs are all available in real-time.

3.4.2.4 ***The use of Other Agreements Authority, Sec.845, did allow savings especially in the area of material selection where the traditional process could be replaced with a commercial***

approach. Relief from Competition in Contracting Act (CICA) was particularly valuable in that Teams could solicit for the best total vendor "package" of cost, technical support, training, upgrades and service life logistics rather than be forced to select the vendor with the lowest acquisition cost. Related benefits include:

(1) allowing Industry to nurture long term relationships,

(2) the numerous benefits, such as reduced non-recurring costs, associated with doing business with a company that is familiar

(3) reduced cost and time associated with contracting, and

(4) benefits associated with providing engineering with more design data at an earlier stage.

It should also be noted, however, that to maintain a favorable negotiating position it may be necessary for industry to protect or conceal their early decision to go with a particular company

3.4.2.5 ***Commercial building practices may be readily applied to naval ship combatant construction.*** Despite the shock, whipping, signature, weapons effects, and other military features built into Arsenal Ship, the Industry Teams demonstrated that it is entirely possible to take advantage of commercial practices, such as:

(1) parallel structure and outfit construction processes,
(2) reduced number of installation parts,
(3) "building block" vice "stick built" products, and
(4) repeatable products and processes.

The approach that seems to maximize the commercial benefits is to start with a 100% commercial criteria and then add military criteria only when required.

3.4.3 ***Design Lessons Learned***

3.4.3.1 ***Similar trade studies conducted by different Teams may lead to results not only quite different from each other, but inconsistent with current Navy thinking.*** Although the Industry Team solutions had many similarities, there were stark differences as well. These differences are a result of differing assumptions, company culture, goals and objectives. One example is the different types of propulsion system chosen by each of the three Teams. In one case, the type of propulsion system seemed inconsistent with that type of system deemed most advantageous based on recent Navy trade studies.

3.4.3.2 ***Industry should have maximum ability to establish those requirements that impact their production process and products.*** To minimize cost and schedule, Industry Teams must be able to optimize use of their unique facilities, personnel, processes and expertise. Freeing Industry affords them that opportunity. Examples include

(1) choosing steel plate sizing that optimizes their production line,

(2) allowing industry to establish construction breaks, block sizes, packaging of equipments, etc. to account for plant layout, lift capability, etc.

3.4.3.3 ***Skid mounting equipment, while affording production efficiencies, actually adds weight.*** Industry trade studies clearly determined that skid mounting equipment provided for significant cost and schedule advantages. It does, however, add weight. This added weight is an important consideration for weight sensitive designs.

3.4.3.4 ***Less dense ships increase design flexibility and lower cost.*** Allowed to generate their own requirements and designs to maximize performance under PAE, all three Teams developed designs more "spacious" than Navy standards. Beneficial results included: areas designated for distributive systems (service trunks), longer and straighter piping runs, simplified vent ducting requiring fewer parts, etc.

3.4.3.5 ***Raft mounting COTS equipment to provide for shock isolation, appears to be technically feasible and provides for significant savings in acquisition, production, and life cycle costs.*** While detailed analyses of the shock mitigating abilities of the rafts was not conducted, preliminary results indicated that it was indeed achievable. The majority of the cost savings are realized through the purchase of COTS equipment.

3.4.3.6 ***The right mix of fire fighting systems was difficult to determine and the subject of considerable design studies.*** Through the conduct of extensive trade studies, industry determined that water mist provided the most effective means of extinguishing fires (in conjunction with foam in machinery spaces), was cost effective, and that acceptable commercial systems were readily available. It is not clear why the Navy is pursuing their own water mist system given the fact that commercial systems are currently in use, especially in cruise ships.

3.4.3.7 ***Acceptable options are available to provide for a low signature, low manning, operationally effective and a safe aviation support facility.*** Through discussions with NAVAIR, NAVSEA and NAWC Lakehurst, the Industry Teams determined that it was indeed possible to design a helicopter landing facility to meet established signature goals.

3.5 -- Missile Launchers

3.5.1 *Background*

3.5.1.1 *Phase I -- Concept Studies*

The CONOPS/SCD specified the need for about 500 VLS cells. This was initially assumed (but not required) by ASJPO to mean MK41 cells, since the cost of a new launcher development was deemed well beyond the scope of the program. Since GFE was eliminated, the launcher was left unspecified primarily to keep in the spirit of free and open trade space given to the Teams. Therefore, at the outset of Phase I, ASJPO wanted to ensure equal access to design data and potential cost reduction modifications for the MK41 VLS Launcher, with Lockheed Martin Aero Naval Systems (LMANS), the sole source Industry Team to PMS-410. LMANS was requested to support all competing Teams which they ultimately did.

None of the Phase II competing Teams chose the current MK41 VLS system. There were at least four business factors that swayed the Teams away from MK41:

(1) since approximately 1/3 of the total ship cost was in launchers, too much of their trade-space would have been locked out if forced to use the MK41 as a fixed cost element,

(2) LMANS, and by inference the whole LM Team, would have too much insight into a competing Team's design,

(3) even if LMANS had been more flexible on trades and pricing, the profit for 1/3 of the ship would go to a non-Team member, and

(4) the potential for launcher profits on SC-21 was even a larger draw than Arsenal Ship for future launcher sales (effectively, Arsenal Ship would be helping support the winning Team develop a launcher to compete with MK41 on the SC-21 program).

As the competing Teams initiated the search for MK41 launcher alternatives, they found they had three options:

(1) develop a new launcher from scratch,

(2) partner with NSWC/Dahlgren on the CCL (Concentric Canister Launcher) program, or

(3) work with UDLP (United Defense) who was a MK41 supplier when there were two manufacturers. All three approaches were taken by various Teams.

3.5.1.2 ***Phase II -- Functional Design***

Tests were needed to reduce development risk to an acceptable level by the end of Phase II and prior to downselect for detail design and construction in Phase III. This necessitated use of representative missile loads. The Government made available spare rocket motors, missile parts and standard VLS canisters at no cost to support the tests. Help from PMS-410 and PMS-422 was indispensable in obtaining these parts and to ensure safe tests.

The fact that launcher tests could be accomplished in the short time period of Phase II is an indication of the maturity of vertical launcher design. With performance essentially ensured, much more concentration could be devoted to the production and cost aspects of the design. All Teams did an in-depth, cost estimate that included materials, labor and process costs.

In parallel with Arsenal Ship Joint Program Office (ASJPO) reviews of the launcher designs, the Weapons Systems Explosive Safety Review Board (WSESRB) conducted a series of safety reviews for the weapon system and its associated systems including the launchers, the computers, software, fire fighting and damage control features affecting explosive safety. ASJPO insisted that all industry designs be driven through the WSESRB, so that the question of weapons firings in Phase IV of the program would not become a contentious issue. The Board advised all Teams (through the ASJPO) on concerns and issues and were very supportive. They were open-minded about the often novel way in which the Teams were developing their designs and focused instead on the desired performance of safety features rather than their specific implementation.

3.5.2 ***Process Lessons Learned***

3.5.2.1 ***New launcher developments were a surprise to the program office.*** Launchers are a major cost item for the ship and were treated as a mandatory new development by each Industry Team to attain a competitive edge. This should be expected on any new competitive procurement where a major item is not specified as Government Furnished Equipment (GFE). It also occurred because the MK41 legacy system was owned by one of the competitors who then would have held too much influence over their cost structure and too much insight into their design.

3.5.2.2 ***Only industry can conduct meaningful cost analysis of such systems as launchers.*** Government does not have the expertise nor data to perform the detailed cost analysis necessary to estimate prices of major weapons systems such as launchers.

3.5.2.3 ***WSESRB is a vital part of the Government's explosive safety program.*** WSESRB's function is to review and comment, not approve or direct. It advises the program manager who is responsible for program safety. Industry Teams are not in a position to accept liability for use of explosive ordnance. Detailed design issues are not normally evaluated by the WSESRB -- it looks at the processes in place and the results.

3.5.3 ***Launcher Technology Lessons Learned***

3.5.3.1 ***VLS launchers have become a mature technology.*** The short development time and low investment cost involved for new launchers development point to a mature technology. Development time for Arsenal Ship launchers would have been about 1/2 that for MK41 with development costs perhaps 1/10.

3.5.3.2 ***Risk reduction on launcher systems requires some live testing.*** Modeling and simulation is not enough because aspects of launcher systems cannot be modeled properly, e.g., ablative erosion.

3.5.3.3 ***There are numerous high quality launcher testing facilities available.*** Several private as well as Governmental launcher test facilities were utilized for the launchers. All had modern measurement equipment, test setups and qualified personnel.

3.5.3.4 ***The Government still needs to invest in survivability features and tests.*** Survivability features of launchers and other systems are still unique needs of the services and are not supported by commercial suppliers. This means that the Antifragment and Explosive Load Reduction (ELR) systems development might have to be borne primarily by the Government, either through a separate effort that was made available to all Industry Teams, or through extra investment in the Arsenal Ship project, with funds going to the single Team entering Phase III. During Phase II, ASJPO invested some funding in testing ELR systems applicable to all Teams' launchers.

3.5.3.5 *Government often has free assets available for testing at no cost.* Older rocket motors (MK104) near the end of their service life were made available for tests. A few R&D motors (MK72) were available because they were not production units. Canisters with defects were also available. They were given to the Teams as no-cost GFE. Significant help from PMS-410 and PMS-422 made this possible.

3.6 -- Operations & Support

3.6.1 *Background*

The subject of Operations and Support encompassed manning, maintenance & repair, operating cycles, forward operating bases, and central operating bases. Issues such as Personnel Tempo of Operations (PERSTEMPO) and dry docking cycles were addressed. All the Teams considered the Blue and Gold crewing concept and the use of civilians for ship operation. The use of military and civilian personnel, both civil service and private industry, was considered to perform maintenance. Phase VI of the program optionally allowed the Teams to propose providing the life cycle maintenance for the Arsenal Ships. All indicated they would propose providing life cycle maintenance.

3.6.2 *Lessons Learned*

3.6.2.1 *The right approach to designing for very low manning is to start with zero manning and justify each billet.* A zero-based approach that looks at whether the function is necessary at all or whether it can be eliminated by more extensive use of technology (e.g., long life hull coatings) and/or automation works! While all of the Teams achieved a very low manning number it was not fully determined whether the required maintenance underway could be achieved, with the issue of corrective maintenance being the most elusive. It was also not clear if housekeeping and food service were adequately addressed.

3.6.2.2 *Manning reductions should be evaluated from a total ownership cost viewpoint.* The military and civilian manning to operate the entire system, from administration, technical support, maintenance support, ship operation and possibly with multiple crews, must be considered. A very small afloat crew with an extensive shore support staff for maintenance is not necessarily less expensive than a larger afloat crew that performs some of its own maintenance. Manpower reductions must reflect the elimination of work rather than the transfer of work to a different venue (although transfer of work ashore may permit it being done at less cost than afloat).

3.6.2.3 *It is important to constantly stress the balancing of operations & support costs and acquisition costs.* There is now little doubt that major crew size reductions are possible. The question is whether the non-recurring engineering (NRE) effort, including software development and modeling & simulation, will be sufficient to allow a few people to operate the ship. Additionally, appropriately reliable equipment and systems must be selected to reduce maintenance workload. The in-service maintenance concept must be worked out in sufficient detail to ensure realistic maintenance cost estimates.

3.6.2.4 ***To achieve proposed manpower reductions, selection of appropriate equipment and systems must be matched to reliability, repair, and maintenance criteria.*** During the early stages of design (e.g., Phases I and II) most equipment have not been selected, yet manning reductions are postulated. The criteria for making future procurement decisions for individual systems or equipment must ensure that the proposed ship availability and crew maintenance workload are not compromised.

4.0 -- Ship Costs

4.1 -- Background

4.1.1 *Philosophy: PAE*

From its initial conception, the Arsenal Ship Program has steadfastly applied Price As Established (PAE) as a major component of its acquisition strategy. Price here means industry cost to manufacture and a reasonable profit (or return of investment to the company). A price goal was established early in the program. All designs were monitored by the contractor teams to ensure an affordable item at a price including all non-recurring and recurring costs to manufacture and a reasonable return on corporate investments or in other words, profit. All other aspects of the government desired capabilities were tradable against the price goal. This is different from Cost as an Independent Variable (CAIV). Cost as independent variable (CAIV) assumes two things; first, the government is monitoring and controlling the trade decisions to ensure affordability, and second, cost is only one of several factors to be traded. Requirements creep is most certain. CAIV in present government literature means a government program manager yardstick to consider trades against requirements.

After several months, all competitive teams were to fix their own price against their design or PAE. During later phases more trades could result. The contractor team should have responsibility for all trade decisions and be encouraged to use the trade space within the government's desired capabilities to ensure an acceptable and capable product that meets PAE. PAE focused the design trades on mission essential performance, without requiring the use of legacy systems as GFE (Government Furnished Equipment). This resulted in designs that were engineered at a total system level, highly integrated, commercially-based, and low cost, while at the same time introducing new systems with some attendant risk. Industry Teams were prepared to meet the goal of $450M for the average production ship cost, about 1/3 less than early Navy estimates for the production ship. Risk reduction programs were key to achieving the PAE goals and worked effectively. Under the Arsenal Ship acquisition strategy, substantial elements of the risk were transferred from Government responsibility to Industry. In short, competing teams concluded "zero risk is unaffordable."

PAE has been applied across all phases of the life cycle. The Phase V Unit Sailaway Price (USP) goal of $450M and the Phase III budget of $389M were established before the start of Phase I. Goals for annual operating and support cost ($13.7M per ship) and total program life cycle cost ($5.7B) were established early in Phase II. The competing Industry Teams were given full flexibility in designing the ships, in controlling the business environment and production process, and specifying the operating and support concept within the Concept of Operations (CONOPS) and Ships Capabilities Document (SCD).

4.1.2 *Government Cost Team Tasks*

The ASJPO cost team openly interacted with the three competing Industry Teams, providing feedback and encouragement, during Phase II. Government tasks included:

- Establish an "aggressive but realistic" goal for the O&S Phase; integrate goals for all program phases into an overall life cycle goal. The cost goal for the development phase (Phase III) and the USP goal for Phase V were established at program initiation.

- Analyze deliverable products and Phase II cost proposals and provide evaluation and recommendations to the Source Selection Evaluation Board (SSEB).

- Develop a consistent but flexible structure for collecting, assessing, and comparing Industry Team cost proposals.

- Provide forthright, but non-directive, feedback to the Industry Teams regarding their cost estimating assumptions, cost modeling approaches, and supporting rationale.

- Through close interaction with industry, develop an understanding of the impact of commercial practices, manufacturing and design process improvement, simplified acquisition practices, and alternative approaches to life cycle support on all aspects of life cycle cost.

- Assess the realism of the Teams' cost proposals in view of the following factors:

  - Experience with performance on current and past programs
  - Commercial practices to the extent incorporated in the Teams' designs
  - Acquisition Process streamlining (Sec.845, etc.)
  - Application of state of the art systems and technology
  - Improved manufacturing and management practices where credibly supported
  - Relevant attributes of the ship concepts

4.1.3 *Industry Cost Teams*

The three Industry cost teams played a major role in the implementation of PAE in their design processes. Each of the three teams had its own individual organization and approach. A number of common characteristics became evident over the course of Phase II. A representative generic process, followed to some degree by all of the teams, included the following tasks:

- Implement PAE as part of the team's system engineering process.

  - Flow down top-level PAE goals to subsystem IPT's.

  - Actively participate in subsystem IPT's to provide real-time cost sensitivity.

  - Define and perform cost trade studies in support of system configuration decisions.

— Act as the "cost police" to ensure continuous emphasis on cost in the design process.

— Flow up and integrate subsystem cost estimates into overall program cost.

- Ensure life cycle considerations were prominently addressed in the system design.

— Participate in the development of life cycle system support concepts.

— Perform RMA and maintenance cost assessments. Motivate designers to address maintainability in the design process.

— Perform manning and training assessments to ensure the consistency and feasibility of the manning and support concepts.

- Improve corporate methods and tools for collecting, managing, and promulgating cost information across the program team.

- Provide realistic assessments of program cost and feasibility to corporate management.

- Interact with unprecedented openness with program office representatives, including cost engineers, during the development of a major acquisition proposal.

— Understand customer concerns re cost visibility and realism.

— Provide necessary backup information and data to make the case for credibility of cost estimates based on commercial practices, new processes, and non-traditional business arrangements.

In summary, the program's approach to PAE and ship cost required an unprecedented degree of cooperation between a Government program office and each of three Industry Teams in a competitive situation. Good will, highly professional attitudes and behavior, and commitment to this new way of doing business by everyone involved were required to make it work

4.2 -- Lessons Learned

4.2.1 *PAE Works.* In a competitive environment, industry will strive to meet aggressive PAE goals.

4.2.2 *Non-traditional acquisition processes offer opportunities for substantial savings.* Expectation of 30-50% (or more) savings vs. traditional process can be supported if industry is given freedom from standard acquisition constraints. Competition is a very powerful motivator and can be used to ensure industry uses this freedom.

4.2.3 *Close interaction between Government and Industry Cost Teams is beneficial to both sides.* Government representatives must be very careful to protect proprietary information of each competitor. Industry personnel must learn to "let down their guard" and be reasonably trusting and open. Government gets unprecedented understanding of Industry's proposal and process and Industry better understands the customer's needs and produces better proposals

4.2.4 ***Early emphasis on reducing total ownership cost (TOC) is required to produce meaningful results.*** Early design choices and equipment selection have a significant effect on procurement and maintenance costs. Making O&S costs an issue encourages the Industry Teams to trade-off for more reliable equipment and to design out some traditional maintenance. Significant maintenance reductions were predicted from re-engineering legacy equipment and performing manufacturer recommended maintenance.

4.2.5 ***The integrated product team (IPT) concept, that combined cost and production engineers with designers in the early stages of the program, led to predicted ownership costs considerably below historical levels.*** Significant process changes in both management and production were required to justify the use of non-traditional cost estimating relationship's (CER's) but when validated, predicted significant savings. Industry cost engineers were the crucial players ensuring that IPT's focused on cost from beginning to end of the process. Production engineers' early involvement was key in the evolution of designs for which a supportable case for lower than historical production costs could be made.

4.2.6 ***Use of cost models based on conventional weight-based methods for unconventional designs can lead to erroneous results.*** In several instances Industry Teams utilized weight-based cost estimating to estimate such areas as labor hours, paint, lagging, etc. A closer look revealed that weight-based cost estimating gave answers that were not accurate, and over inflated.

4.2.7 ***Product design details can be adjusted to improve producibility and substantially reduce cost.*** Designs evolved to utilize repeatable modules, to increase plate thickness and reduce numbers of parts, to employ COTS components where possible, and to tradeoff machinery configurations in an effort to minimize cost. Historically-based CER's can not easily capture these cost impacts; a more detailed and process sensitive methodology is required.

4.2.8 ***Industry offered credible life cycle support under a Full Service Contractor (FSC) approach.*** The FSC approach promises significant cost savings over traditional Navy life cycle support.

4.2.9 ***The key to PAE is to treat cost as another design requirement for the engineer and provide a detail budget breakdown at the outset.*** Although all of the Teams practiced some form of PAE, in the end those that had an early budget in Phase I had a better handle on costs and had more time for total ship trade studies.

4.2.10 ***There is a distinct advantage to use design tools that include real-time cost estimating.*** Industry Teams found it important that cost estimates associated with system design be available in real-time to accommodate faster design cycle times. Use of design tools that did not provide real-time cost estimates necessitated a separate effort to determine cost impacts.

Glossary

A

ACAT	Acquisition Category
AFATDS	Advanced Field Artillery Tactical Data System
AFS/ELR	Anti-Fratricide System/Explosive Load Reduction
AOA	Assessment of Alternatives
ASJPO	Arsenal Ship Joint Program Office
ASN	Assistant Secretary of the Navy
ASOS	Arsenal Ship Off-board Systems
ATWCS	Advanced Tomahawk Weapon Control System

C

CAD	Computer Aided Design
CAIV	Cost As an Independent Variable
CCAT	Combat System C4I Assessment Team
CCL	Concentric Canister Launcher
CD	Carderock Division
CEC	Cooperative Engagement Capability
CER	Cost Estimating Relationship
COEA	Cost and Operational Effectiveness Analysis
COMM	Communications
CONOPS	Concept of Operations
COTS	Commercial Off the Shelf
C4I	Command, Control, Communication, Computers and Intelligence

CS	Combat System
CSS	Coastal Systems Station
D	
DARPA	Defense Advanced Projects Agency
DBOF	Defense Based Operating Fund
DD	Dahlgren Division
E	
EHF	Extra High Frequency
F	
FFRDC	Federally Funded Research and Development Center
FOUO	For Official Use Only
FSC	Full Service Contractor
G	
GFE	Government Furnished Equipment
GFI	Government Furnished Information
GOTS	Government Off-the-Shelf
GSE	Government Specified Equipment
H	
HM&E	Hull, Mechanical & Electrical
HY Series	High strength milspec steels
I	
IMPS	Integrated Magazine Protection System
INFOSEC	Information Security
INSURV	Inspection and Survey

IPDE	Integrated Product Development Environment
IPT	Integrated Product Team
IRAD	Independent Research & Development
J	
JHU/APL	Johns Hopkins University/Applied Physics Lab
JROC	Joint Requirements Oversight Council
JSTARS	Joint Surveillance, Target Attack Radar System
JTIDS	Joint Tactical Information Data System
L	
LAWS	Land Attack Weapons System
M	
MFS	Maritme Fire Support
MNS	Mission Needs Statement
N	
NAVAIR	Naval Air Systems Command
NAVSEA	Naval Sea Systems Command
NAWC	Naval Air Weapons Center
NRL	Naval Research Laboratory
NSWC	Naval Surface Warfare Center
NSWC/PHD/ECO	Naval Surface Warfare Center/Port Division/East Coast Operations
O	
OPTEVFOR	Operational Test and Evaluation Forces
O&S	Operating and Support
OSD	Office of the Secretary of Defense

P

PAE	Price As Established
PARM	PARticipating Manager
PEO	Program Executive Office
PERSTEMPO	Personnel Tempo of Operations
PGM	Precision Guided Munitions
POM	Program Objectives and Milestones

R

RD&A	Research Development and Acquisition
RMA	Revolution in Military Affairs}

S

SBD	Simulation Based Design
SCD	Ships Capabilities Document
SEI	Software Engineering Institute
SETA	Scientific and Engineering Technical Assistance
SPAWAR	Space and Naval Warfare Systems Command
SPO	Special Projects Office
SSA	Source Selection Authority
SSEB	Source Selection Evaluation Board
SSAC	Source Selection Advisory Council
SWBS	Ship Work Breakdown Structure

T

TAD/SD	PEO TAD program office for Ship Self Defense
TADIL	Tactical Digital Information Link
TCP/IP	Transport Control Protocol/Internet Protocol

T&E	Test and Evaluation
TOA	Total Obligational Authority (in the budget)
TOC	Total Ownership Cost
TTO	Tractical Technology Office
U	
UARC	University Affiliated Research Center
USA	United States Army
USD (A&T)	Under Secretary of Defense (Acquisition and Technology)
USMC	United States Marine Corps
USP	Unit Sailaway Price
V	
VGAS	Vertical Gun Advanced Ships
VLS	Vertical Launch System
W	
WD	Weapons Division
WSESRB	Weapon Systems Explosive Safety Review Board

Tab A

Joint Memorandum -- Arsenal Ship Program

Chief of Naval Operations
Assistant Secretary of the Navy (Research, Development and Acquisition)
Washington, D.C.20350-1000
Director, Defense Advanced Research Project Agency
Arlington, VA 22203-1714

Mar 18 1996

Joint Memorandum

Memorandum For: Commander, Naval Sea Systems Command Chief of Naval Research

Subj: Arsenal Ship Program

This joint memorandum establishes the Arsenal Ship Program and provides the Director, Defense Advanced Research Project Agency (DARPA), Commander, Naval Sea Systems Command (NAVSEA) and Chief of Naval Research (CNR) with precepts regarding the basic requirements, goals, and acquisition strategy for the program.

The basic requirement for the Arsenal Ship is to satisfy joint naval expeditionary force warfighting requirements in regional conflicts by providing the theater commander with massive firepower, lone range strike; and flexible targeting and possible theater defense through the availability of hundreds of VLS cells. To meet this warfare requirement affordably, the Arsenal Ship concept and design must be straightforward and simple. Detailed requirements and concept of operations are defined in separate documentation, however, key elements for the Arsenal Ship include --

- Provide approximately 500 VLS calls, with the capability to launch Navy and joint weapons to support the land campaign;
- Integrate the combat system with Cooperative Engagement Capability (CEC) links to serve in, or as, the off-board control;
- Appropriate ship design features for survivability and ship self defense which could be incorporated at a later date if needed;
- Low ownership Costs through the use of innovative maintenance and operational methods, procedures, and technologies;

- Crew size will not exceed 50 personnel. The design objective will be to minimize crew size to the maximum extent below 50 which is technically feasible.

In the face of limited budget levels, the use of acquisition reform initiatives and streamlined contracting methods are paramount to meet the basic requirements of the Arsenal Ship in an affordable manner. To accomplish this, a non-acquisition category demonstrator ship shall be developed that will be convertible to a fleet asset at a future date.

In addition, cost must be viewed as an independent variable, and early industry involvement with the development of a cooperative industry-government team are viewed as key to achieve our goals. To minimize cost, the off-the-shelf systems will be used exclusively. Any development of new systems will require the approval of ASN (RD&A). The cost of acquiring the first ship will not exceed $520 million including the cost of concept development and competition. These funds will be provided jointly by the Navy and DARPA with contributions of $350 million and $170 million respectively. For FY 96, funds will be provided by reprogramming. For FY 97, funds will be requested as part of the budget request. The Program Objectives Memorandum process will be used to provide the funding.

The Director, DARPA; Commander, NAVSEA; and CNR are tasked to establish a plan for a joint Arsenal Ship Advanced Technology Demonstrator Program Office and identify to the ASN (RD&A) a candidate full-time program manager. The program manager will work closely with OPNAV staffs to ensure that requirements are understood and fully met, and with industry in a team approach to ship development and construction. The Arsenal Ship Program Office (ASPO) should operate as a "skunk works" organization, eliminating, or streamlining acquisition procedures, processes, and paperwork. The ASPO shall be comprised of representatives from DARPA, NAVSEA and CNR with a total maximum number of 9 personnel. DARPA will initially have program lead with transition to NAVSEA at an appropriate time during ship production. This program represents a good opportunity to take advantage of DARPA's culture and experience in prototyping to transition alternative business practices into how the Navy buys ships. The ASPO shall be initially located in the National Capitol Region and later co-located at the shipyard chosen to construct the first ship.

DARPA, NAVSEA, and CNR are directed to develop a detailed plan of action, milestones, technology initiatives, acquisition strategy, and budget necessary to execute the Arsenal Ship Program, with the goal to have a demonstrator Arsenal Ship at sea by the year 2000. Specific recommendations and actions necessary to accelerate ship development should be the focus of the plan, eliminating all procedures that are not specifically required by law. The plan should also provide systems integration approach and affordability initiatives to reduce acquisition and ownership costs.

The Arsenal Ship Program is among the highest priority programs within the Navy. All organizations and contractors participating in and supporting the Arsenal Ship Program should view it with priority, and proceed with a sense of urgency to achieve the goal of beginning demonstrator ship at-sea testing of the Navy's first Arsenal Ship in the year 2000.

/Signed/
Larry Lynn
Director, Defense Advanced Research Project

Agency
3/13/96

/Signed/
John W. Douglass
Assistant Secretary of the Navy
(Research, Development and Acquisition)

/Signed/
J.M. Boorda
Chief of Naval Operations
Admiral, USN

Larry Lynn	John W. Douglass	J.M. Boorda
Director, Defense Advanced	Assistant Secretary of the Navy	Chief of Naval Operations
Research Project Agency	(Research, Development and	Admiral, USN

Larry Lynn	John W. Douglass	J.M. Boorda
Director, Defense Advanced	Assistant Secretary of the Navy	Chief of Naval Operations
Research Project Agency	(Research, Development and	Admiral, USN

Tab B

Memorandum of Agreement (MOA)
Joint Navy/DARPA Arsenal Ship Demonstration Program

Purpose:

The purpose of this document is to establish a joint Navy/DARPA agreement as to the objectives, roles and responsibilities, schedule, and funding for the Arsenal Ship demonstration program.

Background:

Arsenal Ship is a high priority program for the Navy to acquire a new capability for delivery of large quantities of ordnance in support of land and littoral engagements. Key to both Arsenal Ship's affordability and operational flexibility is off-board integration of all but the most rudimentary C4I. The ships are conceived to be theater assets that will operate under the authority of the joint Commanders-In-Chief (CINCs) and will receive their targeting along with command and decision information from other assets. Early in Arsenal Ship's life this control will be exercised through an Aegis platform, though as other assets mature, control will transition to aircraft such as AWACS or an E-2 with CEC-like capability and eventually to the Marine or Army shooter on the ground. Thus, the Arsenal Ship will not be fitted with long range surveillance or fire control sensors, but will be remotely controlled via robust data links. The data links will be secure, redundant and anti-jam in order to provide high reliability in the connectivity of the Arsenal Ships in high jamming operational scenarios. The program overall is an attempt to leverage the significant current joint investment in Link 16 and CEC. The Arsenal Ship's survivability will be primarily achieved through passive design techniques. While active systems are not ruled out, they must be consistent with overall cost and manning goals. These design goals will allow the Arsenal Ship to have a very small crew (potentially, none at all) which will be a key ingredient in minimizing its life cycle costs. It is expected that the Arsenal Ship will transit and operate independently but when in a hostile environment, its defense will be enhanced by working cooperatively with other elements of the force. It is envisioned that the Arsenal Ship will be a large hull designed so that the weapons carried onboard are protected from damage and the ship is "virtually unsinkable" if hit by missiles, torpedoes, or mines.

This demonstration program is a non-ACAT program that has been created to evaluate this new capability while minimizing the risks in acquisition of approximately six ships. To ensure that the program remains affordable, a firm cost threshold for the production ships has been established. This program will be conducted using DARPA's Section 845 Agreements Authority so as to allow industry wide latitude in satisfying the Navy's requirements within this threshold. Agreements will be structured to allow tradeoffs between cost and performance. Program success will be judged by the extent to which the Arsenal Ship meets operational requirements.

A second purpose for this demonstration program is to accelerate the Navy's ongoing acquisition reform activities focused on buying improved ships at a lower cost. To this end, the joint program will focus on exploiting DARPA's culture and experience in prototyping system programs. We anticipate the production Arsenal Ship contracts will serve as a model for future streamlining.

Technical Objectives:

The Arsenal Ship is intended to provide a large quantity of (approximately 500) vertical launch systems (VLS) with the capability to launch a variety of weapons for strike, fire support, and area air defense. The exact number of VLS missiles will be determined during the program by optimizing the survivability, performance, sustainablity and costs. The demonstration program will highlight Arsenal Ship's capability as a force multiplier to the Marine Corps, Army, and full array of joint forces. In that regard, it is recognized that certain weapons do not yet exist in the inventory that would allow the full capability to be demonstrated for all missions. No new weapons developments or significant enhancements to weapons are to be pursued as part of this program. Instead, demonstrations should be planned and structured such that significant communications, architecture, and data link functions are evaluated. The goal of the program will be to achieve a balanced design that satisfies the thresholds consistent with the ship's concept of operations (CONOPS).

The demonstration program must show that the production Arsenal Ships are suitable for performing their mission within prescribed cost constraints. To this end, its objectives are to demonstrate:

1. The performance of the mission for 90 days.

2. The architecture, communications, and data link functions to satisfy the Arsenal Ship CONOPS.

3. The capability for remote launch of strike, area air warfare and fire support weapons. It is envisioned that the test program will include:

    a) Salvo launch of up to 3 Tomahawk missiles in 3 minutes

    b) Single SM2 launch using the arsenal ship as a remote magazine for a Cooperative Engagement Capability (CEC) ship

    c) Single Tomahawk launch using the arsenal ship as a remote magazine for air directed and shore based targeting

    d) Single ATACMS launch from a VLS cell in support of a naval surface fire control mission digital call for fire

4. That the proper balance between passive survivability and active self defense will be sufficient for the expected operating scenarios.

Cost Threshold: -- Industry Goal -- $450M/Program Threshold -- $550 M

The acquisition cost threshold is based on the average Navy SCN end costs for the five follow ships acquired after this demonstration program, expressed in FY98 dollars. The costs of the weapons are not included.

Life Cycle Costs:

Industry will be tasked to perform the life cycle cost analyses to demonstrate the operating and support costs for their Arsenal Ship design over a 20 year life. This will ensure that the tenets of the program including reduced manning and innovative operating concepts remain focused on minimizing life cycle costs.

Schedule:

The goal of the demonstration program is to have the ship in the water and ready to start meaningful testing in the year 2000. The program manager will maintain a detailed schedule toward this end and present the plan for approval by the Steering Committee. The basic acquisition strategy for this program is to maximize industry involvement through a competitive multi-phase approach to encourage the maximum innovation within the limits of the cost thresholds. The Government, through the program office, will coordinate with industry to ensure the availability of information that the industry teams need to make informed trades.

Funding:

The cost of the R&D program for this demonstration Arsenal Ship will not exceed $520 million including the cost of concept development and competition. These funds will be provided jointly by the Navy and DARPA as follows:

(Dollars in Millions)

	FY96	FY97	FY98	FY99	FY00	FY01	Total
Navy	$4.0	$25.0	$141.0	$90.0	$80.0	$10.0	$350.0
DARPA	$1.0	$15.0	$47.0	$50.0	$36.0	$21.0	$170.0

The Navy will provide its share of the funds to DARPA at the beginning of each fiscal year.

Roles and Responsibilities:

This joint Navy/DARPA demonstration program will be conducted under the auspices of DARPA's Section 845 Agreements Authority. DARPA will lead the demonstration program and will transition the leadership to the Navy in the later stages of the program, upon mutual agreement of the parties.

The program will be managed by a joint Navy/DARPA program office with the Program Manager reporting to DARPA. A small program office is envisioned. DARPA, Naval Sea Systems Command (NAVSEA), and the Office of Naval Research (ONR) will initially each provide two billets. It is expected that the program office will grow to a maximum of three billets each as the program grows to maturity. Additionally, Director, Surface Warfare (N86) and Marine Corps will designate program office representatives as CONOPS, Requirements, and Fleet Introduction Officer and Naval Surface Fire Testing and Evaluation Engineer, respectively.

The Navy shall develop a concept of operations (CONOPS) for the program that will be reviewed and considered for update as the program develops. The program office will use the CONOPS to guide the trade studies to be conducted by industry.

The Program Manager will develop a program plan including major decision milestones, and the development of a program transition plan. The Steering Committee will approve the initial program plan and thereafter will conduct quarterly reviews to assess progress and provide guidance to the Program Manager.

The Steering Committee will be as follows:

Director, TTO -- DARPA Chairman
Deputy Assistant Secretary of the Navy (DASN, Ships)
Assistant Director, TTO for Maritime Programs -- DARPA
Director of Surface Warfare -- OPNAV (N86)
Director for Expeditionary Warfare -- OPNAV (N85)
PEO for Surface Combatants
Office of Naval Research (ONR33)

An Executive Committee consisting of:

Assistant Secretary of the Navy (RD&A)
Deputy CNO for Resources (N8)
Director, DARPA
Commander, NAVSEA
Chief of Naval Research

will review the program at major decision milestones to evaluate the validity of program cost thresholds and provide re-direction as necessary.

Term of Agreement:

It is expected that this MOA shall remain in effect for the duration of the demonstration program. Early termination of the program due to funding unavailability, lack of legal authority or other reason beyond the control of the parties shall be a basis for termination of this MOA. Any termination shall be preceded by consultation among the parties.

Approved

/Signed/
Larry Lynn
Director, DARPA
5/7/96

/Signed/
John W. Douglass
ASN, (RDA)
24 May 1996

/Signed/
Daniel J. Murphy
RADM USN
Director, Surface Warfare Division
28 May 1996

Larry Lynn	John W. Douglass	J.M. Boorda
Director, Defense Advanced	Assistant Secretary of the Navy	Chief of Naval Operations
		Admiral, USN

Larry Lynn	John W. Douglass	J.M. Boorda
Director, Defense Advanced	Assistant Secretary of the Navy	Chief of Naval Operations
		Admiral, USN

Larry Lynn	John W. Douglass	J.M. Boorda
Director, Defense Advanced	Assistant Secretary of the Navy	Chief of Naval Operations
		Admiral, USN

Tab C

Arsenal Ship -- Concept of Operations (CONOPS)

23 May 1996

The arsenal ship concept is a direct outgrowth of the Navy's shift in focus from the open ocean to the littoral. It is fully consistent with "Forward..from the Sea", and "Operational Manuever from the Sea", and addresses current as well as anticipated future requirements for more decisive, responsive and varied naval support to the land battle. Through concentration of massive firepower, continuous availability and application of netted targeting and weapons assignment, the arsenal ship will increase dramatically the scope and relevance of surface strike and fire support. Tailored specifically to meet the heavy support challenge in the opening days of conflict, arsenal ship will bring firepower to bear in support of Unified CinC's and ground commander plans and schemes of manuever as well as provide significant leverage during the early phases of crises response and control.

Overview

As the foremost world power, the United States will continue to maintain global interests, and therefore must be able to influence and respond to events with credible military presence and power projection capabilities. In the face of steadily decreasing overseas basing and a shrinking military budget, the United States must maintain the ability, in concert with allies, to execute timely combat operations across the spectrum of conflict. Naval forces, sustaining forward presence, will be key to successful introduction as well as early employment of ground forces.

Arsenal ship represents an affordable and much needed enhancement to our existing force of carriers and land attack capable combatants and submarines. It is not a replacement for these or for land-base air. Instead, it is part of the whole-just as the Battleship was a part of the whole for nearly a century. Operating under the control and umbrella of regularly deployed Aegis combatants, arsenal ship will supply substantial firepower, early: giving unified Commanders-in-Chief (CinCs) the capability to halt or deter invasion and, if necessary, enable the build-up of coalition land-based air and ground forces to achieve favorable conflict resolution. With a force totaling about six, arsenal ships will be stationed continuously forward, always available for rapid movement upon receipt of even ambiguous or limited strategic warning. Much like our maritime pre-positioning force, they will remain on station in support of a Unified CinC for indefinite periods without dependence on host nation support or permission.

Operational Concept

With about 500 missiles and space for future extended range gun systems, arsenal ships will be capable of launching many current and planned Department of Defense weapons across the warfare spectrum. Arsenal ship can be positioned to destroy the enemy's critical infrastructure at or near the inception of hostilities. Using precision guided missiles equipped with advanced penetrating warheads and sub munitions, this ship will serve as an additional maneuver element in the landing force or ground force commander's plan by isolating, immobilizing, or destroying enemy forces, including enemy armored fighting vehicles, as well as providing fires in direct tactical support of ground forces.

Employing the Cooperative Engagement Capability (CEC) "remote magazine" launch concept, the Arsenal Ship will provide additional magazine capacity for Theater Ballistic Missile Defense (TBMD) and Air Supremacy missiles. This concept allows for remote missile selection, on-board missile initialization and remote launch orders, and provides remote "missile away" messages to the control platform.

To meet mission goals at affordable cost, ship design will be based on commercial practices and rely extensively on automation in engineering, damage, ship and weapon control systems to achieve a crew size of no more than 50. Berthing spaces for special evolution detachments will enhance operational flexibility.

Arsenal ship and associated weapon control systems will have the flexibility to be responsive to multiple commanders and to conduct simultaneous Long Range Strike, Naval Surface Fire Support, and Theater Air Defense missions. Tables 1 and 2 are representative of the type of capabilities desired.

	Halt Invasion	**Long Range Strike**	**Battlespace Dominance**	**Surface Fire Support**
Complex Adaptive Armed Forces	Air Land Maneuver Battle Groups (e.g., OMGs)	National/Regional C4I Space Control	Manned A/C TBMs, UAVs Cruise Missiles SAM/AAA	Long-Range Artillery TBMs Logistics Assets
Armored Mech Armed Forces	Armor-Heavy Comb. Arms Formations Divisions/BDEs	National and Regional C4I	Manned A/C TBMs SAM/AAA	Long-Range Artillery
Infantry Based Armed Forces	Armor/Mech "Pure" units (BDEs/BNs)	Military Region District C4I	Manned A/C SAM/AAA	Medium-Range Artillery Logistics Assets
Internal Security Light Force	Transportation Railroads Trucking, Light Vehicles	National CMD Authority Military Concentrations	OP Bases Light A/C Coastal Patrol Craft	Logistics Assets Economic Asset Local Forces

Table 1. -- Target Sets to be Countered by Arsenal Ship

	Halt Invasion	Long Range Strike	Battlespace Dominance	Surface Fire Support
Complex Adaptive Armed Forces	SM-2/ATACMS-BAT SLAM	TLAM TLAM-BAT TLAM-C	ATACMS TLAM-C/D SM-2 Blk III A/B and Blk IVA SM-2 LEAP	ATACMS, SLAM, STRIKE-SM TLAM-C/D NAVAL GUNFIRE (VGAS/SCRAM)
Armored Mechanized Armed Forces	SM-2/ATACMS-BAT TLAM-BAT SLAM STRIKE-SM	SM2/ATACMS-BAT TLAM-BAT SLAM STRIKE-SM	ATACMS TLAM-C/D SM-2 Blk III A/B and Blk IVA	ATACMS, SLAM, STRIKE-SM TLAM-C/D NAVAL GUNFIRE (VGAS/SCRAM)
Infantry Based Armed Forces	ATACMS SLAM STRIKE-SM	TLAM-D ATACMS-ER	ATACMS	ATACMS, SLAM STRIKE-SM TLAM-C/D NAVAL GUNFIRE (VGAS/SCRAM)
Internal Security Light Force	NAVAL GUNFIRE (VGAS/SCRAM)	TLAM-C	ATACMS NAVAL GUNFIRE (VGAS/SCRAM)	ATACMS NAVAL GUNFIRE (VGAS/SCRAM)

Table 2. -- Weapons to Counter Target Sets

Operating Assumptions

Arsenal Ship is a firepower multiplier that, in conjunction with other naval forces, increases decisively the options available to the theater CinC. The operational concept for Arsenal Ships is based on the following assumptions.

CinC Requirements. Arsenal Ships will be assigned to theater CinC to provide:

- Conventional Deterrence against regional aggression inimical to U.S. interests,
- Flexible response for demonstration of power independent of diplomatic limitations,
- Credible forward firepower support to joint and coalition land forces early in a regional contingency if deterrence fails. The forward theater arsenal ship weapon loadout will be robust, flexible and tailorable to CinC requirements in order to expand CinC options for use of assigned joint forces.

Joint Warfighting. Arsenal ships will be fully integrated into the joint warfighting force structure. The ships will be capable of firing a variety of weapons in support of a land campaign, including Long Range Strike, Invasion Stopping, Fire Support to Joint Ground Forces, Tactical Ballistic Missile Defense and Air Superiority.

Forward Operations. Arsenal Ships will be stationed, operated and supported in forward theaters for conventional deterrence and to provide immediate responsiveness upon onset of hostilities. The three forward theaters currently envisioned for arsenal ships are:

- Central Command (Southwest Asia/Persian Gulf),
- Pacific Command (Western Pacific)
- European Command (Mediterranean).

Command and Control

Arsenal ships will operate in both peace and war as integral fleet units within the chain of command under Joint Combatant Command (COCOM). Peacetime Operational Control (OPCON) will normally be exercised by numbered fleet commanders. Within a Joint Task Force structure, OPCON will normally be exercised by the Joint Force Maritime Commander. Tactical Command (TACON) will normally be assigned to a naval commander.

Joint Connectivity Mission Planning and Targeting

Key to both arsenal ship's affordability and operational flexibility is off-board integration of all but the most rudimentary C4I. Joint connectivity, including targeting, mission planning, and weapons control will be provided to arsenal ship through the existing fleet of Aegis cruisers and destroyers. Employing an advanced, CEC-like weapons link, the wide array of joint connectivity needed for netted operations will be hosted through an assigned control ship. The role of target and user integration will similarly be performed off-ship, thereby significantly reducing arsenal ship manning, cost and developmental risk; while leveraging the extensive joint C4I investment (Link 16, CEC, etc.) already programmed for the majority of the Surface Navy. The complexity of varied tasking will be reduced to highly reliable, jam resistant targeting, weapons, and launch orders.

Survivability

Though arsenal ship will operate in any threat environment under the protective umbrella of battle force combatants, it must be survivable against 21st century anti-ship missiles, torpedoes, and mines. Passive defense should capitalize on the benefits of mass (tonnage), innovative applications of multiple hull integrity, and signature reduction. Active self defense if required should be roughly equivalent to that of a combat logistics force ship.

Maintenance

Arsenal ships are to be forward deployed for the large part of their operational lives. Low maintenance and high reliability must be engineered into ship design to assure high operational availability.

Tab D

Arsenal Ship Capabilities Document (SCD)

23 May 1996

This ship capabilities document (SCD) compliments the Arsenal Ship Concept of Operations (CONOPS) and provides definition of technical attributes that have evolved as part of ongoing study efforts. This SCD describes functions and capabilities for the Arsenal ships that should be treated as goals when conducting trade studies against the cost thresholds.

1.0 -- Design Philosophy

1.1 ***Arsenal Ship.*** The Arsenal ships are to be delivered fully equipped for fleet operations. They are to have maximized system performance consistent with the CONOPS and the SCD within the cost constraint. The Arsenal Ships should achieve commonality with current Navy systems whenever possible. Innovative approaches that leverage existing DoD investments are strongly encouraged.

1.2 ***Arsenal Ship Demonstrator.*** The Arsenal Ship Demonstrator may not initially have the full capability of the Arsenal Ships. The demonstration program must show that the Arsenal Ships are suitable for performing their mission within the price thresholds. To this end, its objectives are to demonstrate:

1. The performance of the mission for 90 days.

2. The required architecture, communications, and essential data link functions to support the Arsenal Ship CONOPS.

3. The capability for remote launch of strike, area Anti-Air Warfare (AAW) and fire support weapons. It is envisioned that the test program will include:

    a. Salvo launch of up to three Tomahawk missiles in three minutes

    b. Single SM2 launch using the Arsenal Ship as a remote magazine for a CEC (Cooperative Engagement Capability) ship

    c. Single Tomahawk launch using the Arsenal Ship as a remote magazine for air directed and shore based targeting

    d. Single weapon launch from a VLS Cell in support of a naval surface fire control mission digital call for fire.

4. That passive survivability will be sufficient for the expected operating scenarios.

The Arsenal Ship Demonstrator is to be capable of being converted to full mission ship capabilities and configuration and used as a fleet asset.

2.0 -- Warfighting Capabilities

2.1 *General.* The Arsenal Ship should be capable of firing a variety of weapons in support of a land campaign, including Long Range Strike, Invasion Stopping, Fire Support to Joint Ground Forces, Tactical Ballistic Missile Defense and Air Superiority.

2.2 *Launching System.* The ship should have about 500 VLS (vertical launch system) cells capable of launching current and planned vertical launch weapons. The actual number of VLS cells is to be recommended by optimizing the survivability, performance, sustainability and cost.

The ship is to have space, weight and support system capacity reservations for future installation of an extended range gun system.

2.3 *Connectivity.* Targeting, mission planning and command/decision functions will be offboard. The Arsenal Ship is to be connected to command platforms using the CEC "remote magazine" concept or an equivalent data link. An OTH satellite link capability is also to be provided. The ship is to be capable of full time communications with other ships, aircraft, satellites, and shore stations by means of responsive, reliable, clear and secure voice, tactical information distribution and recorded communications. Redundant links may be necessary to achieve robust interconnectivity. It is important that the Arsenal Ship be able to connect to existing joint force communications with minimum impact.

2.4 *Survivability.* The Arsenal Ship is required to be highly survivable in the entire littoral environment. Furthermore, consistent with the objectives for the Arsenal Ship to be an inexpensive platform with low life cycle costs, its' survivability should be achieved through passive means to the extent practicable. Passive techniques to be considered include the use of signature control and countermeasures to make it difficult to detect, target and hit the ship, design/systems that will protect the VLS from damage if the ship is hit, and considerations of ship designs such that the ship will be virtually "unsinkable".

It is expected that the offeror will perform analyses to consider a range of current and future threat systems in performing trade-off studies to develop appropriate levels of survivability that can be achieved within the USP. The threats should include sub-surface, surface and airborne systems. These analyses/trades shall lead to determinations if and where limited active self defense systems are needed to augment the passive design considerations, consistent with minimizing crew size and cost constraints.

The ship shall be able to operate in a chemical-biological-radiation (CBR) environment.

Ship features shall be provided to contribute to the ship's ability to stay afloat and resist further damage including: fire fighting systems, inherent ship stability in damaged conditions, redundant electrical and other support systems.

2.5 *Mobility.* The ship is to be capable of a sustained speed (80% of installed power) of at least 22 knots. The ship is to carry sufficient fuel to conduct a 90 day mission. The ship shall be capable of continuous, precise navigation under all conditions, day or night, independent of geographic location, weather and visibility.

2.6 *Stowage Space.* The ship shall be capable of storing of consumables and repair parts parts for a 90 day mission consistent with the maintenance concept.

3.0 -- Design Standards.

The design life of the ship is to be 35 years.

3.1 *Life Cycle Considerations.* The ships are to be manned, if at all, by a Navy crew to be as small as practicable, but in any event not to exceed 50 people.

The ship shall be ready to perform its missile launch mission when called upon. Availability is the measure of readiness selected for the ship systems. The ship shall be designed, constructed, and integrated with a total ship inherent availability goal of 0.95.

Equipment and material selection, equipment arrangement, built-in-test equipment, redundancy, equipment reliability, manning, logistics facilities, transportation, replenishment, on-board storage, training, and use of off-board support teams and spares pools are to be developed so as to minimize life cycle cost. The maintenance concept shall be developed to achieve the availability goals but a a minimal life cycle cost. The maintenance concept shall be consistent with the Forward Operating Base Concept of the CONOPS.

Material selection, equipment arrangement, built-in-test equipment, redundancy, equipment reliability, manning, logistics facilities, transportation, replenishment, on-board storage, training, and use of off-board support teams and spares pools are to be developed so as to minimize operating and support (O&S) costs and be consistent with the CONOPS.

3.2 *Buoyancy and Stability.* The Ship is to have sufficient reserve buoyancy and stability to withstand flooding as a result of underwater damage. The ship is to withstand grounding or weapons damage that causes a leakage length of 15% of the hull waterline length, assuming the worst combination of flooded and non-flooded compartments within the overall damaged length. The undamaged ship is to have adequate stability to withstand the effects of 100 knot winds and accompanying seas. Stability is to be satisfactory both in full load departure and light load returning condition.

3.3 *Design and Building Margins and Service Life Reserves.*

Design and Building Margins are the responsibility of the offeror.

Service Life Reserves are ship and system capacities in the ship as completed that allow the ship to accept normal growth, planned and unplanned, during fleet service. The following margins are goals for service life of the ship after fleet acceptance of the Arsenal Ships:

- 20% electric power reserves
- 20% air conditioning capacity reserves
- 10% full load displacement growth
- 1 ft of full load center of gravity rise

The Service Life Reserves are exclusive of any margins for items specifically identified as space and weight capabilities.

3.4 ***Regulatory Capabilities.*** The ship design is to comply with 1972 COLREGS for-International-Inland and shall satisfy all the capabilities necessary to obtain certification for transit of the Suez Canal and Panama Canal. Rules-of-the-road equipment may be retracted or covered during low signature military operations.

3.5 ***Standardization.*** Standardization philosophy is be to maximize system performance at the lowest life cycle cost while achieving commonality with current Navy systems wherever possible.

3.6 ***Fuel.*** The propulsion plant and ship service auxiliaries is to be designed to use Diesel Fuel Marine (DFM), corresponding to NATO Code F-76.

3.7 ***Electric Plant Subsystem.*** The ship service generating units are to be of a rating and number such that with one unit inoperable, the remaining installed capacity are able to carry the worst case electric load. At least two sources of electric power are to be provided to all mission critical components.

3.8 ***Underway Replenishment.*** The ship is to be able to be refueled while underway from standard Navy auxiliary ships. Vertical replenishment of provisions is required. Re-arming of VLS cells at sea is not required.

3.9 ***Aviation Support.*** The ship is to be provided with helicopter facilities that meet day and night operations, Visual Meteorological Conditions, landing area with limited service facilities certification for SH-60, V-22 and CH-46 aircraft.

3.10 ***Environment.*** The ship is be capable of operating between latitudes of 70° (North and 60° South. The ships shall not be operated in pack ice. All equipment and machinery installed in exposed locations are to retain full system capability in-40° F to 120° F air temperatures with simultaneous winds up to 40 knots true. All ship systems are to retain full system capability in 28° F to 95° F sea temperatures. All ship systems are to retain full capability with external relative humidity of 0% to 100%.

3.11 ***Machinery Rating Temperatures.*** Rated propulsion power and electric capacity shall be available with 100° F air temperature at prime mover inlet(s).

3.12 ***Performance in a Seaway.*** The ship is to meet the following capabilities:

1. Sea state 5: replenish and strikedown underway

2. Sea state 6: continuous efficient operation (other than replenishment)

3. Sea state 7: limited operation, and capability of continuing its mission without returning to port for repairs after the sea subsides

4. Sea state 8 and above: survivability without serious damage to mission-essential systems.

All structure and fittings are to be designed to withstand dynamic forces produced by motion of a ship in a seaway without operation of any ship stabilization system

3.13 ***Environmental Loading.*** Environmental loading for ship, ship structure, and exposed equipment for design purposes are as follows:

1. Wind loading on vertical projected area, 30lb./sq. ft.

2. Snow and ice loading on horizontal projected area, 7.5lb./sq. ft.

3. Wave slap load on equipment expected to be exposed to green water, 500lb./sq. ft.

3.14 ***Pollution Control.*** The ship is to meet all applicable Federal and International environmental regulations.

3.15 ***Personnel and Equipment Safety.*** The ship is to be designed and constructed to meet internal airborne noise capabilities appropriate to a compartment's function. All installed equipment shall maintain operational effectiveness when exposed to electromagnetic fields as follows: 200v/m for topside mounted equipment, 10v/m for below decks equipment, and 3 Oersteds from below deck equipment. The installed equipment shall satisfy the capabilities for the prevention of Hazards of Electromagnetic Radiation to Personal (HERP) and Ordnance (HERO). The installed equipment shall satisfy the capabilities for the prevention of Hazards of Electromagnetic Radiation to Personnel (HERP) and Ordnance (HERO).

Tab E

Memorandum of Law
Scope of Section 845 Prototype Authority

Memorandum Of: Law Oct 24 1996

Subject: Scope of Section 845 Prototype Authority

Section 845 of the National Defense Authorization Act for fiscal year 1994 [1] authorized the Defense Advanced Research Projects Agency (DARPA) to conduct prototype projects of weapons or weapons systems under the authority of 10 U.S.C.2371. Section 804 of the fiscal year 1997 National Defense Authorization Act [2] amended section 845 extending the authority to the Military Departments and other Department of Defense components. It also extended the authority until 1999. This memorandum reviews the background, purpose and scope of the statute. It is intended to aid in considering whether a project is a candidate for execution using section 845 and contains a preliminary discussion on the implications of utilizing the authority.

1 ***National Defense Authorization Act for Fiscal Year 1994, Pub.L.No.103-160, 107 Stat.1547 (1993).***

2 ***National Defense Authorization Act for Fiscal Year 1997, Pub.L.No.104-201, 1 10 Stat.2422 (1996).***

Background

In June 1986 the President's Blue Ribbon Commission on Defense Management (Packard Commission) made its final report to the President. Among the distinguished members of the Commission was Dr. William J. Perry the current Secretary of Defense. The summary final report contained these findings:

> All too often, requirements for new weapon systems have been overstated. This has led to overstated specifications, which has led to higher cost equipment. Such so-called goldplating has become deeply embedded in our system today. The current streamlining effort in the Defense Department is directed at this problem.
>
> Developmental and operational testing have been too divorced, the latter has been undertaken too late in the cycle, and prototypes have been used and tested far too little.
>
> In their advanced development projects, the Services too often have duplicated each other's efforts and disfavored new ideas and systems. The Defense Advanced Research Projects Agency has not had a sufficient role in hardware experimentation and prototyping. [3]

3 ***President's Blue Ribbon Commission On Defense Management, Final Report (1986) (reprinted in Defense Acquisition: Major U.S. Commission Reports (1949-1988) vol. I, Committee on Armed Services U.S. House of Representatives (100th Cong., 2d Sess., Nov. 1, 1988), 878) (hereinafter referred to as the Packard Commission).***

Corollary recommendations were also contained in the report. They were:

> Rather than relying on excessively rigid military specifications, DoD should make greater use of components, systems, and services available "off the shelf" It should develop new or custom-made items only when it has been established that those readily available are clearly inadequate to meet military requirements.
>
> A high priority should be given to building and testing prototype systems and subsystems before proceeding with full-scale development. This early phase of R&D should employ extensive informal competition and use streamlined procurement processes-It should demonstrate that the new technology under test can substantially improve military capability, and should as well provide a basis for making realistic cost estimates prior to a full-scale developmental decision. This increased emphasis on prototyping should allow us to "fly and know how much it will cost before we buy."
>
> The proper use of operational testing is critical to improving the operations performance of new weapons. We recommend that operational testing begin early in advanced development and continue through full-scale development, using prototype hardware. The first units that come off the limited-rate production line should be subjected to intensive operational testing and the systems should not enter high-rate production until the results from these tests are evaluated.
>
> To promote innovation, the role of the Defense Advanced Research Projects Agency should be expanded to include prototyping and other advanced development work on joint programs and in areas not adequately emphasized by the Services. [4]

4 ***Id. at 880-881.***

DARPA's charter was amended to include "a greater emphasis on prototyping in defense systems by conducting prototype projects and, on request, assist the Military Departments in their own prototyping programs." [5]

5 ***32 C.F.R. Part 358 (1989); DODD 5105.41 (Jan. 25, 1989).***

Prototyping is not new to DARPA. Prime examples include the Pegasus launch vehicle which transitioned directly into operational use. In fact, the first operational launches for the Air Force were conducted under the same contract DARPA had used in the development effort and which was transferred to and modified by the Air Force. Ceramic armor kits for Marine light armored vehicles were prototyped and operationally tested during the Gulf War on seventy-six vehicles. Perhaps the most famous example of DARPA prototyping dates back to 1961 when a DARPA sponsored test program fielded one thousand AR-15 rifles in Southeast Asia for six months. The AR-15 was the prototype of the M-16.

Innovative R&D Agreements

Before discussing section 845, a review of the original "other transactions" authority is appropriate. Prior to 1989, the vast majority of DoD research and development contracting was conducted using procurement contracts as the funding installment. The research offices of the Military Departments, particularly the Office of Naval Research, made significant use of grants. Other elements of DoD rarely, if ever, awarded research grants and there was no consensus within DoD as to the authority to award cooperative agreements. Under the existing DoD policies, grants were awarded only to academic institutions and non-profit research organizations.

In 1989 Congress enacted experimental authority for DARPA to conduct its research and technology development activities by "cooperative agreements or other transactions." [6] Cost sharing, though not absolutely required, was to be used if "practicable." The authority was available only if a "standard contract or grant" was not "feasible or appropriate." Later the authority was made permanent and extended to the Military Departments. [7] Minor amendments have been made without substantive change on subsequent occasions, most recently in the fiscal year 1997 defense authorization. [8] DARPA made extensive use of this authority to conduct technology development efforts not subject to the procurement laws or regulations nor to the regulations governing standard grants and cooperative agreements. [9]

6 ***National Defense Authorization Act for FY90-9 1, Pub.L.No.10 I -- 1 89, 251, 103 Stat.1352 (1989) (codified as amended at 10 U.S.C.2371 (1996)).***

7 ***National Defense Authorization Act for FY92-FY93, Pub.L.No.102-190, 826, 105 Stat 1290 (1991 (codified as amended at 10 U.S.C.2371 (1996)).***

8 ***National Defense Authorization Act for Fiscal Year 1997, Pub.L.No.104-201, 267, 1 10 Stat. 2422 (1996) (codified as amended at 10 U.S.C.2371 (1996)).***

9 ***For information on DARPA's use of this authority, see also United States General Accounting Office, Pub.No. NSIAD-96-1 1, DOD Research -- Acquiring Research By Nontraditional Means (Mar. 1996); Department Of Defense Integrated Process Team, The Services Use Of 10 U.S.C.2371 "Other Transactions" And 845 Prototype Authorities, Final Report (1996); Michael S. Nash, et al., Institute For Defense Analyses, Pub No. D-1793, Participant Views Of Advanced Research Projects Agency "Other Transactions," (1995); Richard N. Kuyath, The Untapped Potential of the Department of Defense's "Other Transaction "Authority, 24 Pub. Con.L.J.521 (Summer 1995); Richard L. Dunn, Testimony before the United States House of Representatives Committee on Science (Nov.8, 1995) (transcript available from the United States Government Printing Office).***

DARPA has found many occasions for using "other transactions" to support science and technology projects. These have included multi-party consortium agreements; agreements for the purpose of establishing standards, reference architecture, or common interfaces; transitioning technology into use; establishing industrial capabilities; technology development agreements with commercial firms, as well as combinations of these or other objectives.

Congress has endorsed DARPA's use of 10 U.S.C.2371 on several occasions in committee reports. Most recently, the conference report for the fiscal year 1997 authorization stated:

> The conferees direct the Services to follow the example of the Defense Advanced Research Projects Agency in the aggressive use of this authority under section 2371. [10]

10 142 Cong. Rec. H9249 (daily ed. July 30, I 996).

There is no definition of an "other transaction." In a sense an "other transaction" is defined in the negative. It is not a standard procurement contract, grant or cooperative agreement. Thus, it is not subject to laws, rules and regulations that govern those instruments. Under 10 U.S.C.2371, "other transactions" can be used to stimulate and support research and development and for other purposes but may not be used for the principal purpose of acquiring goods and services for the direct benefit or use of the Federal Government.

Projects conducted under 10 U.S.C.2371 have typically, but not exclusively, supported dual-use technology developments or technologies with both military and civil applications. Industry's incentive to commercialize the technology serves as the rationale for cost-sharing and reduced government oversight. Cost-sharing and the limitation on the transaction's principal purpose being to acquire goods and services greatly reduces the utility of 10 U.S.C.2371 for the acquisition of military systems.

Prototype Project Authority

After nearly three years of experience using 10 U.S.C.2371 to conduct cost-shared, dual-use technology development projects, a floor amendment to the fiscal year 1994 National Defense Authorization Act, extended DARPA's "other transactions" authority to prototype projects of weapons systems. The text of that amendment was:

Authority of the Advanced Research Projects Agency to Carry Out Certain Pilot Demonstration Projects and Prototype Projects.

(a) ***Authority.*** -- The Director of the Advanced Research Projects Agency may, under the authority of section 2371 of title 10, United States Code, carryout pilot technology demonstration projects and prototype projects that are directly relevant to weapons or weapon systems proposed to be acquired or developed by the Department of Defense.

(b) ***Exercise of Authority.*** --

(1) Subsections (d)(2) and (d)(3) of such section 2371 shall not apply to pilot projects carried out under subsection (a).

(2) The Director shall, to the maximum extent practicable, utilize competitive procedures when entering into agreements to carry out projects under subsection (a).

(c) ***Period of Authority.*** -- The authority of the Director to carry out projects under subsection (a) shall terminate 3 years after the date of-the enactment of this Act." [11]

11 139 Cong. Rec. S 1 1158, S 1 1296 (daily ed. Sept. 9, 1993).

The following colloquy related to the amendment constitutes the principal legislative history of the measure:

Mr. Bingaman. Mr. President, the amendment which I am offering would allow the Advanced Research Projects Agency to use cooperative agreements authority on a pilot basis to execute some of its defense projects. ARPA already has the authority to use cooperative agreements and other transactions to implement its dual-use projects, where industry contributes its own resources and use of contracts would not be appropriate. Indeed, ARPA expects to utilize that authority extensively to implement the programs under the Technology Reinvestment Project.

My amendment would permit ARPA on a pilot basis over the next 3 years to experiment with use of cooperative agreements in carrying out its purely military research and development projects, to which we should not expect industry to contribute its own resources. Use of this more flexible authority is consistent with the thrust of the National Performance review which the Vice President submitted to the President yesterday and with the desire for more flexibility in the defense acquisition system. ARPA led the way in use of cooperative agreements for dual-use projects, such as the high performance computing program. I am sure the agency will make good use of this new authority and urge my colleagues to support this amendment.

Mr. Nunn. This amendment allows ARPA to use the authority in section 2371 of title X, U.S.C. to carry out pilot projects that are directly relevant to weapon or weapons systems. This amendment will allow ARPA to use the cooperative agreements for purely military research as a 3-year test.

The Presiding Officer. The chair, hearing no further debate, without objection, the amendment offered on behalf of the Senator from New Mexico [Mr. Bingaman] is agreed to.

The amendment (No.802) was agreed to.

Mr. Nunn. Mr. President, I move to reconsider the vote by which the amendment was agreed to.

Mr. Warner. I move to lay that motion on the table. The motion to lay on the table was agreed to. [12]

12 139 Cong. Rec. S I 1 158, S 1 1288 (daily ed. Sept. 9, 1993).

As finally enacted the text of the statute reads as follows:

"(a) ***Authority.*** -- The Director of the Advanced Research Projects Agency may, under the authority of section 2371 of title 10, United States Code, carry out prototype projects that are directly relevant to weapons or weapon systems proposed to be acquired or developed by the Department of Defense.

"(b) ***Exercise of authority.*** --

(1) Subsections (c)(2) and (c)(3) of such section 2371, as redesignated by section 827(b)(1)(B), shall not apply to projects carried out under subsection (a).

"(2) The Director shall, to the maximum extent practicable, use competitive procedures when entering into agreements to carry out projects under subsection (a).

"(c) ***Period of authority.*** -- The authority of the Director to carry out projects under subsection (a) shall terminate 3 years after the date of the enactment of this Act. [13]

13 National Defense Authorization Act for Fiscal Year 1994, Pub.L.No.103-160, 107 Stat.1547 (1993).

The text of this version is virtually identical to the original amendment except for technical changes in subsection (b) and the deletion of the words "pilot technology demonstration projects and" before "prototype" in subsection (a). It is clear that no substantive change was intended by the wording change. Indeed the House report on the provision of the current authorization Act which amends section 845 states: "This section would reauthorize and expand to the military services the authority provided by section 845 of the National Defense Authorization Act for Fiscal Year 1994 (Public Law 103-160) to allow additional flexibility in the acquisition of prototype technologies and systems. [14]

14 H.R. REP. No.563, 104th Congress, 2nd Sess., 325-326 (1996).

Section 804 of Public Law No.104-201 reads as follows:

Sec.804. Modification of Authority to Carry Out Certain Prototype Projects

(a) ***Authorized Officials.***

(1) Subsection (a) of section 845 of the National Defense Authorization Act for Fiscal Year 1994 (Public Law 103-160; 107 Stat.172 1; 10 U.S.C.2371 note) is amended by inserting ", the Secretary of a military department, or any other official designated by the Secretary of Defense" after "Agency".

(2) Subsection (b)(2) of such section is amended to read as follows:

"(2) To the maximum extent practicable, competitive procedures shall be used when entering into agreements to carry out projects under subsection (a)."

(b) Extension of Authority. Subsection (c) of such section is amended by striking out "terminate" and all that follows and inserting in lieu thereof "terminate at the end of September 30, 1999."

(c) Conforming and Technical Amendments. Section 845 of such Act is further amended (1) in subsection (b)

(A) in paragraph (1), by striking out "(c)(2) and (c)(3) of such section 2371, as redesignated by section 827(b)(1)(B)," and inserting in lieu thereof "(e)(2) and (c)(3) of such section 2371 "; and

(B) in paragraph (2), by inserting after "Director" the following:

(1) Secretary, or other official"; and
(2) in subsection (c), by striking out "of the Director".

Subsection (e)(2) and (e)(3) of 10 U.S.C.237 1, as amended, are the provisions on cost sharing and limiting use of non-procurement agreements to situations where standard contracts, grants or cooperative agreements are not feasible or appropriate. Both provisions are inapplicable to section 845 agreements.

Expanded Authority

As amended by section 804 of the fiscal 1997 Authorization Act, subsection (a) of 845 now reads as follows:

(a) ***Authority.*** -- The Director of the Defense Advanced Research Projects Agency, the Secretary of a military department, or any other official designated by the Secretary of Defense may, under the authority of section 2371 or title 10, United States Code, carry out prototype projects that are directly relevant to weapons or weapons systems proposed to be acquired or developed by the Department of Defense." [15]

15 Section 845, Pub.L.No.103-160 as amended by section 804 Pub.L.No.103-160.

The authority flows directly to the Director of DARPA, secretaries of military departments, and the Secretary of Defense. Secretaries of military departments may delegate the authority within their departments. The Secretary of Defense may designate other officials (such as Directors of Defense agencies) to exercise the authority. Thus, the authority is potentially available to all elements of the Department of Defense. This is, of course, extremely important to DARPA. In many cases DARPA prototype projects win be conducted jointly with another Department of Defense component and at some point DARPA will want to transition the project to the ultimate user (e.g., the Air Force in the case of high altitude unmanned aerial vehicles, or the Navy in the case of Arsenal Ship). This is now possible. Furthermore, DARPA-funded prototype projects can now be contracted entirely through a Service contracting agent or through another DoD component designated by the Secretary of Defense.

Appropriate Uses of Section 845

Section 845, as amended, authorizes non-procurement agreements ("cooperative agreements and other transactions") to "carry out prototype projects that are directly relevant to weapons or weapons systems proposed to be acquired... Section 845's grant of authority is bounded by the definition of "prototype projects" and "weapons or weapons systems proposed to be acquired."

A principal canon of statutory construction is to apply the plain meaning to words. The standard dictionary classifies "prototype" as a noun and defines it as "an original model on which something is patterned" and also as "a full-scale and (usually] functional form of a new type or design of a construction (as an airplane)."[16] the engineering definition of prototype is "[@j model suitable for use in complete evaluation of form, design and performance."[17]

16 Webster's Ninth New Collegiate Dictionary 947 (1985).

17 McGraw-Hill Dictionary of Scientific and Engineering Arms I 1 83 (1974).

The same conference report which contains the amended section 845 language contains numerous references to "prototypes" such as: "component prototypes for insertion into current undersea weapons"; "operational prototype"; "advanced technology prototype improvements "flyable prototype"; "JASS high band prototype"; "prototype ground-based radar"; "prototype plant."[18] Clearly Congress is aware that the term "prototype" is used in a wide variety of contexts including its dictionary definitions. However, the authority to conduct projects involving "full-scale," "functional," "operational" or "pre-production" prototypes does not limit the authority to conduct projects of lesser scope such as technology demonstrations, sub-system or component prototypes. As noted above, the original version of section 845 expressly included "pilot technology demonstration projects." Furthermore, the legislative history of the amendment to section 845 refers to "additional flexibility in the acquisition of prototype technologies and systems." [19] Thus, "prototype projects" under section 845 includes prototype "systems" but also includes lesser projects involving sub-systems, components, technology demonstrations and technologies.

18 H.R. REP. No.724, 104th Cong., 2nd Sess. (1996).

19 H.R. REP. No.563, 104TH CONG., 2ND SESS.326 (1996).

The breadth of section 845 is also limited by the term "weapons or weapons systems." The statute does not define "weapons." Again, a common sense, plain-meaning approach should be used in interpreting this phrase. A useful reference n-tight be the United States Munitions List published by the Department of State. [20] If the prototype project involves equipment of the types included on the Munitions List, it would be deemed to involve a "weapon." This is not a complete answer, however, since other items of equipment not included on the list may clearly fit the definition of a "weapon."

20 22 C.F.R.121.1 (1994).

It should also be noted that the statute does not require that the prototype project be for the development of a weapon. The statutory requirement is for the project to be "directly relevant" to “weapons or weapon systems proposed to be acquired or developed.” [21] This emphasizes the point made above that sub-systems, components, and technologies are included in the scope of Section 845. Furthermore, this language is broad enough to include training, simulation, auxiliary and support equipment “directly relevant” to weapons or weapon systems.”

21 National Defense Authorization Act for Fiscal Year 1994, Pub.L.No.103-160, 845, 107, Stat 1547 (1993), as amended by the National Defense Authorization Act for Fiscal Year 1997, Pub.L.No.104-201, 267, 1 10 Stat.2422 (1996).

From the plain meaning of its language, it is clear that the scope of section 845 is extremely broad. It would at least include the major categories of equipment contained in the U.S. Munitions list: firearms; ammunition; artillery projectors; launch vehicles, guided missiles, rockets, torpedoes, bombs and mines; explosive propellants and incendiary agents; vessels of war and special naval vessels; tanks and military vehicles; aircraft, spacecraft and associated equipment; military training equipment; protective personnel equipment; military and space electronics; fire control, range finder, optical and guidance and control equipment; auxiliary military equipment; toxicological agents and equipment and radiological equipment; nuclear weapons test and design equipment; and, submersible vessels, oceanographic and associated equipment. Weapons can be either offensive or defensive in character.

Given the trend toward utilizing off-the-shelf components and technologies in defense systems, section 845 prototype projects may often involve the adaptation, testing, or integration of commercial items for military purposes. Indeed it is interesting to note that the Packard Commission recommended greater use of “off-the-shelf” systems and components in the paragraph immediately preceding the recommendations on increased prototyping. [22] A significant “off-the-shelf” content does not preclude a prototype project from being conducted under section 845; indeed, in the future use of off-the-shelf technology, components and systems, as well as prototyping, will probably increase. Furthermore, consistent with the Packard Commission recommendation quoted above, a prototype project could continue through initial low-rate production in order to support operational testing. [23]

22 Packard Commission.

23 Id.

Finally, there is no reason to assume that “proposed to be acquired” means anything more than “if it works it may be the kind of thing we would buy.” Certainly “proposed to be acquired” does not mean that a formal requirement has already been established. The purpose of DARPA has been defined as to “change people’s minds.” Thus a successful prototype may result in the creation of a requirement (e.g., HAVE BLUE). In 1991 Dr. Perry clarified the Packard Commission recommendation by stating that their purpose was not to insert DARPA into the formal acquisition process but to expand DARPA’s role and have the Services act more like DARPA. Thus, prototype projects should not be limited to established requirements or already approved programs.

Effect of Section 845 -- In General

Section 845, as amended, authorizes prototype projects to be conducted under the provisions of 10 U.S.C.2371. Section 2371 in turn authorizes the use of "cooperative agreements and other transactions" for certain research and technology development efforts. Section 2371 efforts will be conducted using cost sharing "to the extent practicable." Section 845 is exempt from that proviso. Section 2371 efforts are authorized only if "a standard contract is not feasible or appropriate." Section 845 is also exempt from that requirement. Competition shall be used in section 845 to the maximum extent practicable.

DARPA has a five year history of interpretation and practical application of section 237 1. DARPA has made numerous reports to Congress on its use of the authority. In addition, Congress has received testimony at committee hearings, reports of the General Accounting Office, and other information on DARPA's use of both the basic 10 U.S.C.2371 authority and section 845 prototype authority. Congress has appropriated millions of dollars to DARPA for projects knowing that they were to be conducted using "other transactions." Statements in floor debate, as well as committee and conference reports, endorse DARPA's interpretation and use of this flexible authority. The interpretation of the effect of "other transactions" authority contained in this section may thus be viewed as ratified not only by Congressional statements but by Congress' act of appropriating millions of dollars in light of DARPA's application and interpretation of the statutes, and by Congress' reenactment of both 2371 and 845 with knowledge of DARPA's interpretation. [24]

24 ***T.V.A. v. Kinzer, 142 F.2d 833, 837 (6th Cir.1944) ("The voting of such appropriations, in the face of the construction placed upon the Act by the Authority, has an effect similar to that resulting from reenactment of a statute the provisions of which had, theretofore, been interpreted they are deemed to have received legislative ratification, and thereby, to have become embedded in the law.") See also U.S. v. Two Tracts of Land, 456 F.2d 264 (6th Cir.), cert. denied, 409 U.S. 887 (1972).***

"Other transactions" conducted under 10 U.S.C.2371 and section 845 are transactions conducted outside the procurement laws and regulations and outside most laws and regulations applicable to assistance relationships. Laws of general applicability such as title VI of the Civil Rights Act, [25] the Trade Secrets-Act, [26] and Conflict of Interest statute [27] are applicable. Laws and regulations specifically applicable to the procurement system are not applicable. Among the laws not applicable are chapters 137, 14 1, and 144 of title 10, United States Code and title 48, Code of Regulations. Neither the Armed Services Procurement Act [28] or other principal procurement laws apply. The Federal Acquisition Regulation (FAR) and the Defense Acquisition Regulation Supplement (DFARS) do not apply. DoD regulations and Military Specifications and Standards applicable to the procurement system do not apply. A number of statutes not falling within the principal title 10 chapters governing procurement are clearly procurement statutes since they reference or use terms defined in chapters 137 or 144 (e.g. 10 U.S.C.2362, 2365 and 2366).

25 ***42 U.S.C.2000(d) (1964).***

26 ***18 U.S.C.1905 (1992).***

27 18 U.S.C.208 (1994).

28 10 U.S.C. Chapter 137.

Contracting under FAR and DFARS is subject to an extensive regulatory system. It might be worth briefly reviewing the 'impact of the procurement regulatory system at this point.

Professor Ralph C. Nash recently reported that a change in DFARS allowing submission of interim vouchers caused him to recall that a similar proposal had been rejected twenty-eight years ago:

> I remembered the incident because I used it in teaching for many years -- as an illustration of the idiosyncrasies of Government procurement. After describing the proposal, which had been made by an industry association, ***The Government Contractor***, in 1968, described the outcome as follows:
>
>> After reviewing this proposal, the ASPR (Armed Services Procurement Regulation] Committee observes that -- although prior DCAA review is in many cases "perfunctory and therefore unnecessary" -- it is nevertheless a control which "should not be lightly discarded." Moreover, it does not appear to the Committee that DCAA reviews cause any significant delay in invoice payment. Accordingly, the Committee has decided not to adopt the proposed procedure.
>
> I always liked this description because it seemed to say something about how the regulatory mind worked. It may be "perfunctory and unnecessary" but it looks good and only does a little harm! That's the regulatory process -- an accumulation of hundreds of those critters. [29]

29 Ralph C. Nash, Payment Procedures: Processing of Interim Vouchers, 10 Nash & Cibinic Report ¶47 (Sept. 1996).

Management consultant Robert C. Spreng compared DoD RDT&E contract awards with the Business Week R&D scorecard and the Fortune 500 Industrials. [30] In addition to a startling concentration of RDT&E dollars in a very small number of firms, Spreng found that 95% of the industry/group leaders that invested the greatest percentage of their sales in R&D received insignificant or no DoD RDT&E awards." [31] These firms were usually on the leading edge of technology developments in their industry. Also 70% of the firms that invested the most total dollars in R&D in their industry/group had insignificant or no DoD RDT&E awards. [32] These 39 leading firms invested $32 billion in R&D in 1993. According to Spreng "a significant share of the most valuable research and product development activity in commercial companies is virtually unavailable to the Federal Government, despite potential benefits to both parties." [33] Spreng's conclusion is that: "Commercial firms will offer the government significantly more of the needed technologies, some right off the laboratory shelf, when the Government can make available adequate protection for commercially oriented intellectual property and the use of existing commercial accounting methods for R&D." [34]

30 Robert C. Spreng, Leading Commercial Firms Shun DOD and R&D (1994).

31 *Id.*

32 *Id.*

33 *Id. at 3.*

34 *Id. at 3.*

Finally, Senator Jeff Bingaman has noted that our government procurement system is one that spends millions to save thousands.

The potential benefits of deregulating the government procurement system are enormous. Deregulation means firms which are unwilling or unable to comply with government rules can be included in government R&D programs. Billions of dollars in privately funded R&D can be leveraged. Government regulations and procedures that do not add value to the process can be abandoned. Contracting techniques can be tailored to take advantage of available opportunities on a case by case or technology by technology basis.

"Other transactions" permit a deregulation of the government R&D system. Existing rules and regulations can be ignored or applied by agreement on a selective basis if deemed to add value. The essence of this non-regulatory system is the principle of freedom of contract.

Freedom of contract should not be a frightening concept. It is the concept upon which the Sales Article (Article II) of the Uniform Commercial Code, as well as many other articles of the Code, is based. [35] The U.C.C. sets up a baseline of rules which apply unless the parties agree otherwise. [36] Under "other transactions" even the baseline is absent and the parties are confronted with a clean slate. DARPA's experience has found that a model agreement expedites the negotiation process by serving as an initial baseline. Likewise use of the model as a baseline facilitates the agreement review process.

35 ***U.C.C. I-102(3) (1962).***

36 ***{T}he idea of flexibility in contracting is embedded in contract theory. Most analyses of commercial contract law simply assume that flexibility exists and should be fostered. The idea that parties are free to choose terms can be justified in a number of ways. It leads to a preference for laws that provide background rules, playing a default or gap-filling function in a contract relationship. A default rule applies if the parties do not agree to the contrary. . . . The second commercial law promise is approached by an effort to identify existing patterns of commercial practice and to follow a presumption that the goal of the drafting is to identify, clarify and, where needed, validate existing patterns of contracting to the extent that these are not inconsistent with modern social policy. Uniform contract laws do not regulate practice. They seek to sustain and facilitate it. The benefits lie in defining principles consistent with commercial practice which, because of their codification and their relevance to actual practice, can be relied on and are readily discernible and understandable to commercial parties." National Conference on Commissioners on Uniform State Laws, Uniform Commercial Code, Article***

2B. Licenses ix-x (Dec. 1, 1995) (Draft available from the National Conference on Commissioners on Uniform State Laws).

As a final note, it is worth pointing out that a non-regulatory system is not unprecedented. Alternative authority to buy experimental and prototype systems without regard to the Armed Services Procurement Act "by contract or other-wise" exists at 10 U.S.C.2373. That provision has been rarely if ever used perhaps exactly because there is no regulatory coverage in FAR. There is no government-wide requirement for the issuance of regulations as a precondition for entering into government contracts. [37]

37 5 U.S.C.553(a)2) (1966).

If the Department of Defense or the Military Departments issue regulations on "other transactions" in the future, they should be baseline guidance embodying the freedom of contract principle and not regulatory mandates.

Effect of Section 84S -- Selected Issues

Competition to the maximum extent practicable is a statutory requirement of prototype projects conducted under section 845. The requirement is not an absolute but obviously by using the term "maximum extent," Congress has made a strong policy statement in favor of competition. The requirement is tempered by the practicability standard. There should be a well documented rationale for initiating section 845 projects in the absence of competition. The statute does not specify the type or characteristics of the competition. Thus, there is plenty of room to introduce innovative forms of competition in prototype projects. The competition can be modeled on the competitive proposal procedures of FAR 6.102(b) and FAR Part 15 or the broad agency announcement (BAA) technique authorized by FAR 6.102(d)(2) and FAR 35.016. Under the FAR, a BAA would normally not be used for a weapons systems prototype development, but under section 845 it can be. Furthermore, entirely new forms of competition can be utilized. Combinations of proposals and oral presentations could be used. Final award selection can be made only after the agreement(s) have been negotiated in final form. Specially tailored techniques can be adopted for special circumstances. Note that the Packard Commission recommended "informal competition." [38]

38 Packard Commission.

During the competition phase of a prototype project a procedure for handling objections or "protests" should be developed and made known to the competitors. This is necessary because the General Accounting Off-ice protest systems applies only to procurements conducted under the procurement statutes. [39] The GAO w inquire into whether a non-procurement instrument was properly used. If it finds use of a non-procurement instrument proper, GAO has no further role. A protester can always go to court but the court's review is based on the Administrative Procedure Act. [40] This essentially requires the plaintiff to show that the Government's action was arbitrary, capriciou s, an abuse of discretion or unlawful. [41] Since relatively few laws apply to section 845 actions, review would typically be focused on whether the action taken by the Government was arbitrary, capricious or an abuse of discretion.

39 ***Energy Conversion Devices, B-260514, June 16, 1995, 95-2 CPD 1121.***

40 ***5 U.S.C.701-706 (1994). See also Scanwell Laboratories v. Schaffer, 424 F.2d 859 (D.C. Cir.1970).***

41 ***5 U.S.C.706 (1994).***

Socio-economic policies implemented specifically through the procurement system do not apply to section 845 projects. For example in the area of equal opportunity while title VI of the Civil Rights Act, applicable to any Federal program applies, Executive Order 11246, applicable to procurement contracts, does not.

As noted by the comments of Robert Spreng, government-required accounting rules and cost principles cause many firms to avoid government R&D contracts. Government-required accounting and purchasing systems are often costly and labor intensive. In the long run the government itself pays for these systems but their initial cost is a serious barrier for commercial firms considering seeking government contracts. Under section 845, there is no requirement to Implement government systems. In fact, cost reimbursement contracting is a rarity in private contracting. DARPA is experimenting with making payment based on technical progress rather than for incurred costs, as well as using multi-factor positive and negative incentives. This is another area open for innovation.

Another major concern of private industry is the protection of intellectual property. Neither the rights in technical data provisions of tide 10, U.S. Code nor the Bayh-Dole Act [42] governing patent rights are applicable to "other transactions" under section 845 thus the government can agree to intellectual property rights tailored to each project.

42 ***35 U.S.C. Chapter 18 (1982).***

The Procurement Integrity Act (41 U.S.C.423) applies only to procurement actions not to "other transactions," however, government employees who are involved in source selection, handling sensitive information, and making programmatic decisions are still subject to 18 U.S.C.1905 (relating to trade secrets and sensitive business 'information) and 18 U.S.C.208 (relating to conflicts of interest). Furthermore, the practice at DARPA has been to have all personnel, whether government or non-government, who have access to source selection and sensitive information sign non-disclosure agreements, submit financial disclosure forms, and sign statements concerning their relationships with competitors. False statements are, of course, subject to the provisions of 18 U.S.C.1001.

Many prototype projects involving systems will be conducted with a view toward an eventual decision to go into production. Although the statutes governing survivability, lethality, operational and other test requirements do not apply to section 845 projects, they nonetheless serve as indications of the type of information which a prototype project should generate in order to support a decision to go into production. Section 845 prototype projects should be planned to address test issues in a manner functionally equivalent to statutory test requirements, while avoiding unnecessary bureaucracy and non-value added documentation.

Current legal authority under section 845 does not extend to full-scale production. If it is intended to transition a system from a prototype project directly into production, the project will have to generate, in addition to survivability and operational test data, life cycle cost and other data sufficient to support a special Defense Acquisition Board, or other authorized review, prior to a production (Milestone III) decision.

If a section 845 systems project involves innovative business and contracting practices, advanced planning must be done to obtain appropriate waivers and exemptions for business practices that will be carried over to the production program. This might include having the project designated a pilot acquisition program in order to obtain expanded waiver authority.

Obviously both program decision-making and production contracting would be greatly facilitated by additional statutory authority allowing an approved section 845 prototype project to transition directly into production on the same contracting basis upon which the prototype project was conducted. DARPA is seeking just such legislative authority.

Conclusion

Section 845 prototyping authority allows for flexibility and innovation in military systems development projects as well as in technology demonstrations and prototyping of subsystems and components. It should allow traditional defense contractors to consider new ways of doing business and permit strictly commercial firms to do business with DoD without changing their existing business practices. Section 845 moves away from an R&D regulatory system to freedom of contract. This is both a challenge and an opportunity. Suggestions on effective ways to implement this new authority are welcomed and should be addressed to the undersigned at the address on the letterhead.

/Signed/
Richard L. Dunn
General Counsel

This Memorandum of Law modifies and supersedes a memorandum for internal DARPA distribution on the same subject dated September 27, 1996.

Tab F

Memorandum -- 10 U.S.C.2371, Section 845, Authority to Carry Out Certain Prototype Projects

Memorandum For: Secretaries of the Military Departments
Directors of Defense Agencies

Subject: 10 U.S.C.2371, Section 845, Authority to Carry Out Certain Prototype Projects

Section 845 of the National Defense Authorization Act for Fiscal Year 1994, Public Law 103-160, authorizes the Director of DARPA to enter into transactions (other that contracts) under 10 U.S.C.2371, for certain prototype projects. The authority under section 845 was to terminate November 30, 1996. Section 804 of the National Defense Authorization Act for Fiscal Year 1997, Public Law 104-201, extends the authority of section 845 through September 30, 1999, and makes it available to the Secretaries of the Military Departments and any other official designated by the Secretary of Defense. The Secretary of Defense delegated his authority and assigned his responsibilities, under the amendments made to section 845 by section 804, to me. I designate the Directors of the Defense Agencies as having the authority to use section 45 as authorized by subsection 804(a).

Section 845 authorizes use of instruments that are alternatives to contracts for the covered prototype projects. Section 845 requires, to the maximum extent practicable, use of competitive procedures when entering into "other transactions." To the extent that a particular statute or regulation is limited in its applicability to the use of a contract, it would generally not apply to an "other transaction." Attachment 1 is a list of statutes that apply to procurement contracts, but thatare not necessarily applicable to "other transactions." The list is provided for guidance only, and is <u>not</u> intended to be definitive. To the extent that a particular requirement is a funding or program requirement or is not tied to the type of instrument used, it would generally apply to an "other transaction." Each statute must be looked at to assure it does or does not apply to a particular funding arrangement using an "other transaction." Use of section 845 authority does not eliminate the applicability of all laws and regulations. Thus, it is essential that counsel be consulted when an "other transaction" will be used.

Section 845 may be used to carry out prototype projects that are directly relevant to weapons or weapon systems proposed to beacquired or developed by the Department. When a prototype project, under the authority of section 845, is used as a precursor to a major defense acquisition program, I expect to be advised of the transition strategy for follow-on contracts at least 30 days prior to award of the "other transaction" that initiates the prototype project. The transition strategy must also address how the DoDD 5000.1 and DoD 5000.2R requirements will be applied to the acquisition process.

I believe it is essential that section 845 instruments incorporate good business sense and appropriate safeguards to protect the government's interest. This includes assurances that the cost to the government is reasonable, the schedule and other requirements are enforceable, and the payment arrangements promote on-time performance.

Annual congressional reporting requirements are established in 10 U.S.C.2371(h), as modified by section 267 of Public Law 104-201. These requirements also apply to prototype projects under section 845. Each military department and defense agency using section 845 authority shall submit a report to OUSD(A&T)/DDP annually, by November 15, that includes all information required by 10 U.S.C.2371 (h), and that summarizes section 845 lessons learned for each prototype project. The format and content for this submission are at Attachment 2. IN addition, users of section 845 may submit lessons learned at any time for inclusion in the Defense Acquisition Deskbook. The Deskbook will also be used to provide discretionary guidance and further information.

I encourage you to use the flexibility provided by this authority and to issue any further guidance you deem necessary. If you delegate authority to use section 845, I expect it will be to officials whose level of responsibility, business acumen, and judgment enable them to operate in this relatively unstructured environment. If we use this authority wisely, I will request that it be extended or made permanent by the Congress.

Attachments

/Signed/
Paul G. Kaminski

Attachment 1

Statutes Inapplicable to Other Transactions

1. Competition in Contracting Act, Pub.L.No.98-369 (1984), as amended-Promotes the use of competitive procurement procedures and prescribes uniform government-wide policies and procedures regarding contract formation, award, publication, and cost or pricing data (truth in negotiations). See DoDD coverage generally at chapter 137 of title 10, United States Code, particularly sections 2301-2305.

2. Contract Disputes Act, Pub.L.No.95-563 (1987, as amended, 41 U.S.C.601 et seq.- Provides for the resolution of claims and disputes relating to government contracts.

3. Procurement Protest System, Subtitle D of Competition in Contracting Act, Pub.L.No.98-369 (1984), 31 U.S.C.3551 et seq. -- Provides statutory basis for procurement protests by interested parties to the Comptroller General.

4. Public Law 85-804, 50 U.S.C.1431-1435, Extraordinary contractual relief -- Authorizes such remedies to contractors as formalization of informal commitments, amendments without consideration, and correction of mistakes, and permits indemnification for unusually hazardous risks.

5. 10 U.S.C.2207. Expenditure of appropriations: limitation -- Permits termination of contracts upon a finding that the contractor has offered or given gratuities to obtain a contract.

6. 10 U.S.C.2306. Kinds of contracts -- Prohibits use of cost-plus-a percentage-of-cost system of contracting; requires a covenant against contingent fees paid to obtain contracts; limits fee amount on virtually all cost-type contracts.

7. 10 U.S.C.2313. Examination of records of contractor-Provides agency and GAO access to contractors facilities to audit contractor and subcontractor records and gives the DCAA subpoena authority.

8. 10 U.S.C.2353. Contracts: acquisition, construction, or furnishing of test facilities and equipment (to R&D contractors).

9. 10 U.S.C.2354. Contracts: indemnification provisions-Indemnification authority against unusually hazardous risks for R&D contractors.

10. 10 U.S.C.2393. Prohibition against doing business with certain offerors -- Prohibition with respect to solicitation of offers and contract awards to contractors that have engaged or are suspected to have engaged in criminal, fraudulent, or seriously improper conduct.

11. 10 U.S.C.2403. Major weapon systems; contractor guarantees.

12. 10 U.S.C.2408. Prohibition on persons convicted of defense-contract related felonies and related criminal penalty on defense contractors -- Generally, convicted felons precluded from working in a managerial capacity on DoD contracts.

13. 10 U.S.C.2409. Contractor Employees: protection from reprisal for disclosure of certain information. Whistle blower protection to contractor employees.

14. 31 U.S.C.1352. Limitation on the use of appropriated funds to influence certain Federal contracting and financial transactions -- Prohibits use of funds to influence or attempt to influence government officials or members of Congress in connection with the award of contracts, grants, loans, or cooperative agreements.

15. Antikickback Act of 1986, 41 U.S.C.51-58 -- Prohibits kickbacks in connection with government contracts; provides civil and criminal penalties.

16. Procurement Integrity Act, section 27 of the Office of Federal Procurement Policy Act, 41. U.S.C.423 -- Imposes civil, criminal, and administrative sanctions against individuals who inappropriately disclose or obtain source selection information or contractor bid and proposal information.

17. Service Contract Act. 41 U.S.C.351 et seq., Walsh Healy Act, 41 U.S.C.35-45; Fair Labor Standards Act, 29 U.S.C.201-219 -- Provide protections for contractor employees.

18. Drug-Free workplace Act of 1988, 41 U.S.C.701-707 -- Applies to contracts and grants.

19. Buy American Act, 41 U.S.C.10 a-d. Applies only in part to "other transactions."

Attachment 2

Section 845 -- Annual Report Requirements

Awarding Activity: Identify the military department or defense agency, a point of contact, and a phone number.

General Description: Provide a general description and discuss the objectives of the 845 prototype project, including the technology areas for which research is conducted.

Reasons for Use of 845 Authority: Explain the reason for choosing to use the 845 authority and address the expected benefits.

Extent of Cost-Sharing, if any: [Section 845 is exempt from the 10 USC 2371(E)(1)(B) requirement for cost-sharing.]

Technology and Industrial Base Implications: Discuss the extent the use of an "other transaction"

(i) has contributed to a broadening of the technology and industrial base available for meeting DoD needs; and

(ii) has fostered within the technology and industrial base new relationships and practices that support the national security of the United States.

Payments Received and Credited, if any: [10 USC 2371 (d) allows an "other transaction" to require payment to the Department as a condition for receiving support under an "other transaction" and permits any such payment to be credited to support accounts.]

Lessons Learned: Contrast the actual benefits or drawbacks experienced to the expected benefits discussed above, being as quantitative as possible, and provide any additional lessons learned. This section is not required by statute, but will be used to decide whether the Department should pursue extension of the authority prior to its expiration in 1999. This section should be updated annually and reported for all section 845 prototype projects until the prototype project is complete.

Tab G

Advanced Research Projects Agency Letters

April 14, 1997
Ser ASJPO/1 69

Dr. Pat Barry
Electronic Sensors and Systems Division
Northrop Grumman Corporation
P.O. Box 1693
Baltimore, MD 21203

Dear Colleagues:

The joint Navy/DARPA Arsenal Ship Program Office has followed your team's Phase II activities with great interest. Your work to date has convinced us that: significant acquisition reform through the 845 Other Agreements Authority is achievable; that cooperative industry-government teaming to achieve CAIV and life cycle goals is producing the desired results; that your focus on minimum crew size with attendant cost of ownership implications is generating exciting designs; that your focus on ship survivability with design features of multiple hulls, automated damage control techniques, and signatures management will alter survivability philosophy significantly; and that your Arsenal Ship Demonstrator if selected, will be a terrific vehicle to evaluate the concept of massed precision firepower in support of the Land Component Commander, within a joint warfighting context.

At the beginning of the Arsenal Ship Phase IV demonstration with the fleet the

Navy is planning on using the Arsenal Ship Demonstrator, in parallel with its primary tests to evaluate the military utility of the Arsenal Ship, to evaluate various SC-21 technologies. The ASJPO has determined that this testing can be accommodated within current demonstrator designs. This is an advisory message only, so that you are apprised of future plans for this Demonstrator. No changes in the design, goals, or evaluation criteria for downselect at the end of the current phase, are intended.

The SC-21 technologies which require risk reduction at sea testing include: --

- Advanced Integrated Electronic Warfare System (AIEWS), EDM Available FY99
- ADCON-21, EDM Available FY01
- Integrated Power System, EDM Available FY01
- Advanced Human Computer Interface, EDM Available FY01

-	Multi Function Radar (X Band),	EDM Available FY02
-	Lightweight Broadband Variable Depth Sonar Handling System	EDM Available FY03
-	Volume Search Radar (L Band)	EDM Available FY03
-	Vertical Gun for Advanced Ships (VGAS)	EDM Available FY04

Some or all of these may be tested on the Arsenal Ship Demonstrator.

Sincerely,

/Signed/
C.S. Hamilton
CAPT, USN
Arsenal Ship Program Manager

April 14, 1997
Ser ASJPO/170

Mr. Tom Egan
ArShip 21 Team
4301 North Fairfax Dr.
Suite 525
Arlington, VA 22015

Dear Colleagues:

The joint Navy/DARPA Arsenal Ship Program Office has followed your team's Phase If activities with great interest. Your work to date has convinced us that: significant acquisition reform through the 845 Other Agreements Authority is achievable; that cooperative industry-government teaming to achieve CAIV and life cycle goals is producing the desired results; that your focus on minimum crew size with attendant cost of ownership implications is generating exciting designs; that your focus on ship survivability with design features of multiple hulls, automated damage control techniques, and signatures management will alter survivability philosophy significantly; and that your Arsenal Ship Demonstrator if selected, will be a terrific vehicle to evaluate the concept of massed precision firepower in support of the Land Component Commander, within a joint warfighting context.

At the beginning of the Arsenal Ship Phase IV demonstration with the fleet, the Navy is planning on using the Arsenal Ship Demonstrator, in parallel with its primary tests to evaluate the military utility of the Arsenal Ship, to evaluate various SC-21 technologies. The ASJPO has determined that this testing can be accommodated within current demonstrator designs. This is an advisory message only, so that you are apprised of future plans for this Demonstrator. No changes in the design, goals, or evaluation criteria for downselect at the end of the current phase, are intended.

The SC-21 technologies which require risk reduction at sea testing include: --

-	Advanced Integrated Electronic Warfare System (AIEWS),	EDM Available FY99
-	ADCON-21,	EDM Available FY01
-	Integrated Power System,	EDM Available FY01
-	Advanced Human Computer Interface,	EDM Available FY01
-	Multi Function Radar (X Band),	EDM Available FY02
-	Lightweight Broadband Variable Depth Sonar Handling System	EDM Available FY03
-	Volume Search Radar (L Band)	EDM Available FY03
-	Vertical Gun for Advanced Ships (VGAS)	EDM Available FY04

Some or all of these may be tested on the Arsenal Ship Demonstrator.

Sincerely,

/Signed/
C.S. Hamilton
CAPT, USN
Arsenal Ship Program Manager

April 14, 1997
Ser ASJPO/171

Dr. Michael Hughes
Lockheed Martin Corporation
P.O. Box 1027
MS 108-108
Moorestown, NJ 08057-0927

Dear Colleagues:

The joint Navy/DARPA Arsenal Ship Program Office has followed your team's Phase II activities with great interest. Your work to date has convinced us that: significant acquisition reform through the 845 Other Agreements Authority is achievable; that cooperative industry-government teaming to achieve CAIV and life cycle goals is producing the desired results; that your focus on minimum crew size with attendant cost of ownership implications is generating exciting designs; that your focus on ship survivability with design features of multiple hulls, automated damage control techniques, and signatures management will alter survivability philosophy significantly; and that your Arsenal Ship Demonstrator if selected,-will be a terrific vehicle to evaluate the concept of massed precision firepower in support of the Land Component Commander, within a joint warfighting context.

At the beginning of the Arsenal Ship Phase IV demonstration with the fleet, the Navy is planning on using the Arsenal Ship Demonstrator, in parallel with its primary tests to evaluate the military utility of the Arsenal Ship, to evaluate various SC-21 technologies. The ASJPO has determined that this testing can be accommodated within current demonstrator designs. This is an advisory message only, so that you are apprised of future plans for this Demonstrator. No changes in the design, goals, or evaluation criteria for downselect at the end of the current phase, are-intended.

The SC-21 technologies which require risk reduction at sea testing include:

- Advanced Integrated Electronic Warfare System (AIEWS), EDM Available FY99
- ADCON-21, EDM Available FY01
- Integrated Power System, EDM Available FY01
- Advanced Human Computer Interface, EDM Available FY01
- Multi Function Radar (X Band), EDM Available FY02
- Lightweight Broadband Variable Depth Sonar Handling System EDM Available FY03
- Volume Search Radar (L Band) EDM Available FY03
- Vertical Gun for Advanced Ships (VGAS) EDM Available FY04

Some or all of these may be tested on the Arsenal Ship Demonstrator.

Sincerely,

/Signed/
C.S. Hamilton
CAPT, USN
Arsenal Ship Program Manager

Tab H

Chief of Naval Operations Letter

2 June 1997

Dear Mr. Chairman,

The four-year, twelve-ship DDG 51 proposal contained in the FY-98 President's budget offers a rare period of surface combatant program stability. At the same time, it has caused us to accelerate somewhat our normal timelines for out-year surface combatant SCN planning. In particular, the Congress has expressed interest in three areas: possible expansion of the multi-year to a thirteenth destroyer: the role of the Maritime Fire Support Demonstrator in transitioning to DD 21 (the surface combatant to allow the DDG 51 class (; the likelihood of a significant Aegis cruiser modernization effort. The purpose of this letter is to respond to this interest, while acknowledging many programmatic details have yet to be refined.

First, with respect to possible expansion of the DDG 51 multi-year proposal, we have included in our FY-98 budget request as many ships as our funding authority would allow. Clearly, the multi-year approach, when applied to a successful and mature program like DDG 51, contains substantial benefits and savings. Acceleration of an additional DDG 51 award into the multi-year window would increase the total savings while addressing validated surface combatant requirements. Before Congress authorized the multi-year buy, the average constant year cost for a single DDG 51 was $950M. The marginal cost of an additional DDG in FY-98 is $720M -- substantially less than the other three ships in that year. I would welcome such action by the Congress.

Second, with respect to the Maritime Fire Support Demonstrator, we anticipate substantial benefit in terms of acquisition reform and technology advance swill accrue to DD 21, as well as CVX and future ships. Our partnership with DARPA and industry is producing results well beyond initial expectations. It was just this success that prompted us to expand the application of the original Arsenal Ship concept so we could capture more completely the advantage so this unique program. DD 21 must be a ship responsive both to twenty-first century requirements and fiscal limits. We are seeking revolutionary advances in joint connectivity and responsive, precise delivery of ordnance to realize our objective of offensive distributed firepower for the surface force. At the same time, DD 21 must be affordable in the numbers needed to replace existing destroyers and frigates. If we do not break with historical norms of design, construction and life cycle costs, I am convinced we will not be able to build the right capability in the right numbers, at the right time. The Maritime Fire Support Demonstrator holds the key; I urge your active support for its continued funding.

With respect to modernization of the Aegis cruiser force; Quadrennial Defense Review deliberations revealed the need to perform a mid-life conversion of these superb ships. Our objective is to package Theater Ballistic Missile Defense (TBMD), Area Air Defense Commander, Land Attack and Smart Ship improvements into a class-wide upgrade, thereby assuring a preeminent role for Aegis cruisers into the third decade of the next century. This modernization program will enable otherwise unaffordable acceleration of TBMD and Land Attack introduction into our surface combatant force, while providing critical industrial work prior to full rate production of the DD 21. It is our intent to commence feasibility studies in FY-97 to permit and FY-98 modernization program start should congress make FY-98 funds available.

Together, the DDG 51 multi-year, Maritime Fires support Demonstrator, Aegis cruiser modernization and DD 21 programs will set the course for the surface navy of the next century. I look forward to working closely with you to ensure their success.

Sincerely,

/Signed/
Jay J. Johnson
Admiral, U.S. Navy

The Honorable Curt Weldon
Chairman, Military Research and Development
Committee on National Security
House of Representatives
Washington, DC 20515-3807

Copy to:
The Honorable Owen Pickett
Ranking Minority Member

The Honorable Duncan Hunter
2365 Rayburn House Office Building
House of Representatives
Washington, D.C. 20515-0552

The Honorable Ike Skelton
2227 Rayburn House Office Building
House of Representatives
Washington, D.C. 20515-2504

The Honorable Curt Weldon
2452 Rayburn House Office Building
House of Representatives
Washington, D.C. 20515-2507

The Honorable Owen Pickett
2263 Rayburn House Office Building
House of Representatives
Washington, D.C. 20515-4003

The Honorable Floyd Spence
2405 Rayburn House Office Building

The Honorable Strom Thurmond
217 Senate Russell Office Building
United States Senate
Washington, D.C. 20515-4001

The Honorable John Warner
225 Senate Russell Office Building
United States Senate
Washington, D.C. 20510-4601

The Honorable Carl Levin
459 Senate Russell Office Building
United States Senate
Washington, D.C. 20510-2203

The Honorable Edward Kennedy
315 Senate Russell Office Building
United States Senate
Washington, D.C. 20510-2101

The Honorable Ted Stevens
522 Senate Russell Office Building

House of Representatives
Washington, D.C. 20515-4002

The Honorable Ronald Dellums
2108 Rayburn House Office Building
House of Representatives
Washington, D.C. 20515-0509

The Honorable C.W. Bill Young
2407 Rayburn House Office Building
House of Representatives
Washington, D.C. 20515-0910

The Honorable John P. Murtha
2423 Rayburn House Office Building
House of Representatives
Washington, D.C. 20515-3812

The Honorable Bob Livingston
2406 Rayburn House Office Building
House of Representatives
Washington, D.C. 20515-1801

United States Senate
Washington, D.C. 20510-0201

The Honorable Daniel Inouye
722 Senate Russell Office Building
United States Senate
Washington, D.C. 20515-1102

Tab I

DON Memorandum -- Maritime Fire Support Demonstrator (MFSD)

Department of the Navy
Office of Legislative Affairs
1300 Navy, Pentagon

In Reply Refer To:
LA-2:bj
October 24, 1997

Memorandum For: Interested Members of Congress

Subj: Maritime Fire Support Demonstrator (MFSD)

1. Appreciating your interest in the current plans, policies, and programs of the Department of the Navy, please be advised that after reviewing Navy budget priorities, Secretary of the Navy John H. Dalton has reluctantly determined that insufficient funding exists to continue the next phase of the Maritime Fire Support Demonstrator (MFSD) program. The $35 million appropriated by Congress is not enough to enter the next phase of the MFSD development plan, which would include the award of a detailed design and construction contract in 1998. The Navy has reviewed and prioritized its programs and has determined that it is not possible to redistribute money from other areas to sustain the MFSD program. Reallocation of the $35 million appropriated by Congress is currently under review by the Navy.

2. The initial phases of the MFSD program provided a good return for the investment. These phases were highly successful in forming closer partnerships with industry as demonstrated by the teaming of combat system integrators and shipyards. The Navy is 100 percent committed to the fleet modernization that the MFSD program represented: teaming produced innovative concepts for modular integration of combat systems, improved ship survivability, design, radar signature reduction techniques, and reduced manning requirements. These innovations are directly transferable tot he Land Attack Destroyer (DD21). With the cancellation of the MFSD program, the Navy will lose the opportunity and the advantages of integrating these technologies into a single test platform at sea.

3. DD 21 development will continue as planned, using an acquisition strategy patterned after the MFSD. The program turns the systems development process over to industry teams at the earliest stage and challenges industry to develop and design the optimum mix of performance capabilities which can be accommodated within production and life-cycle affordability constraints. This approach has already produced several technological

breakthroughs in the initial stage of MFSD development, and the Navy is confident that industry will deliver a revolutionary ship in DD 21.

4. Planned milestones for the DD 21 program include: December 1997 -- Initiate Request for Proposals; March 1998 -- Award multiple concept development contracts; 2004 -- Award contract for the first DD 21.

5. The Office of Legislative Affairs point of contact is Commander Pete Gumataotao, USN, at (703) 695-1366.

Sincerely,

/Signed/
Jay M. Cohen
Captain, U.S. Navy
Deputy Chief of Legislative Affairs

October 24, 1997 1600

Response to Query regarding Maritime Fire Support Demonstrator

After reviewing Navy budget priorities, Secretary of the Navy John H. Dalton has reluctantly determined that insufficient funding exists to continue the next phase of the MFSD program. The Navy and DARPA jointly requested $150 million in FY 1998 for the MFSD program. The $35 million appropriated by Congress is not enough to enter the next phase of the MFSD development plan, which would include the award of a detailed design and construction contract in 1998. The Navy has reviewed and prioritized its programs and has determined that it is not possible to redistribute money from other areas to sustain the MFSD program. Reallocation of the $35 million appropriated by Congress is currently under review by the Navy.

The initial phases of the MFSD program provided a good return for the investment. These phases were highly successful in forming closer partnerships with industry as demonstrated by the teaming of combat system integrators and shipyards. The Navy is 100 percent committed to the fleet modernization that the MFSD program represented: teaming produced innovative concepts for modular integration of combat systems, improved ship survivability design, radar signature reduction techniques, and reduced manning requirements. These innovations are directly transferable to the Land Attack Destroyer (DD 21). With the cancellation of the MFSD program, the Navy will lose the opportunity and the advantages of integrating these technologies into a single test platform at sea.

DD 21 development will continue as planned, using an acquisition strategy patterned after MFSD. The program turns the systems development process over to industry teams at the earliest stage and challenges industry to develop and design the optimum mix of performance capabilities which can be accommodated within production and life-cycle affordability constraints. This approach has already produced several technological breakthroughs in the initial stage of MFSD development, and the Navy is confident that industry will deliver a revolutionary ship in DD 21.

Planned milestones for the DD 21 program include:
December 1997 -- Initiate Request For Proposals
March 1998 -- Award multiple concept development contracts
2004 -- Award contract for the first DD 21.

Tab J

Assistant Secretary of the Navy Memorandum Termination of Arsenal Ship/Maritime Fire Support Demonstrator Efforts

The Assistant Secretary of the Navy
(Research, Development and Acquisition)
Washington, D.C. 20350-1000

Nov 24 1997

Memorandum For: Director, Defense Advanced Research Projects Agency

Subj: Termination of Arsenal Ship/Maritime Fire Support Demonstrator Efforts

After reviewing Navy budget priorities, the Secretary of the Navy reluctantly determined insufficient funding exists to continue the MFSD program. Accordingly, Phases III and all subsequent phases of the Arsenal Ship/Maritime Fire Support Demonstrator program will not be initiated. The Arsenal Ship Joint Program Office (ASJPO) should complete all Phase II activities and then close down the Joint Program Office. I am deeply disappointed that this truly innovative program that embodied the acquisition reforms and early industry design involvement necessary to affordably procure our future surface combatants was not supported, as requested in the President's Budget.

Phase II payable milestones 5 and 6 are due to be delivered on November 14, 1997, by the three industry consortia. As previously agreed between ASJPO, Deputy Assistant Secretary of the Navy for Ship Programs, the Office of the Chief of Naval Operations (N86), and the SC 21 Program Office, a copy of the Functional Design Report, Life Cycle Cost Analysis, and Production Plan will be provided to the SC 21 Program Office on a "for Government use only" basis. This information will be used by the SC 21 Program Off ice to improve and support the acquisition process for DD 21.

FY98 funding to support payable milestones 5 and 6 was transferred to DARPA, along with sufficient funds, to support smooth and complete transfer of MFSD lessons learned to DD 21 between now and October 1, 1998. ASJPO has agreed to conduct a lessons learned study of the Phase I and II Arsenal Ship efforts, which will be completed prior to 31 December 1997. Management of the Project Office should shift to PMS 500 by January 1, 1998, and funding support for the office should shift to the SC 21 Program Office by October 1, 1998

The completed phases of the Arsenal Ship/MFSD program have provided an excellent return for the investment. These two phases successfully demonstrated that industry, involved early in the ship design process, could develop an optimum mix of performance capabilities that could be accommodated within affordability constraints, successfully demonstrated teaming between combat system integrators and shipyards, and the introduced innovative concepts in reduced manning, automated damage control, topside integration, and modular design. This investment has, and will continue to facilitate DD 21 development, and will pay significant dividends in other joint warfighting arenas.

/Signed/
John W. Douglass

Tab K

Memorandum of Agreement Between the Naval Surface Warfare Center and DARPA, Tactical Technology Office

Objective

The objective of this Memorandum of Agreement (MOA) is to facilitate the transfer of the substantial national investment in technology at the activities of the Naval Surface Warfare Center to the industry teams participating in the Arsenal Ship program while maintaining the competitive integrity of the program.

Background

The Arsenal Ship program was established to develop and deliver up to six ships for the delivery of large quantities of ordnance in support of land and littoral engagements. The process is being conducted under the Defense Research Project's Agency's (DARPA) Section 845 Agreements Authority. The program is currently structured to be conducted in phases:

I Trade-off studies and design concept development

II Functional design

III Detail design and construction of the Arsenal Ship Demonstrator

IV Demonstrations

V Option for construction of the subsequent Arsenal Ships and conversion of the Arsenal Ship Demonstrator to a Fleet asset

VI Service life support

Because the ships will be developed from the earliest design stages in the private sector, traditional roles and relationships of government organizations must be reconsidered. The industry teams selected for Phase I have approached NSWC activities and DARPA with specific requests for NSWC involvement. This MOA clarifies the role that NSWC will play in Phase I of the Arsenal Ship program. Since this role may change during each phase of the program, this MOA may be modified as the program proceeds.

Agreement

1. Warfare Center products and services will be made available for a fee to all industry teams on a non-exclusive basis. Industry requests and a list of delivered products will be provided to the ASJPO.

2. Warfare Center products and services will be limited to the use of test facilities and the provision of objective data, (e.g., results of calculation procedures or tests, models, and simulations,).

3. Source selection evaluators will not have participated in services to industry.

4. Agreements, including funding arrangements, will be made directly between industry teams and the Warfare Center.

/Signed/
K.K. Paige
Commander
Naval Surface Warfare Center

/Signed/
Dr. David Whelan
Director
Tactical Technology Office, DARPA

Tab L

Memorandum of Agreement Between the Naval Surface Warfare Center and the Arsenal Ship Joint Program Office

Objective

The objective of this Memorandum of Agreement (MOA) is to facilitate the transfer of the substantial national investment in technology at the activities of the Naval Surface Warfare Center to the industry teams participating in the Arsenal Ship program while maintaining the competitive integrity of the program. This Phase II MOA supersedes the Phase I MOA for the same program.

Background

The Arsenal Ship program was established to develop and deliver up to six ships for the delivery of large quantities of ordnance in support of land and littoral engagements. The process is being conducted under the Defense Advanced Research Projects Agency's (DARPA) Section 845 Agreements Authority. The program is currently structured to be conducted in phases:

I Trade-off studies and design concept development

II Functional design

III Detail design and construction of the Arsenal Ship Demonstrator

IV Demonstrations

V Option for construction of the subsequent Arsenal Ships and conversion of the Arsenal Ship Demonstrator to a Fleet asset

VI Service life support

Because the ships will be developed from the earliest design stages in the private sector, traditional roles and relationships of government organizations must be reconsidered. This MOA clarifies the role that Naval Surface Warfare Center (NSWC) will play in Phase II of the Arsenal Ship program, which is greater in scope than the Phase I role. This role may change again during subsequent phases of the program. Accordingly, this MOA may be modified as the program proceeds.

Agreement

1. Warfare Center products and services will be made available to the two industry teams on a non-exclusive basis. Industry requests and a list of delivered products will be provided to the Arsenal Ship Joint Program Office (ASJPO).

2. Warfare Center personnel may provide technical expertise, upon request, to industry teams in the form of consultation and recommendations or opinions, in addition to performing objective calculations, testing, modeling and simulation. Support available includes but is not limited to the following technical areas:

   Hydrodynamics
   Vulnerability & Survivability
   Signatures
   Weapon Systems and Explosive Safety
   Combat Systems Integration
   Ship, Weapon and Combat System Test & Evaluation Weapon Systems Design
   Chemical, Biological & Radiological Defense
   Electromagnetic Environmental Effects
   Warfare Analysis
   Shipboard Environmental Quality (military mission unique requirements)
   Shipboard Physical Security
   NAVSEA Warfare Center owned (unique) models and simulations

3. Warfare Center personnel will at all times protect the industry teams' proprietary information.

4. Source selection evaluators will not have participated in services to industry teams on this program.

5. Agreements, including funding arrangements, will be made directly between industry teams and Warfare Center activities with the concurrence of Warfare Center Headquarters.

/Signed/
K.K. Paige
Commander
Naval Surface Warfare Center

/Signed/
Dr. David Whelan
Director
Tactical Technology Office, DARPA

Tab M

Biographies

Captain Charles S. Hamilton II, USN

Captain Hamilton attended Duke University, graduating in May 1974 with a Bachelor of Science in Zoology, and commissioned in the Navy through the NROTC Program at Duke. Captain Hamilton's early sea tours included: USS HAWKINS (DD 873) [Electronics Material Officer and Combat Information Center Officer]; USS COONTZ (DDG 40) [Fire Control Officer and Missile Officer]; USS CALLAGHAN (DDG 994) [Operations Officer]; and USS Fox (CG-33) [Executive Officer]. In July 1991, he became Commanding Officer USS OBRIEN (DD 975) where he served until June 1993.

Captain Hamilton's educational assignments included: Naval Postgraduate School, Monterey, Ca. where he graduated with distinction with a Masters of Arts in National Security Affairs; and National War College where he graduated with distinction and was awarded a Master of Science in National Security Strategy.

Captain Hamilton's shore duty assignments included: Office of the Chief of Naval Operations as Anti-Submarine Warfare Program Analyst, Administrative Assistant to Director, Program Resource Appraisal Division (OP-91), and VCNO's speech writer; CNO Staff as Head AEGIS Destroyer Section (OP-355F) and Financial Coordinator, AEGIS Cruiser Destroyer Branch; and Military Staff Specialist for Naval Warfare in the Office of the Under Secretary of Defense (Acquisition & Technology).

Captain Hamilton became Program Manager for the Arsenal Ship in May 1996 and remained in place through 31 December 1997.

Mr. J. Christopher (Kit) Ryan

Mr. Ryan obtained his undergraduate degree from Webb Institute of Naval Architecture in 1967, followed by a Master Degree in Naval Architecture from MIT in 1969.

Mr. Ryan initially worked for Litton Systems, Advanced Marine Technology Division Los Angeles on the DD-963 project. Upon completion of the DD-963 design effort, Mr. Ryan joined the Department of the Navy as a member of the Naval Ship Engineering Center (NAVSEC), Hyattsville, Md. Advanced Design Group. He had major assignments on the FFG-7 project, the Sea Control Ship project, the nuclear strike cruiser (CSGN), Aegis crusier (CG-47) and several aircraft carrier projects such as the CVV.

In the mid-1980's, Mr. Ryan took on the managment of the Computer Supported Design project within NAVSEA 03 to develop specialized ship design software, which ultimately led to the introduction of commercial CAD/CAM systems in NAVSEA, and the progressive integration of the vast analysis software library with CAD geometry. Recent assignments have included leading the Design, Acquisition and Construction process improvement effort at NAVSEA;

being the acting head of Preliminary Design; and being the NAVSEA lead for Simulation-Based Design.

Following graduation from Defense Systems Management College in 1995, Mr. Ryan joined the Navy's Arsenal Ship concept exploration team, where he remained until May 1996 when he became the Technical Director for the Arsenal Ship Joint Program Office.

Ms. Cindy Shaver

Ms. Shaver has been a Contract Specialist for the Naval Sea Systems Command (SEA 02) since 1989. She served as an contracting intern in the Surface Systems and Undersea Warfare Contracts Divisions, where she received exposure to various major weapon system programs. Beginning in 1991, she has served as a senior contracts negotiator for the Navy's Shipbuilding Division and has been responsible for the award and/or administration of seven major shipbuilding programs. She has experience with both research and production programs in all stages of the system acquisition process, from cradle to grave.

Program experience includes: the NSSN New Attack Submarine construction contract negotiation; the AGOR 24/25 and NOAA Oceanographic Research Ship Construction contract award; the settlement of the terminated T-AO 191/192 Fleet Oiler litigation and assignment of the T-AGOS 23 Oceanographic Surveillance Ship Construction contract to another shipbuilder; the WAGB 20 Coast Guard Polar Icebreaker Ship Construction contract award; the CG 47 AEGIS Class Cruiser, DD 963 SPRUANCE Class Destroyer and DDG 993 KIDD Class Destroyer Combined Planning Yard contract award; the and Arsenal Ship Program Section 845 Other Transaction awards.

Ms. Shaver holds a B.S. degree in Business Administration from George Mason University. She is a graduate of the Department of the Navy's Contracting Career Intern Program and is a member of the Department of the Navy's Acquisition Professional Community with level III certification.

Mr. John A. Grizzard

Mr. Grizzard obtained his Bachelor of Science Electrical Engineering degree from Virginia Polytechnic Institute and State University in 1985, followed by Masters in Business Administration from Florida Institute of Technology in 1994.

Mr. Grizzard initially worked for Martin Marietta Orlando Division, Orlando, Florida, where he developed the Master System Test Plan for Supersonic Low Altitude Target (SLAT). Mr. Grizzard joined the Department of the Navy as a member of the Naval Surface Warfare Center Carderock Division, Low Observables group. He served as the Radar Imaging Measurement System (RIMS) Project Manager; the Deployable Signature Measurement System (DSMS) Project Manager; and the Manager, Ship Low Observable Technology Integration office. He establishd the Navy's Topside Integration Working Group including participation as the ONR representative.

Mr. Grizzard joined the Arsenal Ship Joint Program Office as the Survivability Manager in May 1996.

Mr. John G. Frink

Mr. Frink obtained his Bachelor of Science Physics degree from Michigan State University in 1968, followed by a Masters Degree in Computer Science from the University of Maryland in 1970.

Mr. Frink initially worked with the Naval Surface Weapons Center, White Oak, Md. as the group leader for minicomputer/microprocessor applications, the MK 48 torpedo, Captor Mine, and early Navy High Energy Laser efforts. Mr. Frink served four years as a commercial consultant, where he developed the hardware architecture and real-time control programs for the initial series of computerized Versatran industrial robots. In 1978 Mr. Frink joined the Johns Hopkins University, Applied Physics Laboratory where he held various supervisory and technical positions in the Navy Warfare Analysis Department and the Fleet Systems Department. Mr. Frink is a specialist in distributed systems and was awarded a patent for "Intercomputer Communications Based on Message Broadcasting with Receiver Selection". He had major assignments on the following programs: LEAP; CEC Test Coordination oversight; internal Applied Physics Laboratory initiatives in DIS, IW; HiPerD; Aegis; and EA6B/EF111 EW ADCAP upgrades. Mr. Frink recently taught a graduate level Advanced Technology Seminar in the Technical Management curriculum of the JHU Whiting School of Engineering. He is a current member of the IEEE, the Computer Society, and the Association of Old Crows.

Mr. Frink joined the Arsenal Ship Joint Program Office in February 1997 as the C4I Manager.

Mr. Richard L. Dunn

Mr. Dunn has been the General Counsel of the Defense Advanced Research Projects Agency since 1987. From 1979 to 1980 he was an associate in the Washington law firm of Sullivan and Beauregard. From 1980 to 1987 he was with the National Aeronautics and Space Administration where his assignments included serving as counsel to the Space Commercialization Task Force and as Deputy Associate General Counsel.

At the Defense Advanced Research Projects Agency, Mr. Dunn has pioneered innovative techniques to support science and technology projects. Mr. Dunn was responsible for the enactment of legislation which authorizes DARPA to enter into cooperative relationships with commercial companies or "partnerships" of companies and other organizations. In response to DARPA's successful use of this authority Congress substantially increased DARPA's budget and encouraged it to sponsor R&D consortia. Most recently Mr. Dunn obtained legislative authority for DARPA to conduct military prototype projects outside the laws and regulations applicable to the procurement system. Several projects using this authority are currently underway at DARPA.

He is a graduate of the University of New Hampshire (B.A. cum laude, 1966), the University of Maryland (J.D.1969), and George Washington University (LL. M with Highest Honors 1976). Mr. Dunn was on activity duty as a Judge Advocate in the U.S. Air Force from 1970 to 1979.

Ms. Diane M. Sidebottom

Ms. Sidebottom is Assistant General Counsel for the Defense Advanced Research Projects Agency (DARPA) in Arlington, Virginia. Her duties include providing counsel for DARPA's Other Transactions for Prototypes and Research agreements in all aspects of agreement formation and administration. She also advises the agency on intellectual property and Internet-related issues. Prior to joining DARPA, Ms. Sidebottom was an attorney-advisor at the Defense Logistics Agency (DLA) specializing in the suspension and debarment of Government contractors. She also served on the task force supporting the Section 800 Panel, which examined over 800 procurement and acquisition laws and provided recommendations to streamline the acquisition process. She is a graduate of the University of Colorado and Thomas M. Cooley Law School

Mr. John H. Ablard

Mr. Ablard is a Professional Contracts Manager with thirty years of federal government experience. He has worked in all branches of the military services, with experiences ranging from small purchases and local procurement supporting a military base to World-wide communications and ADP systems. The last twenty (20) years have been in what is now called Science and Technology contracting. This is the portion of the R&D cycle where basic technologies are discovered and first exploited. His previous assignment was as supervisor and manager of a large contracting operation for the Naval Research Laboratory, where he received the Navy Meritorious Civilian Service Award in 1990. At DARPA he is furthering the development of new technologies through utilization of new contracting instruments, such as, "Other Transactions" under the authority of 10 USC 2371 and commercial R&D contracts. He received the DoD Exceptional Civilian Service Award in 1997.

Mr. Ablard is involved in the High Altitude Long Endurance Unmanned Aerial Vehicle Projects (Tier II+ and Tier III-Programs) and the joint DARPA/Navy Arsenal Ship Program where these new contracting instruments are being utilized. He has a B.A. degree in History and Political Sciences from Ottawa University, Ottawa, Kansas and a Master of Sciences in Management from the University of Utah. He is a member of the National Contracts Management Association.

Mr. John J. Turner

Mr. Turner is Senior Vice President of Syntek Technologies, Inc. since 1993. He is a registered Professional Engineer with forty years experience in ship design, construction and operations. He is a retired Engineering Duty Officer with assignments in shipyards, design activities, and acquisition and fleet support projects. Before joining Syntek he was Executive Vice President and Chief Operating Officer of JJH, Inc. For the previous 10 years he filled senior management and technical positions at NKF Engineering, Inc. He is a graduate of the U.S. Naval Academy and received an M.S. and the Naval Engineers degrees from M.I.T. He is a member of Society of Naval Architects and Marine Engineers, American Society of Naval Engineers and other professional organizations.

Dr. Robert S. Johnson

Dr. Johnson has been a Vice President of Syntek Technologies, Inc. since 1994. He has 39 years experience in ship design, systems engineering and management in which he consults for Naval Commands, the Defense Advanced Research Projects Agency, and commercial clients. He has served in the Naval Sea Systems Command, as Deputy Chief Engineer, Total Ship Engineering; Director, Field Activities Support Group; Director, Field Operations Subgroup; and Technical Director, Ship Design. He was President of the Ship Engineering and Design of Advanced Marine Enterprises; and Manager, Ship Design, Westinghouse Defense & Electronics Center. He is active in professional societies: a Fellow of the Society of Naval Architecture and Marine Engineers, past member of its National Papers Committee, Editor of *Marine Technology*, and a Life Member, American Society of Naval Engineers (former Member of Council and By-Laws Committee Chairman). He holds the B.S. from Webb Institute, M.S.E. from George Washington University, and D. Eng from the University of California at Berkeley.

Mr. Ronald K. Kiss

Mr. Kiss has been Vice President of Syntek Technologies, Inc. for Systems Development since 1996. He has 35 years experience in ship design and ship acquisition. He has served as the Deputy Assistant Secretary of the Navy for Ship Programs; Executive Director for Amphibious, Auxiliary, Mine and Sealift Ships at the Naval Sea Systems Command; Acting Associate Administrator for Shipbuilding and Ship Operations, Director of the Office of Ship Construction, and Chief of the Division of Ship Design at the Maritime Administration. He has been selected to become the President of Webb Institute and will assume that position in 1999. He is active in professional societies: past President of the Society of Naval Architects and Marine Engineers, member of its Executive Committee and Council, and Chairman of its Nominating Committee, past Vice President of the American Society of Naval Engineers, and member of the Royal Institute of Naval Architects. He holds the B.S. from Webb Institute, M.S. from the University of California at Berkeley, completed the Harvard Business School and Sloan School (M.I.T.) Executive Programs.

Mr. Robert C. Percival

Mr. Percival has been Vice President of Syntek Technologies, Inc. for DARPA Programs since 1996. He has thirty two years experience in ship design, acquisition management, construction and operations. He is a retired Engineering Duty Officer with assignments in research activities, design activities, and acquisition and fleet support projects. Before joining Syntek he was Director of Surface Ship Design and Systems Engineering, Director of Research and Development, Director of Concept Design, and Director of Signature Control at the Naval Sea Systems Command. Previously he was an Acquisition Mannager in the Aegis Program Office delivering the SPY-1B and SPY-1D radar systems. He is a graduate of the U.S. Naval Academy and received an M.S. and the Ocean Engineers degrees from M.I.T. and the Executive Management Program at Penn State University. He is a member of Society of Naval Architects and Marine Engineers, American Society of Naval Engineers and other professional organizations.

Mr. Bruce P. Dyer

Mr. Dyer has been Vice President of Syntek Technologies, Inc. since 1996. He has thirty two years experience in prorgram management and military operations. He is a retired Naval Special Warfare Officer who qualified as an Acquisition Professional. Before joining Syntek he was a Program Manager at the Defense Advanced Research Projects Agency and the Program Manager for a sensitive classified Major Defense Acquisition program at the Naval Sea Systems Command. Previously he commanded the Naval Special Warfare Task Group, Middle East during Earnest Will operations which included Patrol Craft, Army Helicopters, intelligence collection assets, as well as special forces personnel. He is a graduate of the U.S. Naval Academy and received an M.S. in Systems management from the University of Southern California. He completed the Executive Development Program at Cornell University. He is a member of the Institute of Electrical Engineers, U.S. Naval Institute and other professional organizations.

Mr. Michael Yermakov

Mr. Yermakov is Director of Strategic Planning at Syntek Technologies, Inc. since 1997. He has 12 years experience in ship design, systems engineering and management in which he consults for Naval Commands, the U.S. Coast Guard, the Advanced Research Projects Agency, and commercial clients. He has served in the Naval Sea Systems Command as the Chief Systems Engineer for the SC-21 Program, a Ship Design Principal Naval Architect and lead the development and deployment of special systems for anti-terrorist operations. He managed the assessment of new ship acquisition and R&D programs in the Office of the Chief of Naval Operations and served as Chairman of NATO Group 6. He is a member of the Society of Naval Architecture and Marine Engineers, member of its Advanced Naval Vehicles Committee, and a Member of the Association of Scientists and Engineers. He holds the B.S from the University of California at Berkeley and graduated from the Acquisition Program Manager's six month course at the Defense Systems Management College.

Mr. David Schwiering

Mr. David Schwiering has been the Program Manager for Vail Research and Technology providing support to the Arsenal Ship Joint Program Office since 1997. He is a retired Surface Warfare Officer with command assignments afloat and ashore. He served as Head of Programs and Budget for both Surface and Expeditionary Warfare and was the lead analyst for the Research and Development Appropriation in the Office of the Navy Comptroller. He holds a B.A. from Waynesburg College and an M.S. in Financial Management from the Naval Post Graduate School.

www.ingramcontent.com/pod-product-compliance
Lightning Source LLC
LaVergne TN
LVHW080924210926
848485LV00004B/720
9781608881178